MARINE OUTFALL SYSTEMS
Planning, Design, and Construction

ROBERT A. GRACE

Department of Civil Engineering
University of Hawaii at Manoa
Honolulu, Hawaii

PRENTICE-HALL, INC., *Englewood Cliffs, New Jersey 07632*

Library of Congress Cataloging in Publication Data

Grace, Robert A
    Marine outfall systems.

    Includes bibliographies and index.
    1. Waste disposal in the ocean.   2. Ocean
outfalls—Design and construction.   I. Title.
TD763.G64   363.6'1   77-20980
ISBN   0-13-556951-6

© 1978 by Prentice-Hall, Inc.
Englewood Cliffs, New Jersey 07632

Printed in the United States of America

10 9 8 7 6 5 4 3 2 1

PRENTICE-HALL INTERNATIONAL, INC., *London*
PRENTICE-HALL OF AUSTRALIA PTY. LIMITED, *Sydney*
PRENTICE-HALL OF CANADA, LTD., *Toronto*
PRENTICE-HALL OF INDIA PRIVATE LIMITED, *New Delhi*
PRENTICE-HALL OF JAPAN, INC., *Tokyo*
PRENTICE-HALL OF SOUTHEAST ASIA PTE. LTD., *Singapore*
WHITEHALL BOOKS LIMITED, *Wellington, New Zealand*

To the memory
of my father,

Archibald John Grace

Contents

ACKNOWLEDGEMENTS   *xix*

PREFACE   *xxi*

1 THE MARINE DISPOSAL OF WASTEWATERS   *1*

1-1 Orientation   *1*

1-2 The Regional Wastewater Plan   *2*

1-3 Decision to Build an Outfall   *4*

1-4 Protection of Man and the Marine Biota   *6*

1-5 Data on the Receiving Water and its Margins   *8*

1-6 Outfall Design   *9*
   Introduction   9
   Preliminary design   10
   Final design   10

1-7 Outfall Construction   *11*

1-8 Operation, Maintenance, and Monitoring   *13*
   Operation and maintenance   13
   Monitoring outfall performance   13

1-9 Outfalls in Southern California   *15*

1-10 References   *15*

1-11 Bibliography   *15*

2   THE MARINE ENVIRONMENT   17

2–1 Some General Concepts   *17*
        Introduction   17
        Basic ocean chemistry   17
        Variations with depth   19

2–2 Life in the Sea   *21*
        The plankton   21
        The nekton and benthic animals   23
        Sustaining the life of the sea   25
        Larger marine plants   27

2–3 Physical Processes on a Global Scale   *27*
        Introduction   27
        The Coriolis force   28
        Surface air movements   29
        Circulation of surface waters in the ocean   30
        Tides   31

2–4 Physical and Geological Concepts   *33*
        Wind waves   33
        The coastal zone   38
        Coastal currents   42
        Seismic sea waves   45
        Storm tides   46

2–5 References   *46*

2–6 Bibliography   *48*

3   WASTEWATERS:
CHARACTERISTICS, TREATMENT, AND DISPOSAL   52

3–1 Raw Wastewaters   *52*
        Flow Rates   52
        Physical characteristics   53
        Chemical characteristics   54

3–2 Indicator Organisms   *57*

3–3 Pretreatment   *58*
        Introduction   58
        Screening and shredding   59
        Grit removal   60
        Other types of pretreatment   60

3-4 Physical and Chemical Treatment *60*
      Primary sedimentation 60
      Flotation 62
      Chemical treatment 62
      Disinfection 63

3–5 Secondary Treatment Processes *65*
      Introduction 65
      Trickling filters 65
      Activated sludge process 69

3–6 Sludge Treatment and Disposal *70*
      Introduction 70
      Thickening, dewatering, and conditioning 71
      Anaerobic digestion 72
      Aerobic digestion 74
      Sludge disposal 74

3–7 Advanced Waste Treatment *75*
      Introduction 75
      Nutrient removal 76
      Filtration 77
      Organic removal 77
      Removal of dissolved salts 78
      Physicochemical treatment 78

3–8 Wastewater Disposal by Dilution *78*

3–9 Other Wastewater Disposal Methods *80*
      Introduction 80
      Land disposal 80
      Groundwater supplies 82
      Los Angeles experience 83
      Direct and indirect reuse 83
      Wastewater aquaculture 84

3–10 References *86*

3–11 Bibliography *89*
      General 89
      Treatment 90
      Disposal 91
      Reclamation and reuse 93

4   PROTECTION OF THE MARINE ENVIRONMENT   94

4–1 Introduction   94
Raw sewage effluents   94
Beneficial uses and their impairment   96
Chapter plan   97
Marine disposal of wastewaters via outfalls   98

4–2 United States Experience   98
Years after World War II   98
National environmental policy act   99
Public Law 92-500   100

4–3 Water Quality Plans   103
Introduction   103
"Water quality control plan
   for ocean waters of California (1972)"   104
"Pollution control objectives
   for municipal type waste discharges in British Columbia
   (1975)"   107
Specific industry plans   109

4–4 Toxicity   109
Introduction   109
Bioassay   110
Procedures   112
Results of acute toxicity bioassays   115
Chronic toxicity of wastewater discharges   117

4–5 Source Control   118
Introduction   118
Experience of Los Angeles City   119

4–6 Biostimulation and the Minimum Requirement
   of Secondary Treatment   123
Eutrophication   123
Nutrient inputs to marine waters   123
Secondary treatment requirements   124

4–7 Public Health Considerations   126
Introduction   126
Coliform disappearance   126
Pathogens and indicator organisms   129
Coliform standards   131

4–8 Permits and Environmental Impact Statements   132
Introduction   132

Other federal permits   134
State and local permits   136
The environmental assessment
    and the environmental impact statement   136
Review of draft environmental statements   138
Public hearings   140

4–9 References   *140*

4–10 Bibliography   *147*
    General   147
    Public health   150
    Pollutant effects on marine biota   152
    Thermal pollution   156

5 OBTAINING PERTINENT
   PHYSICAL OCEANOGRAPHIC DATA   *157*

5–1 Introduction   *157*
    Orientation   157
    Oceanographic instruments   158

5–2 Wind and Tides   *160*
    Wind data   160
    Tidal variations   160

5–3 Moored Buoy Systems   *161*
    Introduction   161
    The lifetime of moored buoy systems   164
    Notification regarding moored buoy systems   167
    Deploying moored buoy systems   168
    Vessels for deploying and retrieving
        moored buoy systems   169
    Retrieving moored buoy systems   170

5–4 Waves   *172*
    Introduction   172
    Wave measuring systems   173
    Data processing and presentation   179

5–5 Currents   *180*
    Introduction   180
    Lagrangian current-measuring approaches   181
    Eulerian flow measurement   186
    Presentation of current data   191

5-6 Diffusion and Dispersion *193*
   Introduction 193
   Use of drogues 194
   Fluorescent dyes and fluorometers 195
   Dye study field methods 197
   Organizing results of dye studies 199

5-7 Primary Physical Water Column Parameters *201*
   Introduction 201
   Obtaining water samples 202
   Reversing thermometers 204
   Salinities from water samples 205
   Ocean-going systems 205
   In situ systems 207

5-8 Additional Considerations Involving
   Boats and Field Studies *207*

5-9 References *209*

5-10 Bibliography *215*
   General 215
   Remote sensing 216
   Boat operations and moorings 217
   Waves 217
   Currents 218
   Diffusion and dispersion 220
   Oceanographic studies for proposed outfalls 220

6 COLLECTING OTHER PERTINENT MARINE DATA   *222*

6-1 Geophysical Investigations *222*
   Bathymetry 222
   Bottom and subbottom conditions 225

6-2 Obtaining Engineering Properties
   of Sea Floor Materials *228*
   Introduction 228
   Sampling from the surface 230
   Work involving divers 234
   Complete field studies for outfalls 236
   Laboratory analyses 237

6-3 Shoreline and Sea Floor Dynamics *239*
   Introduction 239
   Sediment transport rates and beach profiles 241
   Sea floor dynamics 241

6–4 Marine Flora and Fauna   *242*
 Introduction   242
 Plants   242
 Animals   243
 Diversity   244
 Fish characteristics   246

6–5 Sampling the Benthos and Nekton   *247*
 Introduction   247
 Trawls and trawling   247
 Survey by divers   251
 Sampling the benthos with the grab sampler   254
 Sampling by camera   256

6–6 Gathering Data on Plankton   *257*
 Sampling in the field   257
 Laboratory work   260
 Other approaches   261
 An integrated marine biological study   262

6–7 Obtaining Water-Column Information   *263*
 Orientation   263
 Collecting samples   263
 Chemical parameters   264
 Water clarity   266

6–8 References   *268*

6–9 Bibliography   *273*
 Biological   273
 Quantitative approaches in biology   276
 Geological   278
 Other   282

7 **THE HYDRAULICS OF OUTFALLS,
DIFFUSERS, AND EFFLUENTS   *284***

7–1 Marine Disposal of Wastewaters   *284*
 Wastewater outflow and mixing   284
 Wastewater discharge through outfalls   286

7–2 Hydraulics Background   *288*
 Outflow   288
 Pipe flow   290

7–3 Internal Hydraulics of Diffusers   *290*
 Introduction   290
 Basic equations   292

Flow analysis   292
Computations by computer   295
Risers   298

7–4 Pumping and Hydraulic Transients   300
Driving head   300
Hydraulic transients   300
Surface wave effects on diffuser operation   301

7–5 Initial Dilution of Effluents
in Homogeneous Receiving Water   *303*
Introduction   303
Surface dilution of slot plumes   305
Dilution of round jets and plumes   307
Effluent with the same density as ambient liquid   308

7–6 Initial Dilution of Effluents:
Stagnant Receiving Water of Variable Density   *309*
Introduction   309
Round buoyant jet discharged horizontally   312
Slot buoyant jet discharged horizontally   315

7–7 Secondary Dispersion Process
in Coastal Waters   *317*
Introduction   317
Diffusion in the ocean   318
Coastal currents   320
Brooks' model   321
Another unidirectional current model   326
Computer programs   327

7–8 References   *327*

7–9 Bibliography   *329*
General   329
*Flow in pipes, losses, pumping, and hydraulic
    transients   330*
Internal hydraulics of diffusers   330
Diffuser analysis and design   330
*Jets and plumes: theoretical, numerical, and laboratory
    studies   331*
Buoyant spreading   334
Diffusion   334
Oceanic diffusion   334
Secondary dispersion   335
Field studies of outfalls   335
Thermal effluents   336

Wastewater disposal in estuaries   337
Wave effects on effluent dispersion   338
Wave effects on pipe flow   338

8   THE TECHNICAL ASPECTS OF OUTFALL DESIGN   *339*

8–1 Introduction   *339*
   Outfall pipes   339
   Sand Island No. 2 outfall   340
   Orientation   343

8–2 Outfall Pipe   *344*
   Introduction   344
   Cast iron pipe   344
   Steel pipe   346
   Concrete pipe   348
   Plastic pipe   351
   Cathodic protection   354
   Construction technique influence
      on materials selection   355

8–3 The Spectrum of Potential Trouble
   for an Outfall   *355*
   Introduction   355
   Problems involving ships   356
   Wave-related problems   357
   General failure mechanisms   359

8–4 Outfall Location and Length   *362*
   Introduction   362
   General considerations regarding pipe and route   364
   Foundation stability and settlement considerations   366

8–5 Design Water Levels and Currents   *366*
   Water levels   366
   Currents   371

8–6 Outfall Design with Respect to Currents   *372*
   Introduction   372
   Forces on exposed outfall:
      current perpendicular to pipe   372
   Forces on exposed outfall at angle to current   375
   Sliding stability of pipe   376
   Stabilizing an outfall   376

8–7 Waves   377
     Wave generation and decay   377
     Shoaling, refraction, and reflection of waves   379
     Observed conditions   380
     The design wave   381

8–8 Wave-Induced, Near-Bottom Water Motion   382
     The theoretical basis   382
     Wave-induced kinematics in context   385

8–9 Unburied Outfall Design Against Waves   386
     Introduction   386
     Wave forces: pipe parallel to wave fronts   387
     Wave forces: pipes at angle of attack   391
     Pipe stability against wave forces   392
     Breaking wave forces   392

8–10 Pipe Burial and Ballasting   393
     Backfilling with native material   393
     Backfilling with rock   394
     *The Coastal Engineering Research Center approach
       and its verification   396*
     Another approach   397
     Other considerations   399

8–11 Other Design Issues   400
     Buried pipes   400
     Earthquake design   400

8–12 Appurtenances   401
     Bends   401
     Manholes   401
     Egress structures   403
     Flapgate structures   403

8–13 References   403

8–14 Bibliography   412
     General   412
     General submarine pipeline design   413
     Materials, coatings, and corrosion   414
     Storms and storm surges   415
     General waves   416
     Wave energy dissipation   417
     Wave forces and related studies   418
     Local scour   418
     Foundation stability   419
     Earthquakes and tsunamis   420
     Fluid mechanics   421

9 THE BACKDROP FOR CONSTRUCTION **422**

9–1 Introduction *422*
  Project design 422
  The designer and the construction contractor 422
  The construction manager 424
  Orientation 425

9–2 Details Concerning Plans and Specifications *425*
  Introduction 425
  General provisions in specifications
    for ocean outfalls 426
  Detailed specifications in specifications
    for marine outfalls 427

9–3 Budget Estimate Preparation *427*
  Introduction 427
  Outfall construction costs 429

9–4 Bid Preparation, Bidding,
    and Construction Contract Award *432*
  Introduction 432
  Contract forms 433
  Bidding and construction contract award 435

9–5 Planning, Scheduling, Controlling,
    and Monitoring Construction *437*
  The general concept of CPM 437
  CPM in practice 440
  Example of CPM use in outfall construction 440

9–6 Various Topics *442*
  Insurance and safety 442
  Change orders 442
  Value engineering 443

9–7 References *444*

9–8 Bibliography *444*

10 CONSTRUCTION OF OUTFALL **447**

10–1 Introduction *447*

10–2 Trenching *447*
  Introduction 447
  Mechanical dredges 453

Hydraulic dredges   455
Blasting   458
Additional considerations   462

10–3 Installing the Outfall: Both Inshore
       and Offshore Waters   *464*
       Introduction   464
       Bottom pull method   464
       Flotation method   469

10–4 Installing the Outfall: Offshore Waters   *470*
       Lay barge method   470
       Floating crane   471
       Davy Crockett and the Horse   472
       Towers   477

10–5 Installing the Outfall: Inshore Waters   *478*
       Trestle   478
       Spider   485
       Outfalls across tidal flats   487

10–6 Bedding and Protecting the Outfall   *488*
       Placing bedding stone   488
       Armor or ballast rock   489

10–7 Underwater Operations   *494*
       Orientation   494
       Safety   495
       Diver decompression   496
       Diver pay   497

10–8 Post-Laying Excavation   *498*

10–9 Additional Considerations   *499*
       Orientation   499
       Potential problems with reinforced concrete pipe   499
       *Testing joints in double-gasket reinforced
          concrete pipe   501*
       Miscellaneous   502

10–10 References   *503*

10–11 Bibliography   *508*
       *General submarine pipeline design
          and construction   508*
       Dredging   510
       Blasting   511
       Constructing specific outfalls   512

Constructing specific submarine pipelines
other than outfalls   513
Diving   514
Related to design of trestles   514
General construction   515

**11 CONSTRUCTION EFFECTS
AND POST-CONSTRUCTION ACTIVITIES   516**

11–1 Construction Effects on the Marine Biota   *516*

11–2 Outfall Inspection and Repair   *519*
Introduction   519
External inspection and repair   519

11–3 Outfall Operation and Maintenance   *521*

11–4 Monitoring   *523*
Introduction   523
Effluent monitoring   525
Monitoring reveiving water quality   525
Benthic animals as pollution indicators   527
Southern California coastal water research
project monitoring experience   528

11–5 References   *529*

11–6 Bibliography   *531*
Construction effects   531
Operation and maintenance   533
Monitoring   533

APPENDIX A   ACCURATE POSITIONING OF MARINE VESSELS   *539*

A–1 Introduction   *539*

A–2 The Earth   *540*
Introduction   540
Coordinates on the earth   540
Direction   541

A–3 Charts   *542*

A–4 Positioning and Aligning   *543*
Introduction   543
Electronic positioning methods   544
Position along a line by using lasers   545

Other approaches   546
Use of floating aids-to-navigation
to fix a navigational position   547

A–5 Radar   *548*

A–6 References   *549*

A–7 Bibliography   *549*

APPENDIX B   DIVING   *551*

B–1 Introduction   *551*

B–2 General Concepts   *552*
Pressures and gases   552
Oxygen   552
Nitrogen   553
Decompression   554
Breath-holding   555
Long-term effects of diving   555

B–3 SCUBA Diving   *556*
Standard SCUBA diving   556
Equipment for SCUBA diving   557

B–4 Synthetic Mixed Gases for Breathing   *559*

B–5 Tethered Divers   *559*
Introduction   559
Hard hat diving   560
Band mask and hat diving   563

B–6 References   *565*

B–7 Bibliography   *565*

APPENDIX C   GLOSSARY   *566*

INDEX   *576*

Acknowledgements

The material in this volume comes from three sources: the substantial literature that relates to the planning, design, construction, operation, and maintenance of marine outfalls; my own personal experience; and the verbal input of others based on their own learning and experience. In the latter regard I am particularly grateful to the following for the extensive information they shared with me: Ellis Cross, Fred Delaney, John Goode, John Singleton, Scott Vuillemot, Joseph Peters, Marcus H. Stearns, Richard Vlach, R. Q. Palmer, Patrick Wolter, Ken Knott, Robert W. Morris, Donald Chapman, William W. Tinniswood, G. E. B. Wilson, Gerry Byrne, Fred J. Shumas, Richard Wagner, and William Herder.

The personnel of various organizations, governmental agencies, and companies also provided a great deal of assistance in gathering information for this book. I extend my thanks to: the Southern California Coastal Water Research Project; the U.S. National Oceanic and Atmospheric Administration, Sea Grant Programs; the U.S. Environmental Protection Agency, Region X; the City of Los Angeles, Department of Public Works, Bureau of Sanitation; the County Sanitation Districts of Orange County; the City of San Diego, Utilities Department; the City and County of Honolulu, Department of Public Works, Sewers Division; the Greater Vancouver Sewerage and Drainage District; R. M. Towill Corp.; Morrison-Knudsen Co. Inc.; Hawaiian Dredging and Construction Co.; Sunn, Low, Tom, and Hara, Inc; J. M. Montgomery, Consulting Engineers, Inc.; Brown and Caldwell, Consulting Engineers; and Associated Engineering Services, Ltd.

In my research work for this book, I was aided by Elizabeth Leis, Norman H. Brooks, Tareah Hendricks, John S. Corbin, William J. Kimmerer, Michael Rayfuse, Joseph Castiel, Steven A. Nicinski, Harold D. Pritchett, Roger W. Lindquist, Charles Schuster, Kent K. Reinhard, James T. Sands, William E.

Woodward, Geoffrey Slack-Smith, N. V. Lawson, James Washburn, Richard W. Grigg, Edward Noda, and Michael Isaacson.

I have benefitted from the reviews of individual chapter drafts by S. Arthur Reed, Richard M. Smith, Edward L. Beckman, Hans-Jurgen Krock, Takeshi Yoshihara, E. Alison Kay, Brent S. Gallagher, Gordon L. Dugan, Danforth G. Bodien, Donald Baumgartner, Reginald H. F. Young, John McCain, Graham McBride, John A. Williams, Frederick M. Casciano, Philip J. W. Roberts, and Ralph Bowers. Wayne Tomita and James Fung reviewed several draft chapters as did F. L. Vuillemot who also contributed materially to my researching for the volume.

Six University of Hawaii graduate students reviewed the entire book in draft form: David B. Bills, Lawrence E. Brower, Donn T. Fukuda, Rich Rocheleau, Adrian H. Rowland, and Gabriel T. Y. Zee. I profited a great deal from their thoughtful and probing work.

Through the various drafts of the chapters in this volume, typing assistance was received from Elaine Shimabukuro, Amy Fujishige, Mary Kamiya, Deborah Douma, and Doreen Brom. I am very grateful for their help as well as the assistance of the individuals, firms, and organizations that supplied photographs. These sources are listed in the appropriate figure titles. Photographs without credits were taken by the author.

I owe my greatest debts to two people. Sylvia Khong did much of the necessary typing and most of the drawings. Arthur T. Shak twice reviewed different stages of the book plus assisted in the final collation. His careful attention to detail and his comments were of vital importance.

Preface

The Department of Civil Engineering of the University of Hawaii at Manoa has offered a semester-long graduate course entitled "The Marine Disposal of Wastes" since 1971. Virtually all of the course has been concerned with wastewaters and their disposal through outfalls. The course has no prerequisites other than that the students have graduate standing, and class members have included students from sanitary engineering, ocean engineering, and marine biology as well as Honolulu professionals working in these same fields. Two of us have been co-responsible for the course and have done the bulk of the lecturing. However, we have made frequent use of guest lecturers drawn from the University system, consulting engineering firms, governmental agencies, and elsewhere.

This book is an outgrowth of the course and contains all the material considered in it with the exception of basic concepts in fluid mechanics. I think it is safe to say that there is no other book with this topical outline and guiding focus. The depth of coverage is such that the entire contents of the book can be dealt with in one semester. In addition to teaching this material, we have required one or two term papers focusing on timely issues related to wastewater disposal in the marine environment or to topics not detailed in this volume such as quantitative methods in marine biology. A quarter course could be made out of this book by leaving out the term papers, requiring students to have some background in both oceanography and sanitary engineering, and using Chapters 1, 4–11 as well as the appendices.

It is to be stressed that the material covered is by no means restricted to a graduate course. I have been frequently asked about my intended audience, and in addition to the mix of students we have attracted to the course in the past, I believe the book material is suitable for the following: those drafting legislation, rules, and regulations regarding wastewater disposal in the marine environment; those involved in formulating regional water supply and wastewater plans for coastal areas; engineers involved in outfall design, construction, construction management, and inspection; oceanographers involved in securing physical, chemical, biological, and geological data as a part of pre-outfall-design or monitoring studies; environmentalists who seek a broad understanding of the legal and technical aspects of wastewater disposal into marine waters through outfalls; researchers involved in effluent dilution and dispersion investigations; and divers involved in pre-outfall-design reconnaissance, outfall construction work, and inspection during and after construction.

Although the planning, design, construction, operation, and maintenance of a marine outfall do not require a cast of thousands, a surprisingly large number of disciplines and individuals do come together in this overall area. One of my major reasons for writing this volume, apart for its use in the course mentioned earlier, was to make the individual or team working in one area aware of what is involved in companion areas. I became convinced of the necessity for such a book one night at a hearing concerned with disposal of a thermal effluent. An engineer for a power company argued that withdrawing ocean water through a shoreline inlet, and then discharging the heated water at the shoreline to mix and move away in the top meter of the sea, did corals no harm. After all they were on the bottom, well below the mixed, heated effluent. He was, of course, apparently unaware that the coral in its larval, or planula, stage is a constituent member of the floating zooplankton. But the purpose of the book goes beyond acquainting those in one field with interfacing matters in other fields. I intended as well to provide those in specific fields with new insights into their own disciplines. As an engineer, I can only hope to have done this adequately in engineering-related matters. There is much to be learned from the experiences of others in marine outfall planning, design, construction, operation, and mainte-nance. Many such clues are contained in scattered written material; I have attempted to collate these. Most clues are tucked away in a fertile brain somewhere; I have endeavored to locate those brains and "pick" them. The resulting sessions have had an unexpected result as I can now drink beer without difficulty!

I have been fortunate to have written this book where and when I did. Offshore from my island home, four separate, major outfalls were in various stages of planning, design, construction, or operation, during this two-year period. These projects have provided me an unusual experience to "go to school" both above and below water. Many examples used throughout the book are drawn from these outfall construction projects, and many of the photographs are of them. A leave of absence at Oregon State University allowed me to gain still more perspectives.

Although I owe a personal debt to so many people for assistance in bringing this book to completion that I have used a complete, separate section on Acknowledgements, I would like to acknowledge the debt owed to three particularly important publications. Professor Erman A. Pearson's "An Investigation of the Efficacy of Submarine Outfall Disposal of Sewage and Sludge" California State Water Pollution Control Board, *Publication No. 14*, 1956, was years ahead of its time. *Ocean Outfall Design* written by Hyperion Engineers, 1957, was another early and very valuable contribution. Charles E. Warren's *Biology and Water Pollution Control* (W. B. Saunders Co., 1971) served as an excellent learning book on biological matters.

As in many fields too, much of what is written on the general subject of marine disposal of wastewaters comes from the pens and typewriters of academicians. The practitioner often has a tale to be told, and I would like to enter a personal plea to those outside academia to organize their thoughts and experiences into paper form—for the benefit of all of us who lack necessary practical seasoning. The paper by Douglas L. McKay et al. in *Civil Engineering*, October 1974, and entitled "Large Diameter Submarine Steel Pipeline Crossings" serves as an excellent example.

The matter of units is always a problem as various countries change from the English system to the SI metric system. Since the English system is still heavily used by engineers in the United States, I have felt that this system could not be dropped at this time. Thus in engineering-related matters I have used the English system as the primary units with the corresponding SI units in parentheses. This arrangement makes for ragged reading unless one notes only the entry in one's own units, and I would urge readers to adopt this technique, simply skipping the system that is not being used. Even in the United States the metric system is used as the standard in some fields, for example physical oceanography and marine biology. The areas within the book wherein such matters are discussed employ only the metric system. Virtually all units used in the book are summarized in Appendix C.

Rather than summarize the contents of the book, chapter by chapter, in the Preface, I have chosen to do so in Chapter 1. This has been done since it is easier to explain the book following the presentation of some background material.

Finally, a concerted effort has been made to thoroughly check all information presented in this text. If errors remain, I would encourage readers to bring these to my attention.

University of Hawaii
Honolulu, Hawaii

ROBERT A. GRACE

1

The Marine Disposal

of Wastewaters

1-1 ORIENTATION

Wastewater is a general term that refers to water that has been used in homes, commercial enterprises, or by industries and then discarded. The fraction of wastewater contributed by domestic sources is commonly known as *sewage*. An *outfall*, or *outfall sewer*, is a pipe that carries wastewater from the land out into a water course, such as a river, or into a water body such as a lake or the sea. These recipients of wastewater discharges are collectively known as *receiving waters*, and singled out for consideration in this volume are *marine* receiving waters. These include ocean waters found along open coasts and in embayments, but also include the areas where rivers meet the sea, namely *estuaries*. Although the focus of the book is unquestionably on nonfreshwater receiving waters, much of the material presented can be applied conceptually to lake discharges. However, there are fundamental differences in lake and marine wastewater disposal practices that will be outlined at various points in the text.

The simple expression "marine outfall" indicates merging of two bodies of knowledge that are normally considered entirely separate, namely sanitary engineering and oceanography. Fundamental to the discussion of the various planning, design, construction, and operational aspects of marine outfalls to be presented in this volume is a thorough grounding in the basics of these two fields of study. Since there are numerous people who enter the arena of marine wastewater disposal without having studied either or both of these fields other than superficially, I consider an introductory chapter related to such fields an essential part of this volume. Physical, geological, chemical, and biological oceanographical factors are presented in Chapter 2, with information on the characterization of wastewaters, their collection, treatment, and disposal placed in Chapter 3. I urge readers whose background is weak in either oceanography or sanitary engineering or both to take the time to review Chapters 2 and/or 3 before plunging ahead. Chapters 4 through 11 assume a familiarity with all concepts presented in the two background chapters.

In the latter part of Chapter 3, it is stressed that there are various wastewater disposal and reuse alternatives for coastal communities or industries —including marine disposal by outfall. Since this volume is concerned solely with outfalls, it is implicitly assumed that this disposal method has been chosen as being conceptually superior to the various alternatives confronting a community or industry. Such a decision is by no means a random one. For communities, it is often the outgrowth of a thorough study of the general area resulting in a Regional Wastewater Plan. Such a plan, if properly executed, permits constituent communities to make ordered and relatively long-range plans for wastewater collection, treatment, and disposal facilities.

The following section is devoted to a summary of the components of a comprehensive Regional Wastewater Plan. This is presented for two reasons. First, it provides background for the decision to proceed with a marine disposal approach for wastewater rather than an alternate method. Second, it will show the types of wastewater data to be gathered for such a plan. The assumption has been made in writing this book that all data on past, present, and future constituents and flow rates required for the design of a marine disposal scheme were obtained during the preparation of the Regional Wastewater Plan.

1-2 THE REGIONAL WASTEWATER PLAN*

The general divisions and the subdivisions for a complete Regional Wastewater Plan are presented below in point-by-point form. It is assumed that the region considered borders on marine waters such that one wastewater disposal option is the topic of this book.

1. The natural situation
    a) Environmental conditions (spatial and temporal variations in air temperature, wind, rainfall, cloud cover and solar radiation)
    b) Land characteristics (geology, topography, soil characteristics, drainage, groundwater, runoff, flora and fauna)
    c) General marine conditions (bathymetry, sea floor, coastline, currents, flora and fauna) [†]

---

*In the parlance of the United States Environmental Protection Agency and the Federal Water Pollution Control Act Amendments of 1972, to be discussed in Chapter 4, this would be referred to as an "areawide" or "208" plan. Since U.S. water pollution control approaches are changeable and because this book is intended for a wider audience than the United States, this concept will be described here in general terms.

[†]This topic will be enlarged upon in Chapters 2, 5 and 6.

d) Other receiving waters (physical, chemical, biological, hydrological characteristics)

2. The general urban and rural situation
   a) Population (distribution, density, growth, projections)
   b) Economics (sources of income base and balance; analysis of industries, industrial opportunities; retail facilities; markets; labor force; forecasts)
   c) Transportation
   d) Public utilities
   e) Land use
   f) Educational and cultural facilities
   g) Political situation
   h) Conservation element
   i) Sociological considerations (knowledge, attitudes, behavior, expectations of populace)

3. Wastewater-related background
   a) General institutional studies (laws, ordinances and regulations; public information and support)
   b) Administrative studies (agency organization, management, and costs)
   c) Financial studies (financing methods such as bonds and grants; incentives; federal and state aid; financial structure, requirements and constraints; sewer service charges)
   d) Wastewater treatment criteria
   e) Water quality standards and criteria (geographical subdivisions; beneficial uses; water quality parameters, standards and objectives) *
   f) Existing wastewater facilities (adequacy, septic tanks, pipes, infiltration, routing, rights of way, treatment plants, pumping stations, holding tanks, local ordinances, disposal techniques)
   g) Wastewater characteristics (volumes, strengths, flow rates; possibility of wastewater reduction; ease of treating industrial part)

4. Wastewater-related specifics
   a) Basis for development of wastewater plan (design period, sites, areas, standards; projections of wastewater volumes and characteristics; infiltration)
   b) Costs (capital-property, equipment, structures; annual-repayment of capital costs, principal and interest, taxes; engineering and legal services; rates)
   c) Special studies (toxicity bioassays; biostimulation) *

*This material will be expanded in Chapter 4.

5. Wastewater reclamation and reuse *

a) Past and present regional situation (agricultural practices; industries; water quality for crops and industry; water use and its patterns; schedules of need)

b) Analysis (effect of agricultural return waters on groundwater quality; problems with viruses)

c) Future possibilities (domestic, agricultural and industrial reuse; institutional constraints)

6. Formulated regional plans

a) Recommended wastewater collection plan

b) Alternate wastewater collection plan

c) Recommended wastewater treatment and disposal plan (short and long term with cost estimate; economic, social, ecological and functional evaluation; regulations)

d) Alternate wastewater treatment and disposal plans (for changing conditions—e.g., legislation; cost estimates)

7. Other recommendations

a) Implementation program (administrative; institutional, financial, engineering design, construction, operation and maintenance, inspection; enforcement) †

b) Research

c) Evaluation of plans and replanning

1-3   DECISION TO BUILD AN OUTFALL

It is important that the route and terminus location of a proposed outfall be determined at an early stage of planning so that the detailed oceanographic and geologic data required in the design process can be obtained close to the proposed alignment. One can begin to narrow down a fairly extensive ocean area, delimited only by such terrestrial constraints as extensive staging areas and access ways, into a rather specific route for an outfall line by first consulting at least three general classes of sources: 1) standard charts of the area, such as prepared by the United States National Ocean Survey, as well as possible charts prepared for marine construction undertakings or for other projects in the past; 2) scientific reports, including relevant portions of the Regional Wastewater Plan, dealing with the beaches and sea floor as well as the physical oceanographical and marine biological aspects of the area; 3) members of the local populace, in particular fishermen, divers, surfers, and boating enthusiasts, as well as, where possible, the authors of consulted reports.

---

*These matters are considered further in Chapter 3.

†This subject will be enlarged upon, for ocean outfalls, in Chapters 7–11.

Charts (see Appendix A*) are of fundamental importance in almost any marine construction endeavor. They provide a moderately detailed picture of the bathymetry of the general ocean area under consideration and provide minor details on the nature of the bottom, i.e., whether the material is sand or coral or rock. If it is required, for example, that the outfall be located in a minimum depth of water, a perusal of a chart would quickly determine the shortest route between an entry point on shore and the required depth. Natural channels may be uncovered; regions of rock outcropping can be pinpointed and avoided. An experienced coastal engineer may be able to indicate those areas in which the bathymetry can cause amplified wave action to occur; these would be areas to be avoided by the chosen alignment.

But basically, an initial appreciation of the dynamics of the local shore/ocean system must go beyond the bathymetric chart—to the reports and individuals mentioned above. Along some coastal regions, there may be many reports dealing with such items as beach stability, wave climate, current regimes, fish and fish larvae, plankton, and benthic plants and animals. All such studies are important to an ocean outfall location. As a general rule, the larger the pile of pertinent and comprehensive reports assembled, the smaller the data-gathering mission and the shorter the delays in completing the design.

It should be stressed that local nonscientists should be consulted early in a project's life. Those who use the water should have a substantial say in what is to be done, and it is far better to have the public's views quietly stated in the beginning than shouted at a public hearing later. Consultations with the public can indicate those areas that are important for a number of marine recreational or occupational activities. There have been those involved in the planning of marine outfalls that have regarded public input as a source of hindrance. Some engineers, for example, have felt that an entire marine area might be classified as precious by the collective opinion of the public, leading to an impasse in outfall planning and design. This is one of the reasons for secrecy in planning, but there is no excuse for a public servant or the consultant to a public agency to ignore the views of the public just to make his job initially easier.

I wish to make the point very strongly that nonscientists should also be consulted from the beginning about the design aspects of a proposed outfall. Local people, particularly fishermen, surfers, sailers, and skin and SCUBA divers, have years of experience with the beaches and waters through which the outfall will have to pass. They may well be able to provide detailed advice on the advantages of a particular route for the line or on other design considerations. During a confrontation between a surfing group and the U.S. Army Corps of Engineers in the late 1960's over replenishing Hawaii's Waikiki Beach, which was being steadily eroded, a representative for the surfing group said in essence, "you may have your Ph.D.s, but we know the water." There is a great deal of

---

*Appendix A is primarily concerned with the positioning of marine vessels, a subject of importance for operations described in Chapters 5, 6, 10, and 11.

truth in this statement. There often has been too much reliance in marine matters on the opinions of the person with advanced academic training whose only link with the ocean is his facility with a complicated mathematics that pertains only to some idealized picture of one aspect of marine phenomena.

In the final analysis, however, it is folly for those planning and designing ocean outfalls to rely solely on the verbal input of others—or on what has been shown on charts and written in reports. As a minimum requirement, planning personnel should go to the beach in question and walk it (with their shoes off); they should go out on the water in a boat and look at the area they're considering; and they should attempt at least some of the activities practiced in the nearshore waters so that they develop some appreciation of what is involved for the individual and the water in which he is immersed. I have heard of a designer who reportedly camped for a week on a beach that had been chosen as the entry point for a marine outfall. He probably came away with a good feel for the area. I know of another case in which no design engineer ever visited the site before the design report was written. The result was construction, redesign, and financial disaster.

Finally there is absolutely no substitute for the designer himself to survey the sea floor in the general area planned for the outfall. An increasing number of governmental agencies and consulting engineering firms are encouraging personnel working with the marine disposal of wastewaters to become qualified for SCUBA diving and to undertake such pre-design surveys as referred to above. Some countries and organizations regard diving as a technician's job. They feel that observations of sea floor conditions can, if needed, be transmitted from a technician-diver to the design engineer. Most engineers who have tried to gather detailed information in this way, from anyone less than a very unusual diver, will agree that it does not work.

1-4 PROTECTION OF MAN AND THE MARINE BIOTA

All wastewaters, regardless of how well treated, contain potential pollutants. *Pollution* has been defined by the Food and Agriculture Organization of the United Nations (Gafford, 1972) as follows:

> "Introduction by man of substances into the marine environment resulting in such deleterious effects as harm to living resources, hazards to human health, hindrance to marine activities including fishing, impairment of quality of sea water and reduction of amenities."

The public's experience with ocean outfalls has often been bad. Industries and municipal agencies have frequently dumped their untreated (i.e., *raw*) wastewaters into coastal marine environments. Figure 1-1 shows the discharge of a raw sewage effluent. The result has been a foul-smelling and foul-looking

Figure 1-1 Raw sewage effluent entering the marine environment (Courtesy, Richard W. Grigg)

marine sore. In such situations, the resulting pollution can spread to adjacent beaches, causing their closing on public health grounds if not from pure aesthetics. Unthinking industries and municipalities must be prevented from attempting to use such an "expedient" way of getting rid of their unwanted wastes. Wastewaters should be treated before disposal, and they should enter the marine environment as subtly as possible, some distance from shore.

Public pressure and subsequent legislative action in various countries of the world have led to strict laws and enforcement procedures designed to protect not only man's use of the marine environment but also its natural inhabitants. Municipal agencies and industries now must demonstrate that their wastewaters can comply with these standards. Their plans are reviewed by governmental agencies, academic and research institutions, and environmental groups before the plans are deemed satisfactory and detailed design can follow. These review groups collectively demonstrate substantial competence in the myriad aspects of marine matters.

The protection of the marine biota requires, at the very least, a knowledge of the maximum concentration of a pollutant or combination of pollutants in the receiving water that is not injurious to the most susceptible marine creatures or plants, either in the short or long term. The tests to investigate such tolerance are called *bioassays*. Tests are also needed to indicate when a wastewater might be *biostimulatory*, i.e., conducive to increased plant life.

Chapter 4 covers various topics related to the protection of the marine environment and man's use of it. A short history of United States' experience with water pollution control will provide background for water pollution control activities in California in general and in the Los Angeles area in particular. Non-United States experience will be illustrated by considering water pollution control measures in the province of British Columbia, Canada.

Wastewater discharge in the United States is controlled through systems of permits. Chapter 4 will discuss such permits as well as the documents, hearings, and procedures involved in securing them. Finally, the chapter also contains material on bioassays and public health considerations in the marine disposal of wastewaters.

1-5 DATA ON THE RECEIVING WATER AND ITS MARGINS

In order to make the decision to proceed with an outfall it is likely that some measure of work was done in the receiving water to gather data of physical, chemical, biological, bacteriological, and geological import before issuance of the Regional Wastewater Plan. The data obtained in the preliminary design of an outfall as part of the recommended plan for an overall sewerage system, however, is generally insufficient for the detailed design. More oceanographic studies are required. The types of studies involved in both the preliminary work and in the final planning stages are detailed in Chapters 5 and 6.

Individual physical oceanographic studies of importance are discussed in Chapter 5. Waves are included because they must be considered in the stability design of the outfall, in the construction operations, and in their influence on the spread of the wastewater effluent after leaving the pipe to enter the receiving water. Currents must be considered since they can bodily transport diluted effluent along or perhaps toward the coast; additionally, currents influence the degree that an effluent initially mixes with the receiving water. Information on the variation of salinity, temperature, and density, with distance below the water surface, is of vital importance in determining whether or not the diluted effluent will form a surface "boil" or whether it will stay submerged. The diffusion characteristics of the receiving water largely govern the total spread of a dilute effluent mass in the absence of currents.

Chapter 6 combines one last physical concept (water clarity) with chemical, biological, bacteriological, and geological factors. At the least, chemical studies (for nutrients and heavy metals) are necessary to establish base levels so that changes caused by the commencement of an outfall operation can be measured.

The same is true of bacteriological factors. Thorough biological studies are needed to identify members of the plankton, nekton, and benthic communities. Ideally, the levels of pollutants in the diluted wastewater effluent should be sufficiently low that the normal life functions of the most susceptible organisms are not jeopardized.

Information of a geological nature is of vital importance to the design of the outfall. For example, sea bed materials must be able to support the pipe. Subbottom conditions must also be known, for in the nearshore zone the outfall will probably be buried to protect it from wave-induced water motion. Whether or not there is bedrock close to the bottom is an important consideration; if there is, expensive blasting may be required. The stability of the bottom material and of the beach area is of great concern, since the natural cut of beach and nearshore sediments under storm conditions, or over the long term, can conceivably bare a formerly-buried outfall.

1-6   OUTFALL DESIGN

Introduction

It has already been mentioned that the prime structure in an ocean outfall system is the outfall itself, the pipe that transports the wastewater from the land to its disposal point, usually thousands of feet from shore in water depths of about 15 to over 300 ft (about 5 to 100 m). It is usual to have on shore, at the upstream end of the outfall, a plant that can have various functions. Often this is a wastewater treatment plant; in most cases it is also a pumping plant, although in some rare cases an energy-dissipating system must be used for the wastewater flow before it enters the outfall. Wastewater treatment plants are discussed in Chapter 2, pumping stations in Chapter 7.

Modern sewer outfalls do not simply terminate at some offshore location with a pipe-diameter opening such as shown in Figures 1-1 and 8-7. Rather, the end of the pipe is capped off and wastewater flow enters the sea through a series of small holes spaced along the sides of the outfall over most of the offshore section. The length of pipe through which effluent leaves the outfall is known as the *diffuser* and is typically a few hundred to few thousand feet (a hundred to a thousand meters) in length. When the wastewater enters the marine environment from the diffuser it is buoyant and tends to rise. This *effluent*, initially travelling horizontally after leaving the holes, or *ports*, in the diffuser, gradually ascends on a more vertical path. Throughout its path, whether largely horizontal or vertical, the effluent mixes with the ambient sea water. What was initially a highly-concentrated wastewater becomes progressively more and more diluted with increasing distance of movement through the sea water.

The mixing referred to here, and dealt with in some detail in Chapter 7, is called *initial dilution*. This stage is considered terminated either when the mixed wastewater reaches the water surface or when the mixed plume becomes stable at some intermediate water depth and levels out parallel to the surface. This is possible, usually in deeper water, because the mixture of wastewater and colder water nearer the bottom develops the same density as the salt water at the depth at which the plume becomes vertically stable.

Once the mixed effluent reaches the sea surface or an intermediate equilibrium level it tends to move with the prevailing currents, or to spread due to buoyancy effects or *diffusion*. The latter can be strongly influenced by the prevailing ocean currents which tend to carry the effluent cloud with them in a process known as *advection*. The spread of an effluent after the initial dilution stage is known as *secondary dispersion*; it, too, is detailed in Chapter 7.

Preliminary Design

It has already been remarked that a preliminary design for an outfall is necessary to appraise its suitability within the regional wastewater planning study. The steps to be followed are the same as those to be sketched for the final design in the following subsection. The information available to the designer is somewhat less in preliminary work and greater uncertainty is associated with the various engineering quantities and parameters that must be used in the design and in associated costs.

Final Design

Within certain broad limits, and subject to exceptions, the first decision in the overall design of an outfall system concerns the location (and depth) of the diffuser. Procedures discussed in Chapter 7 are then used to determine pipe and port sizes so that effluent is efficiently passed at desired flow rates. Hydraulics computations, also outlined in Chapter 7, are used to compute the energy required at the shoreline to transport desired flows to and out of the diffuser.

Chapter 8 deals with all remaining design aspects for the outfall. It discusses the types of pipe materials and joints that can be used and which combinations are suitable for specific types of situations. It discusses the effect of wave-induced water motion and quasi-steady currents on outfall design. Because the stability of an unburied pipe may be uncertain under such conditions, I discuss ballasting the pipe with rock or anchoring it to the sea bed. Chapter 8 also outlines the often practiced protective technique of burying the pipe in areas where instability is indicated. Depths of burial and the type and size of material that should be used to cover the trench in which the pipe is laid are considered. Stability of the outfall system under conditions of earthquake-induced ground motion or earthquake or wave-induced liquefaction of the bottom material are briefly considered.

The final design report for an outfall describes the design and the engineering approaches used in determining that design. This document is insufficient in itself, however. Drawings and specifications and detailed cost estimates are called for before proceeding to the construction stage. These matters begin Chapter 9.

1-7 OUTFALL CONSTRUCTION

The real construction details, i.e., what is being done in the field, will be discussed in Chapter 10, with Chapter 9 serving as the link between the termination of design and the beginning of construction, as well as the source of other construction-related material. Chapter 9 will outline bidding by construction contractors on the outfall project plus contract award. Also included will be material on construction management, inspection, and safety. Material on diving, an activity of great importance not only in outfall construction but also in other ocean outfall work, is presented in Appendix B.

Chapter 10 starts off with a list of some of the numerous marine outfalls in existence. Presented along with the locations of the lines are such characteristics as pipe material, pipe size, outfall length, and maximum water depth. Since, as has been outlined earlier, submarine pipes (outfalls among them) are usually placed in a trench when passing through nearshore waters, part of Chapter 10 is devoted to a discussion of the means of creating such trenches, dredging and blasting. A considerable amount of material follows on the various techniques that can be used in different waters and in different depths to lay the outfall. One of these, for nearshore waters, involves a temporary pier or trestle as shown in Figure 1-2. A walking platform (Figure 1-3) is another possibility.

Frequent references will be made to techniques and developments used during various ocean outfall construction projects. Much of this material will be tied into the 1974–75 construction of the Sand Island No. 2 outfall off Honolulu, where I was able to study first hand both the overwater and underwater operations involved. Chapter 10 concludes with material on covering laid pipe with rock; with information on work divers and inspection (Figure 1-4); and with a review of construction problems.

The first third of Chapter 11 is devoted to a discussion of the effects of outfall construction work itself on the marine environment. I believe there is frequently more harm done to the marine biota over one to two years of outfall construction than over many years of operation of an outfall passing satisfactorily-treated wastewater. This is primarily because of the enormous amounts of silt that enter the water column and then reach the sea bed.

Figure 1-2 Construction trestle for marine outfall

Figure 1-3 Walking platform used in outfall construction

Figure 1-4 Author inspects outfall during construction (Courtesy, Arthur T. Shak)

1-8 OPERATION, MAINTENANCE, AND MONITORING

Operation and Maintenance

Many outfalls give no trouble whatsoever after they are built. Generally each one accepts the wastewater given to it and takes it away—day after day, week after week. But just because the outfall accepts its load without a breakdown does not necessarily mean it is operating properly. I know of several lines that have apparently, as far as the people in the treatment/pumping plant were concerned, been working properly but where, in truth, none of the flow was reaching the diffuser; it was all entering the receiving water through cracks or breaks closer to shore. Thus diver inspections must be a part of outfall operation and maintenance, and this matter is discussed at greater length in Chapter 11.

Some operators of outfalls do, however, encounter problems obvious from their on-shore vantage point. These problems will also be outlined in the final chapter.

Monitoring Outfall Performance

It is now law in the United States that outfall performance must be monitored. This subject occupies the last part of Chapter 11. Continuing studies must be made of the water column characteristics at the effluent point, near the

Table 1-1: Los Angeles Area Wastewater Outfalls

Operator	Location	Number	Date	Extension date	Pipe size in.	Pipe size mm	Material	Approximate length ft	Approximate length m	Approximate discharge depth ft	Approximate discharge depth m	Design capacity Mgd	Design capacity m³/s
Los Angeles City	Hyperion (El Segundo)	1	1894	—	24	610	cast iron	600	200	—	—	—	—
		2	1904	—	30	762	—	950	300	—	—	—	—
		3	1908	—	34	864	wood stave	950	300	—	—	—	—
		4	1918	—	54	1,372	wood stave	2,000	600	—	—	—	—
		5	1925	—	84	2,134	RCP	5,400	1,700	55	17	—	—
		6	1948	—	144	3,658	RCP	5,300	1,600	60	18	245	10.7
		7	1959	—	144	3,658	RCP	27,500	8,400	200	61	—	—
Los Angeles County	Whites Point	1	1937	—	60	1,524	RCP	4,500	1,400	110	33	90	3.9
		2	1947	1953	72	1,829	RCP	6,600	2,000	155	47	150	6.6
		3	1956	—	90	2,286	RCP	8,000	2,400	205	63	260	11.4
		4	1964	—	120	3,048	RCP	11,900	3,600	190	58	—	—
Orange County	Newport Beach	0	1924	—	42	1,067	cast iron	2,000	600	30	9	—	—
		1	1953	—	78	1,981	RCP	7,000	2,100	55	17	100	4.4
		2	1970	—	120	3,048	RCP	27,400	8,400	195	59	290	12.8

discharge, and remote from it. The nature of the bottom at and surrounding the diffuser must also be studied, and extensive data on the various components of the marine biota in the area of the discharge point are also necessary. Monitoring differs somewhat from the oceanographic data-gathering outlined in Chapter 5 and 6 to the extent that considerable attention is paid to effluent-induced changes. These include piles of organic materials on the bottom and pollutants in the bottom material, abnormalities in the local marine creatures, altered plankton biomass and impaired water clarity, and above-normal levels of pollutants in the water column.

1-9 OUTFALLS IN SOUTHERN CALIFORNIA

The preference of coastal cities for ocean disposal of their treated wastewaters can be illustrated by considering the western coast of the United States. In the southern California coastal basin in 1969 there were 174 municipal wastewater treatment plants in operation. The combined total effluent of these plants has been estimated at 1111 Mgd (48.7 m^3/s) [SCCWRP (1973)]—with 75 Mgd (3.3 m^3/s) being reused directly. Another 97 Mgd (4.2 m^3/s) was discharged inland (possibly finding its way to the ocean from that point) with the remaining 939 Mgd (41.2 m^3/s), and 84% of the total, being discharged directly into the southern California coastal waters. This flow, only a minute fraction of the volume flow rates associated with coastal currents, had most of its major inputs (89%) at and near Los Angeles—at Hyperion, 15 miles (24 km) southwest of the city center (36%); at Whites Point, 25 miles (40 km) south of the center of the city (39%); and in Orange County, 35 miles (56 km) southeast of the city center (14%). The San Diego outfall contributed 10%. Table 1-1 presents the individual outfalls involved at one time or another in the Los Angeles area. RCP stands for reinforced concrete pipe.

1-10 REFERENCES

GAFFORD, R. D. (1972): "Automation of Monitoring Equipment for Marine Pollution Studies," in *Marine Pollution and Sea Life*, Mario Ruivo (Ed.), Food and Agriculture Organization of the United Nations, Rome, Conference (December). Fishing News (Books) Ltd., London, England, pp. 491–500.

Southern California Coastal Water Research Project (1973): "The Ecology of the Southern California Bight: Implications for Water Quality Management," *Technical Report 104*, El Segundo, California, (March).

1-11 BIBLIOGRAPHY

BARUTH-YODER, Engineers-Planners (1971): "Wastewater Collection, Treatment and Disposal," report prepared for Mid-Humboldt County, California (July).

METCALF and EDDY, Inc. (1972): *Wastewater Engineering: Collection, Treatment, Disposal*, McGraw-Hill, New York.

SALVATO, J. A., JR. (1972): *Environmental Engineering and Sanitation*, Wiley (Interscience), New York.

"Water Quality Program for Oahu with Special Emphasis on Waste Disposal" (1971), report in ten volumes prepared for the City and County of Honolulu, Hawaii, by Engineering-Science, Inc., Arcadia, California, Sunn, Low, Tom and Hara, Inc., Honolulu, Hawaii, and Dillingham Corp., Applied Oceanography Division, Honolulu, Hawaii (December).

"Waste Water Disposal and Reclamation for the County of Orange, California, 1966–2000" (1966), report prepared by Engineering-Science, Inc., Arcadia, California, and Lowry and Associates, Santa Ana, California, for the Board of Supervisors, County of Orange, California (July).

2

The Marine Environment

2-1 SOME GENERAL CONCEPTS

Introduction

The oceans and seas cover 70.8% of the earth's surface, comprising 60.7 and 80.9% of the northern and southern hemispheres respectively. The world ocean has an average depth of 12,500 ft (3,800 m); most of the ocean (83.9%) occupies the *abyssal* depths (2,000–6,000 m) and *hadal* depths (>6,000 m). Near the coasts the sea is divided into two regions: one is known as the *continental shelf*, a region typified by a gradual slope (typically 0.1°) away from the land and shallow depths; the second region is characterized by a pronounced slope (typically 4°) extending from the edge of the shelf down to abyssal depths. This latter region is known as the *continental slope*, and, with the continental shelf, it makes up the *continental terrace*. The continental shelf, which makes up 7.5% of the world ocean, has an average width of about 48 miles (78 km) but on some coasts may be entirely absent or off other coasts as wide as 900 miles (1500 km). The average depth of the world's continental shelves is 436 ft (133 m), and a general rule of thumb is to assume that the continental shelf (and *coastal depths*) extend to a depth of either 200 m or 600 ft/100 fathoms. The waters over the continental shelf comprise the *neritic zone*.

Basic Ocean Chemistry

The feature that distinguishes ocean water from fresh water is that it is salty. One view is that this salinity is due largely to accumulation of substances carried into the sea by the world's rivers. The Colorado River, for example, annually adds about 14×10^6 tons of mineral salts to the Pacific Ocean.

The concentration of salt in "typical" sea water is 35g/1000g water or $35^0/_{00}$ (parts per thousand by weight, ppt) or 35,000 ppm (parts per million by weight). The relative saline composition of sea water is remarkably constant over most of the world's oceans because of the enormous amount of mixing and circulation that has gone on in the sea since its beginnings. The major constituents of sea water salinity are the following: Cl^- ions (55.0% by weight); Na^+ ions (30.6% by weight); SO_4^{--} ions (7.7% by weight); Mg^{++} ions (3.7% by weight). These are the major constituents, and others are apparent in the following recipe for artificial sea water. True sea water includes at least 60 other *trace constituents* (<1 ppm) that are not listed in Table 2-1, and one *minor constituent* (1–20 ppm), silicon.

Table 2-1:   Recipe for Artificial Sea Water*

NaCl	23.48 g	$NaHCO_3$	0.192 g
$MgCl_2$	4.98 g	KBr	0.096 g
Na_2SO_4	3.92 g	H_3BO_3	0.026 g
$CaCl_2$	1.10 g	$SrCl_2$	0.024 g
KCl	0.66 g	NaF	0.003 g
Add H_2O to form 1000 g of solution.			

*From *Oceanography*, 3rd ed., 1976, page 52, by M. Grant Gross; reprinted by permission of Charles E. Merrill Publishing Co., Columbus.

Dissolved gases also occur in sea water, notably nitrogen, oxygen and carbon dioxide. The latter two gases are the most important, because carbon dioxide is necessary for photosynthesis, and oxygen is required in plant and animal respiration as well as the oxidation of wastes.

Carbon dioxide is present in the sea in a number of forms as it reacts with H_2O: as CO_2 in true solution; as the undissociated carbonic acid H_2CO_3, in minute quantity, or as dissociated carbonic acid; as the slightly soluble carbonates; as the more soluble bicarbonates, HCO_3^-. Most of the CO_2 in sea water is present as the bicarbonate ion.

The uses of oxygen make it difficult to state exactly what the concentration of dissolved O_2 is in the surface or near-surface sea waters. In addition, since the solubility of gases in water varies inversely with temperature, the amount of dissolved O_2 in surface sea water varies with the location and season. As an example, however, a dissolved oxygen level of 6–8 mg/ℓ in sea water might be considered typical in the subtropics.

The salinity (S) and temperature (T) of sea water are the prime determinants (with pressure) of its density in g/cm^3 (see Figure 2-1). Oceanographers use

$$\sigma_t = (\text{density in g/cm}^3 - 1)1000$$

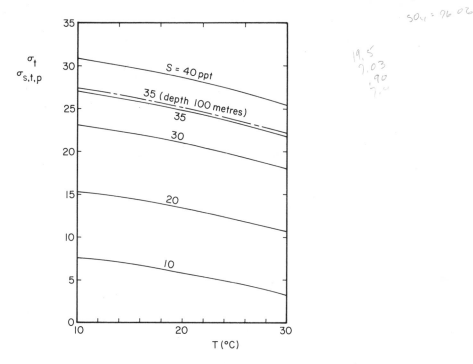

Figure 2-1 Variation of water density with salinity, temperature, and depth

for convenience. Typically σ_t is around 25, equivalent to a density of 1.025 g/cm^3 or a specific gravity of close to 1.025. σ_t always refers to a density for atmospheric pressure; $\sigma_{s,t,p}$ values take the additional effect of pressure into account. The effect of depth (pressure) on σ_t amounts essentially to a 0.1-unit increase for every 20-meter depth increase (see Figure 2-1). For the depths where wastewater outfall diffusers are located, typically 30–250 ft (about 10–75 m), there is little difference between σ_t and $\sigma_{s,t,p}$.

Variations with Depth

Surface waters of the sea are warmed by the sun and thus generally have a higher temperature than deeper waters. Winds and waves agitate the surface waters, resulting in a *mixed layer* of essentially constant temperature. Unless the water is sufficiently shallow that the mixed layer exists throughout the water column, the base of the mixed layer is marked by a *thermocline*, a region through which there is a significant decrease in temperature. In fact, the thermocline is that (more or less) horizontal plane at which the rate of decrease of temperature with depth is a negative maximum. Because of different wind, wave, and solar heating conditions throughout the year, the depth to the thermocline changes seasonally. Below the thermocline the temperature gradually decreases with increasing depth although it may level off in deep water.

Because of precipitation and evaporation, and in some cases the inflow of rivers and the melting of polar ice, salinity is variable over the ocean's surface. The salinity of the lower levels appears moderately constant, and the narrow region through which the surface salinities give way to the deeper salinities is called the *halocline.* Density, lowest near the surface, increases rapidly at the base of the mixed layer through the *pycnocline,* then usually increases slowly with increasing depth. Figure 2-2 depicts the variation of T, S, and $\sigma_{s,t,p}$ with limited depth at a point near the Hawaiian Islands where the water depth was approximately 600 fathoms (1.1 km).

Sunlight penetrates only the surface zone in the sea—the *photic* zone, about 200 m thick in the open ocean. At a depth of 200 m in clear, deep ocean water, less than 0.01% of the incoming radiation remains as visible light. Light at these depths is usually blue since seawater most readily attenuates the red part of the visible light spectrum. Turbidity, resulting from suspended and dissolved materials in the water, may greatly limit the depth of light penetration and change the

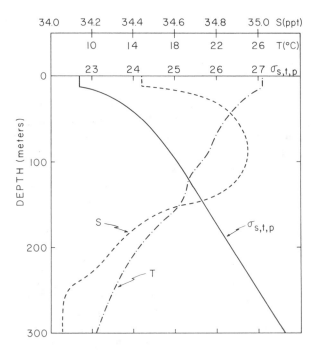

Figure 2-2 Salinity, temperature, and density profiles at station near Hawaiian Islands, June 1967

spectral distribution of the light at that depth. For example, in coastal waters with high turbidity, 99% of the incident light may be absorbed in the top 5 m of the water.

2-2 LIFE IN THE SEA

The Plankton

The sun's light is of enormous importance to life in the sea. The base of the so-called food or *trophic* pyramid in the oceans is occupied by minute, floating plants called *phytoplankton*.* These forms of plankton, which are generally single-celled, utilize the energy of sunlight and readily-available constituents of the sea water in the production of organic material. The process is termed *photosynthesis* and is described by the following equation:

$$6CO_2 + 6H_2O + \text{solar energy} \xrightarrow{\text{chlorophyll}} C_6H_{12}O_6 + 6O_2 \qquad (2\text{-}1)$$

In this process, water and carbon dioxide combine, in the presence of adequate light energy, to form an organic material and release oxygen. Chlorophyll is the green pigment in plants. The organic material is a simple sugar, a fundamental carbohydrate, which is then used by the phytoplankton in two basic transformations. On the one hand it is used as a source of energy so that life processes can be maintained. In this case the energy of the sun has essentially been passed on as chemical energy. On the other hand the simple sugar is used as a basic building block in the formation of other carbohydrates, as well as lipids (fatty substances) and proteins, in processes which do not use solar energy but rather the chemical energy of the simple sugar referred to above. Plants use proteins, for example, in construction of new protoplasm. The term *respiration* concerns the obtaining of energy by the phytoplankton from the photosynthetically-derived simple sugar. This oxidation process can be written as essentially the reverse of the photosynthetic reaction as follows:

$$C_6H_{12}O_6 + 6O_2 \longrightarrow 6CO_2 + 6H_2O + \text{energy (ATP)} \qquad (2\text{-}2)$$

Metabolism refers to the essential chemical processes—all involving energy transformations—of the living body. Movement, growth, and a host of metabolic reactions which constitute living require energy. It is apparent from Equation (2-2) that oxygen is used up in the respiration process and carbon dioxide is liberated. The organic materials synthesized by the plants from the

---

* Diatoms and dinoflagellates are major examples of phytoplankton.

simple sugars are composed of many more elements than simply the carbon, hydrogen, and oxygen. For example, the amino acids which are built into proteins also have nitrogen and sometimes sulfur, magnesium is a part of chlorophyll, and phosphorus (along with nitrogen) are constituents of ATP, which is the chief high-energy compound in plants, storing immediately available energy for the use of the phytoplankton. The nitrogen, phosphorus, magnesium, and sulfur referred to here are, along with various other elements, lumped together under the term *nutrients*. In order to grow and reproduce,* the phytoplankton must have an adequate supply of all the essential nutrients. Frequently one or more of the necessary elements is not available in the same relative abundance as the others; such a *limiting nutrient* then exerts a control over the level of phytoplankton productivity. Nitrogenous compounds and phosphates are the usual limiting nutrients, but silica has also been shown to be limiting in some cases. As an example, plant tissues may contain 2–3% of their dry weight as phosphorus. Phytoplankton are frequently called *producers* or *autotrophs* (meaning self-feeders) because they produce organic materials from inorganic nutrients and the sun's energy.

The amount of organic matter produced within a unit volume of sea water or under a unit area of the sea surface per unit time is known as *production*. The concentration of plankton is normally given either in terms of *standing crop*, the actual number of organisms per unit water volume or surface area, or *biomass* which is the dried weight of living organisms per unit of water volume or surface area.

Solar energy is indispensable to the photosynthetic process. Because this energy attenuates with water depth, phytoplankton at greater distances from the water surface have smaller amounts of solar energy available to them. It is not surprising, then, that not all plants in the sea are net contributors to the supply of free oxygen (available for animals) because in the deeper portions of the photosynthetic zone the plants may consume as much or more O_2 during respiration as they liberate during photosynthesis. The break-even point in the production and consumption of oxygen is known as the *compensation depth*, which may be of the order of 100 ft (30 m) beneath the surface. The region between the sea surface and the compensation depth is known as the *euphotic zone*.

Phytoplankton are consumed by *heterotrophic* ("other-feeders") organisms, which cannot synthesize new organic matter from inorganic substrates but must obtain their nutrients from other organisms. These are animal organisms called *herbivores* or *grazers*, and they are represented among the plankton by the *zooplankton*. The term plankton implies "wandering," and except for some members that possess their own weak means of locomotion, the zooplankton,

---

* Phytoplankton reproduction is asexual. It involves the division of the organism into two more-or-less equivalent parts (*binary fission*) each of which continues to live as a whole organism.

like the phytoplankton, are subject to completely passive dispersion. They drift with the prevailing current systems unless they can move out of such systems by adjusting their buoyancy and rising or falling. But even the movements of the zooplankton with token means of locomotion are dominated by water movements because they are nondirective swimmers.

The types of marine zooplankton have been described, for example, by Coker (1962), Cromie (1966), McConnaughey (1970), and Fell (1974). Almost every phylum of animals is represented in the marine zooplankton which can conveniently be divided into two groups. *Holoplankton* spend all of their lives as plankton, whereas *meroplankton* pass only a fraction of their lives as part of this community.

The microscopic foraminifera and radiolarians are examples of the holoplankton. They are representatives of the phylum Protozoa. Larger permanent plankters are, for example, jellyfish (phylum Coelenterata) and the shrimp-like copepods (phylum Arthropoda) which are the most abundant single members of the animal plankton over the world's oceans.

Larval stages of the *nekton*, which are freely-swimming creatures, and the *benthos*, which are animals that live in or in close contact with the sea floor, constitute the meroplankton. Examples of the latter are the larval stages of oysters and clams (phylum Mollusca), lobsters, crabs, and barnacles (phylum Arthropoda) as well as the larval planula stage of the corals (phylum Coelenterata).

The Nekton and Benthic Animals

Plankton (drifters) and nekton (free swimmers) are not always distinctly separable. Large shrimps or prawns, for example, are on the border line. Most members of the nekton are members of the phylum Chordata, which includes the bony fish, turtles, sharks and rays, dolphins and whales (see Figures 2-3 and 2-4). The nekton feed on the zooplankton or on each other, or they may be herbivorous and feed on the phytoplankton. Some nekton may range over great distances; for example, young tuna tagged off Baja California have been caught in Hawaiian waters.

After benthic animals have reached the bottom upon maturation, they remain in the same area for the rest of their lives. The benthos may be divided into four groups, a very minor one of which consists of creatures like the scallop that can swim very short distances. A second group consists of fixed organisms like sponges, corals, and oysters that live on the bottom (*sessile epifauna*); a third group includes animals such as crabs, starfish (Figure 2-5), urchins (Figure 2-6) and lobsters that can move slowly over the sea floor (*motile epifauna*); in the fourth group (*infauna*) are organisms like clams and worms that burrow in the bottom material and are known as deposit feeders. An example of such worms is the polychaete, or bristle worm, which is used as an indicator of gross pollution

Figure 2-3 School of goatfish

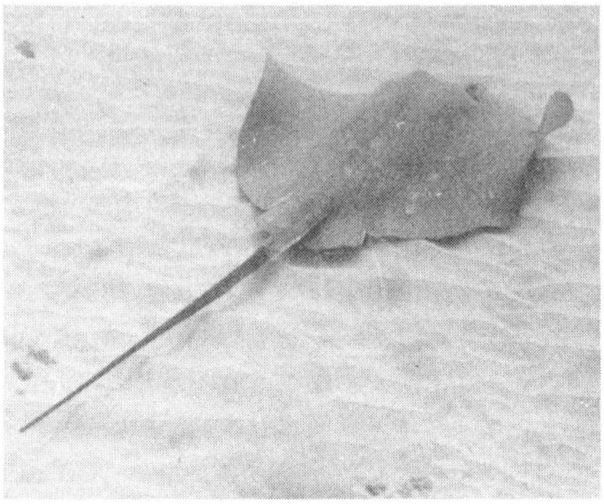

Figure 2-4 Large sting ray

Figure 2-5 Crown of thorns starfish feeding

since some species reach very large concentrations under such conditions. Other major feeding types besides deposit feeders among the benthic animals are suspension and filter feeders.

Marine organisms may concentrate trace elements present in sea water by factors over 10,000 times the element's concentration in the water. Clams, crabs, and sea cucumbers are benthic animals that display this tendency. If the local sea water concentration of certain trace elements is increased due to waste products dumped into the sea, marine organisms may concentrate levels dangerous to themselves and to their consumers. In a similar fashion marine organisms may concentrate man-made chemicals foreign to the natural marine environment, such as pesticides, with, perhaps, unfortunate consequences.

Corals are a particularly important type of benthos in the tropics and subtropics. Reef-building (*hermatypic*) corals flourish where the water tempera-

Figure 2-6 Sea urchins

ture is maintained between 25 and 29°C the year round. They cannot withstand temperatures below 18°C nor elevated temperatures, such as from heated power plant effluents, nor extreme salinities outside the approximate range $27–38^0/_{00}$. Coral polyps live upon calcareous skeletons and obtain their food by filtering plankton from the water. Because they are sessile, a certain amount of water motion is necessary to circulate the food past the organisms and to carry away unwanted clogging (and killing) debris. This is only one of two sources of food for reef-building corals, however. Innumerable small, round, colored plant cells called *zooxanthellae* live within coral tissues. The zooxanthellae use much of the carbon dioxide and nitrogenous wastes of the corals before these waste products enter the water. Coral tissues receive in return the oxygen by-product of photosynthesis and may also receive some carbohydrates from their symbionts.* Reef-building corals thus exist only in water depths where sunlight can penetrate. Corals spread to different areas when the *planulae* finally settle on a solid, stable bottom. Individual coral colonies grow by an asexual reproductive process called *budding* where a small portion of an organism grows out from the parent and eventually develops into another individual or another part of the colony.

Sustaining the Life of the Sea

The amounts of light and mineral salts in sea water vary with the seasons, and generally there are phytoplankton in profusion—plankton *blooms*, in the spring and in the fall. During such blooms of phytoplankton, zooplankton abound. After a bloom of plant plankton has peaked, the population decreases

---

* *Symbiosis* refers to a relationship between two species, principally applied to a relationship in which both species benefit (*mutualism*), such as hermatypic corals and zooxanthellae.

because of grazing by animals, the decrease in available nutrients, lessened light (either because of seasonal changes or suspended matter in the water), or the organisms sink, to the benefit of deeper feeders. It is of note that production can be high but the standing crop low because of consumption of plants by herbivores.

In the trophic pyramid in the sea, there is approximately a tenfold decrease in the amount of protoplasm that can be built from protoplasm of the lower level consumed. Primary consumers feed on phytoplankton; e.g., copepods consume diatoms. Second-order (carnivorous) consumers, such as herring, feed on copepods; third-order consumers (such as tuna) eat herring; and the tuna in turn is devoured by man, a fourth-order consumer. Using the crude factor of 10 mentioned above, this means that it effectively takes about 1000 lb of phytoplankton to add 1 lb to the weight of a tuna. The tenfold decrease at each higher consumer level explains loosely the much larger biomass of herbivorous fish in the ocean than carnivorous.

Dead and decaying forms of marine life, as well as waste products and scraps dropped by herbivores and carnivores, sink into deeper layers of water and, if not consumed by the organisms of those layers, become decomposed liberating chemical materials in a region where they cannot be immediately utilized because of the lack of sunlight. This collection of sinking particles is called *detritus*. Thus, there is a considerable accumulation of organic material in solution, the concentration being greatest in the deepest ocean layers. One may naturally inquire if this enormous amount of dissolved organic matter (estimated to be 300 times the amount of living organic matter in the sea at any time) is irretrievably lost, never to be reused. Since life persists in the open ocean remote from coastal areas where there are always nutrients washing into the sea from the land, one would suspect that there is a means by which the "lost" nutrients are brought back into the mainstream of life. This is brought about where "waters from the deep rise to the surface and become the seat of a large outburst of planktonic life which imparts a distinct tint of green to the water. From these areas of fertility and abundance the waters spread by the currents become progressively poorer in the salts necessary for plant growth, and the large areas of the ocean where the water is of pure blue can only be compared to deserts supporting a minimum of life."* One region of *upwelling* is off the coast of Peru where the prevailing winds and the effect of the earth's rotation drive the surface waters away from the land. The offshore seas around Hawaii are an example of an oceanic desert. An excellent picture of the regions over the world's oceans which are conducive to planktonic abundance is given by Isaacs (1969).

Larger Marine Plants

The phytoplankton are by no means the only plants in the ocean and along its fringes. There are many other species of marine plants, mostly larger

---

*R. E. Coker, *This Great and Wide Sea*, Harper & Row, New York, 1962, with permission.

seaweeds that attach themselves to firm substrates such as rocks. McConnaughey (1970) has described larger marine plants in some detail.

The seaweeds include members of the green, brown, and red algae. A moderately well-known type of brown algae is kelp. These are potentially very large plants and are common off California. They fix themselves to the sea floor by means of a *holdfast* and extend up to the surface supported by gas bladders. Kelp are perennial on rocky bottoms in depths of from about 20–100 ft (6–30 m) especially on the open coast where there is continuous swell. Large kelp beds may serve as wave protection for the adjacent shoreline.

Various types of grasses are suited to growth in the marine environment. Although few animals feed directly on them, sea grasses may contribute more to inshore productivity than planktonic plants. Sea grasses spread vegetatively by the production of long runners or rhizomes either on the surface of the substrate or within the substrate. These rhizomes produce roots that may penetrate deeply into the bottom as well as leaves that form marine meadows. The rhizomes and roots form dense networks in the substrate that bind the bottom sands and muds together, thus greatly reducing the erosion by waves and currents. The leaves trap sediments and detritus and provide extensive surfaces for the attachment of minute sessile plants and animals. The leaves, the attached plants and animals, and the detrital material provide refuge and a natural nursery ground for various types of fish and creatures such as lobsters and shrimp.

The term *mangrove* applies to all woody plants that invade coastal zones covered by water at high tide. Typically, however, one pictures mangrove plants consisting of branching, stilt-like roots that hold the rest of the tree above the water. Mangroves thrive along shallow, protected coastlines in the tropics and subtropics; their roots serve as a base for sediments and detritus, which eventually builds a new land area. It has been reported that an acre of red mangrove drops in the neighborhood of 3 tons of leaves per year. These leaves are quickly coated with microorganisms that are eaten by crabs, worms, insect larvae, shrimp, and small forage fishes. In turn, these detritus eaters are prey for many species of fish juveniles and birds.

2-3 PHYSICAL PROCESSES ON A GLOBAL SCALE

Introduction

There are various oceanic processes with important implications for the disposal of wastewaters in the marine environment that have their origins in global-scale phenomena. The elements in this "large" picture provide the focus of this section, and the first to be considered will be the effect of the earth's rotation on large-scale processes in the atmosphere and in the ocean.

The Coriolis Force

Even though an object is moving at constant speed along a circular path, it still has an acceleration—the centripetal acceleration—directed toward the center of the circle. If an object is accelerating, it is clearly shown by Newton's second law that a force must act in the direction of that acceleration. The magnitude of the force is equal to the product of the acceleration and the mass of the object. If the object is a chestnut being whirled in a plane around a child's head, then the force permitting the acceleration, and that opposing the so-called centrifugal force, would be that provided by the string attached to the chestnut.

Consider a point P on the surface of a sphere rotating about a fixed axis with constant angular velocity w. This point will then have an acceleration directed toward the rotational axis in a plane perpendicular to that axis and containing the chosen point. An object resting motionless at this point has the same acceleration as the point. But if the object moves along the surface with constant velocity U with respect to that point, it turns out that, because of the rotation of the sphere on whose surface the reference point has been located, the object has an acceleration with respect to that point along the surface. This acceleration is given by the equation

$$a = 2 U w \sin \phi \qquad (2\text{-}3)$$

where the angle ϕ is defined in Figure 2-7. The acceleration as given above is known as the *Coriolis acceleration,* and it is oriented perpendicular to the direction of U.

In the case of the Coriolis acceleration, by extension of the argument in the case of the whirling chestnut, for a moving object on a sphere there must be a

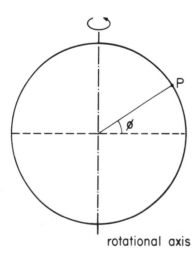

rotational axis

Figure 2-7 Schematic diagram of the rotating earth

force applied perpendicular to its instantaneous velocity—which of course will tend to change its direction of movement.

To a very high level of approximation the earth can be considered as a sphere rotating at constant angular velocity. For the earth this angular velocity is 7.29×10^{-5} rad/s. Thus we expect the reasoning associated with an arbitrary rotating sphere and object to apply to the earth and objects moving along its surface. In this case the angle ϕ in Figure 2-7 becomes the latitude of the object in question. Latitude is discussed in Appendix A.

Consider a 160-lb (712-N) person walking at a comfortable speed $U=4$ fps (1.2 m/s) at the latitude of 30°—which could have him in New Orleans or Cairo. Then the Coriolis force is, from Equation (2-3), approximately 14×10^{-4} lb (0.006 N). Needless to say, this minute force would in no way hamper a person's traveling a straight course and the reader may then question even the mentioning of the Coriolis force.

But even a very small force, if given opportunity to act for a long time, or equivalently over a long distance, can effect noticeable changes. Processes on a global scale, such as atmospheric and oceanic movements, are in such a category, and the Coriolis force is an important consideration. Its tendency, on the earth, is to deflect objects moving in the northern hemisphere to the right and those in the southern hemisphere to the left.

Other perspectives concerning the difficult concept of the Coriolis force are available, for example, in Weyl (1970) and in Gross (1972).

Surface Air Movements

Although the movement of air over the earth's surface is properly a part of meteorology and not oceanography, air movements exert such a significant influence over certain important processes in the sea that they must be considered.

A simplified average sea-level pressure representation for the world would show high-pressure zones circling the earth near latitudes 30°N and 30°S (the *subtropical highs*). High-pressure areas would also be shown at both poles. Low-pressure bands would be depicted surrounding the earth near latitudes 60°N and 60°S (the *subpolar lows*) with another low-pressure zone along the equator. Seasonal alterations result in the zones between the poles shifting somewhat south during the northern hemisphere winter and north in the northern hemisphere summer.

If one follows the normal approach that flow proceeds directly from a zone of high pressure to one of low pressure, it would be expected that there would be movement of air north and south from the subtropical lows to flanking bands of lower pressure as well as air movement south from the North Pole and north from the South Pole. But the fact that the earth is rotating alters the view of small-scale fluid phenomena when one studies processes on a global scale. The Coriolis force causes air movements in the northern hemisphere to be deflected to the right and those in the southern hemisphere to be deflected toward the left. Although if left to its own devices the Coriolis force would ultimately lead to air movements parallel to high- and low-pressure bands, the existence of friction between the moving air and the earth's surface results in the global wind patterns to be described below.

Air moving initially toward the south* from the northern hemisphere subtropical high ultimately blows from the northeast—the so-called northeast tradewinds that dominate the year in the Hawaiian Islands. Air moving initially north from the southern hemisphere subtropical high ultimately blows from the southeast—the southeast tradewinds that blow over Tahiti. The region of calm wind near the equator where these two tradewind systems merge is called the *doldrums*.

The *westerlies* occur in the temperate zone between the subtropical highs and subpolar lows; in the northern hemisphere these winds usually have a southerly component, and in the southern hemisphere, a northerly one. In the middle of this zone lie the "roaring forties" where wind speeds are often elevated.

Although the preceding simplified picture of the global winds is essentially valid in an average sense, there will be at any time on the earth's surface both high- and low-pressure cells as well as *fronts*, lines of demarcation between air masses of different origins, and other meteorological phenomena (see Fig. 8-15). Higher level air flows down and along the earth's surface from high-pressure zones, whereas surface air moves toward low-pressure cells wherein there is rising air. Air does not move directly into or out of such zones again because of the effect of the earth's rotation. In the northern hemisphere air near a high-pressure area (*anticyclonic disturbance*) tends to show a clockwise rotation when seen from above; similarly, a low-pressure zone (*cyclonic disturbance*) shows a counterclockwise rotation of air around its center. The reverse is true in the southern hemisphere.

Circulation of Surface Waters in the Ocean

Details on the circulation of ocean waters are provided in various books, among them Dunlap and Shufeldt (1969), Gross (1972), and Van Dorn (1974). Although the marine interest in this volume concerns coastal waters where wastewaters enter, one cannot completely ignore open ocean current systems because there are localities, such as the southeast coast of Florida, where coastal water circulation can be greatly perturbed by passing large-scale ocean circulation systems (e.g., McAllister, 1968; Lee and Mayer, 1977).

The particular feature of the ocean circulation system that affects Florida coastal waters is the Florida current, part of the Gulf Stream, a mighty flow of

---

*Note that, for explanatory purposes, the usual representation for the wind having a direction from which it blows has been temporarily discarded.

warm water moving swiftly (typically several knots) north in the Atlantic Ocean largely parallel to the east coast of the lower United States. It breaks off at roughly 40°N latitude to move toward Europe as a somewhat diffuse North Atlantic current with various filaments. The trend of circulation along Europe and North Africa is south, and these currents feed the North Equatorial Current flowing westward across the Atlantic between about 10° and 20°N latitude. The North Equatorial Current is a feeder for the Gulf Stream in the Caribbean and Gulf of Mexico. Another feeder for the Gulf Stream is the South Equatorial Current, a westward-flowing current that is part of a current system rotating counterclockwise in the South Atlantic between South America and the southern hemisphere part of Africa.

The winds provide the main driving force for the circulation in all oceans. It is not surprising then that the pattern of oceanic circulation in the Atlantic and Pacific is for clockwise *gyres* (rotating masses of fluid) to exist in the northern hemisphere with counterclockwise ones in the southern hemisphere.

The Gulf Stream has its Pacific Ocean counterpart in the mighty Kuroshio Current that sweeps north up the eastern coast of Asia, breaking to the east off Japan. A southern hemisphere western-ocean-boundary equivalent is the Agulhas Current flowing south along the eastern border of lower Africa. Generally speaking, western-ocean-boundary currents are stronger and more concentrated than those directly opposite them in the same ocean.

At many locations in the ocean, surface currents tend to flow together (a *convergence*), resulting in a sinking (*downwelling*) of surface water; in a *divergence*, water rises to the surface (*upwelling*). Offshore winds can give rise to upwelling near a coast, such as that off Peru, where the Coriolis force also provides a helping hand.

Tides

The gradual rise and fall of the sea surface during a day, called the *tide*, is due to gravitational attractive forces between the sun and moon and the earth, including the water on its surface, coupled with centrifugal effects. The moon, because of its close proximity to the earth, plays the prime role in tide formation; the distant sun, because of its enormous mass, also plays an important role, but its maximum tide-producing force is only about 46% that of the moon. One can visualize long, low mounds of water on opposite sides of the earth forced round and round its surface by these tide-producing forces. Very readable descriptions of the tides are available in Bascom (1964) and Gross (1972); more detailed treatments are contained in Sverdrup et al. (1942) and in Neumann and Pierson (1966).

The tide rises to a maximum height called *high tide* or *high water* (HW), later falling to a minimum level termed *low tide* or *low water* (LW). The time during which there is virtually no change in high or low tidal level is known as *stand*.

The tide-producing forces for any open ocean location have considerable variability because the sun and moon do not describe repeated (apparent) paths along the equator. The sun's declination* varies from 23.5°N (the Tropic of Cancer) to 23.5°S (the Tropic of Capricorn) and back again over a year's time, while the moon shifts between several degrees above the Tropic of Cancer to several degrees below the Tropic of Capricorn and back again in 27.3 days on the average. Because of ellipticity of orbits, the distances of the sun and moon from the earth vary during these cycles. Another complication is that the time taken for one (apparent) complete circle of the sun about the earth is 24 h, whereas that of the moon is 24 h 50.5 min. This difference means that the sun and moon cross an observer's meridian together on an average of about every $29\frac{1}{2}$ days (the *synodic month*). Because of the variability in the sun's and moon's positions, a complete cycle of tides is not realized in a year's time. A better figure for a complete cycle is 18.6 years and a better one yet, 20,940 years (Van Dorn, 1974).

On a month-to-month basis, the sun and moon are most closely aligned on the same side of the earth at new moon and most closely aligned on opposite sides at full moon. Near both such times the tide-producing forces of the two bodies are strongly additive, and the unusually high- and low-water levels that result are known as *spring tides*. When the sun and moon are oriented at approximate right angles, their tide-producing forces do not complement each other, so that abnormally small *tidal ranges*, the differences between maximum and minimum stands of the water, result. This is the time of *neap tides*.

Successive spring tides are not equal because of astronomical variability. Some illustration of this situation is provided by information shown in Table 2-2 for Honolulu, Hawaii, located at 21.3°N latitude, 157.9°W longitude. Here, for the sake of brevity, only data associated with the single-evening passage of the full moon are included. Honolulu local time is delayed 10 h from Greenwich (England) Mean Time, the standard time for astronomical tables.

The type of tide shown for Honolulu in Table 2-2 is known as a *mixed tide*. There are two high tides and two low tides per day, but successive high tides are not equal, nor are successive low tides. In the case of mixed tides the two high tides are distinguished by calling one *higher high water* (HHW), the other *lower high water* (LHW); the two lows are termed *higher low water* (HLW) and *lower low water* (LLW). The average (or mean) LLW (MLLW) for a coastal location having mixed tides serves as zero on nautical charts for that area. This datum differs from that used on land, namely, mean sea level (MSL).

There are two other more usual types of tides. The *semidiurnal tide* produces two highs and two lows per day, each of equal magnitude. In the case of the *diurnal tide* there is only one high and one low per day. In the diurnal tide the highs are separated by 24 h 50 min; in the semidiurnal, 12 h 25 min. In both cases the average low water is taken as the datum for charts.

---

*This is the angle between a line joining the sun to the earth's center and a plane containing the equator.

Table 2-2: Tides Associated with Full Moon at Honolulu, Hawaii, in 1972

Meridian Passage of Full Moon		*East Longitude of Sun at Time of Moon's Meridian Crossing (°)*	*Approximate Declination (°)*		*First HW after Passage of Moon*		*Second HW after Passage of Moon*	
Date	*Approximate Local Time (h)*		*Moon*	*Sun*	*Approximate Local Time (h)*	*Height (ft)*	*Approximate Local Time (h)*	*Height (ft)*
Jan 30	0.7	22.4	N17.5	S17.8	4.3	2.2	16.5	0.8
Feb 29	0.9	20.0	N 3.5	S 7.8	4.3	1.7	16.7	1.2
Mar 30	0.9	17.7	S11.0	N 3.9	4.1	1.2	16.9	1.7
Apr 28	0.3	24.9	S18.6	N14.3	3.5	0.8	16.6	1.9
May 28	0.6	19.7	S25.7	N21.5	3.8	0.5	16.8	2.1
Jun 26	0.3	26.5	S25.7	N23.4	3.6	0.5	16.5	2.3
Jul 26	0.7	20.6	S18.1	N19.4	4.1	0.8	16.6	2.3
Aug 24	0.2	27.2	S10.0	N11.0	3.8	1.2	16.1	2.1
Sep 23	0.5	20.7	N 6.0	S 0.2	4.2	1.9	16.1	1.6
Oct 22	0.0	25.9	N15.1	S11.2	3.9	2.3	15.6	1.1
Nov 21	0.7	15.5	N24.7	S20.0	4.4	2.6	16.1	0.7
Dec 20	0.5	21.7	N25.0	S23.4	4.3	2.7	16.1	0.6

Measurements in the open ocean have shown that rotary currents (when seen from above) result from the propagation of tidal bulges over the sea surface. The water flow is continuous, although not of constant speed, and the direction of rotation is consistent with the Coriolis effect. Typical flow speeds are a knot or less.

Some of the astronomical reasons for tidal variability, dealing mainly with changes with time, have already been outlined. Separate points on the earth's surface differ in position relative to the sun and moon and experience different tide-producing forces and, consequently, unequal tidal excursions. But something out of the realm of the astronomical, the form and orientation of the coastal area being considered, as well as its location, exert a very great influence over the tidal variations experienced.

Some localities have very small tidal ranges, for example, the Mediterranean Sea and Honolulu. Others, such as the Bay of Fundy in Canada, can have spring tide ranges of 50 ft (15 m).

Information on the timing and heights of tides is available for United States waters in the *Tide Tables*. Similar governmental publications exist in many other countries.

2-4 PHYSICAL AND GEOLOGICAL CONCEPTS

Wind Waves

Generally speaking, two things occur when wind blows over a water surface. First, the friction between the moving air and the surface causes surface water to be dragged downwind, causing a wind-driven current considered earlier for the open ocean and to be reconsidered for local conditions in a later section. The wind also causes waves to develop on the water surface.

Wind-generated waves may be visualized as a train of water bumps. Between the *crests* is a lower region that extends below the *still water level*.* This lower region is known as the wave *trough*. Waves are generally represented by the vertical distance from trough to following crest, the *wave height*, and the time taken between the arrival of two successive wave crests at a fixed point, the *wave period*. The horizontal distance between two adjacent crests, the *wave length*, is also used. There is much variation among waves generated in the same general locale by the same wind and one usually focuses on such general wave characteristics as the average height and period. The height and period of a train of waves at a point on the sea surface depends on several factors, namely the magnitude of the wind speed, the duration of the wind, the water depth, and the *fetch*, which is the over-water distance of wind travel prior to the point in question. The wave height and period increase for increases in wind speed and duration and for increase in fetch. The effect of water depth, when it plays a role in shallow waters, is extremely complex.

When waves are being generated, i.e., when they have wind acting on them, they are spoken of as *sea* or *wind waves*. Such surface undulations persist after the wind dies or after they have moved out of the generating area. These surface features are then known as *swell*. When the waves are as large as they can be for a given wind speed, the seas are said to be *fully-developed*. Often such seas are not characterized in terms of an average height and period, for it is difficult, in the considerable confusion of a great storm sea, to speak of such quantities except in terms of *sea state*. Associated with different wind velocities, or wind forces, or *Beaufort numbers*, there are various levels of wave action. These are summarized in Table 2-3, where H_s, the *significant wave height*, is the average of the highest third of all wave heights observed over a certain time and is more often used as a representative wave height than the overall average wave height. $\overline{T}$ in Table 2-3 is an average wave period. The numbers used for sea state and wind force are simply indices.

Although there is a dominant wave period or frequency associated with the waves generated by a particular storm, there is in truth a *spectrum* (or range) of periods in any such seaway. There is a confusion of wave periods within the wave-generating area, but this confusion is largely eliminated when the waves pass out of the generation area. This is because the speed with which a wave advances over the water surface depends on its period; in fact, this speed increases with increasing wave period. Thus there is a natural mechanism, called *dispersion*, by which different frequency components in the wave group are separated.

When swell leaves the generation area, it moves across the ocean along great circles. The height-squared of a wave is a measure of its specific energy,

---

*This would be sea surface if the wave action were to cease.

the energy per unit of water surface area. The specific energy of swell propagating across the deep ocean is slowly reduced by the spreading of the wave train and frictional effects, as well as by adverse winds and currents. In addition, because of nonlinear effects, there can be an energy shift among different wave frequencies. The history of swell passing through deep water* thousands of miles from the generation area begins with long, low waves. The period decreases progressively while the height of the swell gradually rises to a maximum and then diminishes again. Open ocean swells from moderately close, large storms can reach enormous size. In December, 1969, such a storm in the region of the Aleutian Islands generated waves that reached 80 ft (24.3 m) in height in the open ocean 1000 miles (1600 km) north of Hawaii (Rudnick and Hasse, 1971).

Table 2-3: Fully-Developed Wind Wave Characteristics

Beaufort Number	Wind Speed (knots)	Wind Description	Sea State	$\bar{T}$ (s)	H_s (ft)	H_s (m)
4	11–16	moderate breeze	3	6.1	3.5	1.1
5	17–21	fresh breeze	4	6.5	6.0	1.8
6	22–27	strong breeze	5	7.3	9.5	2.9
7	28–33	near gale	6	7.8	13.5	4.1
8	34–40	gale	6	8.3	18.0	5.5
9	41–47	strong gale	6	9.0	23.0	7.0
10	48–55	storm	7	9.4	29.0	8.8
11	56–63	violent storm	8	10.0	37.0	11.3
12	64–71	hurricane	9	10.4	45.0	13.7

In the deep ocean the speed, or *celerity*, of waves theoretically depends only on the wave period. In coastal depths the wave celerity also depends upon the water depth, and in very shallow water the wave speed is, theoretically, entirely dependent on depth. In general, for a given wave period, the wave celerity decreases with decreasing water depth. This situation given rise to a phenomenon called *refraction* when waves (or swell) enter coastal water depths. A long wave crest advancing on a land mass will eventually encounter different water

---

* A rule of thumb in coastal engineering is that "deep water", in a waves context, is any depth that exceeds one-half the length of a wave in question. In "shallow water", the wave length exceeds twenty times the water depth.

depths beneath it along its length. Those regions in shallower water will tend to lag behind those in deeper water with the result that a formerly straight crest line is deformed to face inwards on shoal areas and outwards over depressions. Not only does this refraction tend to change the direction of wave travel but the wave height tends to be augmented in areas where wave energy is concentrated (shoals) and diminished where wave *orthogonals* (lines perpendicular to the wave crests) diverge. In addition, waves tend to increase in height even without refraction because of the decreasing depth. Wave height increases until the wave breaks in a water depth that is approximately 1.2 to 1.3 times the wave height at that point for flat foreshore slopes. The breaking wave height can be greater than the depth in which it breaks for moderately steep to steep foreshore slopes.

There are three major types of breakers: *plunging* breakers (Figure 2-8), wherein the forward face of the wave steepens and the top is thrown forward, to impinge with considerable agitation in the trough at the foot of the wave; *spilling* breakers (Figure 2-9), when foam and turbulent water appear at the wave crest and run down its forward face in a show of white; and *surging* breakers, when the bottom of the wave is thrown forward. Generally speaking, a wave of given deep-water characteristics will break by spilling on a flat slope, by plunging on a moderate slope, and by surging on a steep slope. The area between the seaward limit of breaking wave action and the shoreline is known as the *surf zone*.

Although refraction tends to bring approaching wave fronts into parallelism with the shoreline they are approaching, waves normally break at a small angle

Figure 2-8 Plunging breaker

Figure 2-9 Spilling breaker

to the shoreline. There is then an alongshore component of energy to be accounted for and this leads to longshore, or *littoral*, currents that parallel the beach within the surf zone. Such currents are of enormous importance in some regions, for example, on the ocean-side of Florida, because they transport colossal quantities of sediment along the coast. Under certain local conditions, and with sizable waves, strong littoral currents can be set up.

When the waves break, there is a transport of water from the breaker line in toward the shore. This water, which is renewed as often as the waves break, tends to "pond up" to some degree against the shore. Sufficient head is thereby created to generate a narrow current that flows away from the shore, sometimes straight offshore, but often at an angle to the shoreline. When this *rip current* penetrates beyond the breakers, the flow spreads and decreases in speed. This is the so-called *head* of the rip. Rip currents generated in conjunction with large waves can be exceedingly strong, and the alongshore currents which feed them can also be swiftly moving.

It should be stressed that any wave, before breaking, is a surface feature that has a transient constituency in terms of particular water particles. The behavior of any particular water particle near the surface as a surface wave passes is approximately as follows. As a wave crest approaches, the particle moves in the direction of wave travel, reaching its maximum flow-speed beneath the crest. As the crest moves beyond it the particle gradually slows down, reverses its direction and then travels rearward, contrary to the direction of wave advance, reaching its maximum horizontal rearward flow-speed under the wave trough. The flow-speed is then reduced as the crest approaches and the process is repeated. The flow-path of a near-surface water particle under waves in deep water can be viewed as approximately circular; in shallow water the near-surface water particle orbit is ellipse-like. Close to the bottom in shallow water the flow is parallel to the bottom. Under long waves it is entirely possible for water particles throughout coastal water depths to move (Komar et al., 1972).

The Coastal Zone

Some coastlines are relatively straight, with either unbroken stretches of beach or steep cliffs. But many coastlines demonstrate broad indentations that may be termed either bays or gulfs depending upon whether a semicircle with a diameter equal to the width of the entrance would contain or would not contain the indentation. Long and narrow indentations in a coastline are called *sounds*. An *estuary* is the mouth of a river—its entrance into the sea. Considerable loads of sediment may be carried down a river and into the marine environment through the estuary, over the course of a year. These sediments, plus those derived from erosion of cliffs (*terrigenous* or *physical* sediments) and breakup of shells, shell fragments, and reef debris (*calcarenite* or *organic* sediments) make up the bulk of the unconsolidated material of beaches. A *biogenous* sediment contains at least 30% by volume of particles derived from the skeletal remains of organisms. A *lithogenous* sediment is composed primarily of mineral grains transported into the sea from the continents.

On many coasts, beaches are separated by rocky points or *headlands* such as shown in Figure 2-10. In coastal areas the bottom may be, for example, rock, sand, mud, or coral. Primarily the rock may be igneous (i.e., of direct volcanic origin), such as basalt, or sedimentary. Such sedimentary rocks as shale and mudstone are formed from grains of silt and clay. Shales are laminated, whereas lamination is rare or well-separated in mudstones. Sandstone, siltstone, and claystone are other types of sedimentary rocks with constituents indicated by their names (see Table 2-4). Conglomerate rock is composed of pebbles, boulders, and cobbles in a matrix of finer material and is another sedimentary rock that may appear in the sea bed and adjacent coast. Beachrock, consisting of sand cemented by calcium carbonate, is still another.

The beach is a familiar concept to most people. It is a semiflat area of unconsolidated material extending from the water's edge landward to the furthest point where the beach material has been transported by wave action. The landward extremity of the beach may be marked by dunes (McHarg, 1972) or perhaps by a sea cliff, and the span between this limit and the high tide level is known as the *backshore* (see Figure 2-11). The *foreshore* spans the beach area between the high and low tide levels with the *offshore* extending further seaward. The latter may involve a raised *bar* some distance from the water's edge and separated from it by a *trough*. Flat areas in the backshore are referred to as *berms*. A flat *low-tide terrace* may be involved in the foreshore.

The three beach divisions given above are somewhat vague, and a tighter

Figure 2-10 Rocky headland

Table 2-4: Classification of Unconsolidated Material

		Particle Size Ranges			
		Low		High	
Class	*Name*	*(mm)*	*(inches)*	*(mm)*	*(inches)*
—	clay	—	—	1/256	—
—	silt	1/256	—	1/16	—
—	sand	1/16	—	2	—
gravel {	granules	2	—	4	—
	pebbles	4	2-1/2	64	10
	cobbles	64		256	
	boulders	256	10	—	—

classification system is also employed. In this case the shore between the average of higher high spring tides and the mean level of lower low spring tides, the *intertidal zone*, is divided into *upper littoral, middle littoral,* and *lower littoral* zones. The middle littoral zone covers the range of neap tides with the other two zones adjoining. Immediately above the upper littoral zone is the *splash zone.* Starting at the bottom of the intertidal zone is the *sublittoral zone*; it is regarded as that portion of the continental shelf permanently covered by water but shallow enough to be strongly influenced by wave action, at the least during storms.

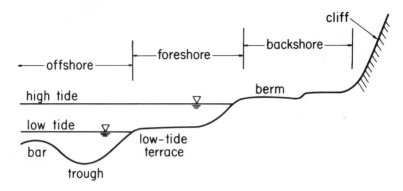

Figure 2-11 Section through a beach

Beach materials vary widely. Most of us can think of at least several distinct beaches composed of completely different particle types and sizes. Granular material can be classified as shown in Table 2-4. The most common beach constituent is sand.

Beach systems are seldom if ever stable. Periods of sand erosion or replenishment at given locations along a beach can extend over several years with the cycle being closed after perhaps a decade. Such has been the case, for example, along Lanikai Beach in Hawaii, even though protected by a barrier reef a mile offshore.

A more noticeable cycle of natural sand erosion and replenishment is an annual one, with small waves (normally during the summer) building up the beach, while the large (usually winter) waves break it down. In the summer case, the wave uprush after breaking is sufficient to transport material up the slope, but some water percolates into the beach and the backwash is too weak to return the sediment to the sublittoral zone. In the winter, on the other hand, large waves during breaking and subsequent runup stir large amounts of material into suspension, and then the strong backwash carries this material down the slope and back into the sublittoral zone. There, still in suspension, the material is carried along the shore by the strong littoral currents that occur in conjunction with large waves, and then part of it moves seaward in a rip current and is dumped in the rip head when the current speed weakens. During times of low wave action, offshore bars move landward and sometimes merge with the beach proper. During times of heavy wave action such bars are displaced offshore.

The volume of unconsolidated material moving downcoast due to littoral transport over a year's time can be enormous. Numbers such as 500,000 to 1,000,000 cubic yards (about 400,000 to 800,000 m³) of sand are not unusual.

During times of even moderate wave action, deposits of unconsolidated materials some distance offshore can suffer extensive changes. In my own ocean research I have seen quarter-acre (0.1 ha) deposits of sand a foot or two in depth in 25 feet (7.5 m) of water completely removed after a period of moderate swell action lasting a couple of days. Under heavy wave action the amount of cut produced in offshore sand deposits may be 10 feet (3 m) or more [Alterman (1962)].

Waves work to change the nature of a coastline even when rock cliffs provide the boundary between land and water. Continuous wave breaking against a cliff face (Figure 2-10) produces an eroded pocket that penetrates further into the cliff as time progresses. At some point the rock overlying the pocket, inadequately supported, crashes into the surf area. Over a period of many years waves break up this debris, moving it largely into the waters fronting the former cliff face. Waves can then work once again on the cliff itself—leading to another cycle of erosion and collapse. After a period of many years the result is a horizontal, flat rock surface, *wave-cut terrace*, at about sea level and extending from the water's edge back to the bottom of the existing cliff, see Fig. 2-12. An example occurs along North Head near Sydney, Australia.

Coastlines in the tropics and subtropics are often separated from the open ocean by a coral reef platform, proving that there are sources of natural coastal structures other than geological processes. These reefs are built up by the hermatypic corals described in Section 2-2 in association with plants such as coralline algae which provide a unifying, cementing action. The bulk of reef platforms is composed of skeletal coral material. There are three types of reefs generally recognized. A *fringing* reef develops in shallow water along a coast and is thus immediately adjacent to it. Breaks occur at points of freshwater outflow from the land. Fringing reefs generally surround the Hawaiian island of Oahu,

Figure 2-12   Wave-cut terrace

for example. *Barrier reefs* form a sort of natural breakwater some distance off a coast. They are separated from the coastline by a long *lagoon* wherein the depths involved are too great for coral growth. The Great Barrier Reef, stretching 1000 miles (1600 km) along the northeast coast of Australia and separated from it by between about 10 and 150 miles (about 15 to 250 km), serves as a prime example. The third type of coral reef is the *atoll*—with a reef surrounding a lagoon having no internal land form. Channels, or *passes*, connect the lagoon to the open ocean. Depth within the lagoon is typically 100 to 150 feet (30 to 45 m). There is strong scientific evidence to indicate that there is a natural long-term progression from fringing through barrier reef to atoll as a volcanic island in tropical or subtropical waters gradually subsides.

The term reef by itself is also used to describe an offshore, raised rock platform that poses a threat to navigation since it lies within 60 feet (18 m) of the surface. The term bar, already mentioned, refers to a hazard consisting of unconsolidated sediment.

When waves break along a reef, water passes from its offshore to onshore side. If the reef is part of an atoll, very swift currents can be created in the passes as this water is returned to the outside under the head built up within the atoll. Similarly, fast currents can occur within the breaks in barrier reefs.

Coastal Currents

The currents resulting from wave action have been examined in the previous two subsections. In this subsection discussion will center on two important types of coastal currents, those due to the dragging action of local winds and those associated with the tides. The large-scale currents forming a part of global oceanic circulation only rarely penetrate into coastal waters.

In the general study of the mechanics of liquids, the term *convection* is used to describe the transfer of some property, say heat, in conjunction with movements of liquid mass. No direction restrictions are placed on the word. In studies of meteorology and oceanography, however, the word convection indicates vertical movements; the term *advection* refers to horizontal motions. In what follows we will consider exclusively the advective type of water motion.

Although there are global and large-scale weather systems over the earth's surface, as described earlier, local effects may play a major role in determining winds at a particular location. Certain localities may, for example, be blocked from the prevailing wind by hills and mountains, and they may experience still conditions or, perhaps, eddying because of the wind blowing past the obstruction. An eddy in this case is a whirling mass of air. Locations set in valleys parallel to the prevailing wind may experience increased wind speeds as the winds blocked by the flanking hills or mountains are forced down the valley. Coastal communities have their own wind systems that may overpower the larger scale systems. Rising air over land heated during the day draws air onshore and results in an onshore breeze. In Honolulu, for example, it fre-

quently happens that 15-knot trade winds (blowing from the northeast-
—offshore) can give way to onshore breezes (from the southwest) in the
afternoon. When the land becomes colder than the sea (or lake) at night, the
cold air adjacent to the earth's surface underflows the warmer offshore air and
causes an offshore wind.

   Throughout the course of a year, the wind at any particular location varies
considerably both in terms of speed and direction. Such changes can be
considered on the basis of many different time scales, from seasonal to daily to
several-minute, for example. *Gusts* involve sudden changes in wind speed (and
direction) over time spans measured in seconds.

   The U.S. National Weather Service measures wind speeds and directions at
many selected stations in the United States. Measurements are taken continuously, but tabulated data reflect the average speed and direction of the wind
over 1 minute's duration at hourly intervals. One way of representing the
accumulated data is in the form of a *wind rose*, an example of which is shown in
Figure 2-13.

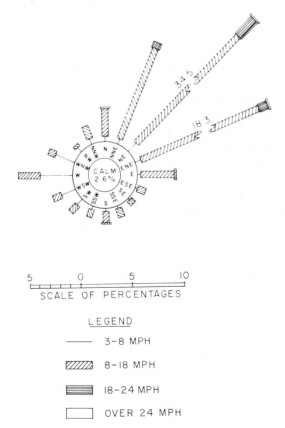

SCALE OF PERCENTAGES

LEGEND

——————   3–8 MPH

▨▨▨▨   8–18 MPH

▤▤▤▤   18–24 MPH

▭▭▭▭   OVER 24 MPH

Figure 2-13 Wind rose for Lihue, Hawaii

The dragging action of local winds on the sea surface causes *wind-driven* currents to be set up, and it is a general rule of thumb that the speed of such currents, if fully established under a fairly steady wind field, will have a surface velocity about 2% of the wind speed. There is a water depth and bottom roughness effect, however. The direction of the surface drift does not necessarily conform exactly with the wind direction because of the effect of the earth's rotation. However, this effect can be considered minor in small-scale wind-current systems and, as a first approximation in such cases, the wind and wind-driven current are assumed to move in the same direction.

As considered earlier, offshore winds can result in surface waters being driven away from the land with colder offshore subsurface waters replacing them.

The rise and fall of the ocean surface in response to the attractive forces of the sun and moon can take place only if there is transport of sea water into and out of a region. The *tidal currents* that result in the open ocean are of a rotary type, as discussed previously, where the flow is continuous with the direction changing continuously through all points of the compass during the tidal period. Close to the coast, or in straits for example, the tidal currents are *reversing currents* lying in a single line. In the latter case, there is said to be a *flooding current* when the tide is rising, an *ebbing current* when the tide is going down. *Slack tide* occurs when there is no horizontal water movement. The direction toward which the tidal current flows is termed the *set*.

Links between tidal currents and tides are discussed in Bowditch (1962), a 1500-page volume that must be the world's greatest book bargain. It is pointed out that no generalities can be put forward to cover the difference in phasing between tides and the associated currents. Even for the same location at different times this is true. It is also shown that the strength of tide-induced coastal currents at different locations in no way reflects differences in tidal ranges. However, it is suggested that it is a reasonable assumption to take tidal current strengths for a given location as proportional to differences in tidal excursions. The tides and associated tidal currents are summarized for different parts of the world in appropriate government publications. For the United States one can consult the *Tide Tables* and the *Tidal Current Tables* published annually in various volumes each covering a separate coastal region.

When a tidal bulge passes an island, such as the Hawaiian island of Oahu, ocean water is forced around the island from the front to the rearward side. The term *divergence* is used to describe the area on the upstream side of the island which appears as the source of the splitting current. The region in which the currents from the two sides reform on the downstream side of the island is known as the *convergence*. This is a very dangerous region for divers since strong currents set offshore for several hours as the bulge passes. The locations of divergences and convergences do not necessarily remain the same from day to day.

The rise and fall of the tides means very swift currents in the passes for many atolls. Such motion, of course, is inland on a rising tide and out of the central lagoon on a falling one. Similarly, tidal action can lead to strong currents at breaks in barrier reefs and other restricted areas such as narrow channels between islands. Examples of the latter are within the Gulf Islands between Vancouver Island and the State of Washington and at the east end of Long Island Sound, New York. The peak tide-induced current listed in the *Tidal Current Tables 1976, Pacific Coast of North America and Asia* (1975), for example, was an incredible 15.8 knots in Seymour Narrows, British Columbia, on November 22.

In many cases it is possible for strong wind-driven coastal currents to completely overpower tide-induced water motion and to materially alter normal tidal flow patterns at and near the water surface. In other cases, such as in the dreaded Alenuihaha Channel in the Hawaiian Islands, even a ferocious head wind is insufficient to turn tide-induced surface currents.

Seismic Sea Waves

Displacements of the sea floor or of the shoreline during earthquakes can lead to the generation of waves. A common name for such waves, although completely a misnomer, is tidal wave; an often-used name for them is *tsunami*, a word taken from the present-day Japanese language. The size of the waves in the immediate vicinity of the seismic activity can be considerable—as in the Good Friday, 1964, earthquake and tsunami along the south coast of Alaska [Wilson and Torum (1968)] or in the Thanksgiving, 1975, earthquake and tsunami on the island of Hawaii.

But the effects of the earthquake can be transmitted to the far corners of the world through the propagation of the seismic sea waves. Although no one has apparently ever measured a tsunami in the open ocean, it is believed that the amplitude of the wave would be of the order of centimeters. Although the wave is very small in height, it has an enormous wave length befitting the inferred speed of such waves in the open ocean (400 knots or so) and the measured range of their periods—10 to 100 minutes, depending upon the magnitude of seismic disturbance involved.

As tsunamis approach land masses and move into shallower depths, they suffer the same transformations as wind waves although this starts in much deeper water than for wind waves, since even in the open ocean the tsunami is classified as a shallow-water wave. The offshore bathymetry and the nature and orientation of the coast can lead, on the one hand, to simply an abnormally high rise and fall of the quasi-mean water surface with the period of the incident tsunami. On the other hand it can cause a catastrophic surge of water that poses a threat to shoreline and coastal water installations and to life and limb. Hilo, Hawaii, has had the distinction of severe tsunami action at various times over the years (e.g., 1946 and 1960) with extensive damage and loss of life. An eye-witness account of the arrival of a destructive tsunami on Oahu, Hawaii, is provided by Shepard (1967).

Storm Tides

A violent cyclonic disturbance* possesses the ability to strongly influence the water level through three separate mechanisms. First, water tends to mound up into the low pressure area within the disturbance. Second, the strong winds force water up abnormally high against the shoreline. An onshore wind component does this directly by applying a shear force on the water surface toward the land whereas an alongshore component can effect the same result due to the Coriolis effect. Third, high wind-generated waves can ride farther inland than normally possible, because of the abnormally-high water level, and bring great pressure to bear against natural and man-made shoreline structures normally well back from the water's edge.

The first two components are lumped together to yield the *storm tide*—also known as meteorological tide or *storm surge*. The effect due to the wind itself is called the *wind setup*. Details concerning the effects of such surges are available in Wiegel (1964). Not only large waves but also swiftly-moving currents can be associated with conditions giving rise to large storm surges. It is not unusual for powerful cyclonic storms to effect 10–15-foot water superelevations or more. An example of this is Hurricane Carla that passed through the Gulf of Mexico and onto the U.S. Gulf Coast in September 1961 (U.S. Army Corps of Engineers, 1973).

2-5   REFERENCES

ALTERMAN, I. (1962): Discussion of "Wave Force Coefficients for Offshore Pipelines" by H. Beckmann and M. H. Thibodeaux, ASCE, *Journal of the Waterways and Harbors Division*, Vol. 88, No. WW4, pp. 149–150.

BASCOM, W. (1964): *Waves and Beaches*, Doubleday, New York.

BOWDITCH, N. (1962): *American Practical Navigator*, U.S. Naval Oceanographic Office, Washington, D.C.

COKER, R. E. (1962): *This Great and Wide Sea*, Harper & Row, New York.

CROMIE, W. J. (1966): *The Living World of the Sea*, Prentice-Hall, Englewood Cliffs, N.J.

DUNLAP, G. D., and H. H. SHUFELDT (1969): *Dutton's Navigation and Piloting*, U.S. Naval Institute, Annapolis, Md.

FELL, H. B. (1974): *Life, Space, and Time: A Course in Environmental Biology*, Harper & Row, New York.

---

* A hurricane is an example. It is usually of tropical origin and covers an extensive area. Peak winds are 64 nautical miles per hour (knots) or higher.

GROSS, M. G. (1972): *Oceanography: A View of the Earth*, Prentice-Hall, Englewood Cliffs, N.J.

GROSS, M. G. (1976): *Oceanography*, 3rd ed., Charles E. Merrill, Columbus, Ohio.

ISAACS, J. D. (1969): "The Nature of Oceanic Life," *Scientific American*, Vol. 221, No. 3 (September), pp. 146–160, 162.

KOMAR, P. D., R. H. NEUDECK, and L. D. KULM (1972): "Observations and Significance of Deep-Water Oscillatory Ripple Marks on the Oregon Continental Shelf," in *Shelf Sediment Transport: Process and Pattern*, edited by D. J. P. Swift, D. B. Duane, and O. H. Pilkey, Chapter 25, Dowden, Hutchinson and Ross, Inc., Stroudsburg, Pa.

LEE, T. N., and D. A. MAYER (1977): "Low-Frequency Current Variability and Spin-off Eddies along the Shelf off Southeast Florida," *Journal of Marine Research*, Vol. 35, No. 1 (February), pp. 193–220.

MCALLISTER, R. F. (1968): "Demonstration of the Limitations and Effects of Waste Disposal on an Ocean Shelf," Florida Atlantic Ocean Sciences Institute, Annual Project Report (March) [NTIS *Publication PB-215-585*].

MCCONNAUGHEY, B. H. (1970): *Introduction to Marine Biology*, The C. V. Mosby Co., St. Louis, Mo.

MCHARG, I. (1972): "Best Shore Protection: Nature's Own Dunes," *Civil Engineering*, Vol. 42, No. 9 (September), pp. 66–70.

NEUMAN, G., and W. J. PIERSON, JR. (1966): *Principles of Physical Oceanography*, Prentice-Hall, Englewood Cliffs, N.J.

RUDNICK, P., and R. W. HASSE (1971): "Extreme Pacific Waves, December 1969," *Journal of Geophysical Research*, Vol. 76, No. 3 (January 20), pp. 742–744.

SHEPARD, F. P. (1967): *The Earth Beneath the Sea*, rev. ed., The Johns Hopkins Press, Baltimore, Md.

SVERDRUP, H. U., M. W. JOHNSON, and R. H. FLEMING (1942): *The Oceans: Their Physics, Chemistry, and General Biology*, Prentice-Hall, Englewood Cliffs, N.J.

Tidal Current Tables 1976, Pacific Coast of North America and Asia (1975): U.S. Dept. of Commerce, National Oceanic and Atmospheric Administration, National Ocean Survey, Washington, D.C.

Tide Tables 1976, West Coast of North and South America, Including the Hawaiian Islands (1975): U.S. Dept. of Commerce, National Oceanic and Atmospheric Administration, National Ocean Survey, Washington, D.C.

U.S. ARMY CORPS OF ENGINEERS, COASTAL ENGINEERING RESEARCH CENTER (1973): *Shore Protection Manual*, Vol. 2.

VAN DORN, W. G. (1974): *Oceanography and Seamanship*, Dodd, Mead, New York.

WEYL, P. K. (1970): *Oceanography: An Introduction to the Marine Environment*, Wiley, New York.

WIEGEL, R. L. (1964): *Oceanographical Engineering*, Prentice-Hall, Englewood Cliffs, N.J.

WILSON, B. W., and A. TORUM (1968): "The Tsunami of the Alaskan Earthquake, 1964: Engineering Evaluation," U.S. Army Corps of Engineers, Coastal Engineering Research Center, *Technical Memorandum No. 25* (May).

2-6 BIBLIOGRAPHY

AMOS, W. H. (1966): *The Life of the Seashore*, McGraw-Hill, New York.

ANIKOUCHINE, W. A., and R. W. STERNBERG (1973): *The World Ocean: An Introduction to Oceanography*, Prentice Hall, Englewood Cliffs, N.J.

BARWIS, J. H. (1976): Annotated Bibliography on the Geologic, Hydraulic, and Engineering Aspects of Tidal Inlets, U.S. Army Corps of Engineers, *GITI Report 4* (January).

BERRILL, N. J. (1966): *The Life of the Ocean*, McGraw-Hill, New York.

BROECKER, W. S. (1974): *Chemical Oceanography*, Harcourt, Brace, Jovanovich, New York.

CARSON, R. L. (1951): *The Sea Around Us*, Oxford University Press, New York.

CARSON, R. L. (1968): *The Sea*, MacGibbon and Kee, London, England.

CUSHING, D. H. (1969): "Upwelling and Fish Production," Food and Agriculture Organization of the United Nations, Rome, *Fisheries Technical Paper No. 84*.

DEACON, G. E. R., editor (1962): *Oceans*, Paul Hamlyn, London, England.

DEFANT, A. (1958): *Ebb and Flow: The Tides of Earth, Air and Water*, The University of Michigan Press, Ann Arbor, Michigan.

DUGDALE, R. C. (1967): "Nutrient Limitation in the Sea: Dynamics, Identification and Significance," *Limnology and Oceanography*, Vol. 12, No. 4, pp. 685–695.

DUXBURY, A. C. (1971): *The Earth and Its Oceans*, Addison-Wesley, Reading, Mass.

DYER, K. R. (1973): *Estuaries: A Physical Introduction*, Wiley (Interscience), New York.

EDMONDSON, C. H. (1928): "The Ecology of an Hawaiian Coral Reef," Bernice P. Bishop Museum, Honolulu, Hawaii, *Bulletin No. 45*, pp. 3–64.

ELTRINGHAM, S. K. (1971): *Life in Mud and Sand*, The English Universities Press, London, England.

FRIEDRICH, H. (1971): *Marine Biology*, translated by G. Vevers, University of Washington Press, Seattle, Wash.

GALVIN, C. J., JR. (1972): "Wave Breaking in Shallow Water," in *Waves on Beaches and Resulting Sediment Transport*, edited by R. E. Meyer, Academic Press, New York.

HANNAU, H. W. (1974): *In the Coral Reefs of the Caribbean, Bahamas, Florida, Bermuda*, Doubleday, Garden City, N.Y.

HENDERSHOTT, M. C. (1968): "Physical and Hydrodynamic Factors," in *Ocean Engineering*, edited by J. F. Brahtz, pp. 202–258, Wiley, New York.

INMAN, D. L. (1971): "Nearshore Processes," in *Encyclopedia of Science and Technology*, 3rd ed., McGraw-Hill, New York.

INMAN, D. L., and B. M. BRUSH (1973): "The Coastal Challenge," *Science*, Vol. 181 (July 6), pp. 20–32.

ISAACS, J. D. (1968): "General Features of the Oceans," in *Ocean Engineering*, edited by J. F. Brahtz, pp. 157–201, Wiley, New York.

KEETON, W. T. (1969): *Elements of Biological Science*, Norton, New York.

KING, C. A. M. (1975a): *Introduction to Marine Geology and Geomorphology*, Crane Russak, New York.

KING, C. A. M. (1975b): *Introduction to Physical and Biological Oceanography*, Crane Russak, New York.

KING, C. A. M., and W. W. WILLIAMS (1949): "The Formation and Movement of Sand Bars by Wave Action," *Geographical Journal*, Vol. 113, pp. 70–85.

KINNE, O., editor (1972): *Marine Ecology*, Vol. 1, *Environmental Factors*, Parts 1–3, Wiley (Interscience), New York.

LAUFF, G. H., editor (1967): *Estuaries*, American Association for the Advancement of Science, Washington, D.C., *Publication No. 83*.

"Manual of Tide Observations," (1965): U.S. Dept. of Commerce, Coast and Geodetic Survey, *Publication 30-1*, Washington, D.C.

MILLER, A. (1966): *Meteorology*, Charles E. Merrill, Columbus, Ohio.

MILLER, R. C. (1966): *The Sea*, Random House, New York.

MILLIMAN, J. D. (1965): "An Annotated Bibliography of Recent Papers on Corals and Coral Reefs," National Academy of Sciences—National Research Council, Washington, D.C., The Pacific Science Board, *Atoll Research Bulletin No. 111* (July 15).

MOORE, H. B. (1958): *Marine Ecology*, Wiley, New York.

MUNK, W. H. (1950): "On the Wind-Driven Ocean Circulation," *Journal of Meteorology*, Vol. 7, No. 2, pp. 79–93.

The Nautical Almanac for the Year 1972 (1970): U.S. Naval Observatory, Washington, D.C.

ODUM, E. P. (1963): *Ecology*, Holt, New York.

ODUM, H. T. (1957): "Primary Production Measurements in Eleven Florida Springs and in a Marine Turtle Grass Community," *Limnology and Oceanography*, Vol. 2, pp. 85–97.

OFFICER, C. B. (1976): *Physical Oceanography of Estuaries (and Associated Coastal Waters)*, Wiley (Interscience), New York.

PARSONS, T. R., and M. TAKAHASHI (1973): *Biological Oceanographic Processes*, Pergamon Press, New York.

RAYMONT, J. E. G. (1963): *Plankton and Productivity in the Oceans*, Pergamon Press, New York.

REISH, D. J. (1969): *Biology of the Oceans*, Dickenson, Belmont, Calif.

RHOADS, D. C., and D. K. YOUNG (1970): "The Influence of Deposit-Feeding Organisms on Sediment Stability and Community Trophic Structure," *Journal of Marine Research*, Vol. 28, No. 2, pp. 150–178.

RICHARDS, F. A. (1968): "Chemical and Biological Factors in the Marine Environment," in *Ocean Engineering*, edited by J. F. Brahtz, pp. 259–303, Wiley, New York.

RILEY, J. P., and G. SKIRROW, editors (1965): *Chemical Oceanography I*, Academic Press, New York.

ROSS, D. A. (1970): *Introduction to Oceanography*, Appleton-Century-Crofts, New York.

SHEPARD, F. P., and H. R. WANLESS (1971): *Our Changing Coastlines*, McGraw-Hill, New York.

SMITH, R. L. (1968): "Upwelling," *Oceanography and Marine Biology—An Annual Review*, Vol. 6, pp. 11–46.

STEEMANN NIELSEN, E. (1975): *Marine Photosynthesis*, Elsevier, Amsterdam.

STEWART, R. W. (1969): "The Atmosphere and the Ocean," *Scientific American*, Vol. 221, No. 3 (September), pp. 76–86.

SUHAYDA, J. N., H. H. ROBERTS, and S. P. MURRAY (1975): "Nearshore Processes on a Fringing Reef," *Preprints*, Third International Ocean Development Conference, Tokyo, Japan, Vol. 5 (August 1975), pp. 63–75.

SUTTON, O. G. (1964): *Understanding Weather*, 2nd ed., Penguin, Baltimore, Md.

SVERDRUP, H. U. (1938): "On the Process of Upwelling," *Journal of Marine Research*, Vol. 1, pp. 155–164.

TAIT, R. V., and R. S. DeSANTO (1972): *Elements of Marine Ecology*, 2nd ed., Springer-Verlag, New York.

United States Coast Pilot 7 (1975): Pacific Coast, 11th ed. (June), U.S. Department of Commerce, National Oceanic and Atmospheric Administration, National Ocean Survey, Washington, D.C.

WEINS, H. J. (1962): *Atoll Environment and Ecology*, Yale University Press, New Haven, Conn.

WEISZ, P. B. (1967): *The Science of Biology*, McGraw-Hill, New York.

WELLS, J. W. (1957a): "Coral Reefs," Geological Society of America, *Memoir 67*, Vol. 1, pp. 609–631.

WELLS, J. W. (1957b): "Corals," Geological Society of America, *Memoir 67*, Vol. 1, pp. 1087–1104.

WILLIAMS, J., J. J. HIGGINSON, and J. D. ROHRBOUGH (1968): *Sea and Air*, U.S. Naval Institute, Annapolis, Md.

WYRTKI, K., and G. MEYERS (1975): "The Trade Wind Field over the Pacific Ocean," Part I. The Mean Field and the Mean Annual Variation, University of Hawaii, Hawaii Institute of Geophysics, *Publication HIG-75-1* (January).

ZENKOVITCH, V. P. (1967): *Processes of Coastal Development*, Oliver and Boyd, London, England.

3

Wastewaters: Characteristics, Treatment, and Disposal

3-1 RAW WASTEWATERS

Flow Rates

Municipal wastewater is composed of various quantities of domestic, commercial, and industrial wastewaters. A nominal figure for the wastewater yield of a municipality is 100 U.S. gallons per capita (person) per day (gpcd) (0.38 m^3pcd). For a population of 464,000 (that of sewered Honolulu, Hawaii, in 1975, including tourists), this converts to a flow of 46.4 million U.S. gallons per day (Mgd) (2.03 m^3/s).

Whereas some municipalities maintain the sewer system for wastewaters separate from the storm-drain system used to carry away runoff from rainfall, others use the same combined system to serve both functions. Even in cases where the systems are separate, it is possible for water not part of the water supply to enter the sewer system through infiltration. Clark et al. (1971) give as a nominal figure 30,000 U.S. gallons per day (gpd) per mile of sewer line (approximately 70 m^3 per day per km) for such infiltration.

Honolulu has a separate sewer system, and a record of the flows for five days in 1975 is shown in Figure 3-1. This particular timespan was chosen because it shows both the dry weather wastewater flows and those corresponding to wet weather. The relatively rapid fluctuations in the record are due to the turning on and off of pumps; but the general dry weather trend is for the wastewater flow to approximately double over the midmorning hours and then to drop off again after midnight. Figure 3-1 shows that the average wastewater

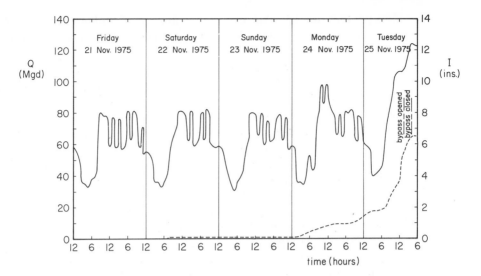

Figure 3-1 Dry weather and wet weather sewage flows at Honolulu,
Hawaii

flow is higher than that determined using the nominal 100 gpcd (0.38 m³pcd)
figure, and the difference is due to infiltration, although the rate is only
approximately half of that suggested by Clark et al. (1971) when one considers
the 1500 miles (2400 km) of Honolulu's sewer system.

The infiltration component of total flow is particularly apparent after the
intense rainfall of November 25. The rainfall curve in the figure (*I*) is given in
terms of the cumulative amount of water.

Physical Characteristics

The term *total residue* (also known by the older expression *total solids*)
refers to all solid matter suspended or dissolved in wastewater. It is determined
by evaporating a well-mixed sample in a weighed dish and drying to constant
weight in an oven at 103–105°C. Total residue is expressed as weight per volume
in milligrams per liter (mg/ℓ). *Fixed residue* remains after ignition for 1 hr at
500–600°C and is used as an index of inorganic constituents. The part of total
residue that burns off in this process, an index of the amount of organic
material, is the *volatile* component.

Total residue includes *nonfiltrable residue* (also known by the older expression *suspended solids*, abbreviated SS) and *filtrable* (dissolved) *residue*. The latter
is material that passes through a standard glass fiber filter disk and remains after
subsequent evaporation at 103–105°C. A portion of the nonfiltrable residue is
settleable matter (or *settleable solids*) expressed in volume-ratio terms of milli-

liters per liter (mℓ/ℓ). This refers to the amount of SS that falls to the bottom of a wastewater sample held in a special container (Imhoff cone) in a specified time (1 hr).

A nominal medium-strength sewage has a total solids of approximately 700 mg/ℓ. On a weight-ratio basis, this is equivalent to 700 ppm (parts per million) since 1 ℓ of sewage weighs very close to 1000 g. A value of approximately 1200 mg/ℓ would be associated with a strong sewage and 350 mg/ℓ with a weak one. The human wastes portion of total solids is approximately 15% by weight with the urine (3.7% solids) contributing twice as much as the feces (23% solids). Of the 700 mg/ℓ of total solids in an average sewage, approximately 30% can be figured to be suspended, with the remainder filtrable.

The amount of total solids contributed to the wastewater by industry depends greatly upon the type, size, and age of the industry involved, as well as the season in some cases. Food-processing plants such as canneries, meat-packing plants, milk-processing plants, and cane-sugar plants may contribute large quantities of solids. Hammer (1975) has indicated that a milk-processing wastewater may have 1600 mg/ℓ of total solids (300 mg/ℓ of SS), that from a meat-packing operation 3300 mg/ℓ of total solids (1000 mg/ℓ of SS), and for synthetic textiles, 8000 mg/ℓ of total solids (2000 mg/ℓ for SS).

Other physical aspects of wastewaters that are of interest are temperature, color, and odor. The important one of these three, for our purposes, is temperature. The wastewaters discharged from certain industrial concerns, such as steel mills and petrochemical companies, as well as from power plants, can have temperatures 20°F warmer than the source water.

Chemical Characteristics

Organic Materials: The principal groups of organic substances in sewage are carbohydrates, proteins, and fats. Metcalf and Eddy (1972) report that the percentage makeup of the organics in sewage by these compounds is 25 to 50% for carbohydrates, 40 to 60% for proteins, and approximately 10% for fats and oils. Urea, the major constituent of urine, is another important organic constituent, and with the proteins adds the bulk of the nitrogen to the sewage. In sewage of medium strength, approximately 75% of the suspended solids and 40% of the dissolved solids are organic.

Three important organic constituents of wastewater, but not of human origin, are surfactants, phenols, and pesticides. A major example of the first of this list is the synthetic detergent used heavily in U.S. homes since 1940. Until 1965, such detergents were essentially nonbiodegradable. Phenols are waste products from a large number of industries that include the following: gas and coke manufacture, synthetic resin manufacture, textiles, tanneries, and tar, chemical, and dye manufacture. Pesticides fall into two large groups: herbicides and insecticides, the former for unwanted plants (weeds) and the latter for unwanted insects. These are principally synthetic organic materials, such as DDT, endrin, or 2, 4-D. They find their way into wastewaters from surface runoff, from return irrigation waters, as spills or wastes from pesticide manufacturing, or from canneries and other food-processing wastewaters.

Biochemical Oxygen Demand: *Aerobic* processes involve reaction with (dissolved, molecular) oxygen. The aerobic decomposition of the organic constituents of wastewaters by resident microorganisms is accompanied by a decline in the dissolved oxygen (DO) in that water. The larger the organic load in the wastewater, the larger the implicit demand on its DO supply. The most common means of indicating the amount of organic material in a wastewater is to specify its *biochemical oxygen demand* (BOD). In terms of a standard test, this refers to the number of milligrams of oxygen required by microorganisms in 1 ℓ of the sewage in the biochemical oxidation of organic matter over (by convention) a 5-day timespan at 20°C. The overall oxidation process is generally only 60–70% complete at this point for readily decomposed substances, but there is a plateau in oxygen usage at about this time, and waiting until the oxidation is 95–99% complete (anywhere up to 20 days) is considered too time-consuming for a standard test. A large portion of this difference involves the oxidation of nitrogenous compounds. The BOD test and other tests provide a measure of the amount of organic matter in wastewater. These tests are detailed in *Standard Methods...* (1976) and presented more briefly in Metcalf and Eddy (1972). A subscript on BOD refers to the length of the test in days.

A typical figure for the per capita municipal production of BOD per day is 0.17 lb (0.76 N), which is based on an average wastewater BOD_5 of approximately 200 mg/ℓ and a wastewater output of 100 gpcd (0.38 m^{3}pcd). A BOD_5 range from 100 mg/ℓ (for a weak wastewater) to 300 mg/ℓ (for a strong wastewater) would not be unusual. Some industries, notably those that contribute largely to the total solids of the wastewater, also provide a large BOD input. Using the same three industries mentioned earlier, a milk-processing wastewater ranges approximately 300–2000 mg/ℓ, that from meat packing approximately 600–2000 mg/ℓ, and a synthetic textile plant effluent approximately 1500 mg/ℓ of BOD_5. Distilleries can have enormous BODs—to 30,000 mg/ℓ (McGauhey, 1967).

Inorganic Substances and Gases: Certain inorganic features of wastewater are important, e.g., the pH, which is the negative logarithm of the hydrogen ion concentration. The pH scale runs from 1 (extremely acidic) to 14 (strongly basic); a neutral substance has a pH of 7. Wastewater has a pH range of from about 6.5 to 7.5. The concentration of various salts may also be of note. Of particular concern are the amounts of nitrogen (N) and phosphorus (P) present.

The bacterial decomposition of organic fecal matter and the hydrolysis of urea form ammonia nitrogen. Depending upon the pH of the solution, the nitrogen exists in aqueous solution as either the ammonium ion (NH_4^+) or as ammonia (NH_3). Further bacterial action oxidizes the ammonia nitrogen to the nitrite (NO_2^-) and thence to the nitrate (NO_3^-) form. This process is called *nitrification*. Nitrites are relatively unimportant in wastewaters because they are unstable and readily oxidized to the nitrate form. The phosphorus and sulfur in wastewaters can also be found in various forms.

Some heavy metals are natural components of water supplies, but specific industries are noted for releasing certain elements. The metal-finishing and electroplating industries can contribute chromium, cadmium, copper, nickel, silver, and zinc to their wastewater, hence to that of the municipality. Paint manufacturing can contribute chromium and lead, and lead can originate here as well as in battery- and gasoline-manufacturing processes. Although sewage itself contains metallic ions, these are customarily in minute concentrations compared to industrial wastewaters.

Eckenfelder (1970) discusses the nature of wastes from the following types of industrial establishments: breweries, fruit and vegetable canneries, dairies, meat-packing plants, petroleum refineries, plating companies, pulp and paper mills, steel mills, tanneries, and textile product companies. Some of the above (e.g., dairies) have wastewaters with high BOD loadings; others (e.g., plating companies) have negligible levels of BOD but contain metallic ions (e.g., cadmium, chromium) at high concentrations that are released throughout the year. The text by Nemerow (1971) contains a wealth of material on wastewaters from industrial concerns organized as follows: apparel, food-processing, materials, chemical, energy, and those with radioactive wastes. The U.S. Environmental Protection Agency has issued a series of reports on various types of industry and their wastewaters (e.g., Hallowell, 1973a, b).

There are five gases commonly found in wastewaters in notable concentrations. One of these, ammonia, has already been considered. Carbon dioxide, the second gas, is the product of carbonaceous oxidation processes under both aerobic and anaerobic conditions. The third gas is hydrogen sulfide formed in part from the decomposition of organic material containing sulfur under anaerobic conditions. Methane is the fourth gas and is the principal by-product of anaerobic decomposition of organic matter in wastewater.

The principal gas to consider is dissolved oxygen (DO) in the wastewater. DO is required for the respiration of aerobic microorganisms as well as other aerobic life forms. To ensure that the decomposition processes in the wastewater are aerobic, such that odors resulting from *anaerobic* conditions are largely prevented, there is a requirement for a minimum level of DO in the wastewater. However, it is not unusual for the DO to be virtually zero in raw wastewaters.

3-2 INDICATOR ORGANISMS

Bacteria are simple, single-cell (unicellular) organisms that use soluble food and are generally capable of self-reproduction without sunlight. This reproduction is by binary fission (splitting in two). These bacteria are so minute that they can be seen only through a microscope, and as a result are known as microscopic organisms (microorganisms).

Bacteria by the billions occupy the lower intestine of man and animals, and vast quantities of these bacteria pass out of the body with the feces. The average daily excrement of fecal matter per person in the United States is about 20 g dry weight, of which approximately 4 or 5 g are bacterial bodies half of which are alive if the feces are fresh. The total number of live bacteria in fresh fecal matter is estimated to be 2×10^{12} pcd. Of these, approximately 10% are bacteria of the coliform group *Escherichia coli* (*E. coli*). They are useful in the breakdown of organic materials into inorganic materials; they are also essentially harmless to animals and man.

Some bacteria lodged in the lower intestine of man and released with fecal material are not harmless; they are pathogenic, i.e., disease-causing and/or transmitting. Such bacteria, released into receiving waters in a wastewater effluent, can potentially infect others with diseases such as typhoid, cholera, paratyphoid, and bacillary dysentery.

Enteric organisms are those originating in the intestines. One such category involves bacteria, whereas two others include viruses and amoebic cysts. A virus is any of a group of ultramicroscopic infectious agents that reproduce only in living cells. These agents can also lead to waterborne diseases, notably poliomyelitis and infectious hepatitis.

The number of viable (live) *E. coli* in normal fresh domestic sewage during dry weather (i.e., when the amount of storm water contribution to the flow is minor) is approximately 10^8–10^9/100 ml. Because of this large concentration, approximately four to five orders of magnitude (factors of ten) larger than that of viruses, and because *E. coli* are relative easy to test for, their presence in water has been normally used to indicate recent fecal pollution. Such bacterial tests are generally preferred to chemical ones. Excess amounts of ammonia, nitrites, and chlorides result from fecal introduction, but since such excesses can result from many effects other than fecal inputs, tests involving such aspects do not provide positive evidence of fecal contamination. However, the use of coliforms per se among various bacterial alternatives may be more due to tradition than to any other factor (Bonde, 1975). It has been said that fecal streptococci might be used as indicator organisms, in part because they have a slower "die-away" rate than coliforms.

Total coliforms are comprised of two major groups: *E. coli*, described above, and *Aerobacter aerogenes* (a saprophyte of the soil). Saprophytes are specialized bacteria involved in the decomposition of organic matter. Many agencies only test for total coliforms in water, whereas others conduct somewhat laborious tests to differentiate between *E. coli* and *A. aerogenes*; however, a newer trend is to segregate coliforms into total coliforms (cultured at approximately 35°C) and fecal coliforms (cultured at approximately 45°C). Fecal coliforms are considered to be primarily composed of *E. coli*.

One method of determining the number of coliform organisms present in a wastewater is the *membrane-filter technique*. A filter of minute pore size is used to retain the bacteria formerly contained in a known volume of wastewater, and bacterial colonies are counted after being allowed to grow in the presence of abundant nutrients. Besides this direct method, there is an indirect, statistical technique that has been used for many years. This is called the *most probable number* (MPN) approach and involves tests on equal volumes of different dilutions of the wastewater and the simple positive or negative indication of coliform presence in each sample.

Since pathogenic microorganisms are few and far between and somewhat difficult to isolate, the presence of nonpathogenic coliforms in water has been taken as indication that pathogens might be present. For this reason, coliforms are known as *indicator organisms*. Water purity standards, to be discussed in Chapter 4, assume implicitly that as the population of viable coliforms wanes because of effects such as bacterial die-off due to natural causes or disinfection, or due to wastewater dilution, the population of pathogenic organisms decreases proportionately. This is likely a satisfactory assumption for most pathogenic bacteria, but it may not be true for pathogenic viruses or cysts that are highly resistant to disinfection by chlorination. In addition, viruses have been shown to die off far less slowly in salt water than coliforms. Despite some disadvantages, fecal coliforms are still generally preferred as an indicator of recent fecal contamination, and will be used as such in this book. Further discussion of indicator organisms and pathogens is contained in Chapter 4.

3-3 PRETREATMENT

Introduction

There is a sizable gulf between the quality of raw wastewaters and the quality of wastewaters that must be maintained for various disposal schemes. The transformation of a raw wastewater into one acceptable for disposal according to criteria presented in Chapter 4 is termed wastewater treatment. There is such a wide array of components in wastewater that are undesirable in terms of ultimate disposal that treatment processes have taken many forms. The following section considers physical and chemical treatment processes, leaving for the ensuing section discussion of biological treatment processes. Means of treatment in which physical forces predominate are known as unit processes. This section deals with the steps usually taken to prepare a wastewater for primary sedimentation (considered in the next section). Together, pretreatment and primary sedimentation comprise *primary treatment*.

Screening and Shredding

The first step in the wastewater treatment process is to remove coarse materials that would hamper the efficient operation of subsequent elements in the process. This is done by passing the wastewater through a structure where such materials will be retained. Such a structure, consisting normally of openings of uniform size, is called either a rack, if parallel bars or rods are employed, or a (fine) screen in the case of wire cloth or perforated plates. Bar racks are also known as coarse screens. The material that accumulates on both types of screens (coarse and fine) can be removed either mechanically or by hand. At the present time, the latter technique is practiced only in small plants.

One arrangement for the screens is to be permanently in place and cleaned by rakes or teeth that pass between the bars. Another possibility is for screens to be stationary when in operation but removable for cleaning purposes. The screens also may be in continuous movement when in operation and cleaned as they move.

Coarse screens are used to remove larger floating materials from wastewater (items such as coarse sewage solids, rags, roots and branches, and large items such as rocks). The removal of rags, for example, is important since such items can clog pumps further along in the plant. The major importance of coarse screens, however, is to reduce the amount of floating material in subsequent settling tanks. The clear opening of coarse screens is normally 0.5–3 in. (about 12 to 75 mm). In some cases, fine screens remove the larger suspended solids from the wastewater, items such as fiber, strings, and plastics. The openings in fine screens are usually 1/32–1/4 in. (0.8–6 mm).

The quantity of screenings depends upon the type of wastewater, the flow discharge, the size of the screen openings, and the frequency of cleaning. However, the range of values for coarse screens would be approximately 0.5–6.0 ft^3/Mg (approximately 0.004–0.045 mℓ/ℓ) with approximately 1.5–3.0 ft^3/Mg (approximately 0.011 to 0.022 mℓ/ℓ) for fine screens.

Screenings can be dumped into suitable containers and then buried on the wastewater treatment plant grounds, hauled away to landfill sites, or fed into the municipal refuse and garbage disposal system. Another possibility is for the screenings to be dewatered and then incinerated. However, a preferred technique for the highly *putrefactive* (foul-odor-producing) screenings is shredding or grinding to at least 1/4-in. size (6 mm) and return to the wastewater flow for subsequent removal.

Shredding devices cut or grind the coarse solids, which then pass to downstream treatment units for removal. There are several different types of shredders; a common one is the *comminutor*.

Grit Removal

The term *grit* covers material that has a specific gravity substantially larger than the organic components in wastewater. Examples of such material are sand, gravel, and cinders, as well as seeds, rice, and coffee grounds. Their removal from the wastewater is important since they can cause increased wear on moving mechanical equipment such as pump impellers; they can also settle out in objectionable quantities in pipes and channels, and can impede the efficient operation of other components in the wastewater treatment sequence. Particle sizes up to 0.2 mm (a fine sand) are commonly handled by *grit chambers*.

Various schemes can be used to remove the accumulation from grit chambers. Two of these are the belt conveyor, fitted with buckets or scrapers, or the screw conveyor. The quantity of grit in a wastewater varies widely depending upon the geographic location involved and weather conditions. A range of $1/3$–24 ft^3/Mg (0.002–0.18 mℓ/ℓ) of wastewater treated is typical (Metcalf and Eddy, 1972). In some cases, grit can be incinerated. It can also be hauled from the treatment plant by truck and used as fill.

Other Types of Pretreatment

One form of pretreatment is *preaeration*, wherein air is added to the wastewater. This step can be combined with the grit-removal process, in particular for so-called *aerated grit chambers*, in part by increasing the standard retention time in the grit chamber. The process of preaeration accomplishes various ends that include the promotion of BOD removal.

In two other pretreatment processes, oil and grease can be removed from the surface wastewaters in *skimming tanks*, and the removal of very fine suspended matter in the later sedimentation process can be promoted by adding appropriate chemicals to the water and providing some stirring by means of paddles, so that *flocculation* occurs. In this process, there is an aggregation of fine material into a united whole, or *floc*, that has a settling speed of the same order as that of other (coarser) suspended material in the wastewater.

3-4   PHYSICAL AND CHEMICAL TREATMENT

Primary Sedimentation

The term *clarification* can be applied to this process in which the continuously flowing water is detained for sufficient time that much of the heavy suspended material settles out (*sludge*) and light (*floatable*) materials rise to the surface. If properly designed, 50–65% of the suspended solids can be removed from the wastewater and with that approximately 30–40% of the BOD. Primary sedimentation may be used as the sole treatment step, in which case the detention time will be 90–150 min; or it may be on the front end of a further chain of treatment processes, in which case the detention time (the first time through) will be 30–60 min. *Detention time* is the tank volume divided by the flow discharge. Recirculation of treated wastewater within the plant can be used to control the detention time. Very long detention times for sewage are undesirable both because of the odor problem under anaerobic conditions and the possibility of a "rising sludge" condition as a result of gas formation in the anaerobic sludge.

Ideally, the primary sedimentation tank is designed so that the detention time of wastewater in the tank is just sufficient to allow an arbitrarily small-diameter particle to pass from the surface to the floor of the tank before its associated water mass exits from the tank. In real cases, such computations are not trivial, however. There is a distribution of particles of many sizes and specific gravities, and there are complicating effects like flocculation, concentration of suspended matter, and wide variations in flow.

It is customary to speak of the *surface-loading rate* for a primary sedimentation tank. This refers to the flow rate per unit plan area of the tank, and typical numbers vary from 400 to 1000 gpd/ft² (1.9–4.7 m³/s/hectare) with 800 gpd/ft² (3.7 m³/s/hectare) a typical design value (Van Note et al., 1975).

Primary tanks can be of either rectangular or circular design with the latter the currently preferred shape. Such tanks are shown in the foreground of Figure 3-2. Typically, such tanks are 40–100 ft (12–30 m) in diameter with a maximum

Figure 3-2 Part of large wastewater treatment plant

of 200 ft (about 60 m). Wastewater depths 7–12 ft (approximately 2–3.5 m) are usual. The normal procedure in circular tanks is for the wastewater inflow to enter the tank at its center and to leave the tank by means of a continuous overflow weir surrounding the tank. Typical weir flow rates are 10,000–15,000 gpd/ft (0.0014–0.0021 $m^3/s/m$). As the wastewater moves imperceptibly radially outward, the suspended matter may move upward or downward depending upon its specific gravity. Both the floating material and sludge are collected by means of arms that slowly rotate around the tank. The surface scum passes over a central weir, whereas the bottom material is scraped toward the center of the tank, where it is collected in a bottom hopper for later removal through a pipe. The frequency of such removal must be such that there is not undue buildup of sludge in the tank. The disposal of such sludge will be considered in a subsequent section. Suspended solids that neither sink to the bottom or float to the surface remain in the mainstream for subsequent treatment or disposal.

Flotation

Fine particles, although heavier than water, can be made to rise to the surface by means of a process called *flotation*, wherein air rises through the wastewater. Air is initially introduced into the wastewater by pressurizing the latter to 2–4 atm in the presence of air. The wastewater is then returned to atmospheric pressure, and under turbulent flow conditions the air comes out of solution. The bubbles become attached to and enmeshed in the suspended particles. This procedure can be aided through addition of chemical additives to the water. The rising of the fine air bubbles through the wastewater also tends to bring greases and oils to the surface where they can be skimmed off. This air flotation process combined with sedimentation has also been called *advanced primary treatment*.

Chemical Treatment

Two classes of chemical treatment will be considered: chemical precipitation and disinfection. Chemical precipitation used in conjunction with the sedimentation process can cause improved results in terms of suspended solids and BOD reduction in the effluent. However, these results are inferior to those possible with biological treatment. Typical numbers for suspended solids reduction are 60–85%; for BOD the range is 40–70%.

The process involves the addition of floc-forming chemicals to sewage to both increase and hasten sedimentation. The precipitants used react with substances present in the wastewater or with other substances added for this purpose. Colloidal particles are removed by coagulation.

Chemical treatment has the advantage that it can be readily adapted to changes in flows and to wastes that are highly toxic to biological growth. However, skilled operator attendance is required to properly effect such necessary changes. Disadvantages of this process are the cost of chemicals involved, and the fact that the chemicals must be stockpiled to ensure continuous plant operation. Also, substantial quantities of chemical-biological sludge must be handled. However, chemical treatment has returned to favor recently, at the least because of the development of proper mechanical equipment to handle such substances and better knowledge of the process.

The mixing of the chemicals can be done hydraulically, e.g., in baffled chambers. It can also be done using violent agitation, a propeller driven by a vertical-shaft motor being a frequent choice. Detention of the wastewater should not be unduly long, since prolonged mixing may impair floc formation. A time of 1–2 min is typical. Following rapid mixing to disperse the coagulant, slow mixing is involved to aid in floc formulation. Chemical treatment causes a volume of sludge in the settling tank that is close to double that in the same tank without prior chemical treatment. The precipitants used form part of this sludge.

Disinfection

Disinfection refers to the selective destruction of disease-causing organisms, although the process may kill nonpathogens as well. All of the organisms are not killed in the process, and this distinguishes disinfection from *sterilization*, which is the destruction of all organisms. The common method of evaluating the efficiency of the disinfection process is to compare the concentrations of live *E. coli* upstream and downstream of the disinfection point.

Wastewater disinfection is best carried out by using chemical agents such as chlorine, bromine, iodine, or ozone. Chlorine is by far the most commonly used disinfectant, but interest in ozone has increased partly because of expense and the marked poisonous (*toxic*) nature of chlorine compounds for aquatic biota occupying wastewater receiving waters.* Ozone leaves little or no residual. However, the brief discussion that follows deals exclusively with chlorine. Alternate methods, particularly involving ozone, are discussed by Hoehn (1976), Nebel et al. (1976), Katzenelson and Biedermann (1976), and Peleg (1976). Thorough references on disinfection by chlorination are Metcalf and Eddy (1972) and Collins et al. (1974).

For discussion purposes, disinfection by chlorination has been placed in this section, implying somewhat that it takes place along with primary treatment. This is not true, since chlorination can take place at various stages along the treatment sequence. On one end of the scale, there is *prechlorination* that occurs on the untreated wastewater. This is often used where the wastewater is to be given only primary treatment. On the other end of the scale, there is chlorination following all levels of treatment. Although only disinfection by chlorination is

---

*The toxicity of chlorinated wastewaters to aquatic biota can be substantially reduced, and in fact made less than that of the unchlorinated wastewater, by using *dechlorination*, or the removal of chlorine. See, for example, Collins et al. (1974) for details of dechlorination practices.

being considered herein, chlorine can be added to a wastewater in the collection system to reduce slime growth, and it can be added along the line for odor control. But the reason for chlorination, as involved here, is to yield a moderately "safe" wastewater.

Chlorine can be added to wastewater as chlorine gas or hypochlorite salts such as sodium or calcium hypochlorite [NaOCl or Ca(OCl)$_2$]. Chlorine gas (Cl$_2$) reacts with water (H$_2$O) to produce, in part, hypochlorous acid (HOCl). In turn some of the HOCl ionizes to form, in part, hypochlorite ions (OCl$^-$). The degree to which this takes place depends heavily upon the pH of the wastewater, with lower pH favoring more HOCL. The quantity of HOCl and OCl$^-$ present in the water is called the *free available chlorine*, and the partitioning between the two is important for disinfection purposes since HOCL is a far more effective bactericide.

However, free chlorine reacts rapidly with wastewater constituents. The nonbactericidal chloride ion forms when HOCl, a strong oxidizing agent, reacts with various readily oxidizable ions such as ferrous, manganese, nitrite, and sulfite, as well as hydrogen sulfide and organic nitrogen. But the major reaction for our purposes is the reaction of free chlorine with ammonia in the wastewater to form the *chloramines*. These are still disinfectants, but slow-acting ones. For example, monochloramine at the same concentration as hypochlorous acid takes about 250 times as long to effect the same 99% elimination of *E. coli* (Collins et al., 1974).

The iodometric method of analysis for chlorine (*Standard Methods...*, 1976) indicates both free chlorine and the chloramines—the *total chlorine residual*. As has already been indicated, the disinfectant properties of the chloramines are clearly inferior to those of free chlorine. However, for wastewaters (as distinct from water supplies), it would be infeasible to introduce chlorine in such amounts as would react with the various ions and still leave an excess to act in the free state.

Mixing of the chlorine with the wastewater should be very rapid (on the order of several seconds), and then there should be adequate contact time in a *contact basin*. Suggested minimum contact times for peak flow are 15 (Metcalf and Eddy, 1972) or 30 min (e.g., Collins et al., 1974). Contact basins can be designed to be a part of the treatment plant, but a long outfall pipe used to carry a wastewater to receiving waters is an excellent contact chamber.

A typical disinfection dose of chlorine to the effluent from a primary sedimentation tank is 20–25 mg/ℓ or higher. The chlorine residual used depends upon the contact time; figures between 0.5 and 1.0 mg/ℓ are typical for sewage treatment. Standard chlorination can result in a 98–99% reduction in bacterial populations, but combined chlorine residual levels in the neighborhood of 1 mg/ℓ are completely unacceptable for adequate viral destructions (Lothrop and Sproul, 1969). Substantially higher chlorine concentrations are necessary—and a free chlorine residual.

3-5 SECONDARY TREATMENT PROCESSES

Introduction

The effluent from the primary treatment processes considered in the previous section serves as the influent to secondary or *biological treatment processes*. Such processes effect a further reduction in the levels of BOD and SS, as well as other constituents in the wastewater, as will be explained in this section. In the United States, Public Law 92-500 of 1972 prescribed a minimum level of secondary treatment for all wastewaters by July 1, 1977.

The two major classes of biological treatment are the (complete mixing) *activated sludge* and (high-rate) *trickling filter* processes, which will be considered in following subsections. A third class, stabilization ponds, will not be described. Details on such ponds are given in Metcalf and Eddy (1972) or Battelle-Pacific Northwest Laboratories (1974). Use of stabilization ponds is generally limited to small wastewater flows. Such ponds are sometimes used in conjunction with outfalls.

Both of the two major forms of secondary treatment processes have their own distinct advantages and disadvantages. Advantages of the trickling filter are ease of operation and the capacity to accept shock loads and overloading without much impairment in process efficiency, let alone complete failure. An activated sludge unit can, however, fail due to overloading. Two distinct advantages to the activated sludge system are the ability to treat high-strength wastewater and adaptability for future use in plant conversion to tertiary treatment. High-BOD removals constitute another advantage of the activated sludge unit, in contrast to trickling filter effluents which would have to be recycled to achieve the same final effluent BOD.

Trickling Filters

Introduction: The expression *biological filtration* is sometimes used to describe this process. The inclusion of the word "biological" is important because secondary treatment using trickling filters has nothing to do with the physical concept of filtration. The fact that biological slimes form on fixed surfaces when domestic wastewater is sprinkled over them can be used to advantage to rid a primary effluent of a large part of its remaining BOD. The slime consists chiefly of bacteria as well as plant and animal organisms that feed on the organic material within the wastewater. In addition, some of the organisms extract oxygen from the wastewater and give off carbon dioxide. To be an efficient system for eliminating BOD from the wastewater, the flow should be evenly distributed over the surface, and this surface area available for slime growth should be maximized within the constraint of assuring that reaeration of the flow can be realized. Clogging, for example, could lead to prevention of wastewater reaeration. Three adaptations of the same theme will be considered below.

Standard Trickling Filter: The first type is the most familiar form of trickling filter. In this case, the medium on which the slimes grow is rock surface. The rocks preferred for use are 1.5–4 in. (38–102 mm) in diameter; smaller rock sizes would provide larger surface areas but would clog more readily. The 1.5-in. (38-mm) size is used for weak wastewaters. The rock must be durable, insoluble, and resistant to *spalling*.* This rock is contained in a large circular tank, up to 200 ft (about 60 m) in diameter, with a usual rock depth of 4–7 ft (1.2–2.1 m). Over the bottom of the tank is an underdrain system, typically consisting of special vitrified-clay blocks, that carries away the water that reaches the bottom of the rock bed and permits circulation of air since the underdrain system is designed to flow only partly full.

The wastewater is applied to the upper rock surface through nozzles in a horizontal, rotating distributor driven slowly either through the dynamic reaction to the water flowing from it or by an electric motor (see Figure 3-3). The primary effluent reaches the distributor through a vertical pipe riser at the turning point for the distributor in the center of the tank. The speed of the outer ends of distributors range from approximately 3 to 10 fpm (approximately 1–3 m/min).

The organisms in at least the upper 2–3 feet (0.6–0.9 m) of the rock filter are plentifully supplied with food and oxygen. Adsorbed organic matter is about 10% oxidized in the slimes. Excess slime growth is sloughed off the surface and passes downward through the filter into the underdrain system. To rid the wastewater of this matter, a final (secondary) clarifier is used with the trickling filter. Recirculation of effluent from the final clarifier back to the wet well at the start of the wastewater treatment chain or to the trickling filter is used for modern high-rate trickling filters (see Figure 3-4).

The BOD load on a trickling filter is determined using the BOD in the primary effluent applied to the filter without considering whatever BOD was contained in recirculated flow. The concentration of BOD is usually less than 200 mg/ℓ. A common design figure for the BOD loading in high-rate filtration systems is 45 lb/1000 ft^3 (720 g/m^3) of rock filter per day, although from 30 to 90 lb/1000 ft^3/day (480 to 1440 g/m^3/day) is possible. To assure that the bed voids are not filled with biological growths to the detriment of flow aeration, a minimum hydraulic flushing to prevent clogging is required. High-rate systems make use of from approximately 10–30 million gallons (Mg) of untreated and recirculated flows per acre of top surface of the filter per day (9–28 m^3/m^2/day).

---

*This involves the chipping or flaking off of parts of the rock surface.

Figure 3-3 Two small trickling filters (Courtesy, Lakeside Equipment Corp.)

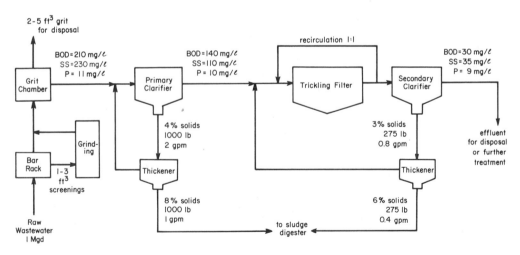

Figure 3-4 Schematic diagram of trickling filter plant operation

BOD removal is enhanced by passing the flow through the trickling filter more than once. A common fraction comparing the discharge of raw wastewater to that recirculated is 0.5–3.0. In some cases, underflow from the trickling filter is returned directly to the initial wet well, whereas in others sufficient flow is returned to the wet well from the final clarifier to carry settled solids so that they can be removed in the primary clarifier. Secondary clarifier sludge is thin (0.5% solids) and is best removed by returning to the flow entering the primary clarifier.

The BOD removal efficiency of a trickling filter is a complex function of several variables that include the BOD loading, the recirculation ratio, and temperature. The range of BOD efficiencies is from approximately 60–90%. A similar range applies to removal of suspended solids.

Recirculating advantages in terms of the filter include the following: more oxygen is carried into the bed; organisms are kept operating at optimum capacity; and hydraulic scouring prevents clogging. For the treatment system as a whole (Figure 3-5), flow fluctuations are smoothed.

Other Forms: The biological tower is the second form of biological filter. It was indicated earlier that it is desirable to have a high specific surface (area/volume) for the filter but that it is also important to have adequate void volume for reaeration. Plastic packing can be manufactured that possesses these desirable features and, at the same time, is uniform, strong, light, and chemically resistant. Networks of redwood slats are also possibilities.

Towers made up of stacked, interlocking plastic packing are 20 ft (6 m) or more high and the flow is applied from fixed distributors. Allowable organic

Figure 3-5   Trickling filter wastewater treatment plant

loadings range from 25–150 lb of BOD/1000 ft^3/day (400–2400 g/m^3/day) with hydraulic loadings up to 2 gpm/ft^2 (0.08 m^3/min/m^2).

The third type of biological filter consists of multiple, corrugated plastic discs, mounted coaxially, that rotate in the wastewater and are approximately 40% immersed in it. When the disc voids move above the wastewater surface, water is replaced by air. When a certain part of a disc is submerged, the organisms attached to that part are exposed to the organic material in the wastewater. Excess growth is sloughed off the discs and is carried to a final clarifier for gravity separation.

Activated Sludge Process

The dissolved and colloidal solids in a wastewater are predominantly of organic origin. Such solids cannot be removed by physical settling, and the *activated sludge* process, also called *biological aeration*, involves the conversion of such solids into bacteriological cell tissue that will readily settle under quiescent conditions.

Primary effluent enters a large basin with a supply of suspended, colloidal, and dissolved organic matter that serves as food for bacteria in the basin. These bacteria grow as they consume the organic matter and dissolved oxygen in the wastewater and give off carbon dioxide. Some bacteria are in turn consumed by other microorganisms. Aerating of the wastewater in the basin facilitates the removal of organic matter from solution, through oxidation and synthesis into microbial cells, by replenishing the dissolved oxygen supply.

The term *mixed liquor* is used to describe the liquid suspension of micro-organisms in the wastewater, and the biological growths are referred to as *mixed liquor suspended solids*. The mixed liquor is constantly moved on to a final clarifier for removal of the biological floc and discharge of the resulting effluent. Typically, wastewater detention times in the aeration tank are between 4 and 8 hours. Organic loadings are normally in the range of 20–40 lb of BOD/1000 ft^3/day (320 to 640 g/m^3/day) for conventional-type activated sludge processes.

There are two major methods of aerating the wastewater in the conventional process. Either air (or perhaps oxygen) is introduced through submerged porous diffusers (small bubbles), or air nozzles (large bubbles), or the wastewater is agitated mechanically to promote solution of air from the atmosphere. Blowers are used to supply air to the diffusers which are often located in a longitudinal row along one side of the tank in order to induce a spiral flow. A nominal air flow of from 0.2–1.5 ft^3/gal of wastewater (1.5 to 11 m^3/m^3) is required—or roughly 1000 ft^3 air/lb BOD (about 1 m^3 air/16 g BOD). Larger bubble diffusers can handle a wider range of air flows. Small bubble diffusers have better oxygenation characteristics, but air filters and extensive system maintenance are required. Typical tank sizes for air-diffusion aeration are as follows: length 50–400 ft (15–120 m); width 15–30 ft (5–10 m); depth 10–15 ft (3–5 m).

The tank for the mechanical aeration is often circular, 20–40 ft (6–12 m) in diameter with a depth between 8 and 18 ft (2.5 and 5.5 m). The mixing devices include submerged paddles or turbines, or revolving blades at the surface. For the surface devices, oxygenation is from induced surface aeration. In the subsurface cases, air is discharged from a pipe or ring beneath the rotating blades. Energy requirements are from 12 to 20 horsepower/Mgd (3300 to 5500 joules/m³ of flow). The period of aeration in such tanks is between 8 and 18 h.

Only a small portion of the organic matter is actually consumed in the aeration tank by the microorganisms (between 10 and 20%). A settling period following detention in the aeration tank permits settling of heavy sludge and overflow of clear liquor. Oxidation of the sludge material continues in the final clarifier.

The diffused air type of biological aeration results in a concentration of suspended solids in the mixed liquor of 1200–3000 mg/ℓ; the figures for the mechanical aeration process are 600–1500 mg/ℓ. The greater the solids content, the greater the floc area available for adsorption of organic matter from the wastewater. Also, the greater the solids, the more the air required, since solids represent biological growths that must be sustained.

Activated sludge settling tanks are either circular or rectangular. The usual range for the diameter of circular tanks is 30–100 ft (10–30 m), and this diameter does not normally exceed 10 times the tank depth. The length of rectangular tanks normally does not exceed 10 times the depth. In both types of tanks, a common system is for the accumulated sludge to be scraped along the bottom of the tank to a central hopper. From that point, some sludge can be returned to the inlet to the aeration tank to assure the perpetuation of the activated sludge process, whereas the excess can be sent to the inlet to the primary clarifier. It is then ultimately removed with the primary sludge. The effluent from the final clarifier has typically a BOD less than 20 mg/ℓ. Activated sludge process efficiencies are up to about 95% in terms of BOD, 90–95% for suspended solids.

An adaptation of the activated sludge process is *step aeration*, in which wastewater is added at two or more points along the aeration tank. The increased flexibility of operation results in an improved BOD removal. Other variants are described in Clark et al. (1971).

3-6 SLUDGE TREATMENT AND DISPOSAL

Introduction

The effluent from the final clarifier in a wastewater treatment chain is essentially all liquid phase. Contrary to what one might suspect, sludge is by no means chiefly solids. For example, the sludge in a primary clarifier (without recirculation) averages approximately 95% moisture, that in an activated sludge settling tank approximately 98–99% moisture or greater, and that in a primary sedimentation tank with activated sludge recirculation approximately 96%, the actual range being 94–99.5%. An average figure for the weight of dry solids resulting from the sludge in a primary clarifier (without recirculation) is approximately 120 lb/1000 persons/day (54 g/person/day). This is about 1200 lb/Mg (140 g/m^3) of sewage. But the corresponding real sludge (solids and liquids) weight per million gallons of sewage in this case is some 12 tons. This would mean that a medium city of population 500,000 would generate some 600 tons of primary clarifier sludge each day. In this section, we consider how this sludge can be suitably disposed of after discussion of the means by which this disposal problem can be eased. A source of extensive information on sludge treatment and disposal is Battelle-Pacific Northwest Laboratories (1974).

Thickening, Dewatering, and Conditioning

Concentration is used to thicken sludge. The following approximate equation can be used to relate initial and final sludge volumes V_1 and V_2 to the initial and final percent of solid matter P_1 and P_2 contained in the sludge

$$\frac{V_1}{V_2} = \frac{P_2}{P_1} \tag{3-1}$$

In thickening sludge, it is better not to attain a solids concentration much above 10%, since such sludges are extremely difficult to pump. A 2–10% solids concentration usually results from thickening. Sludge thickening is an important factor in reducing the costs of transporting sludge by truck to a location where it is to be used as a soil conditioner. Thickening is also important to further sludge treatment processes, because it reduces the required capacity of tanks and equipment, the amounts of chemicals needed for sludge conditioning, and the amount of fuel needed in drying and incineration. In the latter process, approximately 30% solids are needed for a self-sustaining operation. In one sludge-thickening system, *gravity thickening* using mechanical thickeners, the sludge is fed into a tank and slowly stirred. Water-escape paths are thereby opened up, and the water flows over an overflow weir; at the same time the bridges among sludge particles are broken, leading to increased settling and compaction.

Most thickeners have a minimum detention period of 6 h and a hydraulic overflow rate of 400–800 gpd/ft^2 (16–32 m^3/day/m^2). Solids concentration of from two to five times is normal. Another possibility for thickening is to use the flotation process outlined earlier as a form of advanced primary treatment. The previous two techniques of thickening and flotation are used before sludges are sent to digesters; the following techniques are implemented after digestion.

Centrifuges are also used in dewatering. Sludge is moved out of the centrifuge by means of a screw conveyor. This *sludge cake* typically has 75–80% moisture. Vacuum filtration is a common means for dewatering wastewater sludges. In this case, a sludge slurry is kept under vacuum by means of a porous medium that retains the solids but allows the liquid to pass. Cloth and steel mesh are examples of this porous medium. A rotary drum made of the porous material is partly submerged (12–60%) in the slurry and rotated. Solid material adheres to the submerged portion of the drum because of the internal vacuum. Air is drawn through this adhering material, called *filter cake*, when it rises above the slurry surface, causing some drying. Finally, a knife edge scrapes the cake from the drum onto a conveyor, and a water spray washes the porous surface before it again dips below the slurry surface. Total cycle time for a filter varies from 1–6 min. In this process, the solids are concentrated by a factor of 5–10.

Dewatering sludge by using a filter press is reportedly common in Europe. Sand bed drying is widely used for smaller sewage plants and some industrial waste-treatment plants. A bed of sand overlying one of graded gravel, equipped with underdrains, receives wet sludge to a depth of 8–12 in. (20–30 cm). Drying extends over many days but may be shortened through the addition of chemicals.

Conditioning of sludge is carried out to improve its dewatering capability. One common method is to add chemicals, of which lime and alum are examples, to promote coagulation and the release of absorbed water. The chemicals are normally added in liquid form. This type of conditioning is used before sludge is moved to vacuum filtration or centrifuge units. The requirement for coagulating chemicals can be reduced (by 65–80%) through the process of *elutriation* that involves mixing of the digested sludge with water and resettling.

Heat treatment is the second common type of conditioning process and involves heating the sludge for short periods of time under pressure. It accomplishes the same end as *chemical conditioning* and is used before vacuum filters or filter presses.

Anaerobic Digestion

Anaerobic sludge digestion is a biochemical action wherein organic matter undergoes anaerobic decomposition (reduction). The requirements for a favorable anaerobic process are exclusion of oxygen, adequate materials, appropriate temperature (90°F or above), pH control, proper seeding with specific organisms, freedom from excessive concentrations of heavy metals and sulfides, and mixing. Although the major purpose of the digestive process is the destruction of putrescible sludge matter leading to a minimum of odor in the final product, there are other aims: to produce a compact, readily drainable and dryable material, to destroy pathogens, to reduce the water content of the sludge, and to produce methane gas. When mixed with approximately 90% air, methane is highly explosive and can be used as a fuel for sludge heaters, heating treatment plant buildings and providing fuel for engines.

The microorganisms responsible for the anaerobic decomposition of organic matter are commonly divided into two groups. The first of these (acid phase) hydrolyzes and ferments complex organic compounds into simple organic acids, of which acetic acid is a prime example. The second group of microorganisms (gas phase) converts the organic acids to methane gas and carbon dioxide. This conversion by anaerobic bacteria is a somewhat delicate process, depending heavily on the pH of the mixture. The solid matter resulting from the process is usually well stabilized biochemically.

Conventional anaerobic digestion takes two forms. The single-stage process, normally limited to wastewater treatment plants with flows less than 1 Mgd (44 ℓ/s), combines the functions of digestion, sludge thickening, and supernatant and gas formation. In the more usual two-stage process, digestion is carried out in a first tank, the other steps in a second tank. An adaptation of the two-stage process to achieve higher processing rates is the high-rate or complete mixing process.

In this case, the first tank is typically heated, and mixing of the sludge throughout the tank is accomplished by means of pumps, turbine or propeller mixers, or the discharge of compressed digester gas. Sludge is pumped to the digester continuously or at intervals of 0.5–2 h. Digested sludge is displaced and passes to the second (holding) tank.

This second tank is used for storage and concentration of the digested sludge as well as for the formation of an overlying supernatant, above which is the digester gas. Approximately half of the total solids in the incoming sludge are reduced and given off as gas. The second tank is usually heated.

The tanks have either fixed or floating covers and are normally 20–115 ft (6–35 m) in diameter with a central depth 25–50 feet (7.5–15 m). The bottom of the second tank is made to slope toward the center so that sludge can be drawn off at the center. Anaerobic digestion tanks are shown in the background of Figure 3-2.

Detention times in conventional digestion systems vary between approximately 1 and 3 months, of which approximately 20–30 days involves digestion, and the remainder involves storage. For such systems, the solids loadings are between 0.03 and 0.10 lb of volatile solids per cubic foot of tank volume per day (480–1600 g/m³/day). For high-rate digesters, loading rates are larger, viz. 0.10–0.40 lb/ft³/day (1.6–6.4 kg/m³/day) and a reduction in the detention time to between 10 and 20 days is possible. The tank volume required per capita varies from about 2 to 6 cubic feet (approximately 0.06–0.17 m³).

The moisture content of raw sludge is 94–98%, that of digested sludge 90–96%. The total solids in the raw sludge are thus 2–6%, those in digested sludge 4–10%. The supernatant withdrawal is usually approximately three times the sludge withdrawal. The supernatant has a normal range of suspended solids of from 0.03–0.1% and a high BOD and nutrients content.

The resulting gas is 60–65% methane, 30–35% carbon dioxide, with the remainder other gases such as hydrogen, hydrogen sulphide, and nitrogen. The digester gas yield for primary plants treating normal sewage is about 0.6–0.8 ft³pcd (17–23 m³/1000 persons/day), or 6–8 ft³/lb (about 0.4–0.5 m³/kg) of volatile solids introduced, with the figure for secondary plants about 1.0 ft³pcd (0.03 m³pcd). The heating value of this gas is about 600 BTU/ft³ (5×10^6 cal/m³). BTU represents British Thermal Unit.

Not only domestic wastewater sludges but also industrial wastewaters high in BOD, such as meat-packing and cannery wastes, can be efficiently stabilized through the anaerobic digestion process.

Aerobic Digestion

This is an alternative to anaerobic digestion especially (e.g., in package plants) for sludges that are at least in part derived from an activated sludge-treatment facility. This type of digestion is normally carried out in unheated tanks similar to those used in the activated sludge process.

Some of the advantages and disadvantages of aerobic digestion compared to anaerobic digestion have been detailed in Metcalf and Eddy (1972). They outlined some of the features as follows: hydraulic detention time of 12–22 days depending upon the particular treatment processes involved in the sludge production; loading of 0.1–0.2 lb of volatile solids/ft³/day (1.6 to 3.2 kg/m³/day); minimum DO in the liquid with the sludge has to be in the range of 1 to 2 mg/ℓ, which is why the process is normally limited to sludges formed with some aeration as mentioned earlier.

Sludge Disposal

Incineration involves combustion of the volatile matter in sludge. It is not, properly speaking, a means of sludge disposal, since the process converts a sludge into an inert ash which itself must be disposed of. The ash is a harmless end product as are the gases CO_2, H_2O, and N_2 given off during the process. However, particulate matter enters the air as well. With adequate dewatering to about 30% solids, the incineration process is self-sustaining once initiated. The ash resulting from the incineration can be disposed of in landfills. Approximately 0.05–0.07 lb (0.2–0.3 N) of ash are produced per capita per day. Large industrial inputs of sludge can raise the per capita generation significantly in certain communities.

The most common form of disposal of the stabilized sludge obtained through biological and/or thermal means is on land. There are several ways in which this can be done, if the sludge is truly stabilized. In one method of land disposal, the sludge can be dumped in large, unused depressions such as abandoned quarries.

Wet sludge can be spread over farm lands and orchards and plowed under after it has dried, to act as a soil conditioner and fertilizer.* Dried sludge can also be used as a fertilizer for golf courses, lawns, or farms. It can be packaged

---

*A fertilizer has nutritive values for the soil, whereas a soil conditioner retains moisture due to its spongelike nature.

or simply hauled to the disposal site. In the latter case, it is not as much of a handling problem as the wet sludge.

Another possibility is for the sludge to be made part of a landfill if it is combined with a municipality's refuse and garbage. After the sludge has been dumped, it is compacted by means of tractors or rollers and then covered with about 12 in. (30 cm) of clean soil. However, landfills often lead to nuisance conditions, such as odors and flies, as well as the chance of contamination of groundwater and a boggy terrain if not properly engineered.

Lagooning is a final form of direct land disposal of sludges. A lagoon is an earth basin into which sludge is dumped. Raw sludge should be avoided since objectionable odors may result as the sludge decomposes. Sludge solids settle to the bottom and accumulate. In the process, sludge stored in the lagoon can be dewatered from 95% moisture to 55–60% moisture in a period of 2–3 years. In general, lagoons can be considered where large land areas are available and the installation will not be a nuisance for the surrounding area. In some cases, the supernatant liquid is returned to the plant for treatment. The lagoon is abandoned when it has been filled with solids.

Two final means of sludge disposal are composting and ocean disposal. *Composting* involves the transformation of sewage sludge (and municipal refuse) into a humus that is valuable as a soil conditioner and nutrient source for plants. The City of San Diego, California, which maintains an ocean outfall for its treated wastewater, pumps its processed sludge north almost 10 miles (16 km) to Fiesta Island in Mission Bay where it is used as a soil conditioner. The idea of composting has numerous supporters, but there remain unknowns concerning the procedure, such as the survival of pathogens in the compost. Composting is discussed in more detail in Dalton et al. (1968), Clark et al. (1971), and Dalton and Murphy (1973).

Some cities have disposed of their sludge, or part of it, by filling barges that are towed to sea and dumped. The New York Bight has received a substantial amount of publicity in this regard, as it has received countless tons of sludge and other solid waste material much to the detriment of the water and bottom in the area. Sludge has also been moved to sea through outfalls. In some cases, as at Hyperion in California, the ocean disposal of sludge took place for some years through a special sludge outfall. In other cases, at least some of the sludge gathered in the treatment plant has been returned to the wastewater for piping out into the ocean.

3-7 ADVANCED WASTE TREATMENT

Introduction

Tertiary treatment (*advanced waste treatment* or AWT) is any form of treatment beyond those discussed for the removal of pollutants not *substantially* removed by standard primary and secondary treatment processes. These pollutants include refractory organic materials,* and nutrients—usually nitrogen (N) and phosphorus (P). Although secondary sewage effluents (i.e., effluents after secondary treatment) may not vary widely in characteristics, wastewaters containing appreciable quantities of industrial effluents may be widely different, reflecting the differing natures of the raw wastewaters. Since secondary effluents cannot always meet stringent effluent standards, tertiary processes may be called for.

Nutrient Removal

Plants, whether land-based or aquatic, require vitamins and nutrients to grow. Required elements for most green plants are carbon, hydrogen, oxygen, nitrogen, phosphorus, potassium, sulfur, and at least another dozen trace elements including magnesium and zinc. The lack of any one necessary nutrient will limit plant growth. There exists for the plant a definite ratio of the essential elements. For example, N:P ratios are usually considered to be in the order of 10:1–15:1. Even if the concentration of one of these in the water or soil is increased, the plant cannot use that additional nutrient supply unless the concentration of the other nutrient is increased proportionately. There are often specific nutrients that are completely used by plants. These are termed *limiting nutrients*, since they prevent further plant growth even in the face of the plentiful supply of other needed nutrients.

Nitrates and phosphates can have their origin in human wastes. More prolific sources of phosphates, however, are detergents. In fact, it has been estimated that 60% of the phosphorus in domestic wastewater has a detergent origin (Hammer, 1975).

All lakes are subject to a natural aging process wherein the gradual accumulation of silt and organic matter causes a slow transformation to a marsh. Cultural or artificial *eutrophication* refers to the accelerated aging of an aquatic environment due to nutrients introduced into that environment by man. High input levels of nutrients, such as nitrogen and phosphorus in sewage, can lead to rapid and excessive algal production. Resulting algal decomposition can tax the oxygen resources of the water body, and in the extreme lead to anaerobic conditions and associated aesthetic problems. Cultural eutrophication is undesirable since it tends to decrease the recreational, municipal, industrial, and agricultural desirability and usability of the body of water concerned.

Although there is no unanimity of opinion, it appears that nitrogen is the limiting nutrient in most coastal water regions, whereas phosphorus is the limiting and controlling nutrient in lakes and rivers. Whatever the case, it is certain that there are situations in which the wastewater concentration of either one or both of these nutrients should be reduced below that provided by

---

*These are organic materials only partially degraded or entirely nonbiodegradable in biological waste treatment processes.

standard primary and secondary treatment measures. Conventional biological treatment provides about 30% phosphorus removal and 40–50% nitrogen reduction. At the present time, it is easier and cheaper to reduce phosphorus concentrations than those of nitrogen.

Phosphorus can be precipitated in the activated sludge process. Consistent and predictable phosphate removals can be obtained by the addition of iron or aluminum salts to the contents of the aeration tank. Another possibility is to add a coagulant such as lime (CaO) either to the aeration tank or to the secondary effluent. The phosphorus concentration can be reduced to about 1 mg/ℓ, a considerably larger reduction than shown in Figure 3-4 for conventional primary and trickling filter processes.

Nitrogen can be considered a pollutant, both because it exerts an oxygen demand in the nitrification process and because of its contribution to the eutrophication of natural waters. One method of nitrogen removal is first to raise the pH (using lime) of a wastewater so that the ionized NH_4^+ form is converted to un-ionized ammonia, then to pass air through the wastewater in a spray tower, leading to substantial removal of the NH_3—N, called *ammonia stripping*.

An alternate method of nitrogen removal called *nitrification-denitrification* is detailed in Metcalf and Eddy (1972). Other methods are also presented in the same reference.

Filtration

The removal of fine suspended matter (and associated nutrients and BOD) can be accomplished by filtration. The efficiency of the filtration process is generally judged from the decrease in wastewater turbidity. *Multimedia beds*, i.e., adjoining beds of different materials such as anthracite (hard coal), sand, and garnet, have been developed as efficient filtration beds. Pressurization of the wastewater input provides faster throughput rates.

Organic Removal

Adsorption is the attraction and accumulation of one substance on the surface of another. Adsorption is primarily a surface phenomenon; the greater the surface area of the adsorbent, the greater its adsorptive power. Carbon, suitably prepared (*activated*) by controlled combustion of hardwood charcoal and other materials, is highly adsorptive because each particle is honeycombed with minute pores.

Activated carbon is effective in extracting from wastewater low levels of dissolved organics in BOD as well as toxic and taste- and odor-producing compounds. This is normally done following filtration. Such treatment is often necessary because of the organic materials such as pesticides that are resistant to biological oxidation. Detergents, once a problem in this context, are now required to be biodegradable in the United States.

Removal of Dissolved Salts

Four major techniques are available for the removal of inorganic materials remaining in a secondary effluent: distillation, ion exchange, electrodialysis, and reverse osmosis. The first three of this list involve high costs. In *distillation*, water vapor is driven off through heating and is later recovered by condensation. The salts are left behind.

In *ion exchange*, wastewater is passed through a filter bed of a particular insoluble material such as glauconite. In the process, ions are displaced from the glauconite by those from solution. In *electrodialysis*, an induced electrical potential and membranes selective to the passage of positive or negative ions serve to isolate ions of different signs.

In *reverse osmosis*, wastewater containing dissolved materials is placed in contact with a special membrane at a pressure in excess of the osmotic pressure of the wastewater. As a result, water with only a minor amount of dissolved material is forced through the membrane leaving a concentration of dissolved materials behind. This process can be used for removal of chromate and phosphates. It usually follows filtration and activated carbon treatment to remove organics that would foul the membranes.

Physicochemical Treatment

It is possible to achieve high levels of BOD and SS removal and yet not make use of biological processes. Physicochemical processes can be applied directly to a primary effluent. Weber et al. (1970) have proposed that a coagulant be added to the degritted raw wastewater, a clarifier be used instead of a primary settling tank, and the effluent then sent to an activated carbon adsorber followed by a dual media filter. The coagulant is largely recovered from the sludge. Tests showed that 97% BOD removal could be achieved, 90% phosphate removal, and 95% nitrate removal. The effluent was "essentially free" of SS with about 5 mg/ℓ of BOD.

Weber et al. also argued that a physicochemical plant could be built in modular and readily expandable form using one-quarter of the land required by conventional processes for achieving approximately the same effluent conditions. Costwise, conventional secondary treatment (with inferior results) was found to be approximately 30% less expensive than the physicochemical approach, with tertiary treatment about 60% more.

3-8 WASTEWATER DISPOSAL BY DILUTION

Discharge of wastewater effluents into water courses, such as streams and rivers or into water bodies such as lakes, estuaries, and seas, is the most common disposal method. The dilution of the effluent in the *receiving-water* and the action of naturally occurring aquatic organisms can complete the treatment started on land in a wastewater treatment plant. The load on the receiving water is reduced if extensive treatment has been given the wastewater before transfer to the water body.

Water-pollution control includes the development of the means by which a water quality consistent with the beneficial uses of the water body can be assured. Beneficial uses will be detailed in Chapter 4; briefly, they include water supply, recreation, and aesthetics. It is also important to protect fish and other aquatic life. There are two ways of pursuing water-pollution control. The first of these is to specify *receiving-water (stream) standards,* and the second *effluent standards.*

In the first case, minimum standards are specified for the receiving waters outside of a prescribed *mixing zone,** and no discharge can create conditions that violate these standards outside the mixing zone. However, it is difficult for regulatory agencies to enforce such standards. It is easier to specify what water quality criteria the discharge itself must satisfy. Conceptually, these effluent standards are naturally set so that at least the minimum desired receiving-water quality is maintained. But this is in itself a problem because conditions vary from place to place.

Different water bodies, and different areas of the same water body, have different beneficial uses and do not require the same standards. A harbor and a bathing beach serve as examples. The specification of receiving water standards takes this into account by dividing receiving waters into classes for each of which different standards are specified.

The setting of standards in the United States is the responsibility of federal, state, and local regulatory agencies. An element in many receiving-water standards is some form of the "five freedoms" that stipulate that all (receiving) waters should be free of the following:

1. Materials that settle to form bottom deposits
2. Floating materials (e.g., debris, oil, scum)
3. Substances imparting taste, odor, color, or turbidity to the water or off-flavor to the flesh of fish or shellfish
4. Substances that are toxic or harmful to human, animal, plant, or aquatic life or in amounts that interfere with beneficial uses of the water
5. Concentrations of substances leading to undesirable aquatic life

Clearly primary treatment at the least, and possibly a minimum of advanced primary treatment, is required to satisfy items 1 and 2. Items 3 and 4 are mainly aimed at industrial wastewater discharges, and they appear to suggest a minimum of municipal tertiary treatment. Much to be preferred, however, is

*This is a region around a waste discharge, in a receiving water, also called an *initial dilution zone,* that is not subject to receiving water standards.

industry's adequate treatment of its own wastewaters before introduction to the municipal wastewater system, so-called *source control,* detailed in Chapter 4. Item 5 refers principally to nutrients, but the degree of treatment is not clear since the nutrient concentration causing undesirable growth of aquatic life varies greatly with many factors. This will be discussed further in Chapter 4.

The five freedoms are only a part of water quality standards, but it is apparent from them that wastewater must be treated before being discharged into a water course. Hopefully, with adequate standards and enforcement we can say goodbye to the days when a raw sewage effluent was merely allowed to pass, unadulterated, into the nearest water body. I have vivid memories, as a youngster in the 1940s, of the foul-smelling fountain of raw sewage entering the Thames River a mile downstream and west of the city of London, Canada, where I lived. Whenever we swam in the river it was on the northeast side of town, upstream of the London discharge but of course downstream of various cities and towns that had also used the river as their personal sewer.

This book is concerned with wastewater disposal by dilution with the focus being such disposal in the sea. Although only the entry of wastewaters into receiving waters through pipes is to be considered, there are other methods of effecting this transfer, notably by barge (e.g., Ketchum and Ford, 1952; Hood et al., 1958).

3-9 OTHER WASTEWATER DISPOSAL METHODS

Introduction

If one attends the public meetings held in conjunction with ocean outfall planning, or if one reads transcripts of such meetings (e.g., E.P.A., 1973a, b), it is clear that certain elements of the public feel very strongly about three matters that relate to wastewater disposal by dilution:

1. We are throwing away a potentially (after advanced treatment) valuable freshwater resource when we dispose of the wastewater in this way
2. We are "poisoning" marine and fresh waters, animals, and plants by dumping pollutant-laden wastewaters into such waters
3. We are throwing away a potential fertilizer by simply disposing of sludge

The following subsections superficially examine alternate uses for treated wastewaters. Sludge disposal was considered in Section 3-6.

Land Disposal

Land application of treated municipal and industrial wastewater has been used in urban areas for the irrigation of parks, sport grounds, golf courses, sewage treatment plant grounds, cemeteries, college grounds, street trees and median strips, and for ornamental fountains and artificial lakes. The irrigation of farms with such wastewater has also been extensively practiced on such crops as grasses, alfalfa, corn, barley, oats, wheat, citrus trees, grapes, nuts, and cotton. Different constituents in even well-treated wastewaters potentially pose problems for wastewater-irrigated crops and for the consumer (e.g., Baier and Fryer, 1973; Dugan et al., 1975; Larkin et al., 1976; Overman and Ku, 1976). As in most things there are extreme positions on the matter (e.g., Kardos, 1970; Egeland, 1973).

Although irrigation for crops may often be the express reason for applying the wastewater to the land, in some cases other reasons may take precedence, for instance the rejuvenation of badly eroded or strip-mined land. As an example, such applications may present viable economic alternatives to treating wastes and discharging them into receiving waters—without causing direct degradation of rivers, lakes, and marine waters. Also, such application may overcome the lack or unavailability of suitable receiving waters and eliminate the substantial costs of long outfall lines to reach suitable disposal points in large bodies of water.

Application rates of treated wastewater for crop irrigation are usually rather low [less than 2 in. (50 mm) per week]. For one acre (0.4 ha) of land, this is equivalent to about 7800 gpd (30 m^3/day) or less. Thus, extensive land areas are required to dispose of large wastewater flows. The limit of the rate of wastewater application is dictated at a minimum by the rate of uptake in the soil of the nutrients in the wastewater. Too ambitious a rate of wastewater application can lead, for example, to unused nitrates moving unimpeded through the soil to the detriment of groundwater supplies.* Underdrain systems for land application undertakings are rare. Properly applied, suspended particles in the wastewater are strained out, colloids and organic matter are adsorbed by the soil particles, and more complex organic materials are decomposed to simple inorganic materials by soil bacteria. In a sense, as the wastewater percolates down through the soil it is given a measure of tertiary treatment.

The quality of wastewater to be used in land application is under the control of state agencies in the United States, but few states have had formal standards (Sullivan et al., 1973). There has been general agreement that disinfection (with specified chlorine residuals) must be used if land application of secondary (minimum) sewage effluents is to be practiced on land either directly used by humans (e.g., playing grounds) or used to grow food crops. Secondary effluents without disinfection can be used for forage crops and watering of farm animals other than producing dairy animals. Maximum coliform and suspended solids concentrations may be given. Some standards (e.g., Florida) have discussed buffer zones separating areas of land application of treated sewage from residential property as well as the mists possibly resulting from such application.

---

*Excess nitrates in a groundwater used as a potable source can cause the blood disorder methemoglobinemia in infants. Even when diagnosed, this disease can cause 7–8% mortality (Kaufman, 1974).

There are various types of land disposal systems. The first of these is the spray irrigation concept. Buried or overland pipe systems can be used with attached risers and spray nozzles, or rotating spray booms can be utilized. In the case of the infiltration technique, the second such system, water is flooded out onto soil (*surface spreading*) that possesses a high percolation rate. Once the resulting shallow ponds have disappeared, a drying period on the order of 2–3 weeks is allowed so that oxidation of the accumulated organic matter can occur (so that the soil permeability is restored). Evaporation ponds can be employed. Another land disposal system, referred to as overland runoff, is used on sloping, relatively impervious slope terrain. Wastewater is applied to the slope at regular intervals. The effluent is renovated as it moves through a maintained vegetative cover on the ground surface, and it is collected in ditches transverse to the slope. A final technique of land application involves standard ridge and furrow irrigation systems.

Groundwater Supplies

Groundwater supplies are replenished by wastewater that percolates downward after being used on the land. A more direct method of replenishing the groundwater supply, however, is through *injection wells*, wherein water is dumped down shafts (e.g., Winar, 1967; Wesner and Baier, 1970). In some geographic areas, surface spreading is unworkable because of lack of spreading basins and the existence of disconnected aquifers. In such cases only well injections are possible for groundwater recharge. The region surrounding Santa Barbara, California, is an example (James M. Montgomery, 1973).

One of the real problems relating to groundwater withdrawals for communities bordering marine water concerns *salt water encroachment*. As fresh water sources are used and not adequately replenished, the fresh water level (head) is insufficient to hold back seawater, with the result that sea water percolates inland through the soil. Unless arrested, salt water encroachment can lead to contamination of groundwater supplies. There is ample evidence in the literature that many urban areas have suffered the intrusion of seawater into formerly fresh water zones (e.g., Bruington, 1969; Cooper et al., 1964; Schmorak, 1967; Todd, 1974). Toups (1974) relates that salt water intrusion in southern California in 1956 amounted to a distance of 4 miles (6.5 km).

There is no sense in replacing seawater contamination of groundwater aquifers by wastewater contamination. From a simple reuse standpoint, the quality of the disposed wastewater must be adequate to assure that, whatever its mode of reaching the groundwater aquifer, it is suitable as a potable source, satisfying applicable standards (e.g., "National Interim Primary Drinking Water Regulations," 1976) with or without (preferably) further treatment. Argo (1976) has discussed the treatment problem with a combination of trickling filter and wastewater reclamation plant facilities. Removal of organics, turbidity, heavy metals, bacteria, and viruses were at the top of the list.

Experience with groundwater replenishment in the state of Arizona, for example, has shown that pathogenic organisms and nitrates posed particular problems when the waters were withdrawn (Schmidt, 1973). Pathogens, of course, pose health problems (e.g., Romero, 1970; Benarde, 1973).

A 15 Mgd (0.66 m³/s) wastewater reclamation plant has been built in Orange County, California. Various tertiary treatment measures are applied to the input to the plant, which is the secondary (trickling filter) effluent from the main Orange County Sanitation District's plant treating both domestic and industrial wastewaters. The reclaimed wastewater is combined with an equal volume of desalted seawater for reduction in total dissolved solids and injected into the groundwater aquifer to provide a barrier to salt water intrusion. At a later time, after the injected wastewater has been further diluted and purified within the aquifer, it is pumped to the surface as part of the water supply. Further details are contained in Heckroth (1973).

Los Angeles Experience

The Los Angeles, California, area imports large volumes of water by aqueduct from the Owens, Feather, and Colorado Rivers. As outlined in Section 1-9, the Los Angeles area (L.A. City, L.A. County, and Orange County) in 1969 disposed of 841 Mgd (36.9 m³/s) of this imported water plus local groundwater supplies, after use, to the ocean through outfalls. This is a considerable flow of water. As a means of comparison, it is equivalent to the average annual flow of the Clyde River at Blairston, England, and the Assiniboine River at Headingley, Manitoba, Canada. It is half the average annual flow of the South Esk River at Launceston, Tasmania, Australia, as well as the following two U.S. rivers: the St. Johns at De Land, Florida, and the Grand at Grand Rapids, Michigan. It is one-third of the mean annual discharge of the San Joaquin River at Vernalis, California.

It can appear nothing but myopic that enormous volumes of imported water are used once and then discarded. As the population grows and industries expand and proliferate, clearly there is a ceiling on the increased volumes of imported and local groundwater that can be used. For the Los Angeles area, the time has already arrived (Argo, 1976). The severe 1976 and 1977 drought in California, and resultant water rationing, only served to underscore this contention.

Direct and Indirect Reuse

Indirect reuse of wastewater occurs when municipalities draw water from groundwater supplies fed in part by treated wastewaters. The term *uncontrolled indirect reuse* is applied to the withdrawal of river water that has flowed previously through the water supply and wastewater disposal system of an upstream municipality or industry or a string of them. Culp et al., (n.d.) report on a study, carried out by the U.S. Federal Water Pollution Control Administration, involving water use of 155 cities using surface water supplies. Typically, in low flow months, 1 of every 30 gallons of water withdrawn by a city as water supply had been through the water system of an upstream municipality. Industrial wastewaters were not considered, a perhaps significant absentee since, generally, industrial wastewaters are reported by Nemerow (1971) to be three times as plentiful, considering both flow and wastewater strength, as municipal wastewaters.

The use of one's own wastewater, after treatment, as part of one's own water supply is called *controlled direct reuse*. This is a system used with success by the city of Windhoek, capital of South West Africa (Stander and Van Vuuren, 1969; Nupen and Stander, 1973), where the reused component in the water supply comprises about one-third of the total. The thought of drinking water that several days earlier had been flushed down one's toilet is unsavory to most people. As long as the general public is reluctant to accept direct reclamation for reuse of water from wastewater as a potable supply, the most likely immediate use of reclaimed water is agricultural irrigation with a resulting conservation of water by infiltration and percolation. But the strong feelings against controlled direct reuse may have to be overcome in the face of rising population trends and diminishing supplies of relatively pure drinking water. Such reuse of wastewaters for potable supplies must be considered a high probability for many cities (e.g., Work and Hobbs, 1976).

The quality of water used for potable uses must be very high. A considerable degree of treatment of a municipal wastewater must be carried out before direct reuse can be practiced (e.g., Shuval and Gruener, 1973). Direct reuse of industrial wastewater by industry, after suitable treatment, is a not uncommon procedure (e.g., Rey et al., 1971) and will be considered in more detail in Chapter 4. In sewage, a major worry concerns viruses; in industrial wastewaters, the focus is more on heavy metals and pesticides.

Wastewater Aquaculture

Even secondary sewage effluents contain sufficient concentrations of nutrients that undue enrichment of a partly enclosed marine area such as a bay or estuary can result if the sewage is directed into it. It can be imagined that either tertiary treatment of the wastewater may be required to further reduce nutrient concentrations before disposal or else the effluent may have to be directed to an open ocean disposal site.

The latter alternative may be very costly. The city of Kaneohe on the island of Oahu in Hawaii had for many years disposed of its sewage, average 3.4 Mgd (0.15 m^3/s) in 1976, in adjacent Kaneohe Bay, to the great detriment of the Bay.* In 1975 and 1976, lines were built to carry the Kaneohe sewage 3.2 miles

---

*It is of considerable interest, in terms of aquaculture, that fringing Kaneohe Bay are several old Hawaiian fish ponds, the largest of which has a surface area of 125 acres (about 51 ha).

(5.2 km) to join a sewer line at the nearby city of Kailua's sewage treatment plant and then proceed 2.7 miles (4.4 km) to the beginning of an ocean outfall sewer constructed between 1975 and 1977. Pumping stations were built at both Kaneohe and Kailua. The cost of all this construction attributable to Kaneohe was $13.4 million.

It is difficult for many persons to regard sewage as a resource; we are normally devising schemes for quickly getting rid of sewage and sludge. However, the nutrients in sewage can, with proper methods, be turned to man's advantage rather than working to the detriment of desirable forms of plant and animal life in the marine environment. Not only is it possible to save the enormous sums that go into sewer line and outfall systems, but it may also be possible to turn a profit.

Secondary sewage flows can be used in an aquaculture system in which the nutrients are used to desired ends. The experiments of Ryther (1975), begun in 1974, provide a graphic illustration of this fact. Secondary effluent was mixed with seawater; the nutrients in the wastewater were used to grow single-celled marine algae (phytoplankton) in large tanks. Outflow from these tanks, diluted with more seawater, was directed to raceways (channels) containing stacked trays of shellfish (primarily oysters and clams) that consumed the phytoplankton. Water leaving these raceways moved on to others where seaweeds (two species of red algae), suspended in the flow, removed nutrients not initially assimilated by the phytoplankton and those regenerated by excretion of the shellfish and decomposition of their solid wastes. The seaweeds then acted as a final polishing step.

Solid wastes produced by the shellfish as well as uneaten phytoplankton supported dense populations of small invertebrates (e.g., polychaete worms) on the floors of the shellfish raceways. These animals in turn supported marine animals such as the American lobster and a type of flounder (fish).

The twin advantages of such a system are important: virtually complete nitrogen removal from secondary effluents can be assured*; commercially valuable marine organisms (shellfish, seaweeds, lobsters, and finfish) are produced. Thus, this form of aquaculture is effectively tertiary treatment, but with vastly smaller capital and operating expenses than customary tertiary sewage treatment plants, and with a commercially valuable product to defray expenses for the plant, if not to result in a profit. It is stressed, however, that Ryther's system is still being tested. Whether or not results are as good as the concept remains to be firmly established.

Additional material on wastewater aquaculture is available in Allen (1972). The economics of such systems has been considered by Smith and Huguenin (1975). It is to be emphasized that it appears that only small wastewater flows can be handled by such systems of manageable scale.

---

*Ryther was convinced that nitrogen is the nutrient limiting and controlling algal growth in and eutrophication of the coastal marine environment (Ryther and Dunstan, 1971).

3-10 REFERENCES

ALLEN, G. H. (1972): "The Constructive Use of Sewage, with Particular Reference to Fish Culture," in *Marine Pollution and Sea Life*, edited by Mario Ruivo, Food and Agriculture Organization of the United Nations, Conference, Rome (December, 1970), Fishing News (Books) Ltd., London, England, pp. 506–513.

ARGO, D. G. (1976): "Wastewater Reclamation Plants Help Manufacture Fresh Water," *Water and Sewage Works*, Reference Number, pp. R-160–R-170.

BAIER, D. C., and W. B. FRYER (1973): "Undesirable Plant Responses with Sewage Irrigation," ASCE, *Journal of the Irrigation and Drainage Division*, Vol. 99, pp. 133–141.

Battelle-Pacific Northwest Laboratories (1974): "Evaluation of Municipal Sewage Treatment Alternatives," prepared for the U.S. Council of Environmental Quality at Richland, Washington (February) [NTIS *Publication PB-233-489*].

BENARDE, M. A. (1973): "Land Disposal and Sewage Effluent: Appraisal of Health Effects of Pathogenic Organisms," *Journal*, American Water Works Association, Vol. 65, pp. 432–439.

BONDE, G. J. (1975): "Bacterial Indicators of Sewage Pollution," in *Discharge of Sewage from Sea Outfalls*, edited by A. L. H. Gameson, pp. 37–47. Pergamon Press, Oxford, England.

BRUINGTON, A. E. (1969): "Control of Sea-Water Intrusion in a Ground-Water Aquifer," *Ground Water*, Vol. 7, No. 3 (May–June), pp. 9–14.

CLARK, JOHN W., WARREN VIESSMAN JR., and MARK J. HAMMER (1971): *Water Supply and Pollution Control*, McGraw-Hill, New York.

COLLINS, H. F., G. C. WHITE, and E. SEPP (1974): "Interim Manual for Wastewater Chlorination and Dechlorination Practices," California State Water Resources Control Board, *Publication No. 53* (February).

COOPER, H. H., JR., et al. (1964): "Sea Water in Coastal Aquifers," U.S. Geological Survey, Washington, D.C., *Water-Supply Paper 1613-C*.

CULP, G. L., R. L. CULP, and C. L. HAMANN (n.d.): "Water Resource Preservation by Planned Recycling of Treated Wastewater," unpublished manuscript (contained in NTIS *Publication EIS-FL-73-0491-F-2*, pp. 346–377).

DALTON, F. E., and R. R. MURPHY (1973): "Land Disposal IV: Reclamation and Recycle," *Journal of the Water Pollution Control Federation*, Vol. 45, pp. 1489–1507.

DALTON, F. E., J. E. STEIN, and B. T. LYNAM (1968): "Land Reclamation—A Complete Solution of the Sludge and Solids Disposal Problem," *Journal of the Water Pollution Control Federation*, Vol. 40, pp. 789–804.

DUGAN, G. L., et al. (1975): "Land Disposal of Wastewater in Hawaii," *Journal of the Water Pollution Control Federation*, Vol. 47, pp. 2067–2087.

ECKENFELDER, W. W., JR. (1970): *Water Quality Engineering for Practicing Engineers*, Barnes and Noble, New York.

EGELAND, D. R. (1973): "Land Disposal I: A Giant Step Backward," *Journal of the Water Pollution Control Federation*, Vol. 45, pp. 1465–1475.

Environmental Protection Agency, Atlanta, Georgia (1973a): "Ocean Outfalls and Other Methods of Treated Wastewater Disposal in Southeast Florida," Public Hearings on Draft EIS, Transcript of Proceedings held in Lake Worth, Florida, January 24, 1973, Miami Beach, Florida, January 26, 1973, Fort Lauderdale, Florida, January 27, 1973 [NTIS *Publication EIS-FL-73-0491-F-2*].

Environmental Protection Agency, Atlanta, Georgia (1973b): Final Environmental Impact Statement, South Dade County, Florida [NTIS *Publication EIS-FL-73-1490-F*].

HALLOWELL, J. B., et al. (1973a): "Water Pollution Control in the Primary Nonferrous Metals Industry–Vol. I: Copper, Zinc, and Lead Industries," U.S. Environmental Protection Agency, *Report EPA-R2-73-247a* (September).

HALLOWELL, J. B., et al. (1973b): "Water Pollution Control in the Primary Nonferrous Metals Industry–Vol. II: Aluminum, Mercury, Gold, Silver, Molybdenum, and Tungsten," U.S. Environmental Protection Agency, *Report EPA-R2-73-247b* (September).

HAMMER, M. J. (1975): *Water and Waste-Water Technology*, John Wiley, New York.

HECKROTH, C. W. (1973): "Special Report: Ocean Disposal–Good or Bad?" *Water and Wastes Engineering*, Vol. 10, No. 10 (October), pp. 32–38.

HOEHN, R. C. (1976): "Comparative Disinfection Methods," *Journal*, American Water Works Association, Vol. 68, pp. 302–308, with discussion pp. 308–309.

HOOD, D. W., B. STEVENSON, and L. M. JEFFREY (1958): "Deep-Sea Disposal of Industrial Wastes," *Industrial and Engineering Chemistry*, Vol. 50, pp. 885–888.

JAMES M. MONTGOMERY, Consulting Engineers, Inc. (1973): "Ocean Waters Waste Discharge-Technical Report," prepared for Goleta Sanitary District, California (January 15).

KARDOS, L. T. (1970): "A New Prospect," *Environment*, Vol. 12, No. 2 (March) pp. 10–21, 27.

KATZENELSON, E., and N. BIEDERMANN (1976): "Disinfection of Viruses in Sewage by Ozone," *Water Research*, Vol. 10, pp. 629–631.

KAUFMAN, W. J. (1974): "Chemical Pollution of Ground Waters," *Journal*, American Water Works Association, Vol. 66, No. 3, pp. 152–159.

KETCHUM, B. H., and W. L. FORD (1952): "Rate of Dispersion in the Wake of a Barge at Sea," *Transactions*, American Geophysical Union, Vol. 33, pp. 680–684.

LARKIN, E. P., J. T. TIERNEY, and R. SULLIVAN (1976): "Persistence of Virus on Sewage-Irrigated Vegetables," ASCE, *Journal of Environmental Engineering*, Vol. 102, No. EE1 (February), pp. 29–35.

LOTHROP, T. L., and D. J. SPROUL (1969): "High-Level Inactivation of Viruses in Wastewater by Chlorination," *Journal of the Water Pollution Control Federation*, Vol. 41, pp. 567–575.

McGAUHEY, P. H. (1967): *Engineering Management of Water Quality*, McGraw-Hill, New York.

METCALF and EDDY, Inc. (1972): *Wastewater Engineering: Collection, Treatment, Disposal*, McGraw-Hill, New York.

"National Interim Primary Drinking Water Regulations" (1976): *Journal*, American Water Works Association, Vol. 68, pp. 57–68.

NEBEL, C., et al. (1976): "Ozone Provides Alternative for Secondary Effluent Disinfection," *Water and Sewage Works*, Vol. 123, No. 4 (April), pp. 76–78, and Vol. 123, No. 5 (May), pp. 82–85.

NEMEROW, NELSON L. (1971): *Liquid Waste of Industry: Theories, Practices and Treatment*, Addison-Wesley, Reading, Mass.

NUPEN, E. M., and G. J. STANDER (1973): "The Virus Problem in the Windhoek Waste Water Reclamation Project," in *Advances in Water Pollution Research*, edited by S. H. Jenkins, pp. 133–142. Pergamon Press, New York.

OVERMAN, A. R., and H. C. KU (1976): "Effluent Irrigation of Rye and Ryegrass," ASCE, *Journal of the Environmental Engineering Division*, Vol. 102, No. EE2 (April), pp. 475–483.

PELEG, M. (1976): "Review Paper: The Chemistry of Ozone in the Treatment of Water," *Water Research*, Vol. 10, No. 5, pp. 361–365.

REY, G., W. J. LACY, and A. CYWIN (1971): "Industrial Water Reuse: Future Pollution Solution," *Environmental Science and Technology* Vol. 5, pp. 760–765.

ROMERO, J. C. (1970): "The Movement of Bacteria and Viruses Through Porous Media," *Ground Water*, Vol. 8, No. 2 (March–April), pp. 37–48.

RYTHER, J. H. (1975): "Preliminary Results with a Pilot Plant Waste Recycling Marine-Aquaculture System," Woods Hole Oceanographic Institution, *Publication 75-41* (September).

RYTHER, J. H., and W. M. DUNSTAN (1971): "Nitrogen, Phosphorus, and Eutrophication in the Coastal Marine Environment," *Science*, Vol. 171, No. 3975 (March 12), pp. 1008–1013.

SCHMIDT, K. D. (1973): "Groundwater Quality in the Cortaro Area Northwest of Tucson, Arizona," *Water Resources Bulletin*, Vol. 9, pp. 598–606.

SCHMORAK, S. (1967): "Salt Water Encroachment in the Coastal Plain of Israel," International Association of Scientific Hydrology, Symposium of Haifa, Israel, *Publication No. 72* (March), pp. 305–318.

SHUVAL, H. I., and N. GRUENER (1973): "Health Considerations in Renovating Wastewater for Domestic Use," *Environmental Science and Technology*, Vol. 7, pp. 600–604.

SMITH, L. J., and J. E. HUGUENIN (1975): "The Economics of Waste Water–Aquaculture Systems," The Institute of Electrical and Electronic Engineers, Inc., Conference on Engineering in the Ocean Environment, OCEAN '75 (September) Conference Record, pp. 285–293.

Standard Methods for the Examination of Water and Wastewater, 14th Edition (1976): American Public Health Association, American Water Works Association, Water Pollution Control Federation, Washington, D.C.

STANDER, G., and L. VAN VUUREN (1969): "The Reclamation of Potable Water from Wastewater," *Journal of the Water Pollution Control Federation*, Vol. 41, pp. 567–575.

SULLIVAN, R. H., M. M. COHN, and S. S. BAXTER (1973): "Survey of Facilities using Land Application of Wastewater," U.S. Environmental Protection Agency, *Report EPA-430/9-73-006* (July).

TODD, D. K. (1974): "Salt-Water Intrusion and Its Control," *Journal*, American Water Works Association, Vol. 66, No. 3, pp. 180–187.

TOUPS, J. M. (1974): "Water Quality and Other Aspects of Ground-Water Recharge in Southern California," *Journal*, American Water Works Association, Vol. 66, No. 3, pp. 149–151.

VAN NOTE, R. H., et al. (1975): "A Guide to the Selection of Cost-Effective Wastewater Treatment Systems," U.S. EPA, Office of Water Program Operations, Washington, D.C., *Technical Report, EPA-430/9-75-002* (July).

WEBER, W. J., C. B. HOPKINS, and R. BLOOM JR. (1970): "Physiocochemical Treatment of Wastewater," *Journal of the Water Pollution Control Federation*, Vol. 42, pp. 83–99.

WESNER, G. M., and D. C. BAIER (1970): "Injection of Reclaimed Wastewater into Confined Aquifers," *Journal*, American Water Works Association, Vol. 62, pp. 203–210.

WINAR, R. M. (1967): "The Disposal of Wastewater Underground," *Industrial Water Engineering*, Vol. 4, No. 3 (March), pp. 21–24.

WORK, S. W., and N. HOBBS (1976): "Management Goals and Successive Water Use," *Journal*, American Water Works Association, Vol. 68, pp. 86–92.

3-11 BIBLIOGRAPHY

General

BEYCHOK, M. R. (1967): *Aqueous Wastes from Petroleum and Petrochemical Plants*, John Wiley, New York.

CAMP, T. R., and R. L. MESERVE (1974): *Water and Its Impurities*, Dowden, Hutchinson and Ross, Inc., Stroudsburg, Pa.

ECKENFELDER, W. W., JR. (1966a): *Industrial Water Pollution Control*, McGraw-Hill, New York.

EHLERS, V. M., and E. W. STEEL (1965): *Municipal and Rural Sanitation*, 6th ed., McGraw-Hill, New York.

EVANS, D. R., and J. C. WILSON (1972): "Capital and Operating Costs–AWT," *Journal of the Water Pollution Control Federation*, Vol. 44, pp. 1–13.

FAIR, G. M., J. C. GEYER, and D. A. OKUM (1966): *Water and Wastewater Engineering*, 2 vols., John Wiley, New York.

FAIR, G. M., J. C. GEYER, and D. A. OKUN (1971): *Elements of Water Supply and Wastewater Disposal*, John Wiley, New York.

"Manual on Hydrocarbon Analysis," 2nd ed. (1968): American Society for Testing and Materials, Philadelphia, *Special Publication No. 332 A.*

"Manual on Industrial Water and Industrial Waste Water," 2nd ed. (1966): American Society for Testing and Materials, Philadelphia, *Special Technical Publication No. 148-I.*

MICHEL, R. L., A. L. PELMOTER, and R. C. PALANGE (1969): "Operation and Maintenance of Municipal Waste Treatment Plants," *Journal of the Water Pollution Control Federation*, Vol. 41, pp. 335–354.

MITCHELL, RALPH (1974): *Introduction to Environmental Microbiology*, Prentice-Hall, Englewood Cliffs, N.J.

NEMEROW, NELSON L. (1974): *Scientific Stream Pollution Analysis*, McGraw-Hill, New York.

RICH, LINVIL G. (1963): *Unit Processes of Sanitary Engineering*, John Wiley, New York.

SAWYER, C. N., and P. L. MCCARTY (1967): *Chemistry for Sanitary Engineers*, 2nd ed., McGraw-Hill, New York.

Sewage Treatment Plant Design (1959): Water Pollution Control Federation, Washington, D.C.

SMITH, R. (1968): "Cost of Conventional and Advanced Treatment of Wastewater," *Journal of the Water Pollution Control Federation*, Vol. 40, pp. 1546–1574.

STANIER, R. Y., M. DOUDOROFF, and E. A. ADELBERG (1963): *The Microbial World*, 2nd ed., Prentice-Hall, Englewood Cliffs, N.J.

STEEL, E. W. (1960): *Water Supply and Sewerage*, 4th ed., McGraw-Hill, New York.

ZAJIC, J. E. (1971): *Water Pollution: Disposal and Reuse*, 2 vols., Marcel Dekker, New York.

Treatment

BARGMAN, R. D., and W. F. GARBER (1973): "The Control and Removal of Materials of Ecological Importance from Wastewaters in Los Angeles, California, U.S.A.," in *Advances in Water Pollution Research*, edited by S. H. Jenkins, pp. 773–781, with discussion and reply, pp. 783–786. Pergamon Press, New York.

BARTH, E. F., et al. (1965): "Summary Report on the Effects of Heavy Metals on the Biological Treatment Processes," *Journal of the Water Pollution Control Federation*, Vol. 37, pp. 86–96.

CULP, R. L., and G. L. CULP (1971): *Advanced Waste Treatment*, Van Nostrand Reinhold, New York.

ECKENFELDER, W. W., JR. (1966b): "Biological Treatment of Waste Water," in *Advances in Hydroscience*, Vol. 3, edited by V. T. Chow, pp. 154–190. Academic Press, New York.

EISENMANN, J. L., and J. D. SMITH (1973): "Selective Nutrient Removal from Secondary Effluent," U.S. Environmental Protection Agency, *Report EPA-670/2-73-076* (September).

GOLOMB, A. (1973): "An Example of Economic Plating Waste Treatment by Reverse Osmosis," in *Advances in Water Pollution Research*, edited by S. H. Jenkins, pp. 567–577. Pergamon Press, New York.

NEBEL, C., et al. (1976): "Ozone Disinfection of Secondary Effluents," *Journal*, Boston Society of Civil Engineers, Vol. 62, No. 4, (Jan.), pp. 161–187.

NOMURA, M. M., and R. H. F. YOUNG (1974): "Fate of Heavy Metals in the Sewage Treatment Process," University of Hawaii, Honolulu, Hawaii, Water Resources Research Center, *Technical Report No. 82* (September).

POUND, C. E., R. W. CRITES, and R. G. SMITH (1975): "Cost-Effective Comparison of Land Application and Advanced Wastewater Treatment," U.S. Environmental Protection Agency, Washington, D.C., *Technical Report, MCD-17, EPA-430/9-75-016* (Nov.).

ROBINSON, D. J., et al. (1974): "An Ion-Exchange Process for Recovery of Chromate from Pigment Manufacturing," U.S. Environmental Protection Agency, National Environmental Research Center, Cincinnati, *Report EPA-670/2-74-044* (June).

WEBER, W. J. (1972): *Physicochemical Processes for Water Quality Control*, Wiley (Interscience), New York.

WHITE, G. C. (1972): *Handbook of Chlorination*, Van Nostrand Reinhold, New York.

Disposal

BASCOM, W. (1974): "The Disposal of Waste in the Ocean," *Scientific American*, Vol. 231, No. 2 (August), pp. 16–25.

BOUWER, H. (1970): "Ground Water Recharge Design for Renovating Waste Water," ASCE, *Journal of the Sanitary Engineering Division*, Vol. 96, No, SA1 (February) pp. 59–74.

BOUWER, H. (1974a): "What's New in Deep-Well Injection," *Civil Engineering*, Vol. 44, No. 1 (January), pp. 58–61.

BOUWER, H. (1974b): "Renovating Municipal Wastewater by High-Rate Infiltration for

Ground-Water Recharge," *Journal*, American Water Works Association, Vol. 66, No. 3, pp. 159–162.

CHUCK, R. T., and D. LUM (1971): "Disposal of Waste Effluent in Coastal Limestone Aquifers in a Tropical Island Environment," paper presented at the Fifth International Conference, International Association on Water Pollution Research, Reconvened Session, Honolulu, Hawaii (August, 1970) *Proceedings*, edited by S. H. Jenkins, pp. HA1911–1915. Pergamon Press, New York.

CLARK, B. D., et al. (1971b): "The Barged Ocean Disposal of Wastes: A Review of Current Practice and Methods of Evaluation," U.S. EPA, Northwest Region, Pacific Northwest Water Laboratory, Corvallis, Oregon (July).

DAVIS, W. K. (1973): "Land Disposal III: Land Use Planning," *Journal of the Water Pollution Control Federation*, Vol. 45, pp. 1485–1488.

DOMENOWSKE, R. S., and R. I. MATSUDA (1969): "Sludge Disposal and the Marine Environment," *Journal of the Water Pollution Control Federation*, Vol. 41, pp. 1613–1624.

GALLEY, JOHN E., Ed. (1968): *Subsurface Disposal in Geologic Basins–A Study of Reservoir Strata*, The American Association of Petroleum Geologists, Tulsa, Oklahoma, Memoir 10.

GOLDSCHMID, J. (1974): "Water-Quality Aspects of Ground-Water Recharge in Israel," *Journal*, American Water Works Association, Vol. 66, No. 3, pp. 163–166.

GUARINO, C. F. (1967): "Sludge Disposal by Barging to Sea," *Water and Sewage Works*, Reference Number (Nov. 30), pp. R126–R127.

GUARINO, C. F., M. D. NELSON, and S. TOWNSEND (1977): "Philadelphia Sludge Disposal in Coastal Waters," *Journal of the Water Pollution Control Federation*, Vol. 49, No. 5 (May), pp. 737–744.

KOH, R. C. Y. (1971): "Ocean Sludge Disposal by Barges," *Water Resources Research*, Vol. 7, No. 6 (December), pp. 1647–1651.

KOZIOROWSKI, B., and J. KUCHARSKI (1972): *Industrial Waste Disposal*, translated from the Polish by J. Bandrowski, edited by G. R. Nellist, Pergamon Press, New York.

"Marine Disposal of Wastes" (1961): ASCE, *Journal of the Sanitary Engineering Division*, Vol. 87, No. SA1, pp. 23–56.

McGUIRE, J. B., and T. N. LEE (1973): "The Use of Ocean Outfalls for Marine Waste Disposal in Southeast Florida's Coastal Waters," University of Miami, Miami, Florida, Sea Grant Program, *Coastal Zone Management Bulletin No. 2*.

"Ocean Dumping in the New York Bight" (1975): U.S. National Oceanic and Atmospheric Administration, Environmental Research Laboratories, Boulder, Colo., *Technical Report ERL 321-MESA 2* (March).

POUND, C. E., and R. W. CRITES (1973): "Wastewater Treatment and Reuse by Land Application: Vol. I—Summary," U.S. Environmental Protection Agency, *Report EPA-660/2-73-006a* (August).

Task Committee on Underground Liquid Waste Disposal (1975): "Underground Liquid Waste Disposal," ASCE, *Journal of the Hydraulics Division*, Vol. 101, No. HY3 (March), pp. 421–435.

THOMAS, R. E. (1973): "Land Disposal II: An Overview of Treatment Methods," *Journal of the Water Pollution Control Federation*, Vol. 45, pp. 1477–1484.

TODD, D. K. (1959): "Annotated Bibliography on Artificial Recharge of Ground Water through 1954," U.S. Geological Survey, *Water-Supply Paper 1477*.

Underground Waste Management and Artificial Recharge (1973): Preprints of papers, Second International Symposium on Underground Waste Management and Artificial Recharge, New Orleans, Louisiana (September), U.S. Geological Survey and others.

WALLACE, A. T., D. HOWELLS, and A. UIGA (1976): "Let's Consider Land Treatment, Not Land Disposal," *Civil Engineering*, Vol. 46, No. 3 (March), pp. 60–62.

Reclamation and Reuse

Proceedings (1964): Conference on the Problems Associated with Purification, Discharge and Reuse of Municipal and Industrial Effluents, CSIR, Pretoria, South Africa.

SCHMIDT, C. J., and E. V. CLEMENTS III (1975): "Demonstrated Technology and Research Needs for Reuse of Municipal Wastewater," U.S. Environmental Protection Agency, Cincinnati, Ohio, *Publication EPA-670/2-75-038* (May).

SCHMIDT, K. D. (1974): "Nitrates and Ground-Water Management in the Fresno Urban Area," *Journal*, American Water Works Association, Vol. 66, No. 3, pp. 146–148.

"Sewage into Shellfish" (1973): *Newsweek* (December 24).

SHUVAL, H. I., Ed. (1977). *Water Renovation and Reuse*. Academic Press, New York.

"Third Report on the Study of Waste Water Reclamation and Utilization" (1957): California State Water Pollution Control Board, Sacramento, California, *Publication No. 18*.

VAN VUUREN, L. R. J., J. W. FUNKE, and L. SMITH (1973): "The Full-Scale Refinement of Purified Sewage for Unrestricted Industrial Use in the Manufacture of Fully Bleached Kraft-Pulp and Fine Paper," in *Advances in Water Pollution Research*, edited by S. H. Jenkins, pp. 627–636. Pergamon Press, New York.

Protection of the

Marine Environment

> "Hello there. Have a good sleep? Yes? Well, you go down to the stage; you know, the wharf thing made of sticks. And there's a little shack on the shoreward end of it. It's called the fish store and every fisherman has one. You go inside and you'll find a hole just beside the splitting table, where they dump the cod gurry into the water. And, oh yes, better take some paper with you."
>
> Farley Mowat, *The Boat Who Wouldn't Float*,
> McClelland and Stewart Ltd., Toronto,
> Canada (1969), p. 46, with permission.

4-1 INTRODUCTION

Raw Sewage Effluents

There are many magnificent harbors in the world. Port Jackson, the harbor for Sydney, Australia, is one of these. The mile-wide entrance to the harbor is flanked on both sides by steep cliffs that extend to Bondi Beach on the south and Manly Beach on the north. Both beaches are world-famed for their swimming and surfing.

In mid-1972, I flew in a small plane north along the cliffs between Bondi and Manly. During this trip, I was flabbergasted to sight two massive shoreline outpourings of raw sewage, one just north on Bondi and the other just south of Manly. Although both outlets were located where there was no habitation or recreational area, it was easily imagined that perverse currents could thoroughly

94

foul Bondi on the one hand and Manly on the other. This suspicion was later confirmed at Manly through conversations with local swimmers and surfers.

Sydney has by no means been the only metropolitan area with raw sewage outlets at the shoreline. Victoria, British Columbia, Canada, another beautiful place, was such an example until the early 1970s. Many industries and metropolitan areas have done little better than have shoreline outlets. The very readable little book edited by Helliwell and Bossanyi (1975), for example, contains repeated reference to disgusting and sometimes hazardous wastewater-disposal practices in the Solent area of south England.

Until the end of 1976, raw sewage from Honolulu entered the ocean just 4.0 miles (6.4 km) west of Waikiki, Hawaii's tourist mecca. This disgusting flow into 40 ft (12 m) of water was usually of no consequence for the thousands of tourists and local people using the beaches because the trade winds, which blow roughly 80% of the time, moved the effluent to the southwest, away from the land. But this was not always the case.

On the 16th of August, 1976,* three of us working on an undersea research project took our boat out to almost exactly the same water depth as the outflow referred to above, entering the marine environment 1.4 miles (2.3 km) to the west. When we arrived at the site, the water was as clear and blue as I had ever seen it. When we resurfaced after setting up our experiment we found that a south breeze had come up, and over the next hour this freshened and veered to the west. It wasn't long before the smell of sewage-laden seawater came to us on that breeze, and within a couple of hours the diluted effluent was upon us. The smell was now very strong on the boat, and when I dived down through the bits of fecal matter, shreds of toilet paper and, as I imagined, hosts of pathogenic microorganisms, I could still very strongly smell (or perhaps taste) the sewage.

We couldn't just raise anchor and go home when the effluent arrived. We had set up an experiment we wanted to run (one has to take advantage of field data when available) and at the least we were essentially forced to go down to collect the various bits of gear and the cables. I said some rather unkind things that day, both in and out of the water, about the lunacy of sending raw sewage effluents out into the ocean. I meant them then and I mean them now, although I would use better language today.

Our experience on the 16th of August, 1976, was not unique. Many divers have had similar and probably worse experiences. It is understood that divers swimming in excrement-contaminated water can contract disease or suffer infections (Mood and Moore, 1976). Won and Ross (1973) report that there were some acute skin and ear infections among the divers participating in the Tektite II underwater-living experiment where there were no sanitary facilities within the habitat.

---

*This date is over one year after completion of the construction of the Sand Island No. 2 outfall, to discharge into 235 ft (72 m) of water, but it was not yet operative.

Beneficial Uses and Their Impairment

The raw sewage effluent referred to in the first few paragraphs impaired at least several of man's beneficial uses of the immediate ocean area, and one of the primary focuses of this chapter will concern the protection of these uses. The beneficial uses of marine and estuarine waters and their immediate margins have been listed in various places (e.g., McKee and Wolf, 1963; McKee, 1967; Ludwig and Storrs, 1970). For our purposes, the following beneficial uses will be considered:

1. Whole body water contact sports (e.g., swimming, surfing, skin or SCUBA diving)
2. Partial-immersion water activities (e.g., water skiing, wading)
3. Boating (sail, power, paddle, oars)
4. Hiking, jogging, beach picnicking, and sunbathing
5. Commercial and sports (boat and spear) fishing for consumption
6. Shellfishing
7. Harvesting of plants
8. Collecting of marine organisms (e.g., fish for aquaria, shells)
9. Marine research and study
10. Water supply for research, municipal, or industrial uses
11. Aesthetic enjoyment

The word "beneficial" will be regarded in a narrow sense as implying only beneficial to people, putting food in their stomachs, money in their pockets, or giving them pleasure or a "workout."

Early paragraphs in this chapter introduced two very real potential problems associated with the marine disposal of wastewaters, the impairment of aesthetic enjoyment and the risk of infection or disease. As can be readily anticipated, there are also other ways in which such wastewaters can lead to a degradation (pollution) of their receiving waters. The major potential deleterious effects of wastewater effluents will be considered as follows:

1. Impairment of aesthetic enjoyment
2. Risk of infection or disease
3. Biostimulation, the proliferation of marine plants making use of the nutrients in the wastewater
4. Poisonous or inhibitory effects of some constituents of the wastewater (e.g., trace metals, chlorinated hydrocarbons, cyanide, phenols) on the marine flora and fauna
5. Consumption of oxygen in the breakdown of organic material

6. Problems arising from addition of inert particulate matter

7. Effects attributed to addition of waste heat

8. Problems resulting from addition of radioactive wastes

Chapter Plan

Beneficial Uses and Their Impairment: The intent of this chapter is to discuss the protection of the marine environment from pollution. Such protection makes particular reference to the organisms that naturally inhabit the marine environment and to man's beneficial uses of it.

The impairment of man's beneficial uses by pollution was categorized and listed above. Since only the first five in this list will be considered in Chapter 4, it is appropriate to consider the fate of the remaining three. The addition of inert particulate matter to the marine environment and resulting deleterious effects will be considered in Chapter 11. In that context, inert particulate matter entering the marine waters is considered an unwanted effect of outfall construction activities.

There is a voluminous literature concerning *thermal pollution* from the disposal of waste heat in marine waters. Examples of this literature are included in a subsection of the Bibliography for this chapter. Despite the importance of thermal pollution and its prevention, the scope of the book has largely had to be limited to nonthermal wastewaters. Several papers having to do with radioactive wastes are also included in the Bibliography.

Of the five categories to be considered, only three will be involved *explicitly*: public health issues in Section 4-7; biostimulation in Section 4-6; and toxicity in Section 4-4. Because of space constraints, the treatment of biostimulation will be very brief.

Legal Considerations: The protection of the marine environment and man's beneficial uses of it can adequately be assured only if there are appropriate water-pollution-control laws, if there are adequate funds to assure that needed wastewater facilities can be planned, designed, and built, and if the means of enforcement are established. A major attempt in this direction was made in the United States in 1972 when the Federal Water Pollution Control Act Amendments became Public Law (PL) 92-500.

Section 4-2 sketches the short history of U.S. federal activities in the water-pollution-control area up through and including passage of PL 92-500. This section also outlines the important features of this act. The actual impact of PL 92-500 on the states of the United States will be illustrated in Section 4-3 by considering the state of California. Section 4-5, in turn, will consider steps taken by the City of Los Angeles, an enormous metropolitan area that discharges very large volumes of wastewater into the Pacific Ocean, to conform to the California water-pollution-control plan. Section 4-3 will also briefly consider a water pollution control plan from the Canadian province of British Columbia as a means of comparison with that of California. Section 4-8 will sketch what is involved in securing permits required for the construction and then operation of marine outfalls in the United States.

Marine Disposal of Wastewaters via Outfalls

In the past, man has allowed wastewaters loaded with human wastes and industrial pollutants to enter the marine environment through outfalls. As a result, harm has been done to this environment, and man's beneficial uses of estuaries, bays and coastal waters have been impaired. While there appear to be significant perturbations, not necessarily for the worse, to certain segments of the local marine biota from even highly-treated wastewater flows, I believe that outfall systems can be designed to carry even large amounts of such effluents into at least the open coast of marine waters so that the net result for all of man's beneficial uses, as now perceived, is positive.

4-2   UNITED STATES EXPERIENCE

Years after World War II

The U.S. Government made a tentative step into water pollution control with the Water Pollution Control Act of 1948, PL 80-845. The act was designed primarily to provide loans for sewage plant construction, but very little in the way of funds was ever appropriated to back it up. The act also depended upon the states and localities themselves for its effectiveness in pollution control. In 1956, the act was strengthened by a Democratic Congress, and the resulting Federal Water Pollution Control Act (PL 84-660) became the legal framework for federal-state water pollution control and abatement, which generally carried through another fifteen years.*

The basic elements in the 1956 act concerned the primary role accorded the states, enforcement procedures, a system of grants, and the provision of research programs. However, subsequent support from federal agencies and the public for the program was minor. In 1960, an attempt to further strengthen the measure was vetoed by President Eisenhower. A year later, however, President Kennedy let pass amendments to the 1956 Act that extended the role of the federal government and created some pollution-abatement procedures. The Federal Water Pollution Control Act Amendments of 1961 demonstrated the increased significance accorded a national water-pollution-control endeavor.

---

*Political affiliations are referred to only because of the "politics" involved in some of the water-pollution-control legislation.

Instead of being a Public Health Service program, it was directly administered by the Secretary of the Department of Health, Education and Welfare. Key elements in the 1961 legislation concerned water quality data collection, sewage treatment improvement, classification of receiving waters by use, and standards for the quality of receiving waters.

Not until 1965 was further legislation passed and this only after considerable debate and compromise. The result was the Water Quality Act. This act provided for a water-quality-standards program. It required the individual states to draw up their own standards, which when approved by the Federal Water Pollution Control Administration, the responsible federal agency, became the federal-state standards. Involved in this development were three steps for the states: first, to hold public hearings to decide on classifications for waters within their boundaries; second, to determine appropriate criteria to meet the classifications of water use determined; and, third, to decide upon implementation plans for the standards. The latter involved action to be taken, and its timing, against polluters. Enforcement was the responsibility of the states themselves.

National Environmental Policy Act

Over the next few years, legislation concerned with the water environment continued to be enacted, e.g., the Clean Water Restoration Act (1966) and the Water Quality Improvement Act (1970). However, the most important piece of legislation in a water-pollution-control context was the National Environmental Policy Act (NEPA) of 1969, PL 91-190. Its principal thrusts were a national policy for the environment and the establishment of a basic structure for environmental policy making. In 1970, the 1899 Rivers and Harbors Act ("Refuse Act") was brought back into use as an antipollution tool by an executive order. This act required a permit from the Army Corps of Engineers to dump refuse or waste, except municipal sewage, into navigable waters. Such dumping was a crime otherwise. A permit program for wastewater discharge was instituted to be run jointly by the Army Corps of Engineers and the *Environmental Protection Agency* (EPA), the federal pollution-control agency set up by executive order in 1970 to administer the NEPA. The individual states were also given a say as to whether or not a permit was to be issued or denied until a wastewater discharge received treatment. One element of interest within NEPA was that all agencies of the federal government were required to prepare *environmental impact statements* (EIS) on federal actions significantly affecting the environment.

The EPA had many more responsibilities than water pollution control (solid wastes and air pollution, for example). The EPA was set up as a line agency responsible for administering and conducting federal pollution-control programs. It was to be an enforcement and regulatory agency. In the water-pollution-control context, it offered central administrative support to the individual states in the establishment of their own standards, whereas on interstate and navigable waters it could directly enforce related measures. There was to be a system of "matching grants" to municipalities, for the construction of waste-water treatment and disposal facilities administered by EPA.

The NEPA had still another dimension. It sought to overcome the problems of fragmentation in the overseeing of water-quality efforts, for one thing, and established the Council on Environmental Quality (CEQ) located in the Executive Office of the President. Among the various missions of the CEQ were the following: to study the condition of the environment through research, to monitor existing environmental programs and policies, and to ensure that federal agencies considered environmental questions in carrying out their activities. For the agencies themselves, the latter was specifically to mean that activities should be consistent with the NEPA.

Public Law 92-500

The Ideas: The Nixon administration appeared initially supportive of the EPA and the water quality movement, but this soured in a relatively short time. One problem was the enormous amount of funding required for adequate water pollution control but not provided. In 1972, the Federal Water Pollution Control Act Amendments passed in the 92nd Congress over Nixon's veto. Passage of this legislation, PL 92-500, marked the first time that the federal government had concerned itself with intrastate as well as interstate waters. States still retained primary responsibility to prevent, reduce, and eliminate water pollution, but this was to be done within a national framework. If states did not fulfill their obligations under the law, EPA was empowered to take action.

PL 92-500 laid out a series of specific actions in a water-pollution-control context that had to be taken by federal, state, and local governments, as well as industry. The individual states were to adopt water quality standards for their own waters, then submit these to EPA for approval. EPA was required by the law to set standards for states that did not comply. State standards could be more severe than minimum ones considered by EPA and could extend to concepts, such as mixing zones, not explicitly considered in the federal legislation. Reviews of the state water quality standards, with public hearings, had to be held at least once every three years. The timing and treatment requirements of PL 92-500 usually caused municipalities to apply for federal construction grant monies made available under the law.

The 1972 Amendments brought attention to bear on effluent limitations as a part of water pollution control rather than relying solely on receiving water standards as before. The EPA and responsible state agencies were to work together to establish the effluent-limitations regulations along with defining the best "practicable" and "available" water pollution control techniques. PL 92-500 had stated that, by July 1, 1977, effluent limitations "of the best practicable control technique technology currently available" were to be exercised and that by July 1, 1983, the "best available technology economically achievable" was to be exercised. It was felt that the latter would assure receiving waters clean enough for swimming and other recreational activities and clean enough for protection and propagation of fish, shellfish, and wildlife. PL 92-500 had gone on to state that "it is the national goal that the discharge of pollutants into the navigable waters be eliminated by 1985."

Publicly owned treatment facilities extant on July 1, 1977, or those approved for construction before June 30, 1974, had to meet effluent limitations based on a defined secondary treatment. On August 17, 1973,* this level of treatment was taken to mean the application of the *more stringent* of the following two requirements:

1. A maximum of 30 mg/ℓ (average for 30 consecutive days) of BOD_5 and total suspended solids in the effluent; or
2. A minimum of 85% reduction (average for 30 consecutive days) by treatment of influent BOD_5 and total suspended solids †

Upper limits for 7-consecutive-day average BOD_5 and SS were also specified (45 mg/ℓ) and a range limit of $6.0 \leqslant pH \leqslant 9.0$ was also given. Consecutive 30-day and 7-day geometric average limits on fecal coliforms in the effluents were 200 per 100 mℓ and 400 per 100 mℓ respectively. However, this coliform requirement was later (July 26, 1976) dropped.

The National Pollutant Discharge Elimination System (NPDES) permit program for pollutant discharges, with EPA responsible, was set up by PL 92-500. This replaced the 1899 Refuse Act permit program, but the U.S. Army Corps of Engineers retained authority for disposal of dredged and fill material—but with their actions subject to EPA veto. PL 92-500 required that there could be no pollutant discharge without a permit. Polluters were required to keep proper records, to install and use monitoring equipment, and to sample their discharges. A national surveillance system involving EPA, the states, and local authorities was instituted. The law authorized the federal government to seek immediate court injunctions against serious polluters, and fines and jail sentences were to be meted out to those violating the law.

The idea of the "incompatible pollutant" was put forward. This was a pollutant that would not be materially removed from the wastewater in conventional treatment plants and/or that would interfere with treatment processes.

---

* This appeared in the *Federal Register* of that date. After U.S. federal laws such as PL 92-500 are enacted, it is often necessary for the administering agency to establish rules and regulations to implement the law. The regulations are published in the *Federal Register*, become official policy, and essentially carry the same weight as the law itself.

† Since these removal levels are not achievable with trickling filters, their replacement by activated sludge systems was indicated.

Since industries and commercial enterprises are the major sources of such pollutants, *pretreatment standards* were imposed on them. The idea in this case was that such sources of incompatible pollutants should treat their wastewaters in-house, before introduction to the public system, so that the concentrations of incompatible pollutants were below prescribed levels.

Another concept put forward was that of "user charges," fees paid by users of facilities to cover operation and maintenance costs of the system. All users, i.e., residential, commercial, industrial, were to be charged a proportionate fee based on the wastewater treatment services provided.

The Realities: Federal grant money was included in the 1972 Amendments for 75% of the cost of construction of municipal "treatment works" as well as associated planning and design. "Treatment works" included sewage treatment plants, interceptor sewers, outfalls, trunk sewers, collection sewers, and overflow facilities for combined sewer systems. An amount of $18 billion was earmarked for allocation for fiscal years 1973, 1974, and 1975, with actual spending to extend over nine years. But in 1973, for example, only $2 billion actually went to treatment works construction—with only about $0.5 billion for all other aspects of water pollution control. $12 billion had been allocated as of October 1, 1976.

The insufficient funding, particularly in a time of rapid inflation, meant a slowdown in the water-pollution-control program, but there were other reasons. In many instances local governments applying for grants initially supplied insufficient project data. There were delays as EIS's were prepared. But part of the problem appeared to stem from the slowness of an understaffed EPA, working in new territory, to issue adequate guidelines and regulations—in part to tighten the vague language of the law. There was also the problem that EPA had to cope with laws other than simply PL 92-500.*

There were numerous broadsides leveled against PL 92-500 and EPA's administration of it. The procedures were the focus of much criticism, a point of particular note since the act specifically stated that "the procedures utilized for implementing this Act shall encourage the drastic minimization of paperwork and interagency decision procedures...to prevent needless duplication and unnecessary delays at all levels of government." Another complaint was that EPA modified several of its major programs on the basis of opinions by its General Counsel's office without informing the public about these opinions.

In the context of wastewater disposal by coastal communities, a complaint frequently repeated was that EPA focused unduly on ocean outfalls rather than

---

*Other legislative actions with bearing on wastewater effluents and about the time of PL 92-500, for example, were the Coastal Zone Management Act of 1972 (PL 92-583) and the Marine Protection, Research and Sanctuaries Act of 1972 (PL 92-532). The latter is also unofficially known as the Ocean Dumping Act. Partly amended by PL 93-254, it regulates the ocean disposal of wastes including dredged material.

adequately considering alternatives such as land disposal. It was believed that precious fresh water was being wasted by letting wastewater flow directly into the sea. Another problem concerned the blanket policy of requiring secondary treatment of all effluents. Many persons felt that this was an unnecessary and very costly requirement for wastewaters dumped properly into open coastal waters because of the very great assimilative characteristics of such receiving waters. This point will be referred to later in Section 4-6.

The National Commission on Water Quality recommended various changes to PL 92-500.* One proposal was that the July 1, 1983 goal be postponed 5–10 years. Another involved permitting extensions to the 1977 date on a case-by-case basis, with another calling for certification of the states to exercise full authority of the discharge permit and construction grant program. Proposals concerning funding were also involved. Ward (1975) and Heckroth (1976) present some pertinent details.

4-3   WATER QUALITY PLANS

Introduction

The concepts of effluent and receiving water standards were introduced in Section 3-8. Amplified discussion appears in Warren (1971). Waters are often classified by regulatory agencies according to the actual or potential uses of the water that are to be protected. This enables the agency to specify different levels of standards rather than one blanket list.

This section is concerned with "the small picture" in water pollution control, aimed specifically at the outfall disposal of wastewaters in the marine environment. I have chosen to do this by considering two water quality plans, one from the state of California and the other from the Canadian province of British Columbia. Despite the fact that the water quality plans for specific regions can change (markedly) from year to year in the face of recent data, alterations in (expressed) public opinion, and new legislation, the general ideas and components in past plans are instructive.

It is of interest that water pollution control is not recent, despite the environmental furor growing during the 1960s and 1970s. Wisdom (1975) refers to a water pollution ordinance in Britain dating from 1542, the time of Henry VIII. One can well imagine that any ordinance during the reign of that particular monarch was closely adhered to.

---

*PL 92-500 was actually amended by PL 93-207, 93-243, and 93-592 ("Federal Water Pollution Control Act," 1975) and with such changes is known as the Federal Water Pollution Control Act of 1972. Less officially PL 92-500, with amendments, is known as the Clean Water Act of 1972. It has also been disparagingly called the "fish act."

"Water Quality Control Plan for Ocean Waters of California" (1972)

The "Water Quality Objectives" section of the plan dealt with bacteriological, physical, chemical, biological, and toxicity characteristics of the receiving waters and also made reference to levels of radioactivity in them. Two zones were considered for bacteriological concentrations. One applied to areas in which shellfish might be harvested for human consumption and stated that, throughout the water column, the most probable number (MPN) median total coliform concentration should not exceed 70/100 ml with not more than 10% of the samples surpassing 230/100 ml.* The other zone was considered as recreational, extending out to a water depth of 30 ft (10 m) or a distance from shore of 1000 ft (300 m), whichever the further from shore, and also including any other areas outside this zone expressly used for water contact sports. In this zone, the following standards were imposed throughout the water column: the average of individual samples of water from sampling stations in the zone were to have an MPN of coliform organisms less than 1000/100 ml, provided that not more than 20% of the individual samples at any station in any 30-day period exceeded 1000/100 ml, and further that no single sample, when verified by a repeat sample taken within 48 h, was to exceed 10,000/100 ml.

The discussion on physical characteristics dealt briefly with various items. Floating particulates of grease and oil were not to be visible, with the measurable grease and oil concentrations on the water surface below prescribed levels. Numerical limits on the concentration of floating particulates of waste origin were also given. It was generally specified that the discharge of wastewaters was not to cause an aesthetically undesirable discoloration of the ocean surface nor lead to specified forms of degradation of the benthic communities through influence on the type and rate of deposition of inert materials. Finally, degradation of the light transmittance characteristics of the water, outside the zone of mixing of the effluent with the receiving water, was considered. It was specified that the mean of sampling results for any consecutive 30-day period had to be within one standard deviation of the mean determined for natural levels for the same period. It was, of course, assumed that the water clarity after the introduction of the wastewater would be inferior to that existing beforehand.

The chemical characteristics considered in the plan involved first dissolved oxygen (DO) and pH and stipulated a maximum 10% depression from natural levels for DO and a limit of 0.2 departure from natural conditions for the pH. General limitations on organic materials in sediments, so as not to impair benthic life, and nutrients, so as not to cause objectionable aquatic growth or degrade indigenous biota, were also included. Finally, numerical departures from natural conditions were specified for sulfide concentration of waters in and

---

*The recognition of statistical variability was an important aspect of this plan.

near sediments and for materials listed in Table 4-2 in the sediments. The allowable maximum departure was specified as for water clarity: the mean of sampling results for a consecutive 30-day period was not to be greater than one standard deviation above the natural conditions mean for the same period.

The final three items under "Water Quality Objectives" were biological characteristics, toxicity characteristics, and radioactivity, the latter referring to other standards. The biological considerations involved the nondegradation of all marine communities and the nonimpairment of taste, color, and odor of fish and shellfish to be consumed by humans. The single toxic factor listed was that the toxicity concentration should not exceed 0.05 toxicity units. The concepts of toxicity and toxicity units will be explained in Section 4-4.

The section of the plan entitled "Principles for Management of Waste Discharges to the Ocean" stated the intent of maintaining the indigenous marine life and a healthy and diverse marine community. It followed on with the "five freedoms" given in Section 3-8. It then stipulated the use of wastewater outflow systems (diffusers) that materially reduced, through dilution by ambient water, the concentrations of pollutants remaining in the wastewater after treatment. It was specified that diffusion systems should provide a 100:1 dilution with seawater at least 50% of the time and an 80:1 dilution at least 90% of the time. The final part of "Principles for Management of Waste Discharges to the Ocean" considered the sites of wastewater discharges and stipulated that they be located such that pathogenic bacteria and viruses were not swept into shellfish-harvesting and water-contact-sports areas, natural water quality conditions were not altered in areas designated as being of special biological significance, and maximum protection was afforded the marine environment.

Tables 4-1 and 4-2 set out the effluent quality requirements given in the Plan. In addition, parallel data drawn from the "Pollution Control Objectives for Municipal Type Waste Discharges in British Columbia" (1975), to be discussed later, are also presented. Background to the values presented in Table 4-2 will be given in Section 4-4.

Discharge of sludge directly to the ocean was forbidden by the plan as were the discharge of hazardous substances and the discharge of untreated waste-waters. The prohibition against sludge discharge to marine waters is of particular note.* Most experiences with heavy sludge discharges in the ocean have been poor. The classic example is the New York Bight, the dumping ground for New York City's sludge and solid wastes, where the marine environment has been heavily fouled and perturbed from water surface to sea floor (see, e.g., Sindermann, 1972). It appears, moreover, that the fouling and closing of miles of Long Island beaches in the early summer of 1976 because of sewage-polluted water is attributable to currents moving dumped material to that area from the Bight. Cronin (1969) quotes results of the U.S. Bureau of Sport Fisheries and Wildlife

---

*This limitation, coming with the federal minimum requirement of secondary wastewater treatment, created an enormous problem of sludge disposal (Rapoport, 1976).

Table 4-1: Effluent Quality General Requirements

Condition	Units of Measurement	Concentration Not to be Exceeded More than[‡]		Concentration Not to be Exceeded[§]
		50% of Time	10% of Time	
Grease and oil (hexane extractables)*	mg/ℓ	10.0	15.0	15.0
Floating particles	mg/ℓ dry weight	1.0	2.0	—
Suspended solids[†]	mg/ℓ	50.0	75.0	130.0 (open marine) 60.0 (embayed marine)
Settleable solids	mg/ℓ	0.1	0.2	—
Turbidity	Jackson turbidity units (JTU)	50.0	75.0	—
pH	—	within limits of 6.0 and 9.0 at all times		within limits of 6.5 and 8.5 at all times

*Later California requirements called for freon extraction techniques.

[†]It is noteworthy that the California figures fall short of the level of secondary treatment defined in the previous section.

[‡]"Water Quality Control Plan for Ocean Waters of California" (1972), with permission.

[§]"Pollution Control Objectives for Municipal Type Waste Discharges in British Columbia" (1975), with permission.

that clearly showed that currents could move pollutants from the New York Bight into the waters along Long Island beaches within a month.

The final part of the California Plan went briefly into the requirements for dischargers to carry out self-monitoring programs, a topic that will be enlarged upon in Chapter 11. The plan also set out for wastewater dischargers a time schedule for reports concerning and actual compliance with the immediately effective standards. The technical reports were to outline planned improvements, a time schedule for construction and costs, and data supporting less restrictive effluent quality condition than given in Table 4-2 because of particular local conditions.

Effluent quality requirements less restrictive than those in Table 4-1 were not allowed. However, to protect the beneficial uses of a particular ocean water area, it was possible for the State Water Resources Control Board to impose effluent water quality requirements more restrictive than those in the two tables. For particular wastewater discharges, the effluent quality requirements were translated into "mass emission rates," basically the weight of any constituent entering the marine environment from the outfall per day or month.

Table 4-2: Effluent Quality Requirements

Constituent	Unit of Measurement	Concentration Not to be More than* 50% of Time	10% of Time	Concentration Not to Exceed†
Arsenic	mg/ℓ	0.01	0.02	0.05 (total)
Cadmium	mg/ℓ	0.02	0.03	0.005
Total chromium	mg/ℓ	0.005	0.01	0.1
Copper	mg/ℓ	0.2	0.3	0.2
Lead	mg/ℓ	0.1	0.2	0.05 (total)
Mercury	mg/ℓ	0.001	0.002	0.0006 (total)
Nickel	mg/ℓ	0.1	0.2	0.3
Silver	mg/ℓ	0.02	0.04	0.1 (total)
Zinc	mg/ℓ	0.3	0.5	0.5 (total)
Cyanide	mg/ℓ	0.1	0.2	0.1 (total)
Phenolic compounds	mg/ℓ	0.5	1.0	0.2
Total chlorine residual	mg/ℓ	1.0	2.0	1.0
Ammonia nitrogen	mg/ℓ	40.0	60.0	—
Total identifiable chlorinated hydrocarbons	mg/ℓ	0.002	0.004	—
Toxicity concentrations	toxicity units	1.5	2.0	1.0
Radioactivity		specified in other standards		—

*"Water Quality Control Plan for Ocean Waters of California" (1972), with permission.

†"Pollution Control Objectives for Municipal-Type Waste Discharges in British Columbia" (1975), with permission.

"Pollution Control Objectives for Municipal Type Waste Discharges in British Columbia" (1975)

"Pollution Control Objectives for Municipal Type Waste Discharges in British Columbia" (1975) is somewhat more specific than the California standards and more confusing. This plan differentiated among different types of receiving waters. Streams, rivers, and estuaries were lumped together; lakes by themselves formed a second group. Marine waters were divided into two other groups: open waters and embayed areas. The minimum receiving water dilution for streams, rivers, and estuaries was implied by the plan to be 20:1.

Disinfection was required for either open or embayed marine applications, but dechlorination was not. The specified total chlorine residual in both cases was 0.5–1.0 mg/ℓ.* BOD_5 and SS limitations for the open marine case were both 130 mg/ℓ with 45 and 60 mg/ℓ, respectively, for the embayed one. A total

*All numbers given correspond to the AA standard applied to new outfalls in the plan.

phosphorous limit of 1.5 mg/ℓ, specified for only the embayed case, could be waived if it could be demonstrated for a particular case that such inputs would not lead to an "undesirable degree of increased biological activity" in the receiving water.

An environmental assessment (EA) was required for any discharge exceeding 1.2 Mgd (0.053 m³/s) to streams, rivers, estuaries, and to lakes. For discharges to the sea an EA might be required particularly for "shellfish waters, those used for artificial propagation of fish, for recreation or to confined waters." For marine discharges, rules were given for the minimum distance to shore from the inner end of the diffuser in terms of the diffuser depth, the volume rate of flow, and whether or not the receiving waters were shellfish waters.*

The receiving water standards for British Columbia, outside the initial dilution zone, involved the following considerations: the decrease in DO was not to exceed 10%, the residual chlorine had to be below the detectable limits of the amperometric titration test detailed in *Standard Methods...* (1976), there could be no increase in toxicity above background, as measured in a 96-h TL$_{50}$ static bioassay test.[†] There was to be negligible increase in settleable solids and in floatable solids and scum; there was to be no oil visible on the water surface, and there was to be a negligible increase in heavy metals concentrations. Further considerations for receiving waters involved nutrients, coliforms, and organisms.

With regard to nutrients, there was to be "no detectable increase in site-specific productivity limiting parameters." On the one hand, this was to mean phosphates and/or nitrogen forms, and on the other it was to mean (for the sea or estuaries) presence of sludge beds with reduced species diversity and a restricted range of predominant organisms such as *Capitella capitata*, a polychaete (worm) that thrives in polluted zones. These "nuisance conditions," along with stipulations about productivity, formed the bulk of the regulations concerning marine life. Finally, coliform levels were set out as follows: generally, total coliform levels were not to exceed a median MPN of 1000/100 ml or a fecal coliform median MPN of 200/100 ml; in shellfish areas the waters could not have a fecal coliform level above 14/100 ml and the shellfish meats could not show a fecal coliform MPN above 230/100 g.

The British Columbia (B.C.) plan went on to state that for multiport diffusers the initial dilution zone could extend up to 300 ft (90 m) horizontally from all points of discharge but could not exceed 25% of the width of the water body. The latter is a valid consideration in B.C. waters where outfalls frequently extend into narrow marine passes.

---

*An explicit provision in the "Pollution Control Objectives..." allowed for relaxation of effluent controls (Tables 4-1 and 4-2), to the limit of comminuted raw sewage, if receiving water standards were demonstrably satisfied.

†See Section 4-4.

Specific Industry Plans

In some cases industries with their own wastewater disposal facilities and, perhaps, a history of heavily loading the marine environment with waste materials, may be assigned their own effluent and receiving water quality standards. "Report on Pollution Control Objectives for the Forest Products Industry of British Columbia" (1971) serves as an example of specific industry plans, and governmental actions against pulp and paper polluters are discussed by Skory (1976) and in "6 Pulp Mills Face Counts" (1977).

4-4 TOXICITY

Introduction

In March 1965, thousands of dead fish were found washed up on beaches along approximately 100 km of Dutch North Sea coastline north of Scheveningen. Countless other fish were found in the nearshore waters, swimming about with uncoordinated movements. Tests on both the water and the fish established clearly that the reason for the heavy mortality was a very high concentration of copper in the marine waters, in some places well over 100 times the normal level (Korringa, 1968; Merlini, 1971).

The copper that found its way into these nearshore waters was apparently dumped as copper sulfate on a beach near Noordwijk; it did not have its source in effluent from an outfall (such as the one at Scheveningen). However, many analyses have clearly demonstrated that wastewaters entering outfalls carry along appreciable amounts of copper and other pollutants. Chen et al. (1974), for example, gave median and 10th percentile copper concentrations in the final effluent of 150 and 210 $\mu g/\ell$, respectively, for the Hyperion outfalls in California.

Copper is present in ocean water at concentrations of approximately 3 $\mu g/\ell$. The fact that there is myriad life in the sea demonstrates clearly that copper is not toxic (poisonous) to marine creatures at such low concentrations. In fact, a certain level of copper in such waters is required by organisms (e.g., Corcoran and Alexander, 1964).* A very important question then arises. At what concentration in water does a substance cease being harmless to the marine creatures[†] and, furthermore, at what point does it become lethal, or leading directly to death?

Observations during the Dutch incident in 1965 showed clearly that death

---

*Copper in trace amounts is essential in the development of chlorophyll, for example, in the phytoplankton. Copper also plays a role in the hardening of a crab's exoskeleton.

[†]In truth this level is by no means constant even for a given individual, varying with the condition of the environment and the state of the animal.

resulted for many organisms and impairment of life functions for others. The latter was very probably a prelude to death. But it can be imagined that there are intermediate concentrations of pollutants in marine waters that lead to subtle changes in the organisms. They do not lead to death quickly, directly, and dramatically, but over the long term they probably mean an early demise for the organism. This is due to impairment of its life functions or alterations to its food sources, affecting food production, density, and availability, with attendant changes in the organism's growth. The lingering effect of a pollutant on the organism itself is called *chronic toxicity*, sometimes loosely referred to as *long-term toxicity*. *Short-term toxicity* is a somewhat indefinite term at the other end of the scale that includes *acute toxicity*, referring to a rapid response of an organism to a toxic substance. An acutely toxic substance is usually lethal; a chronically toxic one may be lethal or *sublethal*, i.e., below the level that directly causes death.

Wastewater effluents contain all manner of potential pollutants, and one of the very important considerations in marine outfall design involves assuring as far as is possible and predictable that the concentrations of substances such as copper in the receiving waters are well below levels that would impair the natural functions of the local marine flora and fauna. *Toxicity tests* are indicated.

It is generally agreed that there is no certain way in which even a detailed *chemical* analysis of a particular wastewater or diluted wastewater can determine its toxicity to marine organisms. For example, a chemical analysis indicates conditions only at one time; the organism effectively integrates conditions over time. Thus, the quality of a water is judged from the ability of marine organisms to live in it. Nevertheless, using test or indicator organisms has its own complications, one of which is *synergism*, the combined action of two or more substances resulting in greater biological harm than could be inferred from their separate additive effects.

Although there have been suggestions that plants be used to appraise water quality (e.g., Patrick, 1973), and various important toxicity studies have been carried out using plants (e.g., Burrows, 1971; Brown and Varney, 1972), animals are customarily used in such tests. Usually fish are employed as the test organisms, although benthic animals are also used especially in seawater applications. In the work carried out in conjunction with wastewater discharges near San Francisco (Brown and Caldwell, 1971), seven species of fish were used, four species of crab, three of shrimp, and one each of clam, mussel, snail, and sea urchin.

Bioassay

Introduction: The *bioassay* is a test designed to appraise the toxic strength of (in this case) a wastewater or diluted wastewater by the reaction of an organism or organisms living in it. Different concentrations of an individual pollutant or wastewater are maintained in separate tanks. Customarily, in fish bioassays, 10 different organisms of the same species and equivalent size are kept in each tank. Observations are made of fish mortality in the tanks, and the number of live organisms in each tank at the end of a specified trial period is recorded. A test duration of 4 days is fairly standard, since for most toxicants acute mortality has either occurred by that time or else has ceased.

Consider for illustrative purposes a 96-h bioassay where 10 members of a specific species of fish are maintained in each of five separate tanks of diluted wastewater for 96 h. Let the concentrations of waste substance in the five tanks C (the control), D, E, F, and G be 0, 10, 20, 30, and 40 ppm, respectively, and suppose that, at the end of the 4 days, 10 organisms are still alive in tank C, 8 in D, 5 in E, 2 in F, and 0 in G. The standard way of representing these results is to give the lethal concentration for 50% of the individuals (written LC_{50}).* Since in tank E 50% of the individuals are still alive after the specified duration of the test, then $LC_{50}^{96} = 20$ ppm.

Usually in such tests it is necessary to use the results from all five tanks, since it is not overly likely that exactly half of the individuals would be alive in any one tank at the end of the test period. Then the LC_{50}^{96} value could be determined by interpolating from a curve of number of live organisms remaining after 96 h versus wastewater concentration, or by using more complicated techniques (e.g., *Standard Methods...*, 1976). The LC_{50} value is a convenient reference point for expressing the acute lethal toxicity of a given wastewater or pollutant to the *average* or typical fish. The "safe" concentration that permits the *individual* organism successful reproduction, growth, and all other normal life processes in its natural habitat is lower than this value—probably considerably lower.

Whether or not a marine organism survives in a laboratory test is by no means a sensitive measure of how that organism is faring or will fare in a situation that has been perturbed, if ever so slightly, from the "normal." More sensitive means of judging the organism's fortunes are to study its growth, its development, its behavior (the integrated movements of the whole animal), and its reproduction. This can hardly be adequately investigated in the acute toxicity test where the only options are life or rapid death. The acute toxicity bioassay tells little about the maximum concentration of a pollutant possible before it begins to affect the success of a particular species in the marine environment. But this is the *vital* information required for a properly engineered marine wastewater disposal system.

A proper chronic test should extend through a complete life cycle for the test organism. As an example for fish, the bioassay should begin with the eggs,

---

*The term *median tolerance limit*, TL_{50} to TL_m, is also used in toxicity bioassays and refers to the pollutant or wastewater concentration that half the test organisms can tolerate for the test period. Although TL_{50} and LC_{50} are always the same, TL_{10} corresponds to LC_{90}, TL_{80} to LC_{20}, etc.

or with fry less than 20 days old, and last until the offspring of these fish are at least 30 days old (Stephan and Mount, 1973).

Types of Bioassays: Bioassays fall into two general categories: 1. the *static bioassay* wherein organisms are kept in a tank of standing testwater that may or may not be changed during the test period; 2. the *continuous-flow* (flow-through) *bioassay* (*chemostat*) where the testwater is continuously replaced or renewed with frequent periodic additions. The static, or *batch*, bioassay can be carried out at substantially less expense than the continuous-flow test because less involved equipment is required. Also, since the operation is simpler, set-up time is less, and answers can be derived more rapidly than in the throughflow bioassay. Thus the static bioassay is particularly suited to exploratory types of tests. In general, however, the batch bioassay technique suffers from problems of the wastes of the organisms fouling the containers and a decline in toxicant concentration.

The continuous-flow test has the advantage that the test organisms need not be disturbed if and when the water is changed. In addition, in the flow-through test the organisms can be fed, thus eliminating the additional stress due to standard nonfeeding in the static tests. Generally speaking, the continuous-flow bioassay results in more constant test conditions, such as toxicant concentration and dissolved oxygen. Also, if the material being tested is degradable, volatile, high in oxygen demand, or detoxified rapidly by one means or another, the static bioassay will not yield valid results. The continuous-flow bioassay permits testing for periods long enough to include the lifespan of test organisms, an important consideration when considering long-term toxic effects as mentioned earlier.

In the flow-through bioassay, the pollutant or wastewater concentration in the test can be achieved by mixing continuously metered volumes of the dilutant water and the test substance. The proper discharges can be assured by metering pumps or by constant-head chambers. An objection raised to this type of procedure is that the test organisms receive a heavy dose of toxic material if for some reason the discharge of dilutant water should be stopped. A favorite blending system is to use *diluters* wherein a series of dilution systems automatically dilutes toxic materials to several concentrations, usually five or six. Considerable detail on a successful diluting apparatus, and on an entire continuous-flow bioassay system, is available in Esvelt et al. (1972).

Procedures

General Features: The quality of the dilutant water itself is an important consideration in bioassay tests. Where a specific receiving water is involved, this can be used as the dilutant water. However, in other cases some form of standard but aerated water is required. Tap water has to be carefully handled if it is to be used as makeup water because it often contains minute concentrations of chlorine, which is toxic to aquatic organisms. If tap water *is* to be used, it may have to be deionized as practiced in the San Francisco tests (Brown and Caldwell, 1971). Filtered well water is a better supply. The pH of a dilutant water can have a very great effect on the toxicity of a pollutant. Other factors to consider are hardness, dissolved solids, dissolved organic compounds, dissolved oxygen, salinity, and temperature (Brungs, 1973). In the San Francisco tests referred to above, artificial seawater of set salinity was made up from deionized tap water using a commercial salt product (lobster salt, Leslie Salt Co.) as well as a synthetic seawater formulation given by LaRoche et al. (1970). Another synthetic formulation has been given in Courtright et al. (1971).

A *control* in a bioassay is a tank set up as for the regular bioassay test tanks but using unadulterated dilution water. Ideally, the control should yield no mortality of the test organisms, but as a practical matter this is unlikely because of the occasional particularly weak individual and unnatural conditions. Sprague (1973) suggests that for a valid acute bioassay experiment, 90% or greater of the organisms in the control should remain alive at the end of the test. A high mortality in the control can indicate toxicants in the dilutant water itself, or stress because of holding or test tank conditions, lack of food, and other factors.

Before a bioassay begins, preliminary experiments should be run to establish the toxicity range of the pollutant or wastewater. *Standard Methods...* (1976) suggests a scale of dilutions, evenly spaced when plotted on logarithmic graph paper (e.g., 10.0, 5.6, 3.2, 1.8, and 1.0), to be used in the actual tests thereafter. In the San Francisco work, 10, 50, and 100% effluent concentrations were used. Generally speaking, the concentrations used should run the gamut from one in which there is 0% mortality (the control) to one wherein all organisms are dead at the end of the specified test period.

Commercial collectors are best suited for securing test organisms. Fish traps, crab traps, nets, and trawls can all be used for collecting. Divers can collect benthic animals by hand and take fish with some form of "slurp-gun."

When test organisms are brought to the laboratory they must be gradually conditioned to the dilutant water and the test temperature before being used in the bioassays. A minimum time of overall acclimation is 2 weeks with at least 2 days under 100% dilutant water conditions. The organisms should be fed. There should be no disease in the organisms, although in practice this may be difficult to achieve. For example, in the San Francisco pre-outfall-design toxicity work (Brown and Caldwell, 1971), there was a considerable problem with bacterial fin-rot in the fish used. In this same work, there were problems also of predation among the test shrimp. Sprague (1973) suggests that there should be less than 5% mortality of the organisms under acclimation during the 4 days prior to the test beginning.

The *holding* tanks and accessories should be of nontoxic material. Ideally any such holding tanks should be properly designed so that the lowest levels of water are continuously removed. Wastes should gravitate to the bottom center of the tank for removal and there should be some circular water motion so that the fish can exercise. There should be no outside disturbances, a normal day-night light cycle should be maintained either naturally or artificially, and there should be adequate tank volume for the number and weight of fish (typically a litre of water or more for each 10 g).

The test tanks and accessories should also be made of nontoxic materials. A good material for such tanks is glass. Fiberglass is not optimum but is frequently employed; so is stainless steel. Plastics are not the completely inert materials they were once thought to be; they can release toxicants and absorb organic materials. Consequently, they are not ideal for either the tanks or the piping. Whatever the material, rough inside tank surfaces should be avoided as this can lead to fish damage.

Fish that are small when mature are easier to obtain and handle than are larger fish; they also have shorter generation times than larger fish and may be quicker to acclimate. A problem, however, is that fish important to a particular area are often larger ones. The ideal test fish is shorter than 8 cm in length and between 0.5 and 5 g in weight. One suggestion has been that the longest fish used in bioassays be not more than 1.5 times the length of the shortest one (*Standard Methods...*, 1976).

In decisions regarding the choice of test fish species, other factors being equal, attention should be paid to the ability of the organism to adapt to life in a laboratory for some weeks.* The organisms chosen should be readily available and should be from a common source with a similar past history. Whatever organisms are used should be clearly identified in reports by means of species, size, and weight. Certain standard bioassay test species have evolved that might be considered for use in some circumstances. This could be appropriate when state- or province-wide standards must be enforced rather than purely local ones.

Even within species, responses to toxicants have been shown by numerous investigators to vary greatly with different stages in the life history, with previous nutrition, and with unknown causes. And there would not, of course, be such a thing as a median tolerance limit that meant anything if all ostensibly identical specimens reacted to toxicants is the same manner.

Tests: When fish are being placed in the test tanks, some form of randomization procedure should be employed. As an example, those fish taken first from the holding tanks may be the weaker ones, and if these are all placed in the same test tank, a biased picture of the average fish's response would result. The test tanks, apart from the different concentrations of wastewater or pollutant, should be otherwise identical and in similar situations. The volume of water in (static) test tanks has to be sufficient or the organism can deplete the toxicant, lower the dissolved oxygen, or foul the waters with wastes. All three

---

*A problem with this approach, however, is that particularly hardy species are tested, and the response of weaker species is unknown.

can of course happen concurrently. In static tests, a minimum volume of 2 (better 3) ℓ test water per g fish per test day is considered desirable (e.g., *Standard Methods...*, 1976). In the event that there is insufficient tank volume with existing facilities, either the fish can be transferred from one tank to another (leading to problems of acclimation) or else, as is the case for larger fish typically, a test group can be initially split into two tanks. If the bioassay being run is of the continuous-flow type, then the rate of volume exchange can correspond to that given above and the volume of the tank itself made equivalent to 8 h flow.

Suggested test tank volumes are in part determined by the oxygen demands of the test organisms. Neither aeration or oxygenation of the water in the test tanks should be practiced, since this can degrade or remove the toxicant.

The conditions within each test tank should be checked frequently at first. *Standard Methods...* (1976) suggests a schedule of 1.5, 3, 6, 12, and 24 h after the start of the test and once or twice a day thereafter. The importance of pH has already been mentioned; a ± 0.3 variation is suggested by Sprague (1973) as the maximum allowable. Temperature should be controlled $\pm 1°C$. Salinity, if applicable, should be held relatively constant. Dissolved oxygen should be monitored as well as ammonia nitrogen concentrations. The concentration of the toxicant (if applicable) should be checked.

Results of Acute Toxicity Bioassays

Methodology: As a bioassay progresses, the number of live organisms in each tank can be determined at various times such as those given above. For any such time t, the LC'_{50} value can be determined as follows (Sprague, 1973). Log-normal probability graph paper is selected, and the concentrations are plotted along the logarithmic axis. A logarithmic plotting makes sense because one would suspect that an increase in the concentration of a toxic agent 0.01–0.02 mg/ℓ would have more effect on an organism than one 1.01–1.02 mg/ℓ (Warren, 1971). The percent of organisms killed at time t (or alive at time t) for each concentration is plotted along the other (standard normal probability or "probit") axis. A line is drawn through the observations and the LC'_{50} value is read off where the line crosses the 50% alive or dead probability value. This then indicates the concentration that would lead to the death of an average or typical fish in t h. The plotting of a series of LC'_{50} values against time t leads to a *toxicity curve*. A useful aspect of the toxicity curve is that it permits the experimenter to judge whether or not it is warranted to go on with a particular bioassay for as long as planned *a priori*. But such a curve has another function. When such a plot becomes largely parallel to the time axis (i.e., at a more or less constant value of concentration called the *threshold lethal concentration* or *incipient lethal level*) one can loosely envision the end of the acute lethality stage. This level can be visualized as the level of the agent the species could tolerate indefinitely.

Particular attention is focused on the LC_{50} value at the close of the bioassay. The derivation of statistical confidence limits for such a value is discussed briefly by Sprague (1973). Unless the entire series of LC_{50}' values is presented in graphical or tabular form, three important bits of information about any bioassay are the LC_{50} value at its conclusion, the confidence limits, and the slope of the toxicity curve at the end of the test.

Another approach with mortality data is to plot, for each test tank and its specific concentration c, a percent mortality versus time plot. This is also done on log-normal probability graph paper, and a (straight) line is fitted by eye to the series of results for any particular test container. The *median lethal time* or *median effective time*, written as ET_{50}^c, can be read from the plot wherever one of the equal-concentration lines crosses the 50% probability level.

Units of Toxicity: The *toxicity concentration Tc*, a parameter used to measure the acceptability of waters for supporting a healthy marine biota, is defined as

$$Tc = \frac{\text{concentration of substance or wastewater}}{\text{96-h } LC_{50} \text{ for that substance or wastewater}} \qquad (4\text{-}1)$$

Tc as defined above is expressed in *toxicity units tu*. When it is not possible to measure the 96-h TL_{50} because more than 50% of the bioassay test individuals survive 100% waste, it is suggested in "Water Quality Control Plan for Ocean Waters of California" (1972), that

$$Tc = \frac{\log(100\text{-}PS)}{1.7} \qquad (4\text{-}2)$$

where *PS* is the percent survival in 100% waste. When $PS = 50\%$ in Equation (4-2), $Tc = 1$ *tu* as it should.

It is possible to define other toxicity-related parameters (e.g., Brown and Beck, 1972). The *toxicity emission rate TER* is

$$TER = (Q)(Tc) \qquad (4\text{-}3)$$

where Q is the wastewater discharge in Mgd. The toxicity concentration of the receiving water is

$$(Tc)_r = \left[\frac{Q}{Q + Q_r} \right] Tc \qquad (4\text{-}4)$$

where Q_r is the effective diluting flow rate of receiving water past the dispersion section of the outfall. If, for example, a wastewater with $Tc = 1$ *tu* is diluted by $25:1$, the toxicity concentration of the receiving water is 0.04 *tu*.

Available Acute Toxicity Data: There is a substantial literature dealing with the results of acute toxicity bioassays for many different potential pollutants and many different organisms. Numerous of these references are listed and discussed in other papers, and the reader is referred to the following: Bernhard and Zattera (1975); SCCWRP (1973, 1975); Perkins (1974); Eisler (1973); Portmann (1972); Esvelt et al. (1972); Bryan (1971); Sprague (1969, 1970, 1971); Bargman and Parkhurst (1969); Waldichuk (1969); McGauhey (1968); McKee and Wolf (1963); and Doudoroff and Katz (1953).

Chronic Toxicity of Wastewater Discharges

Since wastewater inflows to receiving waters have existed and will exist for many years, the long-term, low-level exposure of marine organisms to pollutants is particularly important. In some cases, the long-term safe concentration for a pollutant is considered to be the lethal threshold defined earlier. One means of connecting the long-term safe concentration of a pollutant and a specified LC_{50} is through what is called an *application factor*. This factor is assumed to be the same for all species of organism (typically fish) in all waters under all conditions. Strictly speaking, this is a very poor assumption, but basically one that has to be made to render the problem manageable (Warren, 1971; Sprague, 1973). The application factors typically used run from 1–10% ("Ocean Plan Review: Status Report," 1976).

Stephan and Mount (1973) have discussed fish toxicology studies in the light of the protection of marine organisms from pollutants introduced into marine waters from wastewater inflows. They stress the fact that decisions must be made on the basis of incomplete information and add that this is something toxicologists must be prepared to do despite their aversion, like all other scientists, to say anything unless it is enormously well supported by evidence.

The types of problems encountered involve the following considerations. Two species of fish with greatly different survival times at the same (reduced) level of a lethal agent may have essentially the same lethal level. The component(s) of wastes leading to acute toxicity are not necessarily those responsible for any effects the waste may have when further diluted in the receiving water. It is *all* organisms that should survive in the receiving waters, not just 50%. The substance causing acute toxicity to one species may not be the one harmful to another, including food organisms.

The toxicity of a treated wastewater to marine organisms is an important consideration in the selection of a treatment process. Even in the cases where secondary treatment may be required, there are various forms of secondary treatment, each of which can yield an effluent of potentially different toxicity to a particular species. Bioassays using fish and invertebrates representative of the potential receiving waters should be undertaken. These are typically organisms important from commercial, sport-fishing, or other recreational viewpoints, or possibly rare or endangered species. Thought should be given to the advisability of using eggs of these organisms and larval or juvenile stages.* In the work of Brown and Caldwell (1971), the Dungeness crab was determined to be a particularly important local marine species and testing of all life forms from eggs through mature adult was carried out. Since the toxicity of a wastewater is by no means constant from hour to hour or day to day, samples drawn from the wastewater flow at different times should be used in bioassays wherein the toxicity of a particular wastewater is being appraised.

Finally, an issue not yet discussed bears mentioning. Exposures of marine organisms to pollutants even at levels below those necessary to perturb normal life functions may still result in adverse effects. These concern the consumers of such organisms. Both flavor impairment and the accumulation of toxic residues in the organism's tissues are in question. Material discussing these problems is contained in Raymont and Shields (1964), Nitta (1972), Halstead (1972), and Connell (1974).

4-5 SOURCE CONTROL

Introduction

There are many potential constitutents of wastewaters that are toxic to aquatic organisms when present in sufficient concentration. Table 4-3 lists the amounts of certain metallic pollutants reaching the Los Angeles City wastewater treatment facilities at one time. Clearly, although the industrial water *flow* contribution was secondary to that of the hundreds of thousands of scattered domestic sources, the amounts of certain heavy metals contributed by industry far surpassed that of the domestic sources. The two most obvious examples in Table 4-3 are chromium and copper.

Copper has been of particular interest; at the beginning of the previous section, the massive Dutch fish kill due to high copper concentrations in nearshore waters was mentioned. The major industrial source of copper in Los Angeles City (Table 4-3) was the metal-finishing industry.† Some 329 separate plants contributed 80% of the total copper. Electrical parts manufacturers (128) added another 10%, blueprinters and engravers (61) approximately 5%, and metal fabricators (121) another 3%. The remaining 2% had as contributors many

---

*An amplified discussion of these matters is available (Cronin, 1969).

†Metal-finishing industries provide such services as coating, engraving, electroplating, plating, polishing, and anodizing. Pollution from such operations can have three forms: accidental spills, leaks, or overflows; periodic dumping of spent processing solutions (the major source); and rinse water discharge. It was estimated in 1976 that the cost to the U.S. metal-finishing industry of complying with PL 92-500 would be $45 billion by 1985 ("PL 92-500...", 1976).

Table 4-3: Approximate Mass Emission Rates for Certain "Incompatible Pollutants" before Stringent Source Control*

Constituents	Inflow at Hyperion Treatment Plant 344 Mgd (14.8 m^3/s)		Domestic Waste 278 Mgd (12.0 m^3/s)			Industrial Waste 66 Mgd (2.8 m^3/s)		
	lb/working day	*N/working day*	*lb/working day*	*N/working day*	*%*	*lb/working day*	*N/working day*	*%*
Arsenic	43	191	35	156	81	8	35	19
Cadmium	29	129	23	102	81	6	27	19
Chromium (total)	1578	7020	116	516	7	1462	6504	93
Copper	1119	4978	162	721	15	957	4257	85
Lead	86	383	70	312	81	16	71	19
Mercury	10	44	3	13	33	7	31	67
Nickel	861	3830	185	823	21	676	3007	79
Silver	52	231	23	102	45	29	129	55
Zinc	1894	8425	626	2785	33	1268	5640	67

*Adapted from Los Angeles City Board of Public Works (1973), with permission.

industries, including foundries, canneries, laboratories, and chemical manufacturers.

Klein et al. (1974) report that 250 electroplating firms in the New York City area contributed 500 lb (2224 N) of copper daily to the sewage system of that area. There were many other appreciable contributors as well, notably laundries and ice cream and soft drink industries. Copper was also added to the water supply (0.059 mg/ℓ) to control algal growth.

Major sources of other industrial "incompatible pollutant" inputs in Table 4-3 were as follows: cadmium, metal finishing (90%); total chromium, metal finishing (80%); lead, metal finishing (90%); mercury, hospitals (number not specified, 70%); nickel, metal finishing (95%); silver, film processing (339 firms, 60%); zinc, metal finishing (90%); cyanide, metal finishing (95%); phenols, cooling towers (number not specified, 50%).

Experience of Los Angeles City

The City: The California State Water Resources Control Board (SWRCB), as part of its total responsibilities, oversees the discharge of wastewaters into the coastal waters of that state. Nine Regional Water Quality Control Boards are scattered over the state such that the SWRCB has local staff to better cope with water pollution control functions. Six of the nine regional offices contain Pacific Ocean coastlines, and three of these six are responsible for the enormous metropolitan areas of San Francisco, Los Angeles, and San Diego.

Every wastewater discharger in California had to bring its effluent into conformity with the "Water Quality Control Plan for Ocean Waters of California" (1972) discussed in the previous section.* The City of Los Angeles, operator of several outfalls that enter the Pacific at Hyperion, was one such municipal body.

Los Angeles City covers some 460 square miles (1190 km^2) and stretches about 50 miles (80 km) from its northwesterly to southeasterly limits, with more than 6800 miles (11,900 km) of sewers and storm drains. The population of the city itself in 1973 was approximately 3 million. However, a larger group than this contributed to the Los Angeles wastewater flow since Los Angeles City accepted flows from certain adjoining urban areas.

Pollutants: The inputs to the Hyperion treatment of certain pollutants have already been given in Table 4-3. To ensure that the levels of these pollutants in the effluent from the plant conformed to those dictated by the SWRCB, Table 4-2, it was readily apparent that some sweeping changes would be necessary. Rather than assuming the responsibility for removing the heavy metals at the treatment plant after they had been diluted with domestic sewage, the City of Los Angeles embarked on an ambitious program of *source control*. This expression pertains in part to the elimination from an industrial or commercial wastewater of at least the bulk of the enterprise's pollutant addition to the water supply before it enters the municipal sewerage system. It is a form of pretreatment.

For our purposes, the above explanation of source control is largely sufficient. However, such a program, which involves regulation of wastes through physical inspection and chemical testing, is substantially broader than this, as outlined by the following list of objectives in a comprehensive source control program: protect the facilities from damage by deleterious wastes; protect the treatment processes; protect the operating and maintenance personnel; protect the capacity of the sanitary sewer system for sanitary wastes and appropriate industrial wastes; insure the safety and welfare of the public; conserve a reclaimable resource (water reclamation); prevent contamination, pollution, and nuisance; and protect the established beneficial uses of receiving waters. The final entry in the above list is the focus of this section. In the Los Angeles case, this was required by the California Regional Water Quality Control Board, Los Angeles Region.

Perhaps nowhere else is the concept of source control put to such thorough use as in Los Angeles (Bargman and Garber, 1973; Los Angeles City Board of Public Works, 1973, 1975). The city's regulating ordinance stated that the "highest and best use" for the system concerned domestic sewage and stressed that it was a privilege for industrial and commercial enterprises to dispose of

---

*It should be stressed that these standards had not been determined solely by the state. The SWRCB was conforming with the requirements of PL 92-500.

their wastewaters in the city sewerage system, i.e., it was not a right. People or companies had the right to dispose of *domestic* wastewater as long as they paid the appropriate connection fee. For the privilege of introducing process wastes, each such industrial or commercial enterprise had to obtain a permit and also pay a fee dependent upon three components of its effluent: its daily volume of discharge, the suspended solids in that flow (above 250 mg/ℓ, the average), and the BOD_5 (above 230 mg/ℓ). Fees received were put into a sewer construction and maintenance fund.

Limits were set on the amounts of "incompatible pollutants" that could be added to the sewage system by any single industrial or commercial wastewater discharger. These "incompatible pollutants" had the potential to interfere with treatment processes or the sewage system or were not removed to an appreciable degree by conventional municipal treatment processes. These maximum levels, presented in Table 4-4, were set so that Los Angeles City could satisfy the "Water Quality Control Plan for Ocean Waters of California" (1972). As an example, it was judged by the Los Angeles City Board of Public Works (1973) that 90% of the 957 lbs (4257 N) of copper (Table 4-3) formerly reaching the treatment plant each day could be removed at the source.

The Los Angeles City Board of Public Works (1973) determined that, with source control and appropriate city treatment of the wastewaters, they could

Table 4-4: Limits on Discharges of "Incompatible Pollutants" to Los Angeles City Sanitary Sewerage System*

Constituent[†]	Maximum	Minimum
Arsenic	3 mg/ℓ	—
Cadmium	15 mg/ℓ	—
Chromium (total)	10 mg/ℓ	—
Copper	15 mg/ℓ	—
Lead	5 mg/ℓ	—
Mercury	essentially none	—
Nickel	12 mg/ℓ	—
Silver	5 mg/ℓ	—
Zinc	25 mg/ℓ	—
Cyanide (total)	10 mg/ℓ	—
Cyanide (free)	2 mg/ℓ	—
pH	11	5.5
Oil and grease (total)		
dispersed	600 mg/ℓ	—
floatable	none visible	—
Chlorinated hydrocarbons	essentially none	—
Temperature	140°F	—

*Interim values pending EPA specifications. Los Angeles City Board of Public Works (1973), with permission.

†Constituents such as phenol, selenium, fluoride, boron, aluminum, iron, tin, and cobalt were not considered to be "critical."

satisfy California State effluent standards (Table 4-2) for all "incompatible
pollutants" except for total chromium. Thus considerable interest in the effects
of chromium on marine organisms developed, so much so that the Southern
California Coastal Water Research Project (SCCWRP), for example, intensified
its chromium studies. One of the real problems with chromium was that the
domestic loading alone could cause the California State effluent standards to be
exceeded.

Procedure: After a reasonable transition period following the introduction of the industrial wastewater ordinance, no commercial or industrial enterprise was allowed to discharge wastes unless its discharge had been analyzed.
The firm had first to make application; once it was established that the
wastewater satisfied the requirements, an industrial waste permit was issued. The
firm had to provide, on a continuing basis, facilities on its premises for sampling
and flow monitoring.

A "self-monitoring" program was instituted. All firms discharging 40,000
gpd (6.3 m^3/h) or more, and other firms with 15,000–40,000 gpd (2.4 to 6.3
m^3/h) with one or more "incompatible pollutants" (Table 4-4), were required to
have periodical sampling and analysis done by a City-qualified laboratory. The
frequency of such self-monitoring and reporting was once per month for
dischargers of 250,000 gpd (39.5 m^3/h) or more, quarterly for the others.
Sampling was to extend over 24 h for "normal" conditions. If the SS and BOD$_5$
were found to be much different from those upon which the rates of that
particular firm were based, an amended permit with new rates was prepared. If
the level of any of the "incompatible pollutants" exceeded the maximum
allowable (Table 4-4), the discharger had 30 days to submit a written report
detailing the measures taken to remedy the situation.

It was within the power of the Los Angeles City Board of Public Works to
revoke a permit if the discharger was not in compliance with the regulations.
This ended his use of the sewage system, and certainly meant a curtailing if not
a complete end to his activities. When dischargers were found to be in noncompliance, the responsible discharger, if identified, was required to pay whatever
damages he had caused. These rules, and others mentioned, applied to storm
drains as well as to the sewage system.

An important element in the Los Angeles system was that materials
removed from the wastewater by enterprises in-house had to be disposed of
properly. This involved approved waste haulers and disposers, and appropriate
forms, properly signed, had to be used.

Staff: In 1976, the City of Los Angeles had 32 staff members committed to
inspection, monitoring and testing activities, and supervision. This small group
was adequate because of the use of industry self-monitoring. Staff members were
freed to carry out the less routine and more difficult industrial waste control
tasks. The sole purpose of using the self-monitoring approach, incidentally, was

not to reduce City of Los Angeles staff size; it was thought that this approach would make industries more conscious of their commitments with respect to waste disposal.

4-6 BIOSTIMULATION AND THE MINIMUM REQUIREMENT OF SECONDARY TREATMENT

Eutrophication

Lake *eutrophication* involves the increase with time in the lake of the nutrients necessary for the production of algae and other aquatic plants. Eutrophication is a natural part of the successional process for lakes which pass slowly from a nutrient-poor (*oligotrophic*) condition, through a shallow (due to deposition), marsh-ringed condition, with ultimate return to land and terrestrial life.

Eutrophication is materially hastened by the addition of nutrients to lakes through sewage discharges. Waters, once clear, become clouded with plant matter. More plant matter grows around the lake fringes. The increased plant production results in an increasing rain of decomposing organic materials to the lake bottom, resulting in an augmented demand on the oxygen resources of the lake and the very real possibility of (odor-producing) anaerobic conditions in at least deeper waters.

Lake Erie is a classic example of a lake that has undergone nutrient *enrichment* or *biostimulation* because of heavy sewage inputs from the many communities surrounding it and those feeding into the St. Clair and Detroit Rivers immediately upstream of it. I spent several summers by Lake Erie when a small boy in the 1940s, and the decline in desirability of those waters since that time has been marked.

Nutrient Inputs to Marine Waters

Estuaries and marine embayments can also suffer water quality impairment due to biostimulation. Warren (1971) has described the problems with San Francisco Bay, the recipient at that time of the effluents of some 50 wastewater outfalls.

The word *bloom* is used to describe a prolific growth of (marine) plants. Algal blooms can cause many side effects as follows (North et al., 1972): odors; dermatitis; interference with filtration apparatus; biotoxicant release; excessive demands on dissolved oxygen; increases in organic loads; rendering human food organisms poisonous; introducing radioactivity or pesticides into animal food chains; mucilage and slime secretion; affecting water taste.

The protection of the marine environment includes the prevention of water quality degradation associated with nutrient enrichment. When a wastewater discharge to marine waters is in the planning stages, the potential for biostimulation should be investigated. As in Section 4-4, where the possibilities of an effluent's being poisonous or inhibitory were investigated using assay methods, the same general approach is followed when testing for the possibilities of enrichment. See *Standard Methods*... (1976) for details.

It is not unusual that the same general procedures should be used for both toxicity and enrichment tests since water constituents can be biostimulatory at low concentrations and toxic at higher ones. As an example, nutrients can be toxic to the hard-shell clam *Mercenaria mercenaria* (Epifanio and Srna, 1976); however, it has been claimed that such nutrients, derived from an effluent discharge, were responsible for spread of the same clam (Helliwell and Bossanyi, 1975).

PL 92-500 specifies a minimum level of secondary treatment for effluents into receiving waters; it does not differentiate between fresh or marine receiving waters. Pearson (1975) has argued vigorously that this U.S. federal requirement of at least secondary treatment for effluents does not then mean that they are eligible for disposal in estuaries or marine embayments. The potential for biostimulation is still present (Warren, 1971). This threat can be eliminated by piping the wastewaters to an open coast disposal point, where, if anything, the additional input of nutrients can be beneficial. This can be reflected, for example, in increased fish catches (Perkins, 1974).

If then the wastewater is to be disposed of in the open ocean, it should not *need* secondary treatment, as will be discussed in the following subsection. The cost savings of using advanced primary treatment over secondary can easily pay for the open coast outfall and feeder lines (Pearson, 1975).

Secondary Treatment Requirements

Introduction: Secondary treatment is biological treatment. When a primary effluent is dumped into receiving waters, whether fresh or marine, it is given natural biological treatment, without the great expense of the on-land facility and operations. The point has been made by various investigators (e.g., SCCWRP, 1973; Pearson, 1975) that secondary treatment for effluents being properly discharged into open coastal marine waters is unnecessary. SCCWRP (1973) concluded that "typical inland requirements for removals of biochemical oxygen demand, suspended solids, and nutrients from municipal wastewaters have little technical justification or relevance to marine ecological problems." This conclusion applied to the open waters off the California coast, where SCCWRP did its work, rather than estuaries or marine embayments.

The BOD limitation is unnecessary for open waters because of the enormous oxygen resources, the great assimilative capacity, and the extensive natural mixing in the sea. The nutrients somewhat increase production but without the deleterious effects encountered in estuaries or marine embayments.

Open Coast Wastewater Disposal in Hawaii: The State of Hawaii and the City and County of Honolulu felt, despite the directives of PL 92-500, that advanced primary treatment coupled with a deep ocean outfall would be sufficient for disposing of Honolulu's wastewater. The federal senator Daniel K. Inouye from Hawaii requested a hearing on the matter, and on March 18, 1974, the "Muskie hearings" took place in Honolulu. Senator Edmund S. Muskie, incidentally one of those involved in writing PL 92-500, was acting in his capacity as Chairman of the Subcommittee on Environmental Pollution of the Senate Committee on Public Works. The hearing was to "consider the impact of the secondary treatment requirement of the Federal Clean Water Act on community wastes discharged into the ocean."

The hearings, open to the public, were addressed by many people, including the Governor of the State and the Mayor of the City. There were other public officials who took part, various University of Hawaii faculty members, representatives of environmental groups, and others. All endorsed the idea that Honolulu's effluent be allowed to enter the marine environment with treatment less than specified by PL 92-500. It is worthwhile to note in passing that this was so even though Honolulu's experience with sewage disposal for many years had been poor, with a raw sewage effluent entering the ocean only 3700 ft (1100 m) from the coast.

During the hearings, the arguments about the tremendous assimilative capacity of open ocean waters and the inapplicability of inland waters requirements of BOD, SS, and nutrients were repeatedly raised. Other points that were raised concerned the additional capital cost of secondary treatment over advanced primary ($17,000,000), the extra annual cost of operation and maintenance ($600,000), and the extensive extra use of land (20 acres or 8 ha) that would otherwise be used as a portion of a park. Despite the arguments, the minimum treatment requirements of PL 92-500 were not relaxed for the Honolulu discharge.*

Treatment and Outfall Costs: The Southern California Coastal Water Research Project (1975) has presented cost data on treatment plants and outfalls that is instructive (see Table 4-5). These figures clearly show the cost penalty associated with specifying unduly severe treatment levels and demonstrate the relative cheapness of outfalls.

---

*The desire to maintain standards of constant severity across the country can be understood at the least in terms of not giving manufacturers or municipalities in different areas undue advantage.

Table 4-5:   Total Costs of Municipal Wastewater Treatment and Outfalls*

Class	Average Flow Rate, Mgd (m³/s)	1 (0.04)		10 (0.44)		100 (4.38)	
		$/Mg	$/day	$/Mg	$/day	$/Mg	$/day
Treatment	Primary	288	288	143	1430	70	7000
	Advanced Primary	356	356	173	1730	83	8300
	Secondary	630	630	307	3070	150	15,000
	Tertiary	1210	1210	630	6300	299	29,900
Outfall	5000 feet (1500 m)	804	804	101	1010	16	1600
	10,000 feet (3000 m)	1090	1090	138	1380	25	2500
	20,000 feet (6000 m) (with pumping)	1690	1690	220	2200	41	4100

*These costs reflect both amortization of capital cost (bonds over 30 years at 7%) and operational/maintenance cost. SCCWRP (1975), with permission.

The report by Van Note et al. (1975) contains information on both capital costs and operating and maintenance costs for an extensive selection of wastewater treatment plant components. However, there are no data on outfalls.

4-7   PUBLIC HEALTH CONSIDERATIONS

Introduction

The bacteriological criteria outlined in the "Water Quality Control Plan for Ocean Waters of California" (1972) specifically mentioned coliform bacteria. It was outlined in Chapter 3 that coliforms are largely harmless to humans and animals and that concern about the concentration of these microorganisms is actually concern about the pathogenic bacteria and viruses that enter the marine environment with the coliforms. The (fecal) coliforms are then *indicators* of possible public health risks. One objective of this section is to discuss how prudent it is to base all one's judgment on the acceptability or nonacceptability of marine waters for water contact sports on imprecisely measured levels of a specific nonpathogen.

Coliform Disappearance

Ninety Percent Mortality: A *pure death process* is described by

$$N = N_0 e^{-kt} \tag{4-5}$$

where N is the concentration of (live) organisms at any time, N_0 is their concentration at the time $t=0$, and k is a (positive) decay coefficient. One of the assumptions made implicitly in dealing with the dieaway of coliform bacteria in the marine environment is that this dieaway corresponds to the exponential form of Equation (4-5). But herein lies two problems. First, one may naturally enquire into the fit of Equation (4-5) to the dieaway of coliform bacteria in the sea. Then one can pose the even more important question regarding the relative rates of dieaway of a coliform population and of the pathogens whose mortality behavior an indicator organism is supposed to be adequately simulating.

The coefficient k in Equation (4-5) is, by convention in wastewater disposal computations, eliminated by defining t_{90} as the elapsed time when the number of (live) organisms is 10% of N_0; i.e., 90% of the organisms have died. Then

$$0.1 = e^{-kt_{90}} \tag{4-6}$$

or

$$k = \frac{\ln 10}{t_{90}} \equiv \frac{2.3}{t_{90}} \tag{4-7}$$

Thus Equation (4-5) can be rewritten in the form

$$N = N_0 \exp\left\{-2.3\frac{t}{t_{90}}\right\} = N_0 10^{-t/t_{90}} \tag{4-8}$$

where

$$\exp\{-a\} \equiv e^{-a}. \tag{4-9}$$

Although the term "dieaway" is frequently used, and was employed above to describe the decrease in concentration of bacteria (or other organisms) in an undiluted parcel of water, Pearson (1971) and Harremöes (1975) prefer the less restrictive expression "disappearance rate," and this term will be used primarily from here on. The significant environmental factors leading to bacterial disappearance are death or inactivation, coagulation or flocculation, sedimentation, and grazing and/or predation (Pearson, 1971). Mitchell and Chamberlain (1975) as well as Jones and Cobet (1975) discuss the causes of bacterial disappearance in some detail.

Determinations of t_{90}: Pearson (1971) has mapped out a strategy for obtaining t_{90} data. This should be done in the ocean at the location planned for the ultimate wastewater discharge.* Pearson suggests transporting to the site a barge load of sewage of characteristics similar to that to be disposed of at the site. This load should be tagged with a suitable fluorescent tracer (Chapter 5) at a concentration well below that leading to bacterial kill. The large volume of

---

*Although they are crude, field studies are superior to laboratory experiments in determining design t_{90} values because the laboratory tests cannot properly simulate all the factors leading to bacterial disappearance.

well-mixed sewage and tracer is then released rapidly into the water. Several samples taken in the center of that discharge provide, after suitable analysis, the initial concentrations of both dye and indicator organisms. Pearson suggests several samples taken at each of 2–3 ft (0.6–0.9 m) deep and at approximately 10 ft (3 m) below the surface both at the initial drop point and at its subsequent positions, at suggested times of 10, 20, 30, 45, 60, 90, 120, 180, and 240 min. Two drogues, one set at 2–3 feet (0.6–0.9 m) and the other at approximately 10 ft (3 m) can be used to indicate the Lagrangian position of the drop point.

The presence of dye in the water allows the investigator to judge the degree of physical dilution of his sample from the time that the test began. The concentration of bacterial organisms, of course, reflects not only Pearson's "disappearance" but also the effect of physical dilution. The dye readings are then used to correct the time-decay rate of the indicator organisms for dilution. It is particularly important to have adequate samples early in the test since that is a time of rapid change. The determination of dye concentration can be done using a Turner 111 Fluorometer (Chapter 5) or equivalent. Either MPN or membrane filter procedures, outlined in Chapter 3 and detailed in *Standard Methods*...(1976), can be followed for the bacterial concentration.

Since, theoretically, Equation (4-8) is supposed to fit the (adjusted) data on bacterial disappearance, the normalized indicator organisms concentration data and time should be plotted together on semilogarithmic graph paper (with time on the arithmetic scale). The individual mean value for the tiny statistical sample should be shown at each appropriate time plus a bar showing a measure of dispersion, such as the double standard deviation or the 95% confidence interval.

A line, hopefully straight, is then passed through the data. The time at which the ratio of bacterial concentration (adjusted for dilution) is 10% of the initial concentration is noted as t_{90}. There is a sizable subjective element in the results if the points are poorly aligned.

When the t_{90} value is given for a test it is also virtually mandatory that the investigator produce the plot from which this was obtained. The initial concentrations of tracer and organisms should also be included. Other pertinent information concerns the water temperature, the weather, and the sample depth.

Harremoës (1975) has termed the method discussed above to determine t_{90} the "concentration ratio approach." He also described a "mass-balance approach," which involved measuring across various sections of the dispersing sewage-dye cloud and making adjustments in dilution determinations for decay of the tracer.

Variability in t_{90} Values: Tabulated t_{90} results from investigators across the world (e.g., Pearson, 1971; Gunnerson, 1975) look more or less like the output from a random number generator. Of course, much of the variability is due to different locations, different waters, different seasons, decay of the "conservative" tracer, and possibly different ways of interpreting the data. But even at one location under ostensibly identical conditions, and with the same investigator, t_{90}

values are known to show substantial variability (e.g., Mitchell and Chamberlain, 1975; Harremoës, 1975).

The choice of a design value of t_{90} is then largely a guessing game because of the dispersion in sample values typically obtained. Consider the case of work done prior to the design of the Sand Island No. 2 outfall in Honolulu (R. M. Towill Corp., 1972).

One hundred separate t_{90} determinations were made during daylight hours. Thirteen values were undefined because subsequent coliform concentrations were higher than initial levels. The remaining 87 t_{90} values had 50 and 80% cumulative distribution values of 20 and 45 min. When the 13 errant samples were assigned infinite t_{90} values, the 50 and 80% cumulative t_{90} distribution values were 25 and 70 min. What value should be chosen?

Sunlight is clearly bactericidal, and t_{90} values obtained for the same location during the day and at night can differ by a factor of 100 (Gameson, 1975; Gameson and Gould, 1975). Since there are probably few t_{90} field experiments carried out at night (Gameson and Gould's were largely done in the laboratory), the average t_{90} value for a real effluent, which pours out night and day, is probably considerably underestimated. Lukin et al. (1971) give two-season (winter and summer) t_{90} values. These differ by a factor of five, the winter values being larger, conceivably because of sunlight intensity and water temperature differences, as well as other factors.

It has been shown by some investigators (e.g., Won and Ross, 1973; Savage and Hanes, 1971) that addition of some organic material to sea water causes the sea water to become less toxic to total and fecal coliforms. This situation might well be one source of the dispersion in t_{90} data.

Variability in t_{90} values is very important in the context of marine disposal of wastewaters. When an *exponent* can vary tremendously, the resulting concentration ratio is subject to enormous variability. Since, say, the 99% confidence interval for the t_{90} value to be used in design is undoubtedly of the order of that t_{90} value, bacterial disappearance calculations are little more than a stab in the dark. When one compounds this problem with the question of whether or not coliform disappearance rates have anything to do with those of the pathogens, there is a real dilemma. The very great variability in t_{90} values renders senseless a vast expenditure of effort in computing dilutions of effluent entering the marine environment from an outfall (Chapter 7) to supposed great accuracy. Further discussion of this point is contained in Pearson et al. (1967).

Pathogens and Indicator Organisms

Enteric viruses (enteroviruses) are found in the gastrointestinal tract and feces of man and many lower animals. Such viruses along with bacteria of human intestinal origin enter sewerage systems from many sources and pass downstream toward the treatment and disposal facilities.

Various types of pathogenic (disease-causing or transmitting) bacteria and viruses (lumped together and called *pathogens*) pass into the sewage system with fecal material. Included among the pathogenic bacteria are the salmonellae that can cause typhoid fever, paratyphoid fever, or gastroenteritis if taken in orally. Other bacterial pathogens can cause the diseases of cholera or bacillary dysentery or lead to streptococcal or staphylococcal infections.

Pathogenic viruses are as follows: adenoviruses that cause upper respiratory infections particularly in children; Coxsackie viruses that lead to poliolike diseases; the ECHO (enteric-cytopathogenic-human-orphan) virus; the infectious hepatitis viruses; the polio viruses; and the reoviruses, leading to enteric and respiratory infections in children.

The pathogens listed above all transmit their diseases or infections via the water route. When sewage is to be disposed of in receiving waters, it is imperative that the treatment and receiving water dilution be blended so that there is minimal chance of contracting one or more of the associated illnesses either through direct water contact or through consuming raw or improperly cooked shellfish taken from polluted waters.

Risks from polluted shellfish come from bacterial pathogens (e.g., typhoid, dysentery), viral pathogens [e.g., infectious hepatitis (Mason and McLean, 1962)] and resistant cysts and eggs. It is of course important to ensure that pathogens are either absent from shellfish-rearing waters or present only in numbers that will not, after being concentrated by the organisms, have any debilitating effects on consumers.* But therein lies a problem. There is incomplete information regarding the probability of infection or the threshold dose that will induce the disease. Katzenelson and Shuval (1975) concluded only that one virus particle "may cause infection." An extensive discussion of outbreaks of disease caused mainly by consumption of shellfish harboring pathogens is available in SCCWRP (1973).

The whole purpose of indicator organisms is to provide evidence of the possible presence in wastewaters, receiving waters, or in shellfish, of pathogenic organisms. Concentrations of enteric viruses and pathogenic bacteria in treated sewage and in receiving waters are generally very low, and the techniques for isolating, identifying, and enumerating these pathogens are difficult despite the advances in microbiology during the 1960s and 1970s (Dutka, 1973; Scarpino, 1975). These difficulties as well as tradition have been behind the continued use of indicator organisms, coliforms generally, or fecal coliforms specifically.

Alternates to coliform bacteria have been proposed. One of these, for example, concerns the *fecal streptococci*. These are characteristic of fecal pollution, do not multiply in surface waters, and only rarely occur in surface soil or on vegetation not contaminated by sewage. In addition, they display greater resistance to adverse conditions than coliforms (Wood, 1972).

---

*Self-cleaning of shellfish, in a process called *depuration*, can be used before the shellfish reach the consumer. This process involves active shellfish feeding for 2–4 days in clean water.

Some investigators appear to prefer bacteriophages as fecal pollution indicators (e.g., McAllister, 1968; Metcalf, 1975). *Bacteriophages* are viruses of bacteria, depending parasitically upon the bacteria for growth. A bacteriophage attaches itself to a bacterium and after penetrating the cell wall reproduces within the cell. New bacteriophages are released into the solution, to infect other bacteria, when the bacterial cell ruptures. The submicroscopic bacteriophages normally occur in nature whenever their prospective bacterial hosts are found. The bacteriophage of coliform bacteria, for example, is often referred to as *coliphage*. Metcalf (1975) states that coliphages are best, but by no means perfect, for indicating unsanitary conditions in shellfish.

Despite these proposals, however, coliform bacteria remain as the indicator organism used in water quality standards. There is ample evidence that many pathogens are much more resistant to chlorination at the treatment plant than coliform bacteria. The polio viruses as well as staphylococci and fecal streptococci serve as examples. In addition, there is abundant evidence that many pathogens can survive for longer periods in the sea than coliforms and thus may be present in water judged satisfactory on the basis of coliform concentrations. Yoshpe-Purer and Shuval (1972) have shown this for salmonellae, Shuval (1975) for polio viruses. Dutka (1973) has cited various epidemics that occurred even when waters satisfied coliform count criteria.

Coliform Standards

Conformance with coliform standards for receiving waters and shellfish has often been achieved through chlorination of the wastewater combined with a sound diffuser-mixing system for the effluent located at some distance from the particular area or areas within which the standards are to be met.

Brooks (1971) has considered some important side issues associated with chlorination that bear mentioning: large amounts of electric power are required in its production; *mercury* anodes are used in such production; there are hazards of transporting large quantities of chlorine to the sewage treatment plants* and minor hazards to plant personnel applying it; chlorine exhibits toxicity to aquatic organisms; chlorine has substantial cost. Brooks felt strongly that the disinfective advantages of chlorination came far short of balancing the bad features associated with chlorine use.

Outfalls should be designed to achieve (reasonable) bacterial standards in recreational and shellfish waters without chlorination. Among the major southern California dischargers, for example, Brooks (1971) reported that at that time Los Angeles City, San Diego City, and the Orange County Sanitation Districts could meet the standards then in force without chlorination.

---

*The Orange County Sanitation Districts (OCSD) stopped chlorinating their wastewater when their longer outfall went into service in late March 1971. OCSD had been using an average of 15 tons/day of chlorine.

There has been a tendency to decrease the allowable concentrations of coliform bacteria in receiving waters and shellfish in order to decrease the possibility of humans contracting water borne illnesses. But there has been considerable controversy about coliform standards for recreational waters.

The whole idea of coliform standards for recreational waters has been largely attacked by Moore (1975). For one thing he discounted studies relating coliform levels and the incidence of disease; for another he claimed that by judicious choices of times and sampling stations virtually any classification could be assigned beach waters. He mentioned in passing, as have others, that there is no enforcement of air standards in public places so that one does not catch, for example, influenza.

Although many environmentalists would differ strongly with the presentation of Moore in arguing against bathing water bacterial standards, there is in truth scant evidence that high coliform counts *per se* in marine bathing waters are significantly correlated with the incidence of disease. SCCWRP (1973), for example, quotes U.S. Public Health Service studies showing that swimmers have an appreciably higher incidence of all illnesses (eye, ear, nose, and throat ailments, gastrointestinal illness, and skin irritations) than nonswimmers, *regardless of water quality*.* Brooks (1971) reported that there were no known instances of disease transmission to southern California bathers from the several major outfalls discharging along that stretch of coastline.

4-8 PERMITS AND ENVIRONMENTAL IMPACT STATEMENTS

Introduction

The Federal Water Pollution Control Act, as amended, prohibits any discharges of pollutants into a United States waterway from a point source unless the discharge is authorized by a permit. Such a permit is issued by the U.S. Environmental Protection Agency (EPA), which is charged with the administration of the Act, or a state agency that has been approved by the EPA. The application is made on a form entitled the National Pollutant Discharge Elimination System (NPDES) Application for Permit to Discharge Wastewater. There are both long and short forms for such an Application, and each type of form is divided into classes. The short forms are as follows: A—Municipal; B—Agricultural; C—Manufacturing and Mining; D—Commercial, Vessels and

---

*Stevenson (1953) has studied illnesses of swimmers in both fresh (Lake Michigan) and salt waters (Long Island Sound). A combined sample yielded slightly less than 20% gastrointestinal disturbances and somewhat greater than 50% eye, ear, nose, and throat ailments. Skin irritations and various combinations of illnesses comprised the remainder. The separate freshwater and marine results were little different.

Other. The long forms are divided between: A—Municipal; C—Manufacturing and Commercial.

Form A, which would be the usual one for outfalls, is divided into sections as follows:

Section I: Applicant and Facility Description (drawings and maps are included)

Section II: Basic Discharge Description (the following are examples of items listed, both as treatment plant input and output—discharge, pH, temperature, fecal streptococci, fecal coliforms, 5-day BOD, total dissolved solids, NH_3 (as N), NO_3 (as N), NO_2 (as N), DO, heavy metals, pesticides, oil and grease, phenols, and radioactivity)

Section III: Scheduled Improvements and Schedules of Implementation

Section IV: Industrial Waste Contribution to Municipal System

The completed form is sent either to the EPA Regional Office for the state concerned or to a designated state agency in a state with a federally approved permit program. There are 10 EPA Regional Offices in the United States, and these have responsibility for all states and territories of the country. Maryland, for example, is in Region III with head office in Philadelphia, Pennsylvania; Oregon is in Region X with headquarters in Seattle, Washington. For minor new discharges, the submission of the application should provide at least a 6-month lead time before discharge is scheduled to begin. For major facilities, where an EIS is possible, at least 12- to 18-month lead times should be used.

If the EPA has not delegated its responsibility to an approved state agency, and still has responsibility for issuing the NPDES permit, the appropriate state agency still serves in a certification capacity for the federal agency. It is required by PL 92-500 that EPA cannot issue an NPDES permit until state certification has been obtained. If a state agency has been approved by EPA, then that agency issues the NPDES permit and EPA holds veto power.

When EPA or the designated state agency receives the NPDES application, a draft permit is prepared and a public notice is issued and sent to interested governmental agencies, organizations, and individuals for comment. If there is an important conflict apparent, a public hearing can be scheduled. Input received as a result of the notice and the hearing are considered in the drafting of the final permit, which is then issued so long as certification has been received. The granting of the permit implies compliance with the pertinent laws, and the case for compliance may be prepared by the applicant in the form of an environmental assessment (EA) or an environmental impact statement (EIS).

When municipalities apply for 75% federal grants under PL 92-500 to assist them in planning, designing, and constructing treatment works, a "federal action" is involved. This means that NEPA of 1969 is applicable. During the planning stage, a municipality prepares an EA—seeking to outline the engineering, cost, social, economic, and primarily environmental impacts of the proposed action. The EA allows EPA to decide if significant adverse environmental effects are involved—thus necessitating the preparation of an EIS under NEPA. If EPA decides not to call for an EIS, then a "negative declaration" is issued as well as public notices. Otherwise, a "notice of intent" is issued and made public, and EPA starts to work on putting together the EIS from the EA. This is often done using an outside consultant, but under EPA guidance. Many months are taken to prepare a draft environmental impact statement and then to have this reviewed prior to the preparation of the final EIS, and in cases where it is anticipated that significant environmental effects might be involved, the EIS is prepared during the early planning process rather than the EA. Both the EA and EIS will be discussed at greater length later in this section.

The EPA has not felt that PL 92-500 gives them authority to impose NEPA on a NPDES permit application for a new source in a state that has an approved permit program and where federal funding, and a federal action, is not involved. Thus industries that fund their own treatment works facilities in a state with an EPA-approved permit program need not conform with NEPA—and no EA or EIS is necessary before a permit is issued. Some environmentalists have voiced strong criticism of this "loophole" in PL 92-500. They have also pointed out another one, namely, that for a new source from an industry that has no discharge guidelines prepared by EPA there is no EIS required.

Other Federal Permits

On a typical ocean outfall project, permits in addition to that from EPA are usually required. It would be unusual for the Coast Guard to issue a permit for an outfall since they require permits only for bridges, causeways, and overhead pipelines. However, their Aids to Navigation Office should be notified at least a month before the construction is to begin so that an appropriate Notice to Mariners, describing the project, can be issued. In addition, the Coast Guard requires that proper day- and nighttime warning balls or lights be displayed on floating construction equipment.

The Corps of Engineers has responsibility for the "navigable waters of the United States." In the context of the oceans, these waters are considered to extend out to the edge of the continental shelf from either the mean high-water line (Atlantic) or the mean higher high-water line (Pacific).

The Corps of Engineers requires a permit for work or structures in navigable waters. The Corps, in their booklet entitled "Application for Department of the Army Permits for Activities in Waterways" (1974), describes their requirements for the issuance of a permit. Basically, the application must name the applicant, describe the location, identify adjoining property owners, discuss dredging details, and must show structures and activities on appropriate plans. The Corps then issues a Public Notice describing the project in detail. This document is circulated to appropriate federal, state, and local governmental agencies, and to concerned firms, groups, and individuals. The latter might include members of the U.S. Congress from the state concerned. Input is solicited. In this regard, a Corps of Engineers Public Notice concerning an application to construct an outfall contains the following passage.

> "Interested parties may submit in writing any comments that they may have on the proposed work. The responsibility of the Corps of Engineers is the protection of the public's interest in the navigable waters of the United States. The decision whether to issue a permit will be based on an evaluation of the probable impact of the proposed activity on the public interest. That decision will reflect the national concern for both protection and utilization of important resources. The benefit which may reasonably be expected to accrue from the proposal must be balanced against its reasonably foreseeable detriments. All factors which may be relevant to the proposal will be considered; among those are conservation, economics, aesthetics, general environmental concerns, historic values, fish and wildlife values, flood damage prevention, land use classification, navigation, recreation, water supply, water quality, and, in general, the needs and welfare of the people. No permit will be granted unless its issuance is found to be in the public interest. Comments should be forwarded so as to reach this District not later than forty-five (45) days from date of the notice."

The District mentioned in the last sentence refers to one of the many area offices of the Corps, in this case the one responsible for the region where the outfall construction is planned.

If there are consequential comments or objections to the Public Notice received from interested individuals, groups, or governmental agencies, two things are done. The comments and objections are sent to the applicant for his consideration and rebuttal and a public meeting may be scheduled. This is a forum where all concerned persons are given the opportunity to contribute to the store of information that will be used in making the decision concerning the granting of a permit.* Even if there is little input in response to the Public Notice and the Corps feels a permit may be warranted, an EIS may be required if significant environmental impact is bound to occur. It has not been unusual for individuals, groups, and governmental agencies to be approached twice or more for comments on the same project, once in response to the Public Notice, then with respect to the EIS.

A routine Corps of Engineers application takes about 3 months to turn into a permit. In many cases, however, where there are complications such as hearings and an EIS, the process may take many months. Once the Corps of Engineers has issued a permit, it is their responsibility to supervise the work to

---

*The *public meeting* is different from a *public hearing*. The latter is more formal, and it is designed to result in a recommended decision on the permit. Such a process may involve the cross-examination of witnesses.

ensure that conditions of the permit are satisfied. In some cases, however, this supervision has been loosely exercised. The usual maximum duration for work associated with a permit is 3 years from the last day of the year in which the permit was issued.

Finally, it is specified, within the Federal Water Pollution Control Act, "that an applicant for a permit to conduct an activity which may result in any discharge into the navigable waters shall provide certification from the appropriate water pollution control agency that the discharge will comply with applicable effluent limitations, standards of performance, or other prescribed water quality standards.... Thus a certification obtained by an applicant for a permit for construction of an outfall would have to be sufficient for a permit from EPA for the discharge from the outfall."

State and Local Permits

The types of other permits that may be required for an ocean outfall can be illustrated by assuming a proposed line on the Island of Hawaii in the Hawaiian Islands. The Hawaii State Harbors Division has its "Application for Permit for Work in the Shore Waters of the State of Hawaii." In this form, the applicant is named; the nature of the work, its purpose, and location are described; property owners on shore are identified; descriptions are included of fisheries involved, the nature of the shore and bottom, and the amount of dredging.

A permit is required from the State Department of Health for a zone of mixing. This has already been mentioned. It is normal that public meetings or hearings be held to gather input before a decision is reached by the Director of the Department to grant or deny the request for a permit.

The State Department of Land and Natural Resources has its "Conservation District Use Application." The nearshore waters constitute a form of conservation area. In addition, on-shore work may take place on land classified as a conservation area.

The County of Hawaii Planning Department would be involved in assuring the proper zoning for onshore work. They would also be concerned with the granting of variances, if these were necessary.

The Environmental Assessment and the Environmental Impact Statement

"The purpose of the environmental assessment is to comply with the intent of the National Environmental Policy Act of 1969 (NEPA), approved 1 January 1970, and insure that full consideration of environmental consequences is provided in processing permit applications. The intended goal of the environmental assessment procedure is three-fold: to gain foresight into the consequences of a proposed action, to illustrate preventative measures which could mitigate environmental degradation, and to determine the significance of the proposed action in terms of its effect on the quality of the human environment. Should it become apparent that the proposed action is significant, the applicant will be required to provide additional information necessary to prepare an environmental impact statement (EIS)."

This quote is taken from the U.S. Army Corps of Engineers publication entitled Application for Department of the Army Permits for Activities in Waterways (1974) which also spells out the topics to be covered by an environmental assessment. However, these topics essentially form a subset of those usually included in the EIS, which typically include the following (e.g., Environmental Quality Commission, 1975):

1. Summary sheet
2. Background and description of project
3. Description of environmental setting
4. The relationship of the proposed action to land use plans, policies, and controls for the affected area
5. The probable impact, primary and secondary, of the proposed action on the environment
6. Any probable adverse environmental effects which cannot be avoided
7. Alternatives to the proposed action
8. The relationship between local short-term uses of man's environment and the maintenance and enhancement of long-term productivity and beneficial uses
9. Mitigation measures proposed to minimize impact
10. Any irreversible and irretrievable commitments of resources
11. An indication of what other interests and considerations of governmental policies are thought to offset the adverse environmental effects of the proposed action
12. Organizations and persons consulted, with discussion of the problems and objections raised
13. List of necessary approvals

Pertinent documentation may accompany the EIS.

The preparation of a proper EIS is a delicate procedure, and municipal bodies with ocean outfall plans will probably not have adequate, qualified manpower in-house to prepare such a document. Specific organizations were set up after 1970 with the express purpose of writing EISs. Such companies had to be staffed with personnel with various backgrounds since the EIS can enter into many fields of knowledge: e.g., meteorology, marine biology, physical oceanography, geography, civil engineering, sociology, political science, and law.

But even a seasoned staff with broad expertise may not be conscious of or able to visualize some aspects of an ocean outfall (or other) project that will impact the environment. In fact, certain issues may be purposely ignored. Neither can it be assured that the organization preparing the EIS can write the document so that it can be understood by a wide selection of readers. Thus a Draft EIS (or DEIS) is first prepared and distributed for comment to appropriate federal, state, and local governmental authorities, to legislators, to public and environmental organizations, and to other concerned individuals and organizations. The collected input is not only used in rewriting the Final EIS (FEIS) but is also used as an appendix in the FEIS.

Review of Draft Environmental Statements

It is informative to list the federal, state, and local agencies that review EAs or EISs concerned with ocean outfalls and to sketch their concerns. The Barbers Point outfall in Hawaii will be considered shortly in this context. The final EIS (R. M. Towill Corp., 1975) contains the agency input and the replies to each by the City and County of Honolulu, Dept. of Public Works.

The Corps of Engineers (U.S. Army) gave its approval and added that a Dept. of the Army Section 10 permit for construction of the outfall in navigable waters was being processed. The U.S. Navy suggested some minor rewording; the Soil Conservation Service (U.S. Dept. of Agriculture), responsible for soil and water conservation, environmental improvement, and agricultural pollution control, replied but had no comments. The U.S. Army itself also replied but had no comment.

The Fish and Wildlife Service* (U.S. Dept. of the Interior) made specific comments on nine paragraphs in the DEIS. Remarks pertained to the following: chlorine effects on marine biota and monitoring; surfacing characteristics of diluted effluent; lack of fish counts and investigations of local fisheries; incorporation of wastewater nutrients into the food web; bird species affected; alternates to blasting and substantiation of statements that blasting has no significant effect on fish life; backfill and soil erosion at land site; turbid water runoff to marine waters.

There were no replies from the following federal agencies with potential interest: Advisory Council on Historic Preservation, responsible for administering the National Historic Preservation Act of 1966; the National Oceanic and Atmospheric Administration (Dept. of Commerce), responsible for the administering of the Marine Protection, Research and Sanctuaries Act of 1972

---

*This agency evaluates the DEIS under the Fish and Wildlife Coordination Act of 1958 and the Endangered Species Act of 1973.

and the Coastal Zone Management Act of 1972; Department of Health, Education and Welfare; Public Health Service; Department of Housing and Urban Development; Department of State, responsible for work extending into international waters; Coast Guard (Department of Transportation); Atomic Energy Commission; Federal Power Commission; Office of Economic Opportunity.

Various Hawaii state agencies replied. The Department of Social Services and Housing commented that there was insufficient information about the incineration and disposal of solids; the Department of Transportation, Harbors Division, asked about modes of transportation of chlorine and suggested mentioning the impact of the proposed line on a proposed state commercial harbor nearby. The Department and Land and Natural Resources (DLNR) related that it was processing a Conservation District Use Application for the project. The DLNR went on to mention the possibility of flooding in the area of the trunk sewer; it also referred briefly to damage to the marine environment during construction and temporary disturbance to commercial and recreational activities in the area. The DLNR then touched on a local fishery and asked about effects of the effluent on marine birds. The Department of Planning and Economic Development gave its support to the site selected. The (state) Department of Agriculture brought up wastewater reuse for agricultural purposes and stated that no benefit-cost analysis of ocean disposal versus reuse had been presented. It presented some figures of its own to point up the possible advantages of such reuse, showing that the value of the wastewater nutrients on land as crop (and then beef) producers was at least two orders of magnitude larger than that for the nutrients incorporated into fish flesh in the sea. The Department of Health made no reply.

Several City and County agencies submitted comments. The Department of General Planning briefly discussed the project in terms of official plans for development of the area and of the local employment market. This agency also asked for consideration of effects of construction blasting on nearby housing structures. The Department of Land Utilization gave its support to the overall project and added a few nitpicking comments. The Department of Transportation Services and Board of Water Supply replied but had no comments.

Academic institutions are notorious for their negativity and nitpicking. The University of Hawaii faculty channelled its comments through its Environmental Center and Water Resources Research Center. There was a series of questions on wastewater flow rates; it was stated that secondary treatment was unnecessary for the effluent passing to the ocean; submergence of the waste field was questioned; there was a comment on species diversity and some remarks on chlorination and sludge incineration; information on the disposal of excavated sea floor materials was solicited.

There was no input to the Barbers Point DEIS by any federal legislators or by any environmental groups.

Public Hearings

Public hearings concerned with ocean outfall systems can sometimes lead to an enormous response and at other times to virtually none.

The ocean outfall at Barbers Point, Hawaii, discussed on pages 138–139, was by all yardsticks a major one. It consisted of a 78-in. diam (1981-mm) reinforced concrete pipe, approximately 10,500 ft long (3200 m), initially discharging advanced primary effluent into 200 ft (60 m) of water with low construction bid (outfall only) at $11.8 million. The public hearing on March 20, 1975, was chaired by the Director and Chief Engineer of the Department of Public Works of the City and County of Honolulu. After a general introduction by four individuals who had played key roles in the design of the system, only four persons testified. In a cordial and largely informal atmosphere, *nothing* was said about the ocean and its margins. Support for the proposed outfall was indicated with only the suggestion that some wastewater reuse might be appropriate. R. M. Towill (1975) reports details.

By contrast, three related public hearings concerning wastewater disposal in the ocean off southeast Florida (see Environmental Protection Agency, 1973) saw the ocean disposal of wastewater attacked by speaker after speaker but with an occasional tentative supporter. The hearing was run in a formal manner by the EPA, but there was a decidedly turbulent atmosphere in the hall. After the introduction by officials in the three meetings, the numbers of speakers were as follows: Lake Worth, 38; Miami Beach, 11; Fort Lauderdale, 10. Some speakers spoke more than once at one session, and some appeared at two sessions. Some obviously had messages from the heart; for some their appearance seemed to be some form of "trip."

An interesting view of the public's views on the environment, and of its immediate reaction to complain about such matters but not to *do* anything constructive, is contained in Hetrick et al. (1974).

4-9 REFERENCES

"Applications for Department of the Army Permits for Activities in Waterways" (1974): U.S. Army Corps of Engineers, *EP 1145-2-1*, (October 1).

BARGMAN, R. D., and W. F. GARBER (1973): "The Control and Removal of Materials of Ecological Importance from Wastewaters in Los Angeles, California, USA," in *Advances in Water Pollution Research*, edited by S. H. Jenkins, pp. 773–781, Pergamon Press, New York, with discussion and reply pp. 783–786.

BARGMAN, R. D., and J. D. PARKHURST (1969): "Evaluation of Sources and Characteristics of Waste Discharges," in *Background Papers on Coastal Wastes Management*, Vol. 1, Chapter 1, National Academy of Engineering, Washington, D. C. (*NTIS Publication PB-198-032*).

BERNHARD, M., and A. ZATTERA (1975): "Major Pollutants in the Marine Environment," *Marine Pollution and Marine Waste Disposal*, edited by E. A. Pearson and E. De F. Frangipane, pp. 195–300, Pergamon Press, New York. [*Proceedings*, Second International Congress of the International Association on Water Pollution Research, San Remo, Italy (December 1973).]

BROOKS, N. H. (1971): "Statement for Presentation to the Public Hearing of the California Water Resources Control Board, Dec. 2, 1971, at San Diego, regarding Proposed General Principles and Provisions for Discharge to Ocean Waters."

BROWN and CALDWELL, Consulting Engineers (1971): "A Predesign Report on Marine Waste Disposal," 2 vols., prepared for City and County of San Francisco at Walnut Creek, Calif.

BROWN and CALDWELL, Consulting Engineers (1975): "A Predesign Report on Marine Waste Disposal," Vol. 4, prepared for City and County of San Francisco at Walnut Creek, Calif. (October).

BROWN, R. L., and L. A. BECK (1972): "A Study of Toxicity and Biostimulation in San Francisco Bay–Delta Waters—Vol. 1. Summary Report," California State Water Resources Control Board, Sacramento, Calif., *Publication 44*.

BROWN, R. L., and G. VARNEY (1972): "A Study of Toxicity and Biostimulation in San Francisco–Delta Waters," Vol. 8, "Algal Assays," California State Water Resources Control Board, Sacramento, Calif., *Publication 44*.

BRUNGS, W. A. (1973): "Continuous-Flow Bioassays with Aquatic Organisms: Procedures and Applications," in *Biological Methods for the Assessment of Water Quality*, edited by John Cairns, Jr., and K. L. Dickson, pp. 117–126, American Society for Testing and Materials, Philadelphia, *Special Technical Publication 528*.

BRYAN, G. W. (1971): "The Effects of Heavy Metals (other than Mercury) on Marine and Estuarine Organisms," *Journal of the Royal Society of London*, Series B, Vol. 177, pp. 389–410.

BURROWS, E. M. (1971): "Assessment of Pollution Effects by the Use of Algae," *Proceedings of the Royal Society of London*, Series B, Vol. 177, pp. 295–306.

CHEN, K. Y., et al. (1974): "Trace Metals in Wastewater Effluents," *Journal of the Water Pollution Control Federation*, Vol. 46, pp. 2663–2675.

CONNELL, D. W. (1974): "A Kerosene-Like Taint in the Sea Mullet, *Mugil cephalus* (Linnaeus). I. Composition and Environmental Occurrence of the Tainting Substance," *Australian Journal of Marine and Freshwater Research*, Vol. 25, pp. 7–24.

CORCORAN, E. F., and J. E. ALEXANDER (1964): "The Distribution of Certain Trace Elements in Tropical Sea Water and Their Biological Significance," *Bulletin of Marine Science of the Gulf and Caribbean*, Vol. 14, pp. 594–602.

COURTRIGHT, R. C., W. P. BREESE, and H. KRUEGER (1971): "Formulation of a Synthetic Seawater for Bioassays with *Mytilus edulis* Embryos," *Water Research*, Vol. 5, pp. 877–888.

CRONIN, L. E. (1969): "Biological Effects on Receiving Waters," in *Background Papers on Coastal Wastes Management*, Vol. 1, National Academy of Engineering, Washington, D.C., Chapter 13 (*NTIS Publication PB-198-032*).

DOUDOROFF, P., and M. KATZ (1953): "Critical Review of Literature on the Toxicity of Industrial Wastes and Their Components to Fish," *Sewage and Industrial Wastes*, Vol. 25, pp. 802–839.

DUTKA, B. J. (1973): "Coliforms are an Inadequate Index of Water Quality," *Journal of Environmental Health*, Vol. 36, pp. 39–46.

EISLER, R. (1973): "Annotated Bibliography on Biological Effects of Metals in Aquatic Environments," U.S. Environmental Protection Agency, Corvallis, Oregon, *Publication EPA-R3-73-007* (February).

Environmental Protection Agency (1973): "Ocean Outfalls and Other Methods of Treated Wastewater Disposal in Southeast Florida," in Transcript of Proceedings held in Lake Worth, Florida, January 24, 1973, Miami Beach, Florida, January 26, 1973, Fort Lauderdale, Florida, January 27, 1973," Atlanta, Georgia (March) (*NTIS Publication EIS-FL-73-0491-F*).

Environmental Quality Commission (1975): "Environmental Impact Statement Regulations," State of Hawaii, Honolulu, Hawaii (June 2).

EPIFANIO, C. E., and R. F. SRNA (1975): "Toxicity of Ammonia, Nitrite Ion, Nitrate Ion, and Orthophosphate to *Mercenaria mercenaria* and *Crassostrea virginica*," *Marine Biology*, Vol. 33, pp. 241–246.

ESVELT, L. A., W. J. KAUFMAN, and R. E. SELLECK (1972): "A Study of Toxicity and Biostimulation in San Francisco–Delta Waters—Vol. 4: Toxicity Removal," California State Water Resources Control Board, Sacramento, Calif., *Publication 44*.

"Federal Water Pollution Control Act" (1975): *Environmental Reporter*, Vol. 71, pp. 5101–5131.

GAMESON, A. L. H. (1975): "Experiences on the British Coast," in *Marine Pollution and Marine Waste Disposal*, edited by E. A. Pearson and E. de F. Frangipane, pp. 387–399, Pergamon Press, New York. [*Proceedings*, Second International Congress of the International Association on Water Pollution Research, San Remo, Italy (December 1973).]

GAMESON, A. L. H., and D. J. GOULD (1975): "Effects of Solar Radiation on the Mortality of Some Terrestrial Bacteria in Sea Water," in *Discharge of Sewage from Sea Outfalls*, edited by A. L. H. Gameson, pp. 209–219, Pergamon Press, Oxford, England.

GUNNERSON, C. G. (1975): "Discharge of Sewage from Sea Outfalls," in *Discharge of Sewage from Sea Outfalls*, edited by A. L. H. Gameson, pp. 415–425, Pergamon Press, Oxford, England.

HALSTEAD, B. W. (1972): "Toxicity of Marine Organisms Caused by Pollutants," in *Marine Pollution and Sea Life*, edited by M. Ruivo, pp. 584–594 [Food and Agriculture Organization of the United Nations, Conference, Rome (December 1970)], Fishing News (Books) Ltd., London, England.

HARREMÖES, P. (1975): "*In Situ* Methods for Determination of Microbial Disappearance in Sea Water," in *Discharge of Sewage from Sea Outfalls*, edited by A. L. H. Gameson, pp. 181–190, Pergamon Press, Oxford, England.

HECKROTH, C. W. (1976): "PL 92-500: Mid-course Correction Grants Options; Next Move up to Congress," *Water and Wastes Engineering*, Vol. 13, No. 5 (May), pp. 28–32.

HELLIWELL, P. R., and J. BOSSANYI, Eds. (1975): *Pollution Criteria for Estuaries*, Pentech Press, London, England.

HETRICK, C. C., C. J. LIEBERMAN, and D. R. RANISH (1974): "Public Opinion and the Environment: Ecology, the Coastal Zone, and Public Policy," *Coastal Zone Management Journal*, Vol. 1, No. 3, pp. 275–289.

JONES, G. E., and A. B. COBET (1975): "Heavy Metal Ions as the Principal Bactericidal Agent in Caribbean Sea Water," in *Discharge of Sewage from Sea Outfalls*, edited by A. L. H. Gameson, pp. 199–208, Pergamon Press, Oxford, England.

KATZENELSON, E., and H. I. SHUVAL (1975): "Viral Pollution Considerations in Marine Waste Disposal," in *Marine Pollution and Marine Waste Disposal*, edited by E. A. Pearson and E. de F. Frangipane, pp. 125–129, Pergamon Press, New York. [*Proceedings*, Second International Congress of the International Association on Water Pollution Research, San Remo, Italy (December 1973).]

KLEIN, L. A., et al. (1974): "Sources of Metals in New York City Wastewater," *Journal of the Water Pollution Control Federation*, Vol. 46, pp. 2653–2662.

KORRINGA, P. (1968): "Biological Consequences of Marine Pollution with Special Reference to the North Sea Fisheries," *Helgoländer wissenschaffliche Meeresuntersuchungen*, Vol. 17, pp. 126–140.

LA ROCHE, G., R. EISLER, and C. M. TARWELL (1970): "Bioassay Procedures for Oil and Oil Dispersant Toxicity Evaluation," *Journal of the Water Pollution Control Federation*, Vol. 42, pp. 1982–1989.

Los Angeles City Board of Public Works (1973): "Technical Report on Waste Disposal to the Ocean," prepared by the Bureau of Sanitation and Bureau of Engineering for the California State Water Resources Control Board (January).

Los Angeles City Board of Public Works (1975): "Rules and Regulations Governing Disposal of Industrial Wastes to the Sanitary Sewer and Storm Drain Systems of the City of Los Angeles," Los Angeles, Calif. (July 29).

LUDWIG, H. F., and P. N. STORRS (1970): "Effects of Waste Disposal into Marine Waters: A Survey of Studies Carried Out in the Last Ten Years," *Water Research*, Vol. 4, pp. 709–720.

LUKIN, L. D., et al. (1971): "Limitation and Effects of Water Disposal on an Ocean Shelf," Florida Ocean Sciences Institute, Deerfield Beach, Fla., Report (December) (*NTIS Publication PB-226-727*).

MASON, J. O., and W. R. McLEAN (1962): "Infectious Hepatitis Traced to the Consumption of Raw Oysters," *American Journal of Hygiene*, Vol. 75, pp. 90–111.

McALLISTER, R. F. (1968): "Demonstration of the Limitations and Effects of Waste Disposal on an Ocean Shelf," Florida Atlantic Ocean Sciences Institute, Boca Raton, Florida, Report (March) (*NTIS Publication PB-215-585*).

McGAUHEY, P. H. (1968): *Engineering Management of Water Quality*, McGraw-Hill, New York.

McKEE, J. E. (1967): "Parameters of Marine Pollution—An Overall Evaluation," in *Pollution and Marine Ecology*, edited by T. A. Olson and F. J. Burgess, pp. 259–266, Wiley (Interscience), New York.

McKEE, J. E., and H. W. WOLF (1963): "Water Quality Criteria," 2nd ed., State Water Quality Control Board, Sacramento, Calif., *Publication No. 3-A*.

MERLINI, M. (1971): "Heavy-Metal Contamination," in *Impingement of Man on the Oceans*, edited by Donald W. Hood, pp. 461–486, Wiley (Interscience), New York.

METCALF, T. G. (1975): "Evaluation of Shellfish Sanitary Quality by Indicators of Sewage Pollution," in *Discharge of Sewage from Sea Outfalls*, edited by A. L. H. Gameson, pp. 75–84, Pergamon Press, Oxford, England.

MITCHELL, R., and C. CHAMBERLIN (1975): "Factors Influencing the Survival of Enteric Microorganisms in the Sea: An Overview," in *Discharge of Sewage from Sea Outfalls*, edited by A. L. H. Gameson, pp. 237–251, Pergamon Press, Oxford, England.

MOOD, E. W., and B. MOORE (1976): "Health Criteria for the Quality of Coastal Bathing Waters," unpublished manuscript, Yale University, School of Medicine (March 30).

MOORE, B. (1975): "The Case Against Microbial Standards for Bathing Beaches," in *Discharge of Sewage from Sea Outfalls*, edited by A. L. H. Gameson, pp. 103–109, Pergamon Press, Oxford, England.

NITTA, T. (1972): "Marine Pollution in Japan," in *Marine Pollution and Sea Life*, edited by M. Ruivo [Food and Agriculture Organization of the United Nations, Conference, Rome (December 1970), pp. 77–81], Fishing News (Books) Ltd., London, England.

NORTH, W. J., G. C. STEPHENS, and B. B. NORTH (1972): "Marine Algae and Their Relation to Pollution Problems," in *Marine Pollution and Sea Life*, edited by M. Ruivo [Food and Agriculture Organization of the United Nations, Conference, Rome (December 1970), pp. 330–340], Fishing News (Books) Ltd., London, England.

"Ocean Plan Review: Status Report" (1976): California State Water Resources Control Board, Sacramento, Calif. (October).

"PL 92-500 to Cost Metal-Finishing Industry $45 Billion" (1976): *Civil Engineering*, Vol. 46, No. 9 (September), p. 84.

PATRICK, R. (1973): "Use of Algae, Especially Diatoms, in the Assessment of Water Quality," in *Biological Methods for the Assessment of Water Quality*, edited by J.

Cairns, Jr. and K. L. Dickson, pp. 76–95, American Society for Testing and Materials, Philadelphia, *Special Technical Publication 528*.

PEARSON, E. A. (1971): "Guidelines for Conduct of Bacterial Disappearance Rate (T-90) Studies for Marine Outfall Design," unpublished paper (April).

PEARSON, E. A. (1975): "Conceptual Design of Marine Waste Disposal Systems," in *Marine Pollution and Marine Waste Disposal*, edited by E. A. Pearson and E. de F. Frangipane, pp. 297–315, Pergamon Press, New York. [*Proceedings*, Second International Congress of the International Association of Water Pollution Research, San Remo, Italy (December 1973).]

PEARSON, E. A., P. N. STORRS, and R. E. SELLECK (1967): "Some Physical Parameters and Their Significance in Marine Waste Disposal," in *Pollution and Marine Ecology*, edited by T. A. Olson and F. J. Burgess, pp. 297–315, Wiley (Interscience), New York.

PERKINS, E. J. (1974): *The Biology of Estuaries and Coastal Waters*, Academic Press, New York.

"Pollution Control Objectives for Municipal Type Waste Discharges in British Columbia" (1975): Province of British Columbia, Dept. of Lands, Forests, and Water Resources, Water Resources Service, Victoria, B.C., Canada (September).

PORTMANN, J. E. (1972): "Results of Acute Toxicity Tests with Marine Organisms, Using a Standard Method," in *Marine Pollution and Sea Life*, edited by M. Ruivo [Food and Agriculture Organization of the United Nations, Conference, Rome (December 1970), pp. 212–217], Fishing News (Books) Ltd., London, England.

R. M. TOWILL Corporation (1972): "Final Design Report, Sand Island Outfall System," prepared for City and County of Honolulu, Hawaii (September 27).

R. M. TOWILL Corporation (1975): "Final Environmental Impact Statement for Honouliuli Wastewater Treatment Plant and Barbers Point Outfall System," prepared for City and County of Honolulu, Dept. of Public Works, Division of Sewers, Honolulu, Hawaii (June).

RAPOPORT, D. (1976): "The Sludge Nightmare," *Parade* (October 31).

RAYMONT, J. E. G., and J. SHIELDS (1964): "Toxicity of Copper and Chromium in the Marine Environment," in *Advances in Water Pollution Research*, Vol. 3, edited by E. A. Pearson, pp. 283–290, Pergamon Press, New York; with discussion and reply. [Proceedings of the First International Conference, September 1962.]

"Report on Pollution Control Objectives for the Forest Products Industry of British Columbia" (1971): Province of British Columbia, Dept. of Lands, Forests, and Water Resources, Victoria, B.C. (September).

SAVAGE, H. P., and N. B. HANES (1971): "Toxicity of Seawater to Coliform Bacteria," *Journal of the Water Pollution Control Federation*, Vol. 43, pp. 854–861.

SCARPINO, P. V. (1975): "Human Enteric Viruses and Bacteriophages as Indicators of Sewage Pollution," in *Discharge of Sewage from Sea Outfalls*, edited by A. L. H. Gameson, pp. 49–61, Pergamon Press, Oxford, England.

SHUVAL, H. I. (1975): "The Case for Microbial Standards for Bathing Beaches," in *Discharge of Sewage from Sea Outfalls*, edited by A. L. H. Gameson, pp. 95–101, Pergamon Press, Oxford, England.

SINDERMANN, C. J. (1972): "Some Biological Indicators of Marine Environmental Degradation," *Journal of the Washington Academy of Sciences*, Vol. 62, pp. 184–189.

"6 Pulp Mills Face Counts" (1977): *The Oregonian*, Portland, Oregon (April 26).

SKORY, L. D. (1976): "Ontario Companies Feel the Sting of Environment Minister Kerr," *Canadian Pulp and Paper Industry*, Vol. 31, No. 2 (February 5), pp. 28–30.

Southern California Coastal Water Research Project (1973): "The Ecology of the Southern California Bight: Implications for Water Quality Management," *Technical Report No. 104* (March).

Southern California Coastal Water Research Project (1975): "Environmental Effects of the Disposal of Municipal Wastewaters in Open Coastal Waters," El Segundo, Calif. (December).

SPRAGUE, J. B. (1969): "Review Paper: Measurement of Pollutant Toxicity to Fish—1. Bioassay Methods for Acute Toxicity," *Water Research*, Vol. 3, pp. 793–821.

SPRAGUE, J. B. (1970): "Review Paper: Measurement of Pollutant Toxicity to Fish—2. Utilizing and Applying Bioassay Results," *Water Research*, Vol. 4, pp. 3–32.

SPRAGUE, J. B. (1971): "Review Paper: Measurement of Pollutant Toxicity to Fish—3. Sublethal Effects and 'Safe' Concentrations," *Water Research*, Vol. 5, pp. 245–266.

SPRAGUE, J. B. (1973): "The ABC's of Pollutant Bioassay Using Fish," in *Biological Methods for the Assessment of Water Quality*, edited by J. Cairns, Jr. and K. L. Dickson, pp. 6–30, American Society for Testing and Materials, Philadelphia, *Special Technical Publication 528*.

Standard Methods for the Examination of Water and Wastewater, (1976): 14th ed. American Public Health Association, American Water Works Association, Water Pollution Control Federation, Washington, D.C.

STEPHAN, C. E., and D. I. MOUNT (1973): "Use of Toxicity Tests with Fish in Water Pollution Control," in *Biological Methods for the Assessment of Water Quality*, edited by J. Cairns, Jr. and K. L. Dickson, pp. 164–177, American Society for Testing and Materials, Philadelphia, *Special Technical Publication 528*.

STEVENSON, A. H. (1953): "Studies of Bathing Water Quality and Health," *American Journal of Public Health*, Vol. 43, pp. 529–538.

VAN NOTE, R. H., et al. (1975): "A Guide to the Selection of Cost-Effective Wastewater Treatment Systems," U.S. Environmental Protection Agency, Office of Water Program Operations, Washington, D.C., *Publication EPA-430/9-75-002* (July).

WALDICHUK, M. (1969): "Effects of Pollutants on Marine Organisms: Improving Methodology of Evaluation—A Review of the Literature," *Journal of the Water Pollution Control Federation*, Vol. 41, pp. 1586–1601.

WARD, P. S. (1975): "NCWQ Releases Staff Findings," *Journal of the Water Pollution Control Federation*, Vol. 47, pp. 2350–2353.

WARREN, C. E. (1971): *Biology and Water Pollution Control*, W. B. Saunders, Philadelphia.

"Water Quality Control Plan for Ocean Waters of California" (1972): State of California, State Water Resources Control Board, Resolution No. 72–45 (July 6).

WISDOM, A. S. (1975): "Legal Controls and Legislation," in *Pollution Criteria for Estuaries*, edited by P. R. Helliwell and J. Bossanyi, Chapter 2, Pentech Press, London, England, with discussion.

WOOD, P. C. (1972): "The Principles and Methods Employed for the Sanitary Control of Molluscan Shellfish," in *Marine Pollution and Sea Life*, edited by M. Ruivo [Food and Agriculture Organization of the United Nations, Conference, Rome (December 1970), pp. 560–565], Fishing News (Books) Ltd., London, England.

WON, W. D., and H. ROSS (1973): "Persistence of Virus and Bacteria in Seawater," ASCE, *Journal of the Environmental Engineering Division*, Vol. 99, No. EE3 (June), pp. 205–211.

YOSHPE-PURER, Y., and H. I. SHUVAL (1972): "Salmonellae and Bacterial Indicator Organisms in Polluted Coastal Water and Their Hygienic Significance," in *Marine Pollution and Sea Life*, edited by M. Ruivo [Food and Agriculture Organization of the United Nations, Conference, Rome (December 1970), pp. 574–580], Fishing News (Books) Ltd., London, England.

4-10 BIBLIOGRAPHY

General

"All You Need to Know about Sewage Treatment Construction Grants" (1976): U.S. Environmental Protection Agency, Washington, D.C. (August).

"A Summary of Knowledge of the Southern California Coastal Zone and Offshore Areas" (1974): U.S. Dept. of the Interior, Bureau of Land Management, Vol. 1, Physical Environment, Vol. 2, Biological Environment (September).

ARGO, D. G., and G. L. CULP (1972): "Heavy Metals Removal in Wastewater Treatment Processes," *Water and Sewage Works*, Vol. 119, No. 8 (August), pp. 62–65; No. 9 (September), pp. 128–132.

BRAHTZ, J. F. PEEL, Ed. (1972): *Coastal Zone Management: Multiple Use With Conservation*, Wiley, New York.

BUELOW, R. W. (1968): "Ocean Disposal of Waste Material," *Transactions of the National Symposium on Ocean Sciences and Engineering of the Atlantic Shelf* (March), Marine Technology Society, pp. 311–337.

CALVERT, J. T. (1975): "The Case against Treatment," in *Discharge of Sewage from Sea Outfalls*, edited by A. L. H. Gameson, pp. 173–179, Pergamon Press, Oxford, England.

"Canada and Environmental Law" (1976): *Water and Pollution Control*, Vol. 114, No. 4 (April), pp. 14–20.

CARTER, L. (1976): "The Disposal of Wastes to Tidal Waters," *Chemistry and Industry*, No. 19 (October 2), pp. 825–829.

CLARK, J. (1974): *Coastal Ecosystems: Ecological Considerations for Management of the Coastal Zone*, The Conservation Foundation, Washington, D.C.

COLLINS, J. C., Ed. (1960): *Radioactive Wastes: Their Treatment and Disposal*, Wiley, New York.

COOK, M. B., and R. SCOTT (1977): "Special Environmental Laws Slow Clean Water Program," *Public Works*, Vol. 108, No. 3 (March), pp. 62–65.

CRONIN, L. E., and D. A. FLEMER (1967): "Energy Transfer and Pollution," in *Pollution and Marine Ecology*, edited by T. A. Olson and F. J. Burgess, pp. 171–183, Wiley (Interscience), New York.

CULP, G. L., and R. L. CULP (1974): *New Concepts in Water Purification*, Van Nostrand-Reinhold, New York.

DRUMMOND, M. (1975): "The Solent as a Recreational Resource," in *Pollution Criteria for Estuaries*, edited by P. R. Helliwell and J. Bossanyi, Chapter 1, Pentech Press, London, England, with discussion.

DYER, K. R. (1973): *Estuaries: A Physical Introduction*, Wiley (Interscience), New York.

FELDMAN, M. H. (1970): "Trace Materials in Wastes Disposed to Coastal Waters: Fates, Mechanisms, and Ecological Guidance and Control," U.S. Dept. of the Interior, Federal Water Quality Administration, Northwest Region, Pacific Northwest Water Laboratory, Corvallis, Oregon, Working Paper 78 (July).

FETTEROLF, C. M., Jr. (1973): "Mixing Zone Concepts," in *Biological Methods for the Assessment of Water Quality*, edited by J. Cairns, Jr. and K. L. Dickson, pp. 31–45, American Society for Testing and Materials, Philadelphia, *Special Technical Publication 528*.

GRANTHAM, G. R., and T. E. BAILEY (1976): "Water Quality Planning Strategy," ASCE, *Journal of the Water Resources Planning and Management Division*, Vol. 102, No. WR1 (April), pp. 11–22.

"How to Obtain Federal Grants to Build Municipal Wastewater Treatment Works" (1976): U.S. Environmental Protection Agency, Washington, D.C. (May).

"Inventory of Data on Contaminants in Aquatic Organisms" (1976): Food and Agriculture Organization of the United Nations, Rome, *Fisheries Circular No. 338* (February).

JOHNSTON, R., Ed. (1976): *Marine Pollution*, Academic Press, London, England.

KINNE, O., Ed. (1972): *Marine Ecology*, Vol. 1, *Environmental Factors*, Parts 1, 2, and 3, Wiley (Interscience), New York.

KINNE, O. (1976): "Cultivation of Marine Organisms: Water-Quality Management and Technology," in *Marine Ecology: A Comprehensive, Integrated Treatise on Life in Oceans and Coastal Waters*, edited by O. Kinne, pp. 19–300, Wiley, New York.

MACKENTHUN, K. M. (1969): *The Practice of Water Pollution Biology*, Federal Water Pollution Control Administration, Washington, D.C.

MCCONNAUGHEY, B. H. (1970): *Introduction to Marine Biology*, C. V. Mosby, St. Louis, Mo.

MCINTYRE, A. D., and R. JOHNSTON (1975): "Effects of Nutrient Enrichment from Sewage in the Sea," in *Discharge of Sewage from Sea Outfalls*, edited by A. L. H. Gameson, pp. 131–141, Pergamon Press, Oxford, England.

"Methods for Chemical Analysis of Water and Wastes" (1971): Environmental Protection Agency, Analytical Quality Control Laboratory, Cincinnati, Ohio.

MINGES, M. C. (1973): "Ocean Disposal Subject to New Controls," *Journal of the Water Pollution Control Federation*, Vol. 45, pp. 782–783.

MUSZYNSKI, W. J., and T. J. OLENIK (1976): "Operation and Impact of NPDES in Region II," *Water and Sewage Works*, Vol. 123, No. 5 (May), pp. 62–65.

NORRIS, D. P., et al. (1973): "Marine Waste Disposal—A Comprehensive Environmental Approach to Planning," *Journal of the Water Pollution Control Federation*, Vol. 45, pp. 52–70.

NORTH, J. E., J. D. LAWSON, and J. S. ROGERSON (1973): "Environmental Impact of Regional Ocean Outfall Sewers in Victoria," *Preprints*, First Australian Conference on Coastal Engineering, Sydney (May), pp. 167–173.

ODUM, H. T., B. J. COPELAND, and E. A. MCMAHAN (1974): *Coastal Ecological Systems of the United States*, 4 Vols., The Conservation Foundation, Washington, D.C. (June).

OLEXSEY, R. A. (1976): "After Ocean Disposal, What?," *Water and Wastes Engineering*, Vol. 13, No. 9 (September), pp. 59–62, 114.

O'SULLIVAN, A. J. (1971): "Ecological Effects of Sewage Discharge in the Marine Environment," *Proceedings of the Royal Society of London*, Series B, Vol. 177, pp. 331–351.

"PL 92-500 Problems and Solutions" (1976): *Water and Sewage Works*, Vol. 123, No. 4 (April), pp. 47–54.

PRESTON, A., and P. C. WOOD (1971): "Monitoring the Marine Environment," *Proceedings of the Royal Society of London*, Series B, Vol. 177, pp. 451–462.

"Quality Criteria for Water" (1976): U.S. Environmental Protection Agency, Washington, D.C., *Publication EPA-440/9-76-023*.

ROSEN, S. J. (1976): *Manual for Environmental Impact Evaluation*, Prentice-Hall, Inc., Englewood Cliffs, N.J.

SELDEN, M., and L. G. LLEWELLYN (1973): *Studies in Environment, Vol. I, Summary Report*, U.S. Environmental Protection Agency, *Publication EPA-600/5-73-012a* (December).

SELLECK, R. E. (1973): "Evaluation of Floatables of Wastewater Origin in the Vicinity of Marine Outfalls," *Proceedings*, Third Annual Technical Conference on Estuaries of the Pacific Northwest (March), Corvallis, Oregon, pp. 104–112.

SIDWICK, J. M. (1976): "The Discharge of Trade Effluents to the Public Sewer," *Journal of the Institution of Water Engineers and Scientists*, Vol. 30 (May), pp. 116–122.

SMITH, G. J. C., H. J. STECK, and G. SURETTE (1974): *Our Ecological Crisis—Its Biological, Economic and Political Dimensions*, Macmillan, New York.

TAIT, R. V. (1968): *Elements of Marine Ecology*, Plenum Press, New York.

"Toward Cleaner Water: The New Permit Program to Control Water Pollution" (1974): U.S. Environmental Protection Agency, Washington, D.C. (January).

UNGER, S. G., M. J. EMERSON, and D. L. JORDENING (1973): State-of-Art Review: Water Pollution Control Benefits and Costs, Vol. I, U.S. Environmental Protection Agency, *Publication EPA-600/5-73-008a* (October).

U.S. Environmental Protection Agency (1973): "The Economics of Clean Water—1973," Washington, D.C. (December).

VAN DER LEEDEN, F. (1975): *Water Resources of the World: Selected Statistics*, Water Information Center, Inc., Port Washington, N.Y.

VONDRIK, ART V. (1976): "Municipal Problems with PL 92-500: Conformance vs. Performance," ASCE, Natural Water Resources and Ocean Engineering Convention, San Diego, Calif. (April), *Preprint 2645*.

"Water Programs, Secondary Treatment Information" (1973): *Federal Register*, Vol. 38, No. 159 (August 17), Part II.

"Water Quality Criteria" (1968): Report of the National Technical Advisory Committee to the Secretary of the Interior (April 1), U.S. Federal Water Pollution Control Administration, Washington, D.C.

WESTON, A. D., and G. P. EDWARDS (1939): "Pollution of Boston Harbor," *Proceedings*, ASCE, Vol. 65, pp. 383–418.

WILLIAMS, F. P., and A. C. RUCKS (1976): "Water Pollution Control: Crossroads of a Municipal Financial Dilemma," *Public Works*, Vol. 107, No. 6 (June), pp. 86–88.

Public Health

AKIN, E. W., W. F. HILL JR., and N. A. CLARKE (1975): "Mortality of Enteric Viruses in Marine and Other Waters," in *Discharge of Sewage from Sea Outfalls*, edited by A. L. H. Gameson, pp. 227–236, Pergamon Press, Oxford, England.

AUBERT, M., D. PESANDO, and M. J. GAUTHIER (1975): "Effects of Antibiosis in a Marine

Environment," in *Discharge of Sewage from Sea Outfalls*, edited by A. L. H. Gameson, pp. 191–197, Pergamon Press, Oxford, England.

BAALSRUD, K. (1975): "The Case for Treatment," in *Discharge of Sewage from Sea Outfalls*, edited by A. L. H. Gameson, pp. 165–172, Pergamon Press, Oxford, England.

BERG, G., Ed. (1967): *Transmission of Viruses by the Water Route*, Wiley (Interscience), New York.

BERG, G. (1972): "Viruses in Water-Current Status," in *Air and Water Pollution*, edited by W. E. Brittin, R. West, and R. Williams, pp. 45–65, Colorado Associated University Press, Boulder, Colorado.

CARTER, H. H., J. H. CARPENTER, and R. C. WHALEY (1967): "The Bactericidal Effect of Seawater under Natural Conditions," *Journal of the Water Pollution Control Federation*, Vol. 39, pp. 1184–1189.

COOK, D. W., and R. W. HAMILTON (1971): "Factors Affecting the Survival of Pollution Indicator Organisms in Estuarine Waters," *Journal of the Mississippi Academy of Sciences*, Vol. 16, pp. 3–10 (*NTIS Publication COM-72-10426*).

"Florida's Sewage Treatment Debate" (1972): *Enfo Newsletter*, Environmental Information Center of the Florida Conservation Foundation, Inc., Winter Park, Fla. (November).

FLYNN, M. J., and D. K. B. THISTLETHWAYTE (1965): "Sewage Pollution and Sea Bathing," *Advances in Water Pollution Research*, Vol. 3, edited by E. A. Pearson, pp. 1–14, Pergamon Press, New York, with discussion and reply, pp. 15–25.

GELDREICH, E. E. (1976): "Fecal Coliform and Fecal Streptococcus Density Relationships in Waste Discharges and Receiving Waters," *Critical Reviews in Environmental Control* (C. R. C. Press), Vol. 6, No. 4, pp. 349–369.

KABLER, P. W. (1954): "Water Examination by Membrane Filter and Most Probable Number Procedures," *American Journal of Public Health*, Vol. 44, pp. 379–386.

KAISER, C. (1976): "More Pollution at Jones Beach," *The New York Times* (July 24), p. 46.

KETCHUM, B. H., J. C. AYERS, and R. F. VACCARO (1952): "Processes Contributing to the Decrease of Coliform Bacteria in a Tidal Estuary," *Ecology*, Vol. 33, pp. 247–258.

KOTT, Y. (1975): "Effluent Quality of Chlorinated Sewage Discharged from Sea Outfalls," in *Discharge of Sewage from Sea Outfalls*, edited by A. L. H. Gameson, pp. 155–163, Pergamon Press, Oxford, England.

METCALF, T. G., C. WALLIS, and J. L. MELNICK (1974): "Environmental Factors Influencing Isolation of Enteroviruses from Polluted Surface Waters," *Applied Microbiology*, Vol. 27, pp. 920–926.

MITCHELL, R. (1969): "Biological Control of Intestinal Bacteria and Viruses," ASCE, *Proceedings*, Civil Engineering in the Oceans II, Miami Beach, Fla. (December), pp. 1241–1249.

MITCHELL, R. (1971): "Destruction of Bacteria and Viruses in Sea Water," ASCE, *Journal of the Sanitary Engineering Division*, Vol. 97, No. SA 4, pp. 425–432.

MITCHELL, R., and H. W. JANNASCH (1969): "Processes Controlling Virus Inactivation in Seawater," *Environmental Science and Technology*, Vol. 3, pp. 941–943.

MOSLEY, J. W. (1975): "Epidemiological Aspects of Microbial Standards for Bathing Beaches," in *Discharge of Sewage from Sea Outfalls*, edited by A. L. H. Gameson, pp. 85–93, Pergamon Press, Oxford, England.

ORLOB, G. T. (1956): "Viability of Sewage Bacteria in Sea Water," *Sewage and Industrial Wastes*, Vol. 28, pp. 1147–1167.

Public Health Activities Committee (1963): "Coliform Standards for Recreational Waters," ASCE, *Journal of the Sanitary Engineering Division*, Vol. 89, No. SA 4, pp. 57–94.

SLANETZ, L. W., C. H. BARTLEY, and T. G. METCALF (1965): "Correlation of Coliform and Fecal Streptococcal Indices with the Presence of Salmonellae and Enteric Viruses in Sea Water and Shellfish," *Advances in Water Pollution Research*, Vol. 3, edited by E. A. Pearson, Pergamon Press, New York.

STEIN, J. E., and J. G. DENISON (1967): "Limitations of Indicator Organisms," in *Pollution and Marine Ecology*, edited by T. A. Olson and F. J. Burgess, pp. 323–335, Interscience Publishers, New York.

THOMAS, H. A., JR., R. L. WOODWARD, and P. W. KABLER (1956): "Use of Molecular Filter Membranes for Water Potability Control," *Journal of the American Water Works Association*, Vol. 48, pp. 1391–1402.

U.S. Environmental Protection Agency (1973): "Pretreatment Standards," *Federal Register*, Vol. 38, No. 215 (November 8), pp. 30981–30984.

U.S. Environmental Protection Agency (1977): "Preparation of Environmental Impact Statements—New Source NPDES Permits," *Federal Register*, Vol. 42, No. 7 (January 11), pp. 2449–2459.

VIND, H. P., J. S. MURAOKA, and C. W. MATHEWS (1975): "The Survival of Sewage Bacteria at Various Ocean Depths," U.S. Navy, Civil Engineering Laboratory, *Technical Note N-1396* (July).

WALKER, J. D., and L. J. GUARRAIA (1975): "Other Factors Determining Life Expectancy of Microorganisms in the Marine Environment," in *Discharge of Sewage from Sea Outfalls*, edited by A. L. H. Gameson, pp. 221–226, Pergamon Press, Oxford, England.

Pollutant Effects on Marine Biota

"A Statistical Evaluation of the Relationship between Toxicity and Species Diversity Index" (1972): "A Study of Toxicity and Biostimulation in San Francisco Bay–Delta Waters," Vol. 2, California State Water Resources Control Board, *Publication 44*.

BAPTIST, J. P. (1966): "Uptake of Mixed Fission Products by Marine Fishes," *Transactions of the American Fisheries Society*, Vol. 95, pp. 145–152.

BARADA, W. R. (1972): "A California Horror Story—Pollution Produces Cancer in Fish," *Skin Diver* (September), pp. 12–15.

BEST, R., et al. (1973): "Effect of Sewage Sludge on the Marine Environment: A Case Study in Liverpool Bay," *Proceedings*, The Institution of Civil Engineers, Vol. 55 (March), Pt. 2, pp. 43–66.

CALDWELL, R. S. (1970): "Toxicity of Pulp Mill Wastes to Selected Marine Organisms and the Characterization of the Biological Communities in the Vicinity of a Proposed Ocean Outfall for Pulp Mill Wastes at Coos Bay, Oregon," Oregon State University, Corvallis, Oregon, Dept. of Fisheries and Wildlife (November 6).

CHEN, C. W. (1970): "Effects of San Diego's Wastewater Discharge on the Ocean Environment," *Journal of the Water Pollution Control Federation*, Vol. 42, pp. 1458–1467.

CHEN, C. W., and R. E. SELLECK (1969): "A Kinetic Model of Fish Toxicity Threshold," *Journal of the Water Pollution Control Federation*, Vol. 41, pp. R294–R308.

COPELAND, B. J. (1966): "Effects of Industrial Waste on the Marine Environment," *Journal of the Water Pollution Control Federation*, Vol. 38, pp. 1000–1010.

DANIEL, D. A., and H. K. CHADWICK (1972a): "A Study of Toxicity and Biostimulation in San Francisco Bay–Delta Waters," Vol. 3, "Acute Toxicity of Discharged Wastes," California State Water Resources Control Board, *Publication 44*.

DANIEL, D. A., and H. K. CHADWICK (1972b): "A Study of Toxicity and Biostimulation in San Francisco Bay–Delta Waters," Vol. 7, "Effects of Wastes on Benthic Biota," California State Water Resources Control Board, *Publication 44*.

DAVIS, C. C. (1972): "The Effects of Pollutants on the Reproduction of Marine Organisms," in *Marine Pollution and Sea Life*, edited by M. Ruivo, pp. 305–311 [Food and Agriculture Organization of the United Nations, Conference, Rome, December 1970], Fishing News (Books) Ltd., London, England.

DEGUCHI, Y., et al. (1975): "Influence of the Treated Sewage on Marine Organisms," The Third International Ocean Development Conference, Tokyo, Japan (August), *Preprints*, Vol. 4, pp. 93–105.

DOMENOWSKE, R. S., and R. I. MATSUDA (1969): "Sludge Disposal and the Marine Environment," *Journal of the Water Pollution Control Federation*, Vol. 41, pp. 1613–1624.

ELDER, D. L., and J. P. VILLENEUVE (1977): "Polychlorinated Biphenyls in the Mediterranean Sea," *Marine Pollution Bulletin*, Vol. 8, No. 1 (January), pp. 19–22.

"Fisheries and Pollution" (1970): *Undercurrents*, Vol. 3, No. 2 (April), pp. 14–16.

HARRIS, R. C., D. B. WHITE, and R. B. MacFARLANE (1970): "Mercury Compounds Reduce Photosynthesis by Plankton," *Science*, Vol. 170, No. 3959 (November), pp. 736–737.

HUME, N. B., C. G. GUNNERSON, and C. E. IMEL (1962): "Characteristics and Effects of Hyperion Effluent," *Journal of the Water Pollution Control Federation*, Vol. 34, pp. 15–35.

JERNELOV, A. (1975): "Heavy Metals in the Marine Environment," in *Discharge of Sewage from Sea Outfalls*, edited by A. L. H. Gameson, pp. 115–122, Pergamon Press, Oxford, England.

JOHANNES, R. E. (1972): "Coral Reefs and Pollution," in *Marine Pollution and Sea Life*, edited by M. Ruivo, pp. 364–375 [Food and Agriculture Organization of the United Nations, Conference, Rome, December 1970], Fishing News (Books) Ltd., London, England.

KLEIN, D. H., and E. D. GOLDBERG (1970): "Mercury in the Marine Environment," *Environmental Science and Technology*, Vol. 4, No. 9 (September), pp. 765–768.

KNAUER, G. A., and J. H. MARTIN (1972): "Mercury in a Marine Pelagic Food Chain," *Limnology and Oceanography*, Vol. 17, pp. 868–876.

KROCK, H-J. (1974): "Toxicity and Biostimulation of Domestic Wastewater in San Francisco Bay," University of Hawaii, Water Resources Research Center, Seminar Series No. 3 (June), pp. 19–27.

KROCK, H-J., and D. T. MASON (1972): "A Study of Toxicity and Biostimulation in San Francisco Bay–Delta Waters," Vol. 6, "Bioassays of Lower Trophic Levels," California State Water Resources Control Board, *Publication 44*.

MASON, J. W. II, and D. R. ROWE (1969): "How Pesticides are Threatening Our Oysters," *Ocean Industry*, Vol. 4, No. 7 (July), pp. 60–62.

McDERMOTT, D. J., D. R. YOUNG, and T. C. HEESEN (1975): "Polychlorinated Biphenynyls in Marine Organisms off Southern California," Southern California Coastal Water Research Project, El Segundo, Calif., *TM 223* (November).

McNULTY, J. K. (1970): *Effects of Abatement of Domestic Sewage Pollution on the Benthos, Volumes of Zooplankton, and the Fouling Organisms of Biscayne Bay, Florida*, University of Miami Press, Coral Gables, Fla.

MEARNS, A. J., et al. (1976): "Chromium Effects on Coastal Organisms," *Journal of the Water Pollution Control Federation*, Vol. 48, pp. 1929–1939.

MENZEL, D. W. (1977): "Summary of Experimental Results: Controlled Ecosystem Pollution Experiment," *Bulletin of Marine Science*, Vol. 27, No. 1, pp. 142–145.

MILEIKOVSKY, S. A. (1970): "The Influence of Pollution on Pelagic Larvae of Bottom Invertebrates in Marine Nearshore and Estuarine Waters," *Marine Biology*, Vol. 6, pp. 350–356.

MITROVIC, V. V. (1972): "Sublethal Effects of Pollutants on Fish," in *Marine Pollution and Sea Life*, edited by M. Ruivo, pp. 252–257 [Food and Agriculture Organization of the United Nations, Conference, Rome, December 1970], Fishing News (Books) Ltd., London, England.

NEUSHUL, M. (1970): "The Effects of Pollution on Populations of Intertidal and Subtidal

Organisms in Southern California," *Proceedings*, Santa Barbara Oil Symposium (December), pp. 165–172 (*NTIS Publication COM-73-10317*).

NORTH, W. J. (1964): "An Investigation of the Effects of Discharged Wastes on Kelp," California State Water Quality Control Board, Sacramento, Calif., *Publication No. 26*.

PARKER, R. R., and J. SIBERT (1973): "Effect of Pulpmill Effluent on Dissolved Oxygen in a Stratified Estuary," *Proceedings*, Third Annual Technical Conference on Estuaries of the Pacific Northwest (March), Corvallis, Oregon, pp. 134–145.

PHELPS, D. K., G. TELEK, and R. L. LAPAN, JR. (1975): "Assessment of Heavy Metal Distribution within the Food Web," in *Marine Pollution and Marine Waste Disposal*, edited by E. A. Pearson and E. de F. Frangipane, pp. 341–348, Pergamon Press, New York [*Proceedings* of the Second International Congress of the International Association on Water Pollution Research, San Remo, Italy (December 1973).]

PORTMANN, J. E. (1975): "Persistent Organic Residues," in *Discharge of Sewage from Sea Outfalls*, edited by A. L. H. Gameson, pp. 123–130, Pergamon Press, Oxford, England.

PRINGLE, B. H., et al. (1968): "Trace Metal Accumulation by Estuarine Mollusks," ASCE, *Journal of the Sanitary Engineering Division*, Vol. 94, pp. 455–475.

RAYMONT, J. E. G. (1972): "Some Aspects of Pollution in Southampton Water," *Proceedings of the Royal Society of London*, Series B, Vol. 180, pp. 451–468.

REISH, D. J. (1970): "The Effects of Varying Concentrations of Nutrients, Chlorinity, and Dissolved Oxygen on Polychaetous Annelids," *Water Research*, Vol. 4, pp. 721–735.

TASLAKIAN, M. J., and J. T. HARDY (1976): "Sewage Nutrient Enrichment and Phytoplankton Ecology along the Central Coast of Lebanon," *Marine Biology*, Vol. 38, No. 4, pp. 315–325.

TEMPLETON, W. L., and A. PRESTON (1966): "Transport and Distribution of Radioactive Effluents in Coastal and Estuarine Waters of the United Kingdom," in *Disposal of Radioactive Wastes into Seas, Oceans, and Surface Waters* [Proceedings of the Symposium, Vienna, Austria (May)], International Atomic Energy Agency, Vienna.

"The Perils of PCB's" (1976): *Time* (May 10), p. 75.

TSAI, C-F. (1973): "Water Quality and Fish Life below Sewage Outfalls," *Transactions of the American Fisheries Society*, Vol. 102, No. 2 (April), pp. 281–292.

U.S. Environmental Protection Agency (1974): "Effects of Pesticides in Water," Report, Washington, D.C.

VERNBERG, F. J., et al., Eds. (1977): *Physiological Responses of Marine Biota to Pollutants*, Academic Press, New York.

WARREN, L. M. (1977): "The Ecology of *Capitella Capitata* in British Waters," *Journal of the Marine Biological Association of the United Kingdom*, Vol. 57, No. 1 (February), pp. 151–159.

ZIEBELL, C. D., et al. (1970): "Field Toxicity Studies and Juvenile Salmon Distribution in Port Angeles Harbor, Washington," *Journal of the Water Pollution Control Federation*, Vol. 42, pp. 229–236.

Thermal Pollution

"Bibliography on Thermal Pollution" (1967): ASCE, *Journal of the Sanitary Engineering Division*, Vol. 93, No. SA 3, pp. 85–113.

CAIRNS, J., JR. (1970): "Ecological Management Problems Caused by Heated Waste Water Discharge into the Aquatic Environment," *Water Resources Bulletin*, Vol. 6, No. 6 (November–December), pp. 868–878.

COLES, S. L., P. L. JOKIEL, and C. R. LEWIS (1976): "Thermal Tolerance in Tropical versus Subtropical Pacific Reef Corals," *Pacific Science*, Vol. 30, No. 2, pp. 159–166.

COUTANT, C. C. (1970): "Biological Aspects of Thermal Pollution. I. Entrainment and Discharge Canal Effects," Chemical Rubber Co., *Critical Reviews in Environmental Control*, Vol. 1, No. 3 (November), pp. 341–381.

DE SYLVA, D. P., and A. E. HINE (1972): "Ciguatera—Marine Fish Poisoning—A Possible Consequence of Thermal Pollution in Tropical Seas?" in *Marine Pollution and Sea Life*, edited by M. Ruivo, pp. 594–597 [Food and Agriculture Organization of the United Nations, Conference, Rome (December 1970)], Fishing News (Books) Ltd., London, England.

DROST-HANSEN, W., and A. THORHAUG (1974): "Biologically Allowable Thermal Pollution Limits," U.S. EPA *Publication No. EPA-660/3-74-003* (May).

FRYE, J. (1970): "Thermal Pollution," *Sea Frontiers*, Vol. 16, No. 2 (March–April), pp. 85–95.

HEINLE, D. R. (1976): "Effects of Passage through Power Plant Cooling Systems on Estuarine Copepods," *Environmental Pollution*, Vol. 11, pp. 39–58.

KRENKEL, P. A., and F. L. PARKER, Eds. (1969): *Biological Aspects of Thermal Pollution*, Vanderbilt University Press, Nashville, Tenn.

NAYLOR, E. (1965): "Effects of Heated Effluents upon Marine and Estuarine Organisms," *Advances in Marine Biology*, Vol. 3, pp. 63–103.

PANNELL, J. C. M., A. E. JOHNSON, and J. E. G. RAYMONT (1962): "An Investigation into the Effects of Warmed Water from Marchwood Power Station into Southampton Water," *Proceedings*, The Institution of Civil Engineers, Vol. 23 (September), pp. 35–62.

PARKER, F. L., and P. A. KRENKEL, Eds. (1969a): *Engineering Aspects of Thermal Pollution*, Vanderbilt University Press, Nashville, Tenn.

PARKER, F. L., and P. A. KRENKEL (1969b): "Thermal Pollution: Status of the Art," Vanderbilt University, School of Engineering, *Report No. 3* (December).

5

Obtaining Pertinent Physical Oceanographic Data

5-1 INTRODUCTION

Orientation

Proper planning for and design of an ocean outfall require extensive physical, chemical, geological, and biological marine data. How such information is used in the various components of a marine wastewater disposal project has been sketched in Chapter 1. Both this chapter and the following one will be devoted to an in-depth discussion of the securing of required oceanographic and related information. This chapter discusses physical data gathering; the chemical, geological, and biological aspects are presented in Chapter 6. The final organization and use of the data will be divided among Chapters 5, 6, and 8.

The gathering of marine data for an outfall project can be handled in one of two ways: the owner or designer can issue a contract to a marine survey specialist, or he can carry out the work himself. In the latter case the owner or designer can lease needed equipment and instruments. Another possibility is that he can use equipment and instrumentation that he has on hand and/or buys for express use in the current project. Then (assuming no losses) such items would be available for future gathering of similar marine data.

There are numerous companies in the United States and in other countries that specialize in marine surveys. It is not always necessary to contract with companies to obtain such information, since many universities have staff members eminently qualified to obtain and process marine data. In fact, for marine data of a biological nature it is probably more usual for university personnel to be retained because companies involved in marine surveys usually concentrate on the physical and geological oceanographic information, such as waves, currents, and sea floor conditions.

It is my strong belief that a representative of the owner or designer should remain in very close contact with his consultant. By "very close" I mean that the representative (1) participate in some of the field work; (2) periodically inspect both at-sea and onshore installations; (3) review the raw data from time to time; and (4) keep abreast of the data reduction processing. The first purpose of a close relationship between the owner/designer and the contractor is that the former can "go to school." He can learn about the equipment and instruments involved, the operations, and the data analysis and presentation to the benefit of his own understanding and the success, with in-house expertise only, of future projects of a similar nature.

The second purpose is to check on the data. Data gathering in the ocean is fraught with so many uncertainties and difficulties that marine surveys often end up with considerably less data than was contracted for—and in some cases with no field information. Properly organized marine survey organizations make allowances in their bidding or negotiations for such contingencies, recognizing that they may have to redo part of the program. However, in some cases the lack of data is not mentioned in the final contract report, or the organization simply states that it was unable to secure some of the information desired, or data are invented.

Some companies and manufacturers will rent marine instruments to prospective users. The instrument manufacturer Hydro Products (San Diego, California), for example, leases various types of instruments along with support from the field operations staff. For organizations with a "one-shot" data need and some in-house technical competence, the instrument-leasing possibility has real promise. In some cases manufacturers offer a lease-purchase plan. Using rented instruments may be the only reasonable way to proceed if time is short.

Oceanographic Instruments

The variable or condition to be quantified is measured by a *sensor*. The value obtained is reflected in the end by the amount of movement of a needle on a dial or meter ("readout"), by the degree of displacement of a writing pen on a chart recorder, or through the code placed on a punched paper tape or magnetic tape unit. If the quantity being measured is varying quickly or moderately rapidly with time, the readout has its limitations. These are not insurmountable, as values can be rapidly read into a tape recorder, for example. But when the element of time becomes important, it is more usual to go to a recorder where the chart or tape being used moves past the writing pen, paper punch, or encoding head.

The chart recorder is an *analog* instrument; it makes a continuous trace of the quantity measured by the sensor. On the other hand, the punched paper tape representation is a *digital* description of the data; the values are given at discrete intervals of time. Since the familiar computer of the present day is a digital machine, the digital type of data presentation is directly compatible with it. Where the computer is to be used to process (large amounts of) data there are enormous advantages in having one's data in a compatible form.

The data gathering–processing–reporting sequence can be comfortably divided into various discrete steps. The first of these is the specification phase where exactly what is to be measured and how the data are to be used are thoroughly thought out. At some point the financial constraint phase is introduced; the ceiling on expenditures for instruments is specified.

During the next phase, descriptive (and price) information on instruments is obtained. Manufacturer's documents are predictably biased presentations on the characteristics and operating capability of their products. McAllister (1968) makes the following appropriate comment: "... like many other commercially available marine instruments (it) needed gross modification for successful use at sea." Comments from users of various instruments are most valuable. Another source of unbiased, pertinent information on marine instrumentation was the U.S. National Oceanographic Instrumentation Center (NOIC), and its publications, during the time it was in operation.

NOIC had various functions some of which were as follows: to operate facilities for the evaluation of oceanographic instruments; to perform laboratory and field testing and calibration of oceanographic instruments for government, academic, and industrial interests; to collect and disseminate data on instrument performance and deterioration; to develop ocean measurement instruments when these were not otherwise available; and to develop equipment needed in the testing and calibration of oceanographic instruments.

For example, the NOIC "Index to National Oceanographic Center Publications" (1976) listed tests on and reports issued as follows: acoustic positioning devices (2); corers (2); current meters (12); pressure sensing systems (7); salinity/conductivity systems (13); sound velocimeters (7); sounder-recorders (12); temperature-sensing systems (13); underwater release devices (9); water quality systems (8); wave measurement systems (5); and miscellaneous (3).

Once a decision has been made on the commercially available sensor-recorder combination to be procured, the order is placed and one or more months typically pass before the system arrives. The user is often so excited to have a new toy to play with in the sea that the checkout and instrument calibration and modification phase of the data-gathering sequence is frequently ignored or treated too rapidly. The last place in the world to try to read an instrument manual and fiddle with a complex bit of instrumentation is in a small boat at the mercy of water and weather.

5-2 WIND AND TIDES

Wind Data

Once current data have been obtained (Section 5-5), it is usually necessary to sort such data so that the separate effects of wind and tides on such currents can be determined. For this reason it is frequently important, in conjunction with current measurements, to have access to wind and tide data for the days and hours when the survey was made. Wind data can either be collected as part of the survey or obtained from meteorological stations. In the latter case, the correspondence between wind conditions offshore and onshore at the station should be studied and perhaps concurrent records obtained from time to time for checking purposes.

A very simple and reliable method of making wind measurements during surveys is to carry along an inexpensive hand-held anemometer, such as the Model HH-A manufactured by Martek Instruments Inc. (Newport Beach, California), and to take measurements of wind speed at frequent intervals. The wind direction can be determined through use of a piece of yarn tied off to a stay on the upwind side of the stationary boat and a hand bearing compass. Some larger boats and certainly oceanographic ships have permanent wind anemometers providing a continuous deck readout of wind speed and direction.

Governmental meteorological agencies are often excellent sources of wind data. In the United States, the Environmental Data Service of the National Oceanic and Atmospheric Administration (Department of Commerce) issues a monthly publication entitled "Local Climatological Data" for major airport-based National Weather Service centers. This publication lists the wind speed and direction at 3-h intervals throughout the month and contains, in addition, information on precipitation, temperature, sky cover, relative humidity, and other parameters.

Even for coastal locations remote from major airports it is entirely possible that state or local agencies maintain a meteorological station in the vicinity. Another possible source is a military establishment. The United States Air Force, for example, obtains wind data at its bases.

The National Climatic Center (Asheville, North Carolina), a part of the Environmental Data Service referred to above, collects, processes, stores, and disseminates not only United States weather information but that for the rest of the world as well. Land stations as well as ocean weather stations and moving ships feed data into the system.

Tidal Variations

It is often unnecessary to gather information on tide-induced changes in water level for a particular coastal area because of the availability of Tide Tables that present predicted sequential information on the timing and heights of high and low tidal water levels. Such tables are only for major or semimajor coastal locations, but since it is more than likely than an oceanographic study for a marine outfall is opposite such a location, the tables often suffice. Even for locations that are not specifically referenced, there are listed corrections that can be applied to the tidal histories at the closest reference station that enable one to represent the local conditions (timing and heights) with fair precision. Care should be exercised in equating conditions at nearby points on opposite sides of a peninsula or inside and outside a bay or harbor. From time to time there still exists the need to make measurements of tidal variations, and such methods will be examined later.

The *Tide Tables 1976—West Coast of North and South America including the Hawaiian Islands* contains detailed tidal variation information for 38 reference stations spread from Cape Horn to upper Alaska and also Hawaii. In turn, 1,502 other locations are keyed to one of the reference ones. Still further detail on the Canadian portion of this enormous stretch of coastline is provided in the *Canadian Tide and Current Tables 1976—Pacific Coast* with 14 reference ports and 216 secondary ports. Similar publications exist for most of the world's coastlines.

The "standard" method of making tide observations is to use a vertical pipe connected to the adjoining marine waters by a small-diameter tube that acts to suppress high-frequency signals (waves) from affecting the water level in the stand pipe (e.g., Cross, 1968). A float gage and chart recorder complete the system.

This type of system, normally employed in ports and harbors by official agencies, requires mounting of some parts above the surface, and such parts may be subject to vandalism. A completely submerged, tide-measuring instrument of pressure sensor type may be preferable. An example of such instruments, of proven practical utility, is the compact Bass Engineering (San Diego, California) Model STG/100. There are two modes of operation for such units. Either the power supply and recording system can be self-contained or it can be located on shore, connected to the underwater instrument by a suitable cable. The latter alternative is attractive when a data display is also required. Two recording possibilities are possible with the self-contained unit, either a small strip chart recorder or a digital magnetic tape cassette recorder. In either case the sensor, attached to a suitable frame, can either be lowered to the bottom from a boat or placed on station by a diver.

5-3 MOORED BUOY SYSTEMS

Introduction

Although (Eulerian) current meters, to be described later, can be mounted on weighted platforms set on the sea floor, it is usual that such meters are mounted on a cable extending between an anchor on the bottom and a buoy at or below the surface. Similarly, wave-measuring devices may be mounted on cables or buoys. Since buoy–cable–anchor systems are of great importance for both current meters and wave sensors, this entire section is devoted to a discussion of such systems with only vague reference to the instruments they support.

Figure 5-1 Navigation buoy and its slack-line mooring arrangement

Buoy–cable–anchor systems are either *slack-line* or *taut-line*. In a slack-line mooring the top part of the system is a buoy floating on the surface. The cable is maintained somewhat loose such that the float is not pulled underneath the surface. The anchor line for a boat, for example, could be referred to as a slack-line mooring. So is the chain used with navigation buoys; see Figure 5-1. In a taut-line mooring the upper portion of the system is a positively-buoyant float that is held beneath the water surface by the anchor and cable, and in turn pulls the cable tight. It is this second type of mooring system that is preferred for mounting current meters; it can also be used for supporting wave sensors. But the slack-line mooring arrangement is also used in both cases, and the discussion to follow will deal with both the taut- and slack-line types.

Boat anchors, such as the Danforth shown in Figure 5-2, are not normally used with moored buoy systems since such anchors dig in best when the pull is

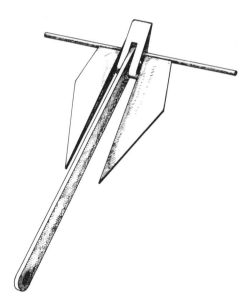

Figure 5-2 Danforth anchor

virtually horizontal.* Certainly in the case of taut-line moors, and often in slack-line moorings, the pull on the anchor is vertical. A very good anchor for such cases, and one that is largely expendable, is one made of concrete, possibly lightly reinforced, with a chain link or other eye extending out of the concrete mass on the top side. Such gravity anchors are often referred to as *clump anchors*. They must, of course, have adequate *submerged* weight to withstand the buoyancy of components in the mooring array with an allowance for environmental forces. Another common anchor for mooring besides the clump consists of links of heavy chain. Still another type that has found some use involves railway car wheels.

The line joining the anchor at the bottom with the buoy at or near the surface is often steel cable, frequently called *wire rope*. In many cases, synthetic fiber rope such as nylon, dacron, polypropylene, or polyethylene is employed. Such lines are light and do not deteriorate appreciably in seawater. Oceanographic measuring devices are either attached to the line being used or onto the float. There are various fittings used in such attachments and in connecting one piece of line to another. The most common is the *shackle*, shown in Figure 5-3.

Since a taut-moored buoy rests below the water surface, it is often necessary to provide a surface marker for the system in the form of a small but visible float connected to the buoy by means of a cord. However, the days of almost

---

*In order to accomplish this, boat anchor lines must have adequate *scope*, the ratio of line length to depth. A figure of 5 : 1 is often quoted as adequate.

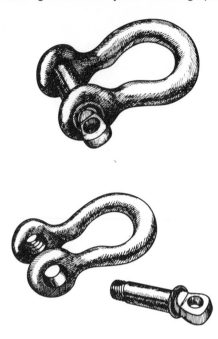

Figure 5-3   Shackle used in mooring cables

any form of surface float are numbered. The likely thing is that someone on a boat will site the float, take a fancy to it, and cut it off.* A coconut moored to a subsurface float by a length of strong, clear fishing leader has been occasionally used successfully in Hawaiian waters as such a marker, a surface marker that no one wants, but one that is difficult to find. Surface floats can be run over and accidentally cut off; they can also be torn off by heavy wave action. Potential problems are such that in many cases in well-traveled waters a taut-moored system is established with no surface marker float. But there are still difficulties.

The Lifetime of Moored Buoy Systems

The quotation from Bathen (1974) in the following paragraph illustrates the types of problems that arise with taut-line current meter stations. In this case there were four such stations in various depths between the 100- and 300-ft (30- and 90-m) contours. No surface markers were used and the top floats lay about 30 ft (9 m) below the water surface. This depth was considered to be a reasonable compromise between the desire to have the float close enough to the

---

*The incident reported in the short article "Precious Equipment Lost" (1973) took place in the Tasman Sea off the Taranaki coast of New Zealand and concerned various buoys marking moored arrays obtaining wave and current data. A trawler went from buoy to buoy and hauled all but one from the water and then departed. A considerable amount of data and some instruments were reported lost as a result.

surface to be easily relocated by a towed diver and the desire to maintain the float well below the draft of local ships.

> Between the months of September and December, 1972, the mooring systems (without current meters) at Stations 2 and 4 disappeared. After exhaustive searches failed to relocate these systems, two more mooring arrays were fabricated, current meters added, were emplaced in early January, and subsequently succeeded in recording data. The problem, however, produced a gap of about twenty days between the installation of meters at Stations 1 and 3 and installations at Stations 2 and 4. A second problem was encountered when the mooring array at Station 1 also vanished during the period of February–April 1973, resulting in no installation of current meters for the last observation period of May, 1973 at Station 1. Following the third deployment period of current meters, the entire array system at Station 3, including meters, was lost. Extensive efforts, including towing of divers to locate anchor, cable, etc., failed to produce the meters. Overall, 65% of the intended data was obtained from the four stations and three depths each.

Whether the problems mentioned by Bathen were due to an inability to relocate the site of a moored float system still intact or were due to malfunctions in the system, leading to the loss of the source of buoyancy and the resultant collapse of the remaining cable and meters onto the bottom, out of sight, is unknown. Many other investigators have lost taut-line mooring systems with or without attached current meters, e.g., Richardson et al. (1963), Wyrtki et al. (1967, 1969), Stalcup and Metcalf (1972), and Halpern et al. (1974).

There are many factors to guard against in keeping a moored buoy system, whether taut- or slack-line moored, intact. These can be summarized point by point as follows:

1. There should be no direct contact between dissimilar metals. Such contact leads to the development of a galvanic cell with possibly rapid erosion of one (the more anodic) metal. The use of sacrificial anodes might be considered in some cases. An electrically isolating element should be used when it is absolutely necessary to couple different metals. It has been reported that a 5/16-in. (8-mm) mild steel shackle coupled directly to type 316 stainless steel will corrode away in a month [*Operation...* (n.d.)]

2. Steel wire rope will corrode on its own with or without a protective zinc coat. Such rope, often used in buoy moorings, is frequently used protected by an outer hard plastic coat, normally nylon or polyethylene with 1/32-in. (0.8-mm) thickness. In such a case the terminations of the rope must be very carefully designed to be 100% watertight. If no protective coating is employed, allowance must be made for strength reduction as a result of corrosion

3. When a steel wire rope goes slack and then is again put in tension, it is possible that loops of the cable, *kinks*, are pulled tight. This creates a weak point in the line. Nylon-covered steel wire rope tends to avoid kinking.

Swivels in the line also tend to guard against this danger as does a *torque-balanced cable* design (Berteaux, 1976)

4. Abrasion of rope as it is constantly pulled back and forth over the sea floor and chafing at points of attachment are both sources of mooring cable failure. One way of avoiding the abrasion problem on the sea floor for slack-line moorings is to use a length of polypropylene line, which is positively buoyant, next to the anchor. Another approach is to place a small float on the rope 10–15 ft (3–5 m) from the anchor or chain. This also avoids having the mooring line wrapped around the clump anchor as the buoy moves about

5. In my experience, I know of more failures due to (threaded) shackle pins backing out under the continuous shaking of a mooring system than any other. Obviously, such a development separates the two parts of the mooring system on either side of the shackle; the buoyant portion drifts away and the lower part collapses to the bottom. Shackle pins, and other threaded connections, if they must be used, should be pinned or wired in place. Attention must then be paid as well to the type of pin or wire being used, for galvanic action could corrode these away in short order. Nylon cord is useful for securing threaded items. "Aircraft" or "safety shackles" have both nuts and cotter pins

6. The cable can snag on the bottom (e.g., around a wreck, under the anchor) and break

7. If a surface buoy is used, or if a subsurface buoy is not sufficiently deep under the water surface, the cable can easily be accidentally cut by a passing ship. If the float buoy is on the surface, it should be equipped with a strong flashing light, such as a strobe. Otherwise, if possible, such a buoy can be placed in water too shallow for shipping, near wreck buoys, or near a manned station such as a light ship or drilling platform, to reduce chances of collision

8. Surface buoys are not infrequently pilfered. If the buoy is actually a valuable item such as a floating wave-measuring device, it is probably wise not to make the reward for turning in such an item too large. Figures used in the Netherlands and Denmark, for instance, are 2 and 4%, respectively, of the value of the buoy

9. Surface buoys have been used by yachts in storms with the result that the overloaded cable breaks.

10. There is sufficient evidence that fish bite buoy cables. It is doubtful that fish bite could injure wire rope, but it could expose plastic-covered steel wire rope to the corrosive sea environment. Fish bite could account for the direct failure of synthetic fiber ropes

11. It is widely understood that numerous materials become weakened when subjected to many cycles of elevated stress. This situation is known as

fatigue. Normally, fatigue is only thought of as a problem when a member experiences of the order of a million stress cycles, but this is easy in the ocean where a million stress cycles can occur because of surface wave action in about two months. Many more cycles may result from vortex shedding. The corrosive nature of sea water adds to fatigue still another dimension called *corrosion fatigue*, in which corrosion gains entry to the interior of a member by way of minute cracks opened up as the member is cycled

Notification Regarding Moored Buoy Systems

The U.S. Coast Guard issues weekly publications called "Local Notice to Mariners" from each of its District offices. The publication is the way in which the Coast Guard notifies the private, commercial and governmental sectors with boating interests of such matters as alterations in the charted positions of buoys, of electronic navigation aids not operational, of changes in charted bathymetric information because of sand shoaling, of construction and regatta activities that could hamper normal boat traffic, and of the placing of new items in navigable waters. The placing of moored buoy systems in the sea, whether the buoys are on or below the surface, falls within the latter category.

The Coast Guard should be notified of the planned installation of a moored buoy system. Two examples taken from the Local Notice to Mariners issued by the Commander, Fourteenth Coast Guard District, Honolulu, Hawaii, in 1975 and 1976, respectively, are the following.

II.   *AIDS ESTABLISHED, DISCONTINUED OR CHANGED*:

HAWAIIAN ISLANDS—KAUAI—BARKING SANDS:

The Pacific Missile Range Facility at Barking Sands, Kauai has advised that an international orange sphere buoy with a six foot antenna mounted on top will remain temporarily established in position 22-12N 169-51W until about 15 January 1976.

Ref: Chart No. 19381 (C&GS 4100)

II.   *AIDS ESTABLISHED, DISCONTINUED OR CHANGED*:

HAWAIIAN ISLANDS—HAWAII ISLAND—KEAHOLE POINT:

The University of Hawaii has advised that underwater buoys will be established for approximately 18 months commencing 20 March 1976. The 3 sets of buoys will be positioned on a line due west (magnetic) of Keahole Point Light (LLNR 3694) at 100 feet, 200 feet, and 2000 feet depths. These subsurface buoys will not interfere with normal vessel traffic but will have small orange surface marker buoys attached by line.

Ref: Chart No. 19327

Deploying Moored Buoy Systems

It was intended that the three subsurface buoys referred to in the preceding paragraph be located about 30 ft (9 m) below the surface. Planting taut-line moored systems so that plans are strictly achieved is difficult as evidenced by what really did happen to those three systems.

The first was installed without much trouble, but in 130 ft (40 m) of water so that the float was 60 ft (18 m) below the surface. When the second system was dropped, the subsurface float ended up more than 100 ft (30 m) below the surface, much too deep for easy locating by divers and dangerously deep for disconnecting an electrical cable mated to a connector on the buoy. The concrete clump anchor for the third buoy was eased over the side but the wire rope allowed to pay out over the gunwale of the Coast Guard ship rather than over a block (pulley). The result in this case was that the wire parted and the anchor and the already paid-out line disappeared into the blue Pacific.

This series of events underlies the difficulty in establishing moored buoy systems as designed. Thus mooring system failures are not limited to developments after the system is put in place, but in addition to the care exercised in placing the system. Failures are also possible during retrieval operations.

The mooring assembly should be put together on a trial basis, if at all possible, before the ship leaves port. In addition, members of both the scientific party and the boat crew should receive a thorough briefing on procedures and their responsibilities and functions.

The deploying of moored buoys and associated equipment can be done in two ways. First, the anchor is placed in the water as the initial step and then the cable and attachments are paid out as the anchor is lowered to the bottom. Second, the top buoy is dropped over the side, the cable and attachments are paid out, with or without the help of a small skiff, as the vessel eases forward, and then the anchor is dropped over the side when the laying vessel arrives at the desired point. The first alternative is classified as an *anchor-first* technique, the latter *anchor-last*. The former generally requires heavier ship equipment and a longer deployment time.

Neither approach is perfect. In the anchor-first technique, movement of the laying ship with prevailing currents can cause the anchor to be set downcurrent, even though placed in the water in the appropriate position, since part of the system is still aboard the ship. Allowance for this tendency can of course be made by dropping over the anchor initially an appropriate distance upcurrent of the desired site. In the anchor-last approach the anchor is released directly over the intended location, and since there is no part of the string in the vessel there is no ship-induced influence. However, wind and currents can make a mess out of cable system spread out along the water surface.

An anchor-last technique used in setting current meter mooring systems during University of Hawaii oceanographic work in which I participated was to make use of a type of teeter-totter set at the stern of the 90-ft (27-m) vessel used in the project. This teeter-totter, a steel box 8 ft (2.4 m) long and 3 ft (0.9 m) wide with the after end open, was used to launch the 1,500 lb (6,700 N) concrete clump anchors used with the arrays. A clump was placed in the forward part of the teeter-totter. Current meters, buoys, and the cable were streamed in the wake of the vessel until a precision fathometer indicated that the proper depth had been reached. Then the forward end of the teeter-totter was raised using a winch and cable through a block on one of the ship's booms, and the clump disappeared into the depths with the remainder of the system in close pursuit.

It is very important to make the drop in the proper depth. If the water is too deep, then it may be virtually impossible for the array to be relocated by towed divers at a later date as noted earlier. On the other hand, a depth too shallow means that the top of the mooring system may not be below possible ship drafts. In the extreme, the top float and current meter may end up on the surface. In such an event divers must usually enter the water to secure a line to the system so that it can be towed into deeper water. A reason for moored buoy systems to be dropped in improper depths may result from the improper value of the average acoustic velocity over the site depth involved being used in the fathometer (see Chapter 6). Another reason is that many sea floor areas are highly convoluted.

Vessels for Deploying and Retrieving
Moored Buoy Systems

The handling and deploying of a $\frac{1}{2}$-ton or more concrete cube with its associated gear, is by no means child's play and requires a ship of 50 ft (15 m) or more, equipped with a boom and winch. The boom, when fully extended, must be capable of handling the snap loading of the clump due to an unruly seaway. The rigging must also be capable of standing up under such a strain.

In Hawaii, we have made frequent use of the type of vessel shown in Figure 5-4, an Alaskan purse seiner (fishing boat). Such vessels are 55 ft (17 m) long, 16 ft (5 m) in the beam, and displace 48 tons. There is ample work space aft, and booms plus a winch for the raising and deploying of heavy gear. Another arrangement used on many work boats, especially those heavily used for oceanographic purposes, is an "A" frame at the stern used in conjunction with a winch behind the pilot house.

A good boat in the hands of a poor skipper may be next to useless; a poor boat in the hands of a good skipper may still be able to get the job done; a good boat in the hands of a good skipper is the way to get things done properly. In addition, a good skipper usually "commands a right good crew."

The vessel to be used in setting a mooring must have the requisite handling equipment referred to earlier. In addition to skipper and crew competence, other elements of importance are the following: usable deck space, electrical power, maneuvering ability (single or twin screw), anchoring capability (bow and stern),

Figure 5-4 Purse seiner used in nearshore oceanographic work

and navigation and safety equipment (e.g., radar, fathometer, radio) in good working order.

Stickland and Bigham (1976) detail design, preparation, launching, and recovery considerations for several types of mooring systems. Their proposals were based on extensive experience with "small vessels" of opportunity, 32–65 ft (10–20 m) in length. The system used for a taut-line mooring for current meters involved a current meter string and ground line, since the system was grappled (dragged) for in the recovery operation. The anchor-last approach was employed in launching.

Although the use of small ships has been proposed in the foregoing, it is possible for some moored current meter systems to be designed so that they can be deployed, serviced, and retrieved by two men in a small boat such as a Boston Whaler. Every reasonable effort should be made to achieve this end, since even small ships command a considerable rental sum.

Retrieving Moored Buoy Systems

There are many ways of establishing a position out on the water as described in Appendix A. In the case of mooring systems, compass bearings on landmarks or lining up two objects on shore, called *ranges* or *lineups*, may work satisfactorily if radar and standard electronic positioning systems are not mounted on the boat being used. Horizontal sextant angles are also valuable, but of course can be used only when there are conspicuous objects on shore. For the relocation of a station out on the water, it is not necessary that the sighting points be shown on charts.

In some cases, seemingly trivial exercises can assist considerably in the relocation of a current meter array. A series of photographs of the shoreline taken at the time of deployment of the mooring system has been used to advantage; so too have pencil sketches of the neighboring coast done at the same instant.

When divers are used in the retrieval of mooring systems, the concrete clump anchor and much of the cable leading from it are considered expendable. A SCUBA diver releases a pelican release hook (Figure 5-5) under the current meter, permitting the float and meter to rise to the surface where they can be picked up and put aboard the attending ship. Since such systems may have considerable buoyancy, the ship should stand well clear so as not to risk being holed by the rapidly rising system or not to damage the system. Current meters in water depths too great for diving can be released from the anchor and cable by time-release or acoustic-release instruments. These devices, many tested by the NOIC, have great utility. Because malfunctions can occur, two release instruments should be used in parallel so that actuation of only one still sends the system to the surface.

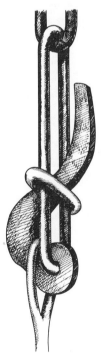

Figure 5-5 Pelican release hook used in mooring cables

The InterOcean Systems, Inc. (San Diego, California) Model 5000T Timed Release is an example of a commercially available buoy system that functions after preset duration. In the case of the Model 5000T these times can be chosen from 1 to 9,999 h (roughly 14 months). The unit operates on standard 9-V dry cells and has a 5,000-lb (22,000-N) tensile load capability. Timed releases can pose problems if for some reason a boat is unable to rendezvous with the surfacing package as planned.

The AMF Sea-Link Systems (Alexandria, Virginia) Models 314 and 255 are releases that function upon receipt of an appropriate acoustic code from a vessel. *Pingers* (underwater acoustic sources) are mounted on the AMF releases to assist in their recovery.

If a submerged float is to be used for a moored system, and is to be recovered in place, the possibilities of using a pinger mounted on the float should be investigated in order to ease the recovery operation. The pinger can operate continuously—but of course this is a drain on the power source—or can be set up to function only when it receives an appropriate signal from the search vessel.

Another way that current meters set in water too deep for diver work can be retrieved is to use a technique described by Bathen (1974). Sliding frames fitting around taut-line cables and handled by other lines can be lowered and raised by divers near the top float. The sliding frame used by Bathen (1974) consisted of two coaxial pipes. The inner one of these was attached to the retrieval cable. The outer pipe, free to rotate about the inner one, was fitted with a bracket having provision for shackling on the General Oceanics, Inc. (Miami, Florida) Model 2010 current meter (Section 5-5). The sliding frame assembly thus allowed the current meter and frame to continuously adopt a downcurrent orientation without the retrieval cable becoming fouled.

5-4 WAVES

Introduction

Information on waves is important in various areas during the planning, design, and construction sequence for a marine outfall. First, as will be discussed in Chapter 7, wave conditions strongly influence the speed with which a cloud of diluted wastewater spreads to neighboring areas. Second, wave conditions dictate the manner in which the outfall is protected on the ocean floor (see Chapter 8). Third, the wave climate at a given location may have a very strong bearing on construction methods as well as construction costs, since work may be impossible for extended periods of time (see Chapter 10).

It is by no means always necessary that the outfall designer set up a system for the gathering of wave data to be used in the outfall design and in the construction planning. Vast amounts of wave data have already been gathered in various coastal locations across the world and on the high seas. A report of the latter type, using data exclusively from ship reports, is Hogben and Lumb (1967), a very valuable compilation of observations for various oceanic areas over the earth's surface. Such observations can be used, with techniques that will be described later, to obtain an estimated wave climate at a location in coastal waters that lies within the area for which the Hogben and Lumb wave data are applicable.

A vast store of wave data has been gathered, and continues to be gathered, for the coastline of the United States by the Coastal Engineering Research Center (CERC) (e.g., Thompson and Harris, 1972; Harris, 1972; Peacock, 1974; Thompson, 1974). Wave data for specific locations along the ocean coastlines of many other countries are being obtained on a regular basis. Examples of such efforts are as follows: Canada—Baird et al. (1971), Baird and Glodowski (1975); Australia—Lawson and Abernethy (1975); Norway—Houmb (1973). Other sources of wave data are referenced in Chapter 8.

Wave Measuring Systems

Introduction: Wave-measuring devices can be separated into four general groups. The first of these involves sensors that look down on the water surface from above. For example, there are radar and infrared devices that can be mounted on offshore platforms. Laser systems can operate from airplanes. This class of sensors would normally be out of the question for any organization unless it had available an offshore platform or suitable plane plus substantial financial resources. An organization meeting these requirements is the U.S. Navy.

The other three groups all are possibilities for gathering wave data for marine outfall-related projects. The first involves sensors that float on the water surface, the second devices that operate from beneath the water surface, and the third, those that extend through the surface. The premier sensor in each group will be discussed in what follows.

Waverider: The most actively used wave measuring system worldwide, the Waverider by Datawell (Haarlem, Netherlands), is the outstanding member of the first class. Externally, the Waverider consists of a stout spherical buoy 0.7 m in diameter designed to float on the surface, a 2.0-m antenna mounted on the top of the buoy, and a mooring arrangement to maintain the buoy on station (see Figure 5-6). The buoy can be operated without a mooring system, but such a situation would be unusual. The mooring system connects to an eye on the underside of the buoy. A swivel at the underside of the buoy connected to the mooring system reduces the tendency of the buoy to spin. Spurious readings and possible breakdown can result from spinning of the buoy.

As the sea surface rises and falls, the Waverider rises and falls along with it. Such motion results in vertical acceleration of the buoy. It is this quantity that is sensed within the buoy housing by a stabilized accelerometer suspended in a plastic sphere placed at the bottom of the buoy. The buoy, hence the water, displacement history can then be retrieved by twice integrating (electrically) the

Figure 5-6 Datawell Waverider buoy (Courtesy, Datawell bv)

acceleration history. The internal workings of the buoy housing are such that horizontal-acceleration effects are minimized, with the horizontal sensitivity reportedly less than 3% of the vertical.

The Waverider has been designed to transmit the information it measures to shore (up to 50 km) via an FM signal. A shore unit provides for recording of the FM signal on analog magnetic tape as well as display of the translated signal on a strip-chart recorder.

The Maritime Services Board of New South Wales maintains a Waverider in 75 m of water offshore from Botany Bay immediately south of Sydney, Australia. This buoy was placed on station April 1, 1971, and up to mid 1976 the system displayed a phenomenal 87% reliability. Wave records are taken four times daily for 20 min, using a 0.5-s sample interval, and the data are punched onto paper tape. The buoy is retrieved twice per year and recalibrated by moving it at a given period around a circular path of set radius in the laboratory. The greatest error ever found in such recalibration was −8% (Lawson and Abernethy, 1975), an excellent result.

The buoy setup used by the Maritime Services Board is shown in Figure 5-7. The overall length of mooring is 1.75 times the water depth. A rubber cord keeps the buoy from being pulled under the surface by all but very strong currents and at the same time possesses enough resilience for the buoy to follow the waves. The length of line running parallel to the rubber cord aids in preventing the buoy from being detached when the line is intercepted by a fishing trawl.* The extra line allows the rubber cord to work but stops excess stretching from pulling out of the terminals under a heavy pull. The preventer line is secured to the rubber cord by light rope. On the average, the Botany Bay buoy is broken loose by trawlers four times per year. When the Maritime Services Board established another Waverider station offshore from Newcastle,

---

*Some details on commercial fishing trawls are provided in Chapter 6.

100 miles (160 km) north of Sydney, they moored it adjacent to a wreck in a successful attempt to keep it out of the path of trawlers.

The largest wave measured by the Waverider off Botany Bay was during a storm on June 21, 1975, when a wave 50 ft (13.7 m) in height passed the station. The period of this wave was 11.7 s.

The Waverider was specified for wave measurements starting in 1974 in the stormy Gulf of Alaska. Extensive details on the systems and approaches, as well as the problems encountered, are available in McLeod et al. (1976) and McLeod (1976). A group of three separate Waveriders was used in various locations, usually in depths of about 100 fathoms (180 m).

Two items from that project are of particular note. First, a 0.9-m buoy was fabricated and tested, since the standard 0.7-m buoy had insufficient buoyancy to overcome strong currents. Second, on fifteen occasions in the first five months of data gathering a buoy and anchor were separated. Corrosion seemed to be a major problem.

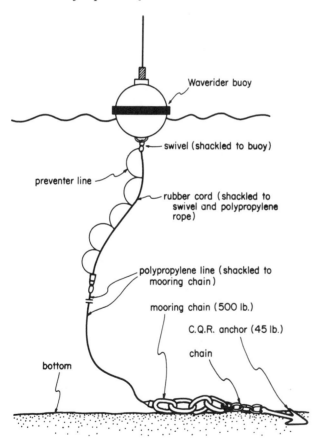

Figure 5-7 Mooring arrangement for Datawell Waverider

Pressure Sensors: Pressure sensors have taken several different forms. One type (vibratron) features a vibrating tungsten wire under tension, whose natural vibrational frequency changes as a function of the pressure-induced axial force applied to it. Walther and Lee (1975) detail the use of such a system. A second type, one I have used in my own research (Fig. 5-8), involves a chamber with yielding walls. Strain gages mounted in these walls sense the expansion and contraction of the chamber as a result of variations in the wave-induced pressure on it. A *strain gage* is a small electrical resistor mounted on a suitable backing material that is in turn cemented to a surface undergoing deformation due to loading. As the gage length changes in concert with the loading, its electrical resistance changes. Such changes can be detected by suitable electronic circuitry and correlated with applied loads.

There are commercially available pressure systems (e.g., Hydro Products, San Diego, California) wherein the pressure data are recorded on a chart within a sealed housing that is a part of the whole subsurface package. Such systems are normally capable of unattended operation for one month. After that time both the chart and batteries require replacement. A disadvantage of such a system is that there is no way of checking that the unit is working properly between site visits, when the instrument is brought to the surface and taken apart. Great care must be taken in properly sealing the unit.

The most usual case combines a pressure sensor with a cable running to recording instruments on shore. The disadvantage of this system is the cable. Even if heavily armored, the lifetime of such systems seldom exceeds two years,

Figure 5-8 Pressure sensor and base for wave mast

particularly in areas of heavy wave action. Corrosion at the point where the cable leaves the sea can be severe. This effect can be somewhat guarded against by tarring that portion of the cable before laying.

A sizable advantage of having the recorder on shore is that it is always possible to determine whether or not the sea unit appears to be operating satisfactorily. Then, too, wave data can be collected or not collected as the conditions warrant.

The major operating advantage of the wave-pressure system is, in my view, that the sensor is on the bottom, out of reach of boat traffic and vandals. Even then, on one occasion, I have seen a pressure sensor system ruined by inquisitive divers. I am convinced, incidentally, that data-gathering systems for the ocean must be as carefully designed against the human element (scientist or vandal) as against the rigors of the sea.

Assume that a pressure sensor is located on or near the bottom with a wave crest directly over it. Let the distance of this crest above the still-water level be η_c as shown in Figure 8-17. The wave pressure-head departure from normal at the sensor is not η_c, because vertical accelerations of the water column invalidate the usual hydrostatic pressure approach. The pressure-head variation under the wave would be less than η_c. How much less turns out to be a severe problem. It is clear from measured data that the Airy wave theory (Chapter 8) does not accurately estimate the ratio between η_c and the associated pressure head at the sensor (Grace, 1970). Nor are other theories adequate. So the major problem with the pressure sensor is that the prediction of a surface wave climate from a wave-pressure record is subject to some unknowns. The curve in Figure 7-8 is provisionally suggested for pressure "height"–surface height computations.

A system of relatively inexpensive wave-pressure stations was initially set up in 1975 by the California State Department of Navigation and Ocean Development and Scripps Institution of Oceanography (Seymour et al., 1976). Ultimate plans called for one hundred or more stations off California, each with three sensors, such that the directional attributes of the waves could be obtained in addition to the usual height and period information.

The per station cost was in the area of several thousand dollars. The individual systems were designed to eliminate undersea connectors, to provide for FM transmission of the pressure information right from the sea end, and to use standard telephone lines for on-land transmission of the signals to a computer that dialed each station periodically—initially once every 10 hours.

The pressure information was digitized at 1-s intervals, and each sensor recorded for about 17 minutes, providing 1024 data points, a convenient number for fast Fourier transform techniques for obtaining wave-pressure spectra.* The pressure spectrum is converted to a surface wave spectrum through use of the

---

*Useful references on spectra are Blackman and Tukey (1959), Bendat and Piersol (1971), and Kinariwala et al. (1973).

linear wave theory curve despite the fact that this curve appears to differ somewhat from reality for both long and short waves.

Resistance Wave Staff: There are various types of vertical rods passing through the sea–air interface that can be used as wave sensors. All of the electrical principles of resistance, inductance, and capacitance have been used, as well as others (e.g., Zwarts, 1975), but only the resistance type will be covered here. This category can be further subdivided into the *continuous resistance* and *step resistance* classes.

I have used a commercially available continuous wave staff in ocean research. This staff consists of a plastic-jacketed high-strength steel cable with a helically-wound nicrome wire wrapped over its length and embedded flush with the plastic insulation surface. Eyes at both ends of the cable permit mounting on compatible fittings. The central cable is also an electrical ground bar, since it comes in direct contact with sea water. The lower end of the nicrome wire is connected to the lower end of the ground bar so that a circuit will be completed down through the nicrome wire and back through the cable. The resistance of the staff in question is about 100 ohms per foot (30 ohms/m) so that the 15-foot (4.6-m) staff used has a loop resistance of approximately 1500 ohms.

As sea water rises along the staff, the electrical circuit through the nicrome wire to ground decreases. Since this decrease is linear with staff immersion, any means of accurately charting the resistance changes can be used to produce a record of water level variations. Various approaches are used to this end by manufacturers. A relatively simple method is the use of a potentiometer setup in which the staff is placed in a circuit with a high-resistance series resistor. Voltage changes across the staff are then used to reflect water level fluctuations.

It was stated earlier that there are real data analysis problems with wave-pressure data because of the largely unknown relationship between the height of a wave and the total pressure head variation caused at some depth of submergence where a wave pressure sensor is located. The wave staff has the enormous advantage that there are no unknown conversion factors to estimate; the staff really measures the surface wave height.

When I began an ocean study of wave-induced forces on a simulated pipeline in 1974, I was well aware of the desirability of having a better method of obtaining individual wave heights than the wave-pressure approach that we had used in earlier projects. But our site was 1400 ft (430 m) from shore, in 37 ft (11.3 m) of water. There was no structure anywhere near and we had neither the funds nor the patience to go through the red tape necessary to build our own platform. I got my wave staff by constructing a tiltable wave mast upon which a staff could be mounted. Half the water depth was occupied by a vertical pipe fixed to a heavy base composed of steel beams (see Figure 5-8). Pinned near the top of the permanent mast section was another length of pipe. A counterweight at one end of this pipe and a buoyancy tank part way along it allowed us to raise the tiltable portion to a vertical position on a data-taking day and to lower it so that the free end rested on the bottom when the day was over. For data-gathering projects that have divers available who don't mind working under conditions of heavy wave action, this tiltable mast idea may have promise.

Our continuous-wire wave staff was mounted at the top of the wave mast, with wires running from it to a four-point-moored boat near the mast. Power supply, instrumentation, and recorders were housed on the boat. The boat served a dual role as a data-recording station and a marker buoy for the mast and staff.

There are no commercially available step resistance gages as far as I know. However, Williams (1969) has very thoroughly described the design (as well as fabrication, calibration, operation, and maintenance) of three different types of step resistance gages, one for fresh water, one for salt water, and the third for either. With the information given, it should not be a difficult task for a competent technician with adequate facilities and supplies to fabricate a staff in a moderately short time.

Available Systems: There are many, many manufacturers of wave-measuring instruments. It is very strongly suggested that those seeking wave data use commercially available devices or use only those self-fabricated devices that have already been tried and tested and for which detailed plans are available (e.g., Williams, 1969). Too many wave-data projects have come to nought because both time and resources were taken up in wave sensor development.

Data Processing and Presentation

Assume that a 20-min wave record is obtained once every six hours. This is the approach of CERC, for example (Thompson, 1974). If an average wave period of 6 s is assumed, then during a year's recording there will be approximately 5×10^6 waves to process. At 4 samples per second, the annual number of data points is roughly 10^8. With such a data reduction load, it is virtually mandatory to use a digital computer for data processing. For this reason, much thought should be given to the complete data measuring and recording system.

Recorders using punched paper or magnetic tape are computer-compatible with or without an intermediate step. The Waverider system, for example, comes with a punched-paper tape recorder. But a digital recording system should never be used by itself. For monitoring purposes, and as a backup in the event of instrumentation malfunction or difficulty in data interpretation, an analog strip-chart recorder is a must. The Waverider system also comes with a chart recording feature.

I believe that there exists a potential danger in the desire to have wave data in a computer-compatible form. So much money may be allocated from a limited budget to obtain a first-line data processing system that there may be insufficient funds to purchase a first-rate measuring system. There can be no data processing without data. An inferior-quality wave measuring device (and cable) should not be obtained in order to free extra money for proper data logging.

A useful data presentation form for wave characteristics is to have joint frequency distributions of height and period listed for various segments of direction of wave approach. This is the system followed by Hogben and Lumb (1967) as well as that used for one type of data presentation by CERC. When information on the direction of wave approach is lacking, a single table of associated height and period information is obtained. Average heights and periods for such tables can be listed as well as standard deviations for both quantities.

The term "frequency," as employed above, refers to the number of times a particular blend of conditions is observed. A cell in the joint frequency table would, for example, be $4.0 < H \leqslant 6.0$ and $12.0 < T \leqslant 14.0$ where H is the wave height in ft or m and T is the wave period in s. The number entered in the particular box would show the number of times during wave measurements that this particular combination of wave characteristics had been obtained. The heights and periods given above could either be taken as those for individual waves or, substantially more often, taken as the significant wave height (Chapter 2) and average wave period for a record of (often) 20 min duration.

Normally the "zero-up-crossing method" is used to extract wave height and period information from wave data in the computer. As the digitized data values rise and fall in step with wave crests and troughs, the zero level is repeatedly crossed. The wave period is taken as the time between two consecutive "upcrossings" of the zero level by the stored values. That particular wave height is then the difference between the highest value within that time span and the lowest value in the preceding time span.

5-5 CURRENTS

Introduction

There are hundreds of reports and papers dealing with currents at specific locations. The logical place to start a program of research for ocean-current data for a prospective outfall location is in a good oceanographic library. However, it should be remarked that oceanographic current data are gathered in most cases beyond the continental shelf and are thus of little direct applicability.

Another source of ocean-current information, albeit somewhat crude, is in governmental nautical publications, for example: *Tidal Current Tables* 1976; *United States Coastal Pilot 7* ...; *Canadian Tide and Current Tables* 1976, and *Sailing Directions* ... .

It is possible, conceptually, to divide current measurements into two main classes. In the *path*, or *Lagrangian*, method a device is placed in the water, and as the water currents carry it with them its position is periodically checked and the between-positions average speed and direction of the current inferred. In the *flow*, or *Eulerian*, method a current sensor is positioned at one location and the variation with time of the current speed and direction is measured. Both techniques will be studied in this section. However, there are other approaches that should be mentioned.

There is a device for measuring currents from a boat underway called the *geomagnetic electrokinetograph* (GEK). The GEK measures the electrical potential induced by the movement of seawater (an electrolyte) through the earth's magnetic field. This is accomplished by towing a pair of electrodes behind a ship and recording the potential between the electrodes as they pass along a course and then again at right angles to it. Details are provided in von Arx (1962).

Remote sensing from aircraft or satellites is discussed by Cameron (1952, 1962, 1965), Klemas et al. (1974), McLeod and Hodder (1976), and Ruzecki (1976).

At the least, data on coastal currents are necessary because dispersion of an effluent in the ocean is so dependent on current speed and direction. The current climate for a desired outfall diffuser location can indicate the likelihood of perverse currents carrying the effluent directly onshore. The current climate is often represented by a *current rose*, the oceanographic equivalent to the wind rose presented in Figure 2-13.

Coastal currents outside the surf zone come mainly under the influence of winds and tidal effects. Both are obviously time-dependent. Thus, there are temporal variations in coastal currents at a given point that must be reflected in current measurements. There can also be marked spatial differences in current velocities at a given instant, and these too must be charted in an investigation of coastal currents. Variations in depth can be extremely abrupt, as I witnessed myself on one occasion when measuring currents at a station in the entrance channel to a harbor. At the time of measurement, the wind was blowing about 25 mph (40 km/h) from the northeast and the tide was rising. The top several feet (about 1 m) of water set off rapidly directly downwind, but under this surface layer there was a sudden direction shift, and the remainder of the flow was headed north into the harbor.

Lagrangian Current-Measuring Approaches

Introduction: The Lagrangian determination of ocean currents can be divided into two separate types of approach. With one approach a current-following device called a *drogue*, is placed in the sea at a known location, depth, and time. As the drogue moves with the prevailing currents it is periodically relocated and its position again established and the time noted. As a result a discrete time series of drogue positions is obtained. By plotting location and time a reconstruction of the drogue's (and, we hope, the current's) detailed movements is possible; the degree of definition depends on the frequency with which the drogue was relocated and the accuracy with which its position was reestablished.

The other type of approach involves a current follower that is released at a known position but not relocated from time to time. Its final position, whether washed up on a shoreline (where it may lie for an undetermined time), caught in a net, or picked up at sea is all that is known. With just the beginning and ending points and times known for this type of current follower, which we will call a *drifter*, no detail can be provided on the movements of the drifter between the two points. All that can be established is that the minimum speed of the drifter, and again, we hope, also the current, is the "great circle" distance between the beginning and end points divided by the time between release and recovery. Many releases are required to provide an accurate picture of the currents when using drifters.

Drifters: Two types of drifters that have been used to indicate surface water movements, *drift bottles* and *drift cards*, will be considered in this section. In addition, bottom drifters, used to follow bottom currents, will be described.

A drift bottle is customarily a long-necked glass bottle of 4- to 6-oz (0.11- to 0.17-ℓ) size. Sand is added to the bottle until about 0.25 to 1 in. (6 to 25 mm) of the bottle neck remains above the water surface. Generally, also within the bottle are a visible (e.g., red) post card and/or instructions (perhaps in two or more languages depending upon the working location) so that the ultimate finder of the bottle can notify the originator of the point of its discovery. A reward ($.50 to $1 in the U.S.) is usually offered for return of the enclosed post card. The bottle must be coded in some way for identification purposes. A drift bottle is stoppered or capped and sealed with wax, and often released in sets of six or twelve. In fact, in some cases, "six-packs" of drift bottles are made up for group release. Although drift bottles are normally released from surface vessels, it is also possible to drop them from airplanes (Harrison et al., 1967). The percentage return of drift bottles is usually very low. Crowe and Schwartzlose (1972) quote 3.4%.

Drift cards can have various forms. One common type features an instruction-return-identification card within a plastic envelope. Another uses a small rectangular slab of solid plastic with identification numbers and instructions stamped into the material. Shannon et al. (1973) describe the latter type of drift card. Both drift cards and bottles must be influenced to some degree by the wind. All that can be hoped is that with proper precautions the effect is secondary.

A *bottom drifter* provides a Lagrangian indication of prevailing average currents near the sea floor. Some form of the Woodhead sea-bed drifter is normally employed (Rehrer et al., 1967). This device features a small (18 to 19 cm diameter) plastic, dished saucer with four 2 cm diameter holes. The saucer is attached at its center to a plastic rod. This rod, usually 54 cm long and 0.65 cm in diameter, terminates in a sharpened point near which there is a weight collar typically weighing 5–7 g. When deployed, the plastic umbrella hovers slightly above the sea floor; the end of the rod rests on the bottom. One technique for sending a group of (five) seabed drifters to the bottom is to use a salt ring that dissolves in an hour or two and then frees the individual drifters (Harrison et al., 1967). Drifters can also be taken to the bottom by divers and released.

Drogues: The paper by Monahan and Monahan (1973) depicts, in one figure, the many types of drogues that have been used for Lagrangian measurements of current velocities. Such drogues have been divided by Monahan and Monahan into four groups as follows: 1) plane, perpendicular to the flow; 2) cruciform; 3) circular; 4) parachute. Models of various types within each group have been thoroughly tested for drag characteristics and stability by Vachon (1973, 1974), and numerous forms are possible for oceanographic work. However, our attention here will be limited to the cruciform and parachute classes, apparently the most heavily used types in field work. Examples in the literature on the use of cruciform drogues are Gaul and Stewart (1960), Bathen (1974), Stevenson et al. (1974), and Cederwall et al. (1975); for parachute drogues see Volkmann et al. (1956), Wyatt et al. (1967), Wyrtki et al. (1967, 1969).

Two representative examples of cruciform and parachute drogues are shown schematically in Figure 5-9. The main part of the cruciform drogue can be made from identical slotted pieces of plywood or masonite. This design can

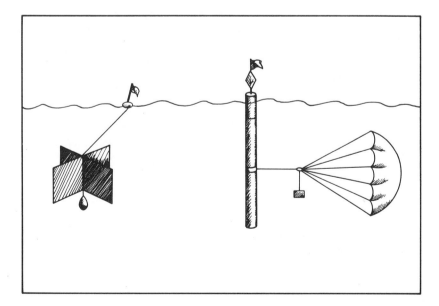

Figure 5-9 Schematic representation of cruciform and parachute drogues

be readily stowed when not in use, is not too bulky to transport, and can be easily assembled. A hinged design is also possible. To strengthen the drogue, one can attach a metal frame on the top and bottom or cross-tie with twine.

A weight keeps the cruciform drogue submerged and a small surface float connected to the drogue by a line provides sufficient buoyancy to keep the entire system buoyant. The length of line is selected so that the drogue is positioned at the desired depth.

The main part of the parachute drogue shown in Figure 5-9 is a pipe, probably polyvinyl chloride (PVC), capped at the bottom and weighted so as to float in a vertical attitude. The parachute and a weight complete the underwater arrangement.

In general terms, the cruciform drogue is usually the smaller of the two and the one used in current measurements from a small boat. Parachute drogues can be large enough to be impossible to deploy from any boat less than about 50 ft (15 m) in length.

Tracking Drogues: Although the two drogues differ considerably in design under the water, they are similar in that a portion of the overall drogue system penetrates above the sea–air interface in order that the drogue can be followed (*tracked*) as it moves with the currents. It is often difficult for those inexperienced in looking for an object on the sea surface to fully comprehend how frustrating it can be to try to locate such an object visually, even if it is within a few hundred yards. A major problem concerns waves. Even an object extending up above the water surface several feet (about 1 m) can be easily obscured for long periods of time by moderate wave action. Thus some means of locating the marker part of a drogue in addition to a visual marker, such as the often-used red flag, is required.

One possibility frequently employed is to mount a small radar reflector (see Appendix A) on the drogue marker pole. Then, at least theoretically, the drogue can be located in any weather, night or day, as long as it is located within the radar horizon of the search vessel. But this immediately implies a relatively big operation, since small boats and many work boats do not carry radar, and at any rate, radar is temperamental and not always properly operating. But given both the required radar-equipped ship and a properly-outfitted drogue, there is always the problem of picking out the drogue's radar echo from the mass of echoes that constitute *sea return*—echoes from suitably reflecting parts of the ocean waves.

Another marker feature used with drogues is a blinking light. Different colors are sometimes used for drogue identification purposes. A flashing strobe light can conceivably be seen up to 10 miles (16 km) away or more. Another possibility is for a radio transmitter to be housed in the drogue's marker mast. Rather than having the light and transmitter work continuously at a considerable drain on available battery power, it is possible to have both actuated by radar signals from the search ship (Walden and Webb, 1964).

It is customary to use a hand-held receiver for homing in on a radio beacon from a drogue. Such receivers, for example the Ocean Applied Research Corporation (San Diego, California) Model FR-206, have several frequency channels, selected by a switch. Thus one receiver can be used with several drogues. The FR-206 has eight channels. The receiver has an integral antenna, and when the system is turned horizontally, a panel meter indicates when the antenna is directed toward the target. I have seen a boat steam right up to a drogue whose light had gone out, in the middle of an inky black night, directed only by the man on the radio direction finder.

Only rarely is a single drogue tracked. It is more usual for several similar drogues to be released and then relocated from time to time by the oceanographic vessel being used in the work. The chart position of the ship, and the buoy beside it, can be obtained by using one or more of the positioning techniques described in Appendix A. Such positioning is not always done with respect to charted features, for it is not unusual that a fixed, taut-line buoy station, or perhaps two such stations, be used as reference points. The buoy stations themselves would, of course, have to be located with respect to charted features at some point.

When drogues are tracked close to the coast it is possible to establish the position of the drogues by means of a theodolite (Murray, 1975).

The systems described above for locating drogues play a role in not permitting the drogue to perfectly follow the current systems of which they are supposed to be a part. This is because the action of wind on the exposed components causes a drag force that can deflect the entire drogue system to a degree dependent upon the relative drag of the exposed portions to wind and the lower portions to water. Then also, there is the drag force exerted by flowing water of current systems on lines extending up through the water to the surface from relatively deeply-submerged drogues (Terhune, 1968).

The above-water portion of the drogue system must be designed to have low drag. Thus the desire for an easy-to-see large marker flag and long pole must be balanced against a tiny flag and short, thin pole which is desirable if the drogue is to faithfully follow the currents. Low wind-resistance radar reflectors are also a must.

Aerial Observations: A cross between a drift card and a drogue can be used when floating devices indicating surface water movements are tracked by airplanes. Yeske and Green (1975) used white posterboards; Stevenson et al. (1974) employed floats made of three plywood squares (one horizontal, two vertical and crossed). Yeske and Green used moored buoys for reference as did Stevenson et al., but in some cases the latter team obtained position by flying a compass heading to landfall. A high-wing plane, such as the Cessna, is best adapted to aerial tracking.

Eulerian Flow Measurement

Introduction: This section will summarize the major types of meters used in the Eulerian determination of current velocities. Older types of meters, such as the Ekman (-Merz), will not be dealt with. Information on such instruments is available in the *Manual of Current Observations* (1950).

The meters to be considered are those either mounted on a moored cable or on a weighted frame set on the bottom. Although it is entirely possible to transmit ocean-current data to a shore recording station either by radio telemetry from a suitable surface buoy or along a submarine cable, I will consider only situations where the data are recorded at the measuring site. Such *in situ* recording is generally intermittent in order to conserve self-contained power.

Current Meters with Savonius or Similar Rotors: The *Savonius rotor* consists of a circular housing with blades (see Figure 5-10). It rotates about an axis perpendicular to the flow at a rate roughly proportional to the flow speed. Used in conjunction with Savonius rotors are vanes to indicate flow direction, and such systems are normally completed by battery-instrument packs so that current speed and direction are recorded on station. Numerous companies manufacture such systems.

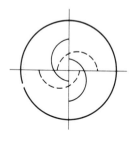

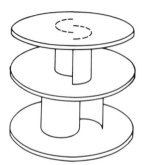

Figure 5-10 Savonius rotor

Halpern et al. (1974) tested three major commercial Savonius-rotor-type or adapted Savonius current meters against each other in ocean experimentation. It was found that these meters, the Aanderaa RCM-4, the Geodyne A850-2, and the AMF vector-averaging all gave equivalent average results. These same meters have also been tested by the NOIC.

Considerable discussion of the Aanderaa meter is available in Dahl (1969). This meter, shown in Figure 5-11, obtains data on temperature, conductivity, depth, as well as current speed and direction. These are all recorded on magnetic tape. This is also the recording system for the Geodyne and AMF units.

Figure 5-11   Aanderaa current meter (Courtesy, AMF Inc.)

Savonius rotors suffer from the same biological fouling problems as all moving parts immersed in shallow-water seas (see Figure 5-12). After some days, the calibration characteristics of the meter can be completely altered due to growth on the rotor and blades or along the shaft. In one unpublished test run by NOIC, a fouled rotor indicated only 36% of the true current speed as measured by the rotor immediately upon deployment and before fouling. The Savonius is also found wanting since it translates mooring motion and wave-induced water motion into net rotations that are logged by the measurement system (Pollard, 1973). Thus, in this particular case, Savonius-rotor meters indicate current speeds that are too high (Karweit, 1974). Other errors, such as those due to tilting of the meter, are reviewed by Kalvaitis (1972). Subsurface rather than surface floats should be used with Savonius sensors to reduce the effects of tilt and mooring motion. University of Hawaii oceanographic practice

Figure 5-12 Fouled Savonius current meter
(Courtesy, William E. Woodward)

has been to assure about 200 lb (890 N) of positive buoyancy in cables supporting Savonius-rotor-type current meters.

Ducted Impeller Meters: The Bendix Corp. (Annapolis, Maryland) Model Q-15 current meter features a 3- or 5-bladed plastic impeller turning inside a 4-in diameter (102-mm) stainless steel duct. Small magnets mounted in the tips of the impeller blades close an electrical reed switch located in the housing alongside the duct whenever they pass the switch. The rate of impeller rotation, hence of switch closing, is directly proportional to the flow speed. Ideally, a long vane attached to the duct orients the meter parallel to the current, but how well this operates in the presence of substantial wave surge is largely unknown.

The pulse history from the reed switch closures is treated, integrated, and printed on one channel of a small, two-channel chart recorder in a battery-instrument pod mounted near the meter. The other channel of the chart receives current direction information.

I have used the Bendix ducted meter a great deal in investigations of wave-induced water motion. The sensor has a tendency to foul rapidly; tube worms take up space on the inside duct walls and minute organisms penetrate into the impeller-shaft bearing areas. I have seen such meters completely frozen, both in my own work and in diving on meters used by others in gathering data on currents. The sensor works well if cleaned thoroughly every few days.

Meters with no Moving Parts: The problems attending moving parts in a current meter used in the ocean point toward the desirability of meters with no moving parts. Two such types of meters are the ultrasonic and the electromagnetic. The former either relates the minute difference in arrival time between upstream- and downstream-propagating ultrasonic pulses to the flow speed of the water, or works on a Doppler-shift basis, sorting out current speed from frequency alterations of pulses moving up- or downstream in the flow. Such meters are commercially available (e.g., Westinghouse Underseas Division, Annapolis, Maryland) but are very expensive and, to my knowledge, have seen no oceanographic use other than wave-induced kinematics.

The electromagnetic current meter operates on Faraday's law of electromagnetic induction by generating a magnetic field in the water and utilizing the water as the electrical conductor. An electrode pair is used to measure the potential produced by the moving water, a potential that is proportional to the velocity of the conductor. Electromagnetic current meters are available commercially (e.g., EG & G, Waltham, Massachusetts and Marsh-McBirney, Rockville, Maryland). Users of electromagnetic sensors make frequent reference to electrical problems.

Various other types of current meters with no moving parts are summarized by Lomas and Marks (1975).

Tilting Current Meters: The concept of a current meter as a device tipped over by a current is at least decades old. The fundamental idea is that the faster the current, the greater the tilt of the meter. The problem has always been the manner of recording, time after time, hour after hour, the speed of the current and its direction.

The General Oceanics (Miami, Florida) Model 2010 is a relatively inexpensive, heavily-used commercial sensor that uses the tilting wand idea. A positively buoyant, plastic (PVC) cylindrical housing, usable to a depth of 50 m,* contains the electronics and mounts twin fins that cause the meter to be oriented stably in a current with the fins downcurrent. A swivel eye at the bottom of the meter attaches to a weight or mounting bracket; either may be used by the current measurement group. The plastic housing is 4.5 in. (114 mm) in outside diameter, 20 in. (508 mm) long, and has a positive buoyancy of slightly over 2 lb (10 N).

The manufacturer's calibration curve is nonlinear in the two tails, but linear in the velocity (angle of tilt) range of approximately 20 cm/s (19°) to 40 cm/s (56°). The meter is advertised as being usable for current speeds of between 0.05 and 1.6 knots (2.5 to 82 cm/s).

A cover, fitting over an O-ring, provides access to the inside of the meter (see Fig. 5-13). The tilt-direction sensor is an ingenious design featuring two concentric spheres. The outer hollow sphere is transparent with a small circular target at its top. The inner sphere, trimmed to neutral buoyancy, floats in water (with wetting agent) within the outer sphere. The surface of the inner sphere is marked with a latitude–longitude representation: the latitude markings indicate the tilt, the longitude scale, activated by a bar magnet inside the inner sphere, indicates the magnetic direction. The bar magnet is mounted low so as to assure the rotational stability of the inner sphere.

---

*A negatively-buoyant model (2011) with an anodized aluminum case is used in deeper waters.

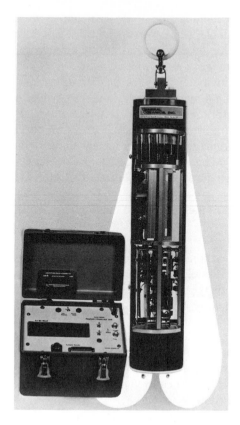

Figure 5-13 Tilting-type current meter (Courtesy, General Oceanics, Inc.)

A record of the current speed and direction is captured on film by a super 8 cartridge camera that advances the film frame-by-frame at intervals that can be set from 1 to 60 min. Power for a bulb to illuminate a calendar-type watch and the tilt-direction sensor, and to run associated circuitry, is provided by batteries. The film has about 3500 usable frames, so the total data-taking time depends on the film advance rate. For a 60-min timing interval, about five months of data are theoretically possible.

Data can be obtained from the developed film by projecting each frame on a wall and reading the date and time as well as the inner sphere's longitude and latitude under the outer sphere's target. This data reduction is very tedious and can best be carried out by two people, one advancing film and writing the data, the other reading off the values into a tape recorder. Finally, the data can be punched onto cards for computer analysis.*

Every current meter has its own set of bad features and the General

---

*The General Oceanics Model 6011, a later development than the Models 2010 and 2011, has a cassette tape recording system for meter-tilt data. This development has, at the least, a decided advantage in terms of data reduction.

Oceanic Model 2010 is no exception. Two chief complaints concern the watch stopping and the inner sphere sticking. It has also been said that the swivel eye is not sufficiently robust and that the twin fins are too small. Field personnel of the Southern California Coastal Water Research Project (SCCWRP) replaced the standard fins with larger ones.

A typical SCCWRP installation in 1975 and 1976 was in 185 ft (56 m) of water offshore from the sewage treatment plant at Point Loma, California. At this station, there were two lobster floats on the surface, separated by about 500 ft (152 m). A light polypropylene line descended from each float to a chain anchor on the bottom, and two 250-ft (76 m) polypropylene grapple lines ran inwards from these anchors to another anchor that held down a taut-line mooring. The subsurface buoy on this line supported a frame to which a Model 2010 current meter was attached. The lines were retrieved using a "lobster puller," a capstan device driven by a gasoline engine.

Presentation of Current Data

Data obtained using Lagrangian techniques of current flow measurement are normally presented on a chart of the area showing the successive positions of the drogues involved, with a notation opposite each position indicating the time that the drogue occupied that location. Using the distance scale on the chart and the times involved, the average current speeds for each segment can easily be obtained.

Lagrangian current information is useful because it shows the track that a cloud of wastewater effluent would likely follow once introduced into the marine environment. How the cloud would spread over the path would not be represented in the drogue tests. Methods by which such spread can be estimated will be presented in the next section.

The diffuser for an ocean outfall lies along a line between two points, and it is useful to have current information that is Eulerian in nature, i.e., that applies to these two points and possibly others between them. Such data are required in computing effluent dilutions and in assessing the stability of the outfall line and its rock protection to current-induced forces.

It has already been remarked that a moderately common method of presenting Eulerian data derived from current measurements is in the form of a current rose patterned after the wind rose depicted in Figure 2-13. One such representation is shown in Figure 5-14. The length of any radial line indicates the relative frequency of occurrence of currents setting in that direction. The length of any segment along that line indicates the relative frequency of currents having the appropriate direction *and* strength. The current rose representation provides an excellent quick-look picture of the current velocities. However, this is necessarily a crude picture because directional and speed classes must be formed, thus causing a loss of detail. Current roses can be plotted on a chart at the station involved or given separately with station identification.

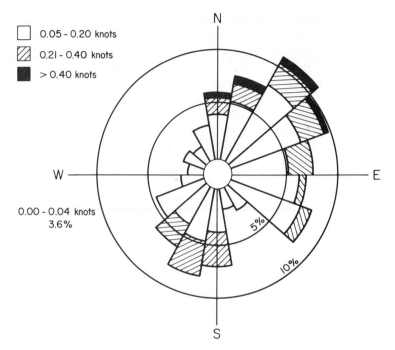

Figure 5-14   Current rose

One not unusual method for presenting current velocity information for a given station and water depth is to separate the speed and direction features and to present these in frequency distribution (*histogram*) form. This approach is inferior to that involving the current rose in that the cross correlation between speed and direction is not shown.

An excellent computer-generated, three-dimensional, graphical presentation of current data uses an observational frequency surface drawn over the speed-direction plane. The method of presenting the data can be other than graphical; a joint frequency table of hourly averages of current speed and direction classes is very useful.

Although the current rose representation demonstrates the cross correlation between current speed and direction, it does not demonstrate the serial correlation within an individual speed or direction time series; i.e., it does not show the time over which the current consistently sets in about the same direction and at, say, approximately the same speed. For example, if the plot of Figure 5-14 shows that a current flows in a northerly direction 25% of the time, then the hour-by-hour extremes would have the current flowing north for an hour once every four hours, or else flowing consistently north for six hours once every day.

There are at least two or three ways by which such serial correlation can be demonstrated. First, hourly averages of current speed and direction can be plotted against time. Second, the *progressive vectors* approach places on a chart,

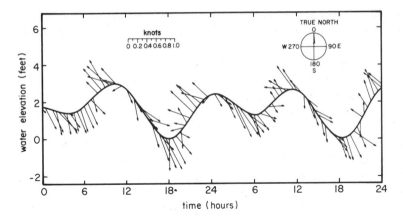

Figure 5-15 Correlation of current velocity and tidal variation

using the station involved as the origin, a head-to-tail string of the hourly average current velocity vectors expressed in terms of distance travelled by the water past the station involved. Such a plot looks like the charted movement of a drogue, but is not. What is shown is the time variation of flow at the given station.

A third graphical technique that implicitly includes serial correlation information is that shown in Figure 5-15. However, the major use of the type of plot in Figure 5-15 is to demonstrate the effect on the currents of the tides. Where tidal effects predominate, this is an excellent representation.

It is well known that wind-induced currents can overpower tide-induced currents in certain localities when the winds are strong. In the Hawaiian Islands, for example, there are very large differences between the current systems that prevail under normal trade-wind conditions and those that occur rarely when the trade-wind system breaks down for a period of several days and purely tidal influences can exert themselves. With this in mind, another useful type of graphical accounting of current conditions is to plot on a chart current vectors, both surface and subsurface, at various stations for one tidal stage—wind condition situation. Various different situations are represented. For instance, one plot could be for a flooding tide and strong northeast winds, another for an ebbing tide and calm winds.

5-6 DIFFUSION AND DISPERSION

Introduction

If a sewage effluent reaches the water surface, it may still be lighter than the ambient surface waters and tend to flow over the surface because of gravitational effects. An effluent that has largely, through dilution, reached the density of the ambient sea water is advected away from the outfall by ocean currents and spread by dispersive mechanisms. These topics are discussed in some detail in Chapter 7.

In this section we are concerned with determining, in the sea, the manner in which an effluent diluted to near sea water density would disperse. Two classes of approach will be described for determining the appropriate dispersion coefficient. Substantial ocean research has established that, for reasons outlined in Chapter 7, the dispersion coefficient is a function of the scale of the process involved. Furthermore, such coefficients are clearly dependent upon wind and wave conditions and upon the ocean water density profile. Thus dispersion measurements are closely related to the various other physical measurements considered in Chapter 5. Proper interpretation of dispersion data can only be done with concurrent data on these physical quantities. Since these change with the seasons, dispersion measurements should properly be gathered at different times of year.

Use of Drogues

Stommel (1949) has set out a practical method by which the eddy diffusivity of a specific length scale can be determined. This approach apparently has its origins in the manner by which Richardson conceptualized the diffusion process in the ocean (Richardson and Stommel, 1948). Stommel's approach was to release floats in pairs at an initial separation l_0, then to measure the separation (l_1) after an elapsed time T chosen to be small enough that the quantity $(l_1 - l_0)$ averaged only a small fraction of l_0. Let the initial and final separations of the ith float pair be represented by l_{0_i} and l_{1_i} respectively. Then the scale of the process for the ith pair can be written

$$l_{*_i} = \tfrac{1}{2}(l_{0_i} + l_{1_i}).$$
(5-1)

If numerous floats are used, many pairs exist. Consider pairs in groups for which l_{*_i} is relatively constant. Then the scale of the process for a group of N pairs is

$$\bar{l} = \frac{1}{N} \sum_{i=1}^{N} l_{*_i}$$
(5-2a)

$$= \frac{1}{2N} \sum_{i=1}^{N} (l_{0_i} + l_{1_i})$$
(5-2b)

The overbar indicates an average value. Stommel has shown that the eddy diffusivity of length scale $\bar{l}$ is given by

$$\epsilon_{\bar{l}} = \frac{\displaystyle\sum_{i=1}^{N} (l_{1_i} - l_{0_i})^2}{2NT}$$
(5-3)

Different groups of floats yield eddy diffusivities of different scales. Also, as time progresses the length scale for a particular group enlarges.

The best means of establishing a quick-look picture of the floats and their positions is from the air as Stommel did. However, if few floats are involved and/or if spacing is not too great, boat and ocean-surface-positioning techniques (Appendix A) can be used.

The Marine-Estuarine Technical Committee of the California State Water Resources Control Board proposed, in 1976, a standardized way of determining eddy diffusivities through use of Stommel's approach. It was proposed that seven drogues should be initially planted so that (ideally) they occupied the vertices and center of a hexagon. The sides of the hexagon were to be about 100 m in length since this distance well approximates the size of a potential diffusing cloud of effluent.

Requirements for the drogues themselves were as follows: they should be set to float at a depth of 7 m; the parachute should be greater than 10 ft (3 m) in diameter; the maximum ratio of elements in the air to those in the water should be $1:200$. The drogues were to be tracked within a 5-mile (8-km) radius or for 24 h, whichever occurred sooner.

There are twelve equal distances in a hexagon with a central point. Thus the seven drogues provide twelve pairs for consideration as one group for the determination of ϵ_j.

Fluorescent Dyes and Fluorometers

Although radioactive tracers have been used to simulate the dispersion of an effluent cloud in the ocean (e.g., Harremoës, 1966), only fluorescent tracers will be described here. The word tracer in this context refers primarily to a way of tagging a water mass and, secondarily, to the representation of dilution phenomena undergone by elements within that water mass.

> Basically, fluorescence is a form of luminescence, a broad term for any emission of light not directly ascribable to heat. Fluorescent substances emit radiation (light) immediately upon irradiation from an external source; emission ceases when the source is removed. The emitted (fluoresced) energy nearly always has longer wave lengths and lower frequencies than the absorbed energy. ... (Wilson, 1968).

The general idea involved in the use of fluorescent tracers is that an aqueous solution of the material is released into the water for which the dispersive properties are to be studied. Although such releases can be of an instantaneous or continuous nature, the former will be considered for the time being. As time passes, the initially very concentrated slug of dye spreads to neighboring areas. The dye concentration at the Lagrangian position of the initial dye drop decreases; dye concentrations along Lagrangian rings surrounding the drop point increase and then decrease as time passes. The dye occupies a larger and larger area. By conservation of mass, the average concentration over that area continuously decreases. The speed with which the dye patch is spread to neighboring Lagrangian areas is a measure of the diffusion coefficient. Quantitative methods for determining such coefficients from dye concentration data will be given later.

A slug of dye released into surface waters is easily visible to the naked eye. However, as dispersion progresses appreciably, the dye is no longer apparent. Thus, some instrumental way of noting the presence of dye is necessary. In addition, even if the eye can discern the dye at a particular location in the water, it can not differentiate between subtle concentration differences. As we shall see later, it is not only necessary to follow the edge of a diffusing dye cloud to obtain diffusion coefficient information but also to chart the transitions in concentration within the cloud.

The type of instrument used to determine the concentration of fluorescent material in water is a *fluorometer*. There are two types of these instruments, one that provides a spectral representation of the emission of the fluorescent substance, the other that does not discriminate among different wavelengths. The latter is called a *filter fluorometer*, or *fluorimeter*. Although there are various manufacturers of such instruments, the G. K. Turner Associates (Palo Alto, California) Model 111 is by far the most widely accepted in the United States.

There are several key elements in most filter fluorometers (Wilson, 1968). An energy source supplies light that reaches the water sample after passing through a primary filter that passes only a selected band of the source's output spectrum. Emission from the sample passes through a secondary filter, then onto a sensor that connects to a readout. There are two ways of handling the sample in the Turner fluorometer. Either a sample in a test tube can be inserted into a special holder or water can be continuously passed through a special "flow-through door" and readings continuously taken. The latter is the technique used in dye dispersion studies. The design and operation of the Turner Model 111 fluorometer are completely described by Wilson (1968) and in *Turner Laboratory Instruments* (1972).

There are many commercially available fluorescent dyes.

> Four dyes, each a variation on the same basic organic structure (xathene), have been used extensively as tracers: rhodamine B, Rhodamine WT, Pontacyl Pink, and fluorescein. Generally speaking, they are good tracers because they are (1) water soluble, (2) highly detectable—strongly fluorescent and easily isolated from background, (3) harmless in low concentrations, (4) inexpensive, and (5) strongly stable in a normal stream environment. Special training is not required for handling the dyes, but care in handling is recommended to avoid undesirable stains resulting from spills. (Wilson, 1968)

The latter is a significant remark, although understated. Rhodamine B, for example, will remain on the skin, despite frequent washings and brushings, for many days. Spilled dye on floors or cabinets means a red stain for anyone venturing near for months, and boat lines once contaminated by rhodamine B can be a source of annoyance for many months as well. An associate of mine, who used to prepare solutions of rhodamine B from powder, always referred to himself as having been "caught red-handed" after he had finished his chore.

Fluorescein is less expensive than rhodamine B but has a very high photochemical decay rate. This means that there is a rapid degradation of the fluorescent intensity of the dye in the water body under the action of sunlight. For this reason fluorescein is considered inferior to rhodamine B for accurate field work. The remaining three all have advantages and disadvantages. Rhodamine B is readily absorbed by suspended particles in the water body and has a moderate photochemical decay rate. Pontacyl pink and rhodamine WT do not suffer from these deficiencies, but they are both more expensive than rhodamine B, particularly pontacyl pink. Rhodamine WT tends to be heavy. Feuerstein and Selleck (1963a, 1963b) concluded that pontacyl pink B was the best quantitative tracer for natural surface waters. However, only rhodamine B will be considered in what follows, as it is very often the choice of those carrying out dye dispersion measurements in the sea, and the only one I have used. Substantial material comparing various attributes of rhodamine B, pontacyl pink, and fluorescein is available in Feuerstein and Selleck with summary and additional material in Wilson (1968).

The fluorescence given off by a water sample containing a fluorescent tracer depends most strongly on the dye concentration. In dilute solutions (solutions in which less than five percent of the exciting light is absorbed), and for a given fluorometer setup, fluorescence is directly proportional to the dye concentration in the sample. Temperature influences fluorescence to some degree; the effect can be represented by the equation

$$F_s = F \exp\left[\alpha(T_s - T) \right]$$

where $\exp[a] = e^a$, F_s is the fluorescence intensity at a standard temperature T_s, F is the fluorescence intensity at a temperature T, and α is a coefficient with value -0.027 $(°C)^{-1}$ for rhodamine B. Fluorescence can also be altered by the action of chemicals in the water. However, for pH values between 6 and 11 the fluorescence is essentially constant. The minimum detectable concentration of rhodamine B in distilled water was found by Feuerstein and Selleck (1963a, b) to be 0.000046 mg/ℓ at 20°C, or approximately 5×10^{-11}, an exceedingly small figure.

Dye Study Field Methods

The dye is released into the surface water either continuously or in a discrete dose, using methods that will be detailed later. As the dye cloud spreads, a boat bearing a fluorometer and ancillary equipment passes back and forth through the cloud, monitoring the spatial variation in dye concentration from one extremity of the cloud to the other at (approximately) a given time.

Generally, water is continuously drawn through an intake pulled behind and to one side of the slowly moving survey vessel. The water passes through an opaque hose to the fluorometer where the fluorescence value is marked on a chart recorder. The water is then discharged overboard on the side away from the intake. There have been various forms for the intake. One (Bailey, 1966) involved a submersible pump mounted on a diving plane; another (Foxworthy, Tibby, and Barsom, 1966) consisted of an open torpedo-like body with adjustable tail fins. An open-ended hose with a weight and hauling line running back to the boat has also been employed. Pumps for the latter two types of intakes are located on board the boat. It is suggested by Cederwall et al. (1975) that the pump used in any continuous water-sampling system be located downstream of the fluorometer to prevent air bubbles from affecting its satisfactory operation. A thermometer for continuously monitoring the sample water temperature should also be upstream of the pump.

The depth of the intake below the water surface can be obtained using the length of towing wire paid out and a wire-angle meter or equivalent. The intake should be submerged at all times.

Sea water outside the dye cloud is first pumped through the fluorometer system in order to establish the "background" level of natural fluorescence. Thereafter, readings shown by the instrument are corrected for the background level.

Corrections must also be made to values shown for temperature variations, and an attempt should be made to compensate for the photochemical decay of the dye. In order to better estimate the effect of such decay on fluorescence values, Cederwall et al. (1975) used two different tracers of different decay rates. One of these was a radioactive tracer and the other a fluorescent one.

It is common for 115 v ac power for the fluorometer to be supplied by a gasoline-powered alternator. Some researchers suggest installing a voltage/frequency regulator between the alternator and fluorometer, since it is known that faulty readings can result for imprecisions in the input power. However, for a properly-designed alternator this should not be a problem.

In order to plot dye concentration contours upon completion of the field work, one must be able to determine the position of the boat out on the water at any time. Any of the methods listed in Appendix A can be used. A workable approach is to lay out a series of straight-line traverses that will cover the cloud. When the boat completes one traverse and turns onto the compass bearing of another, the time is recorded and noted on the fluorometer chart record. Sextant angles are quickly shot, and, if possible, a depth reading is taken from the fathometer as an extra check. The boat skipper then proceeds at constant power along the new traverse until the next turning point is reached and the process is repeated. A running check on position along a traverse to assist in determining when the turning point is reached can be made using the sextant, then quickly establishing position on a chart using a three-arm protractor.

There are various ways of actually making the dye drop when instantaneous discharges are used. Most involve the same boat used in the follow-up sampling work. The density of the dye should be adjusted to that of the ambient sea water, and this is customarily done by adding appropriate amounts of methanol.

As remarked earlier, rhodamine B stains anything it comes in contact with a bright red, and it is undesirable to simply dump a pail of the dye solution overboard from a boat unless the appearance of the boat is of absolutely no consequence. It is desirable to drop the dye into the water at some distance from the boat and to ensure that the container for the dye is truly shut until it has been removed from the deck or immediate vicinity of the survey vessel. The Army type of 5-gal (19-ℓ) gasoline can is an acceptable item. It is filled with dye solution and foam floats are lashed to it. A sea anchor with two lines is put into the water followed by the dye container connected to the boat by two lines, one a tow line connected to the body of the can and the other an actuating line tied to the spring-loaded top. When the actuating line is pulled, the dye flows out of the can. The trip line on the sea anchor is then pulled and the system returned to the boat.

One approach that we have used is to make the dye drop from a helicopter. The aircraft hovers over the drop point, and the glass container or plastic container of dye is lowered down to just above the water surface. The dye can be released either by a second line or by breaking the container, thus releasing the dye. Accomplishing either method from an open helicopter door some distance above the water can be somewhat unnerving.

For continuous point-source experiments involving dye, a very workable technique is to anchor a skiff at the desired point; aboard the boat is a container of dye and a variable-speed, positive-displacement pump discharging dye into the water through a hose ending at the desired discharge depth (e.g., Foxworthy, Tibby, and Barsom, 1966). The sampling boat will work downstream of this point. The continuous discharge approach more closely approximates the actual marine disposal of a wastewater than does the instantaneous discharge technique.

It has been implicitly assumed that the spread of a *surface* dye patch was monitored. The following method of dealing with subsurface dye releases and the resultant dispersion was used in a study in Kaneohe Bay, Hawaii, in October 1976. A diver, dragging a surface float, swims repeatedly around the edge of the cloud. An attending boat follows the float, with an operator frequently checking position by sextant. In the Kaneohe Bay study, the establishment of position was aided by using a grid composed by nine buoys in a square pattern, with 200 ft (61 m) between nondiagonally-adjacent buoys. The dye release had been at a depth of 35 ft (10.7 m) directly under the middle float in the array.

Organizing Results of Dye Studies

For some dye studies, only the migration of the entire dye patch is presented; no details on dye concentration profiles within the cloud are provided. The manner of presenting such gross information usually consists of plotting on a chart the outline of the dye patch as it appeared at the location shown, with notation showing the appropriate time. The point of dye drop is also shown. This particular type of representation is useful for providing an approximate measure of current strength as well as of dispersive tendencies. A useful insert is a sketch of the tidal variation over the study period.

When dye dispersion measurements are carried out in the ultimate receiving waters for a wastewater effluent, the intent is usually to provide numbers that can be used in mathematical dispersion models that simulate the spread of the effluent after its entry to the marine environment. Two such models are discussed in Chapter 7. The link-up between the field data and these models will be sketched below.

One approach is to use the following relationship between the eddy diffusion coefficient ϵ and the variance of the concentration profile (σ^2) perpendicular to the current (e.g., Brooks, 1959).

$$\epsilon = \frac{1}{2} \frac{d(\sigma^2)}{dt} \tag{5-4}$$

It is standard theoretical practice to assume that the distribution of dye concentration across the current is Gaussian, i.e., has a normal distribution. As a practical matter the dye cloud is defined by widths w_1 at time t_1 and w_2 at later time t_2. The axis of symmetry of the cloud corresponds to $w = 0$. The link-up of this approach with Equation (5-4) depends upon what value z of the unit normal probability density function is taken to define the boundary. The general form of Equation (5-4) becomes

$$\epsilon = \frac{w_2^2 - w_1^2}{C_z(t_2 - t_1)} \tag{5-5}$$

where C_z is a constant depending upon the choice of z. In fact $z = \sqrt{C_z/8}$. Numbers for C_z used in practice are 24 and 32 and correspond to $z = \sqrt{3}$ and 2, respectively. The former conforms to Brooks' definition of field width.

If the dye release is a discrete one, determination of ϵ requires that the same Lagrangian cross section be followed and sampled. For a continuous release, however, an assumption of steady-state permits one to determine w_1 and w_2 at approximately the same time at two different stations, and to determine $(t_2 - t_1)$ from the down-current spacing between stations and the current flow speed.

Even in the absence of a current, Equation (5-5) can be used. A lateral eddy diffusion coefficient independent of horizontal direction is obtained.

Foxworthy, Tibby, and Barsom (1966) attempted, in their field work, to establish the axial (maximum) dye concentration as a function of distance downstream of a dye-simulated wastewater source. Their mathematical model, detailed in Chapter 7, did not require diffusion coefficients *per se* but, rather, indices of spread for the diffusing cloud (variances). Inputs to their model besides the axial dye concentration profile were the depth of dye field, the rate of dye supply for the continuous-flow tests, the speed of the current, and the standard deviation of the source-dye-profile perpendicular to the current.

5-7 PRIMARY PHYSICAL WATER COLUMN PARAMETERS

Introduction

One of the most familiar instruments to physical oceanographers is the *bathythermograph* (BT), a torpedo-shaped device that is lowered from a boat or ship and obtains an *in situ* mechanical record of the variation of temperature with pressure, hence depth. The instrument, after use, is brought back aboard using the light cable attached to it and a special BT winch (see Figure 5-16). Conceptually, the bathythermograph can be dropped from a slowly moving vessel or into a moderate current since the amount of line paid out is not used to judge the instrument depth. But operationally, dropping the BT from a ship underway can be difficult (Snodgrass, 1968).

Figure 5-16 Bathythermograph winch and operator (Courtesy, William J. Emery)

The temperature-pressure profile is recorded on a thin, gold-plated glass slide (BT slide) that is inserted into the side of the instrument on deck and then withdrawn after retrieval. The profile appears as a continuous scratch along the surface of the slide. This scratch is made by a stylus connected to a xylene-filled bourdon tube within the instrument. Such a tube is formed in a short helix in one plane and tends to "unlay" as the temperature rises. Xylene has a rather high coefficient of thermal expansion. The pressure is inputed because the slide mounting is placed on a piston that is depressed by unbalanced pressure.

There are special calibrated BT slide viewers that enable the oceanographer to convert the scratch into a moderately accurate picture of temperature change with pressure (depth)—if the instrument is in calibration. BTs go out of calibration fairly readily, in part because of the rough usage they receive on board ship.

In work related to the marine disposal of wastewaters, the variation of temperature with depth is less important than the variation in water density. This is because, for appropriate ambient density differences between the surface and bottom waters, the wastewater effluent can stabilize vertically at some subsurface level. It was explained in Chapter 2 that the temperature and the salinity of the water at any depth can be used to determine the water density. Since density changes are very small, both the temperature and salinity need to be measured very accurately.

A technique used for many years in ocean work is to capture a water sample at a desired depth, for later determination of salinity and other attributes, at the same instant that the water temperature at that location is obtained. The water is obtained by a *sampling bottle* and the temperature is taken by two *reversing thermometers*.

Obtaining Water Samples

It can readily be appreciated that a sampling device should be designed first to capture a water sample from a particular depth and not from various layers of the water column through which it has been lowered, and second, so that the resulting water sample is not contaminated by the material that makes up the interior of the sampling bottle. In order to satisfy the first requirement the bottle, in nonclosed mode, should be of a flow-through type. To meet the second requirement the bottle, or bottle interior, should be made of a relatively inert material, such as plastic.

A sampling bottle is normally clamped onto a wire leading to a winch on an oceanographic vessel.* As the wire is unspooled from the winch, sampling bottles are attached at positions corresponding to whatever sampling depths are being used (see Figure 5-17). The distance of wire runout can be obtained from a counter at the winch or from pre-installed markers on the wire itself. Corrections may have to be applied to actual wire segment lengths to obtain water depths if, because of relative motion between the water and the boat, the line descends at an angle to the vertical. There are wire-angle meters that assist in

---

*Standard sampling bottles can also be taken to a prescribed depth by a diver having a pressure gage and triggered there by the diver.

such determinations. A heavy weight is attached to the bottom end of the wire supporting the sampling bottles in order to make it hang as straight as possible.

An early form of sampling bottle was the so-called Nansen bottle, with a brass housing. One later version of this bottle featured a Teflon liner. It has been found (e.g., Park, 1968) that the Nansen bottle, even if coated, produces incorrect salinity values as well as other incorrect results as shown in Chapter 6. Park suggested that either the NIO (National Institute of Oceanography, England) or Niskin sampling bottles be used for such determinations. However, since Park's paper, an all-plastic (Lexan) form of the Nansen bottle has appeared.

The sampling bottle developed by NIO is widely used and distributed by MPA, Inc. (Houston, Texas). It comes in four sizes—1.35, 2.5, 5, and 7.5 liters. Only the smallest one carries a reversing thermometer frame. This model has a bottle body, thermometer frame, and taps of polypropylene. The hemispherical valve caps are of soft natural rubber, and these are mounted on spring-loaded pivots released by a messenger.

A *messenger* is a device that is released from a boat and slides down the wire supporting the oceanographic instrument. One form of messenger consists of two slotted cylindrical halves that can be turned 90° with respect to each other and locked. The slots in the two halves are lined up for the messenger to be placed on the wire, and then turned at right angles for use.

Figure 5-17 Oceanographers attaching Nansen bottle to wire (Courtesy, William J. Emery)

Oceanographic systems are customarily set up so that a cascaded messenger system can be employed, one messenger not only triggering the closing of a sampling bottle, but also causing the release of still another messenger to travel on down the wire to the next bottle and so on. The second messenger has a line clipped into the sampling bottle frame, and it is this line that is dropped as the sampler is actuated.

A commercial form, General Oceanics Inc. (Miami, Florida) Model 1010, of the PVC Niskin sampling bottle is available in sizes of 1.7, 5, 8, 10, 12, and 30 liters. These all mount reversing thermometer frames. The bottle end stoppers are held open during lowering, and a messenger-operated trigger system releases them. O-ring seals within the end stoppers are insurance against leakage. Water can be drained from the bottles through stopcocks at the lower ends.

The *Van Dorn bottle* is frequently used to obtain water samples to be later analyzed in the laboratory. The major part of a Van Dorn bottle is a cylindrical PVC tube. The ends of the bottle are closed off by two elastically-coupled stoppers, one at either end. The stoppers are pulled out of the mouths of the bottle before it is lowered into the water, and the tripping mechanism is actuated by a messenger dispatched down the support cable from the surface.

Reversing Thermometers

Reversing thermometers were used at an early stage with Nansen bottles. When a messenger was sent down the wire to the open bottle, it struck a catch at the top, releasing the top of the bottle so that it flipped over, closing the end valves and sealing in the water sample. Thermometers maintained in frames mounted on the Nansen bottles were rapidly reversed at the same time.

Reversing thermometers are designed so that the reversing process freezes the temperature reading at the time the messenger arrived. In this way, the bottle and thermometers, when brought back to the surface, provide an indication of the temperature at depth. The bottom end of the thermometer, when in lowering orientation, houses a mercury reservoir. Above this bulb there is a constriction followed by a loop. When the thermometer is reversed the mercury column breaks at the constriction and the column of liquid formerly above the constriction then flows downward into what is now the bottom of the thermometer. A scale marked on the thermometer converts this column height to temperature. A second thermometer adjacent to the first one is provided to correct the column height shown for the expansion of the glass and mercury at the time the reading is taken. The first thermometer, housed in a heavy glass tube, is known as a *protected thermometer* and the second an *unprotected thermometer*. The difference between the temperatures indicated by the two thermometers gives an indication of the water pressure (depth) at the sampling position. The action of pressure on the unprotected bulb forces out an amount of mercury that corresponds to $0.01°C/m$ of depth (Weyl, 1970). This provides a check on the desired sampling depth.

Salinities from Water Samples

Strickland and Parsons (1972) detail two separate laboratory methods for determining the chlorinity of a sea water sample. A mathematical equation, or, equivalently, tables, is (are) then used to determine the salinity. The *chlorinity* is a measure of the chloride content by mass in the sample. Space limitations preclude a detailed summary of the analytical methods involved.

The ability of a water sample to transmit electrical current (*conductivity*) is also used to determine salinity. Sea water conducts electricity far more readily than fresh water.

The Beckman Instruments, Inc. (Cedar Grove, New Jersey) RS7-B Induction Salinometer is an example of a laboratory instrument used to determine the salinity of a sea water sample. This is done, in this case, by determining conductivity with the instrument and then looking up the corresponding salinity in appropriate tables. The salinometer is zeroed by first filling the sample cell with standard Copenhagen sea water.* Thereafter the conductivity of any other (50 mℓ) sample can be determined. The instrument is designed to properly compensate for up to 30°C differences between the temperature of the standard and the field sample. The RS7-B can deal with salinities from 0 to 49$^0/_{00}$ and has a stated accuracy of $\pm 0.003^0/_{00}$.

Manufacturers distributing conductivity probes make available charts relating conductivity, temperature, and salinity. Part of such a table would contain the following information.

Table 5-1: Conductivity (mmho/cm) of Sea Water

Temperature (°C)	Salinity ($^0/_{00}$)		
	32.0	34.0	36.0
10.0	35.14	37.11	39.06
15.0	39.60	41.81	43.99
20.0	44.21	46.66	49.10
25.0	48.97	51.68	54.37
30.0	53.88	56.86	59.81

Ocean-Going Systems

Oceanographic data concerned with the variation of temperature and salinity with depth are frequently gathered by a system that consists of sensors located at the desired depth and a readout and/or recording system on deck. An electrical cable connects the sensors and the deck unit. This cable, and perhaps also a strength cable for lifting purposes, is stored on a winch drum. There are various designs possible for the sensors.

*An international sea water office is maintained at Copenhagen and distributes sealed glass ampules of sea water of specified chlorinity.

A *thermistor* is a semi-conductor in which a slight temperature change causes a pronounced change in electrical resistance. Accurate measurement of this resistance gives direct reading of the temperature at the thermistor position. The thermistor is a fairly standard oceanographic temperature item.

Conductivity is normally measured, rather than salinity itself, by such sensors. Williams (1973) discusses conductivity salinometers in some detail. A figure for conductivity along with the temperature yields not only the salinity, through a table such as Table 5-1, but also allows acoustic velocity and water density to be determined.

There are various commercially available conductivity/temperature/depth systems for use in coastal waters. Usually the sensor package contains, in addition to conductivity, temperature, and pressure probes, pH and dissolved oxygen sensors, as discussed in Chapter 6. These shallow-water systems have standard read-out scales for the individual parameters involved, but in some cases can be used in conjunction with recorders. A suitable arrangement for plotting the variation with depth of water parameters is shown in Figure 5-18.

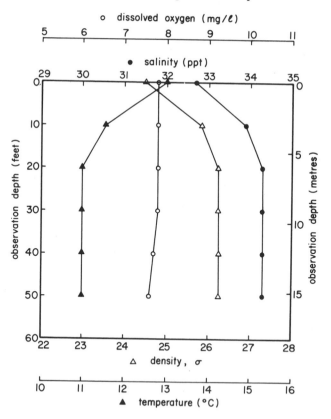

Figure 5-18 Sample plot for depth variation of salinity, temperature, density and dissolved oxygen

The Plessey Corp. (formerly Bissett Berman, San Diego, California) S/T/D is a favorite oceanographic instrument for obtaining salinity-temperature-depth information, but it can be regarded as a high seas, deep-water system. The salinity and temperature data in Figure 2-2 were obtained from such an instrument; the density information shown was calculated from the salinity, temperature, and depth (pressure) information. The Plessey Model 9040 transmits measured data back to recording instruments on the attending ship via an electrical cable whereas the Model 9060 records *in situ*. The latter arrangement has a great advantage in that a standard ship winch, rather than one with slip rings adapted to electrical cables, can be employed. In the Model 9060 only salinity, temperature, and depth are measured, and the data are recorded on a circular chart roll.

Data recording in the case of the Model 9040 can take various forms, the most basic being the multi-channel chart recorder. The Model 9040 has a standard depth range of from 0 to 1500 m with options of up to 3000 or 6000 m, and use of such a complex system in nearshore waters may be regarded as underutilization. The Model 9060 can be ordered for a depth range of 0 to 100 m, however. Whereas the Model 9040 weighs 105 lb (467 N) in air, the Model 9060 (for shallow depths) weighs 53 lb (236 N).

In Situ Systems

Several systems that obtain an *in situ* measure of an oceanographical variable have already been discussed, the bathythermograph, for example. Recording in that case was mechanical. There are systems on the commercial market that can be left at a desired depth for long periods of time to monitor the temperature and salinity histories at the point concerned and to record, in place, these histories. Examples are the Recording Thermograph and Recording Refractometer/Salinometer of Environmental Devices Corporation (Marion, Massachusetts). The latter instrument makes use of the fact that the refractive index of water is a function of salinity.

5-8 ADDITIONAL CONSIDERATIONS INVOLVING BOATS AND FIELD STUDIES

A physical oceanographic study related to the marine disposal of wastewaters demands most of the scientific instruments described in earlier sections of this chapter. In Section 5-3, some details concerning the type of boat to be selected for such a study were included. In this section I will enlarge upon the requirements for a suitable oceanographic vessel and, in addition, briefly discuss operations and nonscientific equipment that are an integral part of field operations. The book by Chapman (1967) provides excellent background on such activities.

The boat selected should, preferably, have a ship-to-shore radio in good repair. As long as the survey work is close enough to shore that sextant angles can be used, the vessel need not have radar. However, there are many regions where closeness to shore does not mean an ability to see shore features because of persistent fog. In such cases radar is a must.

Coast Guard (or equivalent) certification of the boat, plus safety inspection, are minimum requirements. Two anchors of adequate holding capacity for the boat with the existing bottom, shackled and pinned to chain with enough line, in good repair, to provide for 5-to-1 scopes for planned and even unplanned working depths are minimum requirements.

We should also consider the personnel for the survey boat. One at least should be an experienced boat person. Unless the survey area is in well-pro-

Table 5-2: Items for Possible Inclusion in Marine Data-Gathering Expeditions

Positioning and Plotting	Writing	Hardware and Related	Boat and Diving	Miscellaneous
sextant	pencils	bag of tools	boat ladder	extra clothes
hand-bearing compass	erasers plastic slate	binoculars	swim mask	sunglasses
three-arm protractor	grease pencil	camera	swim fins	knife
protractor	underwater-writing paper*	film	snorkel	rope
charts			back pack	twine
dividers	pair of compasses	inverter	SCUBA tank	electrical tape
triangles		alternator	SCUBA regulator	lunch
parallel ruler		voltage transformer	weight belt	sun cream
plastic cover sheet for charts		can of gasoline	life jacket	waterproof watch
			flares	whistle
		anchor	throw ring	hat
		battery	boat hook	container of water
		miscellaneous electrical items	oars and oar locks	
		bolts		
		shackles		

*Appleton Papers, Inc. (Appleton, Wisconsin)

tected waters, individuals without a propensity to get seasick should be selected. Nevertheless, an extra scientific crewmember is always an excellent idea. I have been involved in dye studies where, as it turned out, only two scientific crewmembers were capable of doing any work and they had to carry out the entire multitude of operations involved alone.

Table 5-2 lists items of potential usefulness for a field outing concerned with gathering the types of data considered in this chapter or in Chapter 6. Token comments on major entries in Table 5-2 follow.

Charts for the area concerned are an absolute must, and a table on the vessel selected should be available so that such charts can be laid out. This not only aids in being able to quickly scan the chart for some type of feature or clue but also permits one of the scientific party to quickly work out a position from a pair of sextant angles, using a three-arm protractor. The chart should be thoroughly consulted before disembarking. It is embarrassing and potentially dangerous to suddenly find oneself over a shoal area with a set of big waves looming up outside. The units of measurement used for the depths should be very carefully noted.

Besides the sextant a hand-bearing compass occasionally becomes a useful instrument if, for example, an intended sighting point becomes obscured. With a hand-bearing compass, a sight is taken on a distant (charted) object through a slit and over a pointer mounted on the compass. The magnetic bearing of the sighted object is then read off the compass.

5-9 REFERENCES

BAILEY, T. E. (1966): "Fluorescent-Tracer Studies of an Estuary," *Journal of the Water Pollution Control Federation*, Vol. 38, No. 12, pp. 1986–2001.

BAIRD, W. F., et al. (1971): "Canada's Wave Climate and Field Measurement Program," *Preprints*, Third Annual Offshore Technology Conference, Houston, Texas (April), Vol. 2, pp. 163–170.

BAIRD, W. F., and C. W. GLODOWSKI (1975): "Accelerometer Wave Recording Buoy," *Proceedings*, Civil Engineering in the Oceans III (June), Newark, Delaware, ASCE, pp. 1104–1123.

BATHEN, K. H. (1974): "Results of Circulation Measurements Taken during August 1972 to May 1973 in the Area between Barbers Point and the Entrance to Pearl Harbor, Oahu, Hawaii," University of Hawaii, Look Laboratory of Oceanographic Engineering, *Technical Report No. 34* (July).

BENDAT, J. S., and A. G. PIERSOL (1971): *Random Data*: *Analysis and Measurement Procedures*, Wiley (Interscience), New York.

BERTEAUX, H. O. (1970): "Design of Deep-Sea Mooring Lines," *Marine Technology Society Journal*, Vol. 4, No. 3 (May–June), pp. 33–46.

BERTEAUX, H. O. (1976): *Buoy Engineering*, Wiley, New York.

BLACKMAN, R. B., and J. W. TUKEY (1959): *The Measurement of Power Spectra*, Dover, New York.

BROOKS, N. H. (1959): "Diffusion of Sewage Effluent in an Ocean Current," in *Waste Disposal in the Marine Environment*, edited by E. A. Pearson, pp. 246–267, Pergamon Press, New York.

BURT, W. V., and B. WYATT (1964): "Drift Bottle Observations of the Davidson Current off Oregon," *Studies on Oceanography*, pp. 156–165.

CALDWELL, J. M. (1956): "The Step-Resistance Wave-Gage," *Coastal Engineering Instruments*, Council on Wave Research, The Engineering Foundation, Berkeley, Calif., pp. 44–60.

CAMERON, H. L. (1952): "The Measurement of Water Current Velocities by Parallax Methods," *Photogrammetric Engineering*, Vol. 18, pp. 99–104.

CAMERON, H. L. (1962): "Water Current and Movement Measurement by Time-Lapse Air Photography: An Evaluation," *Photogrammetric Engineering*, Vol. 28, pp. 158–163.

CAMERON, H. L. (1965): "Currents and Photogrammetry," in *Oceanography from Space*, Woods Hole Oceanographic Institution, Woods Hole, Mass. (April), pp. 29–36.

Canadian Tide and Current Tables, 1976 (1975): Environment Canada, Fisheries and Marine Service, Ottawa, Ontario, Canada (in various volumes for different regions).

CARPENTER, J. H. (1960): "Tracer for Circulation and Mixing in Natural Waters," *Public Works* (June), pp. 110–112.

CEDERWALL, K., C. G. GÖRANSSON, and T. SVENSSON (1975): "Subsequent Dispersion—Methods of Measurement," *Discharge of Sewage from Sea Outfalls*, edited by A. L. H. Gameson, pp. 309–319, Pergamon Press, Oxford, England.

CHAPMAN, C. F. (1967): *Piloting, Seamanship and Small Boat Handling*, Motor Boating, New York.

CROSS, R. H. (1968): "Tide Gage Frequency Response," ASCE, *Journal of the Waterways and Harbors Division*, Vol. 94, No. WW3 (August), pp. 317–330.

CROWE, F. J., and R. A. SCHWARTZLOSE (1972): "Release and Recovery Records of Drift Bottles in the California Current Region 1955 through 1971," State of California, Marine Research Committee, Calif. Cooperative Oceanic Fisheries Investigations, *Atlas No. 16* (June).

DAHL, O. (1969): "The Capability of the Aanderaa Recording and Telemetering Instrument," in *Progress in Oceanography*, Vol. 5, edited by M. Sears, pp. 103–106, Pergamon Press, New York.

FEUERSTEIN, D. L., and R. E. SELLECK (1963a): "Tracers for Dispersion Measurements in Surface Waters," University of California, Berkeley, Calif., Sanitary Engineering Research Laboratory, *Report No. 63-1* (February).

FEUERSTEIN, D. L., and R. E. SELLECK (1963b): "Fluorescent Tracers for Dispersion

Measurements," ASCE, *Journal of the Sanitary Engineering Division*, Vol. 89, No. SA4, pp. 1–21.

FOXWORTHY, J. E., R. B. TIBBY, and G. M. BARSOM (1966): "Dispersion of a Surface Waste Field in the Sea," *Journal of the Water Pollution Control Federation*, Vol. 38, No. 7 (July), pp. 1170–1193.

GAUL, R. D., and H. B. STEWART, JR. (1960): "Nearshore Ocean Currents off San Diego, California," *Journal of Geophysical Research*, Vol. 65, pp. 1543–1556.

GRACE, R. A. (1970): "How to Measure Waves," *Ocean Industry*, Vol. 5, No. 2 (February), pp. 65–69.

HALLERMEIER, R. J., and W. R. JAMES (1974): "Development of a Shallow-Water Wave Direction Gage, Measurement and Analysis," *Proceedings of the International Symposium on Ocean Wave*, New Orleans, La. (September), ASCE, pp. 696–712.

HALPERN, D., R. D. PILLSBURY, and R. L. SMITH (1974): "An Inter-comparison of Three Current Meters Operated in Shallow Water," *Deep-Sea Research*, Vol. 21, pp. 489–497.

HARREMÖES, P. (1966): "Prediction of Pollution from Planned Wastewater Outfalls," *Journal of the Water Pollution Control Federation*, Vol. 38, pp. 1323–1333.

HARRIS, D. L. (1970): "The Analysis of Wave Records," *Proceedings*, Twelfth Coastal Engineering Conference (September), Washington, D.C., ASCE, pp. 85–100.

HARRIS, D. L. (1972): "Wave Estimates for Coastal Regions," in *Shelf Sediment Transport*, edited by D. J. P. Swift, D. B. Duane, and O. H. Pilkey, Dowden, Hutchinson and Ross, Inc., Stroudsburg, Pa., pp. 99–125.

HARRISON, W., et al. (1967): "Circulation of Shelf Waters off the Chesapeake Bight: Surface and Bottom Drift of Continental Shelf Waters between Cape Henlopen, Delaware, and Cape Hatteras, North Carolina, June 1963–December 1964," U.S. Dept. of Commerce, Environmental Science Services Administration, Professional Paper 3 (July).

HOGBEN, N., and F. E. LUMB (1967): *Ocean Wave Statistics*, Her Majesty's Stationery Office, London, England.

HOUMB, O. G. (1973): "A Norwegian Wave Climate Study," *Proceedings*, Second International Conference on Port and Ocean Engineering under Arctic Conditions, Reykjavik, Iceland (August), edited by T. Karlsson, pp. 772–779.

Instrument Manual for Oceanographic Observations (1955): U.S. Naval Oceanographic Office, *H. O. Publ. No. 607*, U.S. Government Printing Office, Washington, D.C.

KALVAITIS, A. N. (1972): "Survey of the Savonius Rotor Performance Characteristics," *Marine Technology Society Journal*, Vol. 6, No. 4 (July–August), pp. 17–20.

KANWISHER, J., and K. LAWSON (1975): "Electromagnetic Flow Sensors," *Limnology and Oceanography*, Vol. 20, No. 2 (March), pp. 174–182.

KARWEIT, M. (1974): "Response of a Savonius Rotor to Unsteady Flow," *Journal of Marine Research*, Vol. 32, No. 3, pp. 359–364.

KINARIWALA, B. K., F. F. KUO, and N-K. TSAO (1973): *Linear Circuits and Computation*, Wiley, New York.

KIRWAN, A. D., JR., et al. (1975): "The Effect of Wind and Surface Currents on Drifters," *Journal of Physical Oceanography*, Vol. 5, No. 2 (April), pp. 361–368.

KLEMAS, V., et al. (1974): "Monitoring Coastal Water Properties and Current Circulation with ERTS—1," *Scientific and Technical Aerospace Reports*, Vol. 12, No. 20 (October 23), p. 2423 (abstract only).

LAWSON, N. V., and C. L. ABERNETHY (1975): "Long Term Wave Statistics off Botany Bay," *Preprints*, Second Australian Conference on Coastal and Ocean Engineering, Surfers Paradise, pp. 167–176.

LOMAS, C. G., and C. H. MARKS (1975): "Low Velocity Current Meters—Their Capabilities," *Marine Technology Society Journal*, Vol. 9, No. 7 (August), pp. 17–19.

Manual of Current Observations (1950): U.S. Dept. of Commerce, Coast and Geodetic Survey, *Special Publication No. 215*, rev. ed.

McALLISTER, R. F. (1968): "Demonstration of the Limitations and Effects of Waste Disposal on an Ocean Shelf," Florida Atlantic Ocean Sciences Institute, Annual Project Report (March) [*NTIS Publication PB-215-585*].

McLEOD, W. R. (1976): "Operations Experience with a Wave and Wind Measurement Program in the Gulf of Alaska," *Proceedings*, Eighth Annual Offshore Technology Conference, Houston, Texas (May), Vol. 2, pp. 719–733.

McLEOD, W. R., L. C. ADAMO, and R. C. HAMILTON (1976): "Installed System Redundancy for Wave Measurements—Gulf of Alaska," *Journal of Petroleum Technology*, Vol. 28, No. 4 (April), pp. 482–488.

McLEOD, W. R., and D. HODDER (1976): "Environmental Baseline Study of the Gulf of Alaska by Remote Sensing," *Proceedings*, Eighth Annual Offshore Technology Conference, Houston, Texas (May), Vol. 2, pp. 683–696.

MONAHAN, E. C., and E. A. MONAHAN (1973): "Trends in Drogue Design," *Limnology and Oceanography*, Vol. 18, No. 6 (November), pp. 981–985.

MURRAY, S. P. (1975): "Trajectories and Speeds of Wind-Driven Currents near the Coast," *Journal of Physical Oceanography*, Vol. 5, No. 2 (April), pp. 347–360.

NATIONAL OCEANOGRAPHIC INSTRUMENTATION CENTER (1976): "Index to National Oceanographic Center Publications," Washington, D.C. (January).

"Oceanographic Study Specifications for the City of Watsonville Clean Water Grant Project" (1976): State of California, State Water Resources Control Board (January 27).

Operation and Service Manual for Waverider, Datawell bv, Laboratorium voor Instrumentatie, (n.d.) Haarlem, Netherlands.

PARK, P. K. (1968): "Alteration of Alkalinity, pH and Salinity of Seawater by Metallic Water Samplers," *Deep-Sea Research*, Vol. 15, pp. 721–722.

PEACOCK, H. G. (1974): "CERC Field Wave Gaging Program," *Proceedings*, International Symposium on Ocean Wave Measurement and Analysis, New Orleans, La. (September), ASCE, pp. 170–185.

POLLARD, R. (1973): "Interpretation of Near-Surface Current Meter Observations," *Deep-Sea Research*, Vol. 20, pp. 261–268.

"Precious Equipment Lost" (1973): *The Press*, Christchurch, N.Z. (February 23).

REHRER, R., A. C. JONES, and M. A. ROESSLER (1967): "Bottom Water Drift on the Tortugas Grounds," *Bulletin of Marine Science*, Vol. 17, No. 3 (September), pp. 562–575.

RICHARDSON, L. F., and H. STOMMEL (1948): "Note on Eddy Diffusion in the Sea," *Journal of Meteorology*, Vol. 5, pp. 238–240.

RICHARDSON, W. S., P. B. STIMSON, and C. H. WILKINS (1963): "Current Measurements from Moored Buoys," *Deep-Sea Research*, Vol. 10, pp. 369–388.

RUSSELL, T. L. (1963): "A Step-Type Recording Wave Gage," *Ocean Wave Spectra*, pp. 251–257, Prentice-Hall, Englewood Cliffs, N.J.

RUZECKI, E. P. (1976): "The Use of the EOLE Satellite System to Observe Continental Shelf Circulation," *Proceedings*, Eighth Annual Offshore Technology Conference, Houston, Texas (May), Vol. 2, pp. 697–708.

Sailing Directions—British Columbia Coast (South Portion) Vol. 1, 9th ed. (1974): Dept. of the Environment, Marine Sciences Directorate, Canadian Hydrographic Service, Ottawa, Ontario, Canada.

SERKIN, H. C., and M. KRONENGOLD (1974): "The Effects of Tilt on a Savonius Rotor Exposed to a Turbulent Flow Regime," *IEEE Transactions on Geoscience Electronics*, Vol. GE-12, No. 2 (April), pp. 55–69.

SEYMOUR, R. J., et al. (1976): "Coastal Engineering Data Network," University of California, La Jolla, Calif., Institute of Marine Resources, Sea Grant Publication No. 50 (July) (with monthly updates).

SHANNON, L. V., G. H. STANDER, and J. A. CAMPBELL (1973): "Oceanic Circulation Deduced from Plastic Drift Cards," Republic of South Africa, Dept. of Industries, Sea Fisheries Branch, *Investigational Report No. 108.*

SNODGRASS, J. M. (1968): "Instrumentation and Communications," in *Ocean Engineering: Goals, Environment, Technology*, edited by J. F. Brahtz, pp. 393–477, Wiley, New York.

STALCUP, M. C., and W. G. METCALF (1972): "Current Measurements in the Passages of the Lesser Antilles," *Journal of Geophysical Research*, Vol. 77, No. 6 (February 20), pp. 1032–1049.

STEVENSON, M. R., R. W. GARVINE, and B. WYATT (1974): "Lagrangian Measurements in a Coastal Upwelling Zone off Oregon," *Journal of Physical Oceanography*, Vol. 4, No. 3 (July), pp. 321–336.

STICKLAND, J. A., and R. H. BIGHAM (1976): "Techniques for Mooring Oceanographic Instruments for Small Vessels," Institute of Ocean Sciences, Patricia Bay, Victoria, B.C., Canada, *Pacific Marine Science Report 76-9* (March), unpublished manuscript.

STOMMEL, H. (1949): "Horizontal Diffusion Due to Oceanic Turbulence," *Journal of Marine Research*, Vol. 8, pp. 199–225.

STRICKLAND, J. D. H., and T. R. PARSONS (1972): *A Manual of Seawater Analysis*, Fisheries Research Board of Canada, *Bulletin 167*, 2nd ed., Ottawa, Canada.

TERHUNE, L. D. B. (1968): "Free-Floating Current Followers," Fisheries Research Board of Canada, *Technical Report No. 85* (July).

THOMPSON, E. F. (1974): "Results from the CERC Wave Measurement Program," *Proceedings of the International Symposium on Ocean Wave Measurement and Analysis*, New Orleans, La. (September), ASCE, pp. 836–855.

THOMPSON, E. F., and D. L. HARRIS (1972): "A Wave Climatology for U.S. Coastal Waters," *Preprints*, Fourth Annual Offshore Technology Conference, Houston, Texas (May), Vol. 2, pp. 675–688.

TIBBY, R. B. (1960): "Inshore Circulation Patterns and the Oceanic Disposal of Waste," *Waste Disposal in the Marine Environment*, edited by E. A. Pearson, pp. 296–327, Pergamon Press, New York.

Tidal Current Tables 1976 (1975): U.S. Dept. of Commerce, National Oceanic and Atmospheric Administration, National Ocean Survey, Washington, D.C. (in various volumes for different regions).

Tide Tables 1976 (1975): U.S. Dept. of Commerce, National Oceanic and Atmospheric Administration, National Ocean Survey, Washington, D.C. (in various volumes for different regions).

Turner Laboratory Instruments (1972): G. K. Turner Associates, Palo Alto, Calif.

ULTRAMAR CHEMICAL WATER LABORATORY (1969): "Water Quality Study: Nearshore Waters of the Island of Kauai," Report prepared for State of Hawaii, Dept. of Health, Honolulu, Hawaii (July).

United States Coast Pilot 7, Pacific Coast California, Oregon, Washington, and Hawaii 11th ed. (1975): U.S. Dept. of Commerce, National Oceanic and Atmospheric Administration, National Ocean Survey, Washington, D.C. (June).

VACHON, W. A. (1973): "Scale Model Testing of Drogues for Free Drifting Buoys," The Charles Stark Draper Laboratory, Inc., Cambridge, Mass., *Technical Report R-769* (September).

VACHON, W. A. (1974): "Improved Drifting Buoy Performance by Scale Model Drogue Testing," *Marine Technology Society Journal*, Vol. 8 (January), pp. 58–62.

VAN DORN, W. G. (1959): "Large Volume Water Samples," *Transactions of the American Geophysical Union*, Vol. 37, p. 682.

VOLKMANN, G., J. KNAUSS, and A. VINE (1956): "The Use of Parachute Drogues in the

Measurement of Subsurface Currents," *Transactions of the American Geophysical Union*, Vol. 37, No. 5, pp. 573–577.

VON ARX, W. S. (1962): *An Introduction to Physical Oceanography*, Addison-Wesley, Reading, Mass.

WALDEN, R. G., and D. WEBB (1964): "Methods of Locating and Tracking Buoys," *Transactions of the 1964 Buoy Technology Symposium*, Marine Technology Society, Washington, D.C. (March), pp. 317–323.

WALTHER, J. A., and J. J. LEE (1975): "Measurement of Wave Energy Transmission through the San Pedro Breakwater," University of Southern California, Ocean Engineering Program, *Sea Grant Publication No. USC-SG-1-75* (May).

WEYL, P. K. (1970): *Oceanography: An Introduction to the Marine Environment*, Wiley, New York.

WILLIAMS, J. (1973): *Oceanographic Instrumentation*, Naval Institute Press, Annapolis, Md.

WILLIAMS, L. C. (1969): "CERC Wave Gages," U.S. Army, Corps of Engineers, Coastal Engineering Research Center, *Technical Memorandum No. 30* (December).

WILSON, J. F., JR. (1968): "Fluorometric Procedures for Dye Tracing," U.S. Dept. of the Interior, Geological Survey, *Applications of Hydraulics*, Book 3, Chapter A12.

WYATT, B., et al. (1967): "Measurements of Subsurface Currents off the Oregon Coast Made by Tracking of Parachute Drogues," Oregon State University, Dept. of Oceanography, *Data Report No. 26* (October).

WYRTKI, K., et al. (1967): "Oceanographic Observations during 1965–1967 in the Hawaiian Archipelago," University of Hawaii, Hawaii Institute of Geophysics, *Publication No. HIG-67-15* (August).

WYRTKI, K., V. GRAEFE, and W. PATZERT (1969): "Current Observations in the Hawaiian Archipelago," University of Hawaii, Hawaii Institute of Geophysics, *Publication No. HIG-69-15* (July).

YESKE, L. A., and T. GREEN (1975): "Short Period Variations in a Great Lakes Coastal Currents by Aerial Photography," *Journal of Physical Oceanography*, Vol. 5, No. 1 (January), pp. 125–135.

ZWARTS, C. M. G. (1975): "A Transmission Line Wave Height and Level Transducer," National Research Council, Canada, Division of Mechanical Engineering, Reprint (April).

5-10 BIBLIOGRAPHY

General

COX, D. R., and P. A. W. LEWIS (1966): *The Statistical Analysis of Series of Events*, Wiley, New York.

GRAHAM, M. J., G. R. MILLER, and P. E. SCHAFER (1977): "Comprehensive Oceano-graphic Baseline Study at Barceloneta, Puerto Rico," *Journal of the Water Pollution Control Federation*, Vol. 49, No. 4 (April), pp. 558–567.

JENKINS, G. M., and D. G. WATTS (1968): *Spectral Analysis and Its Applications*, Holden-Day, San Francisco, Calif.

KOONTZ, W. A., and D. L. INMAN (1967): "A Multi-Purpose Data Acquisition System for Instrumentation of the Nearshore Environment," U.S. Army, Corps of Engineers, Coastal Engineering Research Center, *Tech. Mem. No. 21* (August).

KRAMER, W. P., and R. H. WEISBERG (n.d.): "Fortran Graphics Programs for Physical Oceanographic and Time Series Data," University of Rhode Island, *Marine Technical Report* 46.

Oceanography of the Nearshore Coastal Waters of the Pacific Northwest Relating to Possible Pollution (1971): Prepared by Oregon State University, Corvallis, Oregon, for the Environmental Protection Agency, Water Quality Office, Grant No. 16070 EOK, 2 Vols. (July).

PARKHURST, J. D., W. E. GARRISON, and M. L. WHITT (1964): "Effects of Wind, Tide and Weather on Nearshore Ocean Conditions," *Advances in Water Pollution Research*, Vol. 3, edited by E. A. Pearson, pp. 199–212, Pergamon Press, New York.

TELEKI, P. G., F. R. MUSIALOWSKI, and D. A. PRINS (1976): "Measurement Techniques for Coastal Waves and Currents," U.S. Army, Corps of Engineers, Coastal Engineering Research Center, Fort Belvoir, Virginia, *Miscellaneous Report No. 76-11* (November).

VAN HAAGEN, R. H. (1969): "Oceanographic Instrumentation," in *Handbook of Ocean and Underwater Engineering*, edited by J. J. Myers, C. H. Holm, and R. F. McAllister, pp. 3-57 to 3-112, McGraw-Hill, New York.

WEBSTER, F. (1964): "Some Perils of Measurement from Moored Ocean Buoys," Marine Technology Society, *Transactions of the 1964 Buoy Technology Symposium*, pp. 33–48.

Remote Sensing

BADGLEY, P. C., L. MILOY, and L. CHILDS, Eds. (1969): *Oceans from Space*, Gulf Publishing Co., Houston, Texas.

GOWER, J. F. R. (1972): "A Survey of the Uses of Remote Sensing from Aircraft and Satellites in Oceanography and Hydrography," Dept. of the Environment, Marine Sciences Branch, Pacific Region, Victoria, B.C., Canada, *Pacific Marine Science Report 72-3*.

HARRIS, G. P., R. P. BUKATA, and J. E. BURTON (1976): "Satellite Observations of Water Quality," ASCE, *Transportation Engineering Journal*, Vol. 102, pp. 537–554.

MAGOON, O. T., J. W. JARMAN, and D. W. BERG (1971): "Use of Satellites in Coastal Engineering," *Proceedings*, First International Conference on Port and Ocean Engineering under Arctic Conditions, Norway, Vol. 2, pp. 1234–1255.

MAUL, G. A. (1977): "Recent Progress in the Remote Sensing of Ocean Surface Currents," *Marine Technology Society Journal*, Vol. 11, No. 1, pp. 5–13.

TELEKI, P. G., and D. A. PRINS (1973): "Photogrammetric Experiments on Nearshore Mixing and Diffusion," *Proceedings*, Second International Conference on Port and Ocean Engineering under Arctic Conditions, Reykjavik, Iceland (August), edited by T. Karlsson, pp. 251–265.

TELEKI, P. G., J. W. WHITE, and D. A. PRINS (1973): "A Study of Oceanic Mixing with Dyes and Multispectral Photography," *Proceedings of the ASP Symposium on Remote Sensing in Oceanography* (October), Lake Buena Vista, Fla., American Society of Photogrammetry, pp. 772–787.

Boat Operations and Moorings

McLOAD, K. W., and W. E. BOWERS (1964): "Torque Balanced Wire Rope and Armored Cables," *Transactions of the 1964 Buoy Technology Symposium*, Marine Technology Society (March), Washington, D.C., pp. 341–357.

MEALS, W. D., et al. (1969): "Rigging, Tackle, and Techniques," in *Handbook of Ocean and Underwater Engineering*, edited by J. J. Myers, C. H. Holm, and R. F. McAllister, pp. 4-32 to 4-90, McGraw-Hill, New York.

NATH, J. H. (1971): "Dynamic Response of Taut Lines for Buoys," *Marine Technology Society Journal*, Vol. 5, No. 4 (July–August), pp. 44–46.

SMITH, C. E., T. YAMAMOTO, and J. H. NATH (1974): "Longitudinal Vibration in Taut-Line Moorings," *Marine Technology Society Journal*, Vol. 8, No. 5 (June), pp. 29–35.

SNYDER, A. E. (1969): "Winches and Deck Machinery," in *Handbook of Ocean and Underwater Engineering*, edited by J. J. Myers, C. H. Holm, and R. F. McAllister, pp. 4-90 to 4-123, McGraw-Hill, New York.

SNYDER, R. M. (1969): "Buoys and Buoy Systems," in *Handbook of Ocean and Underwater Engineering*, edited by J. J. Myers, C. H. Holm, and R. F. McAllister, pp. 9-81 to 9-115, McGraw-Hill, New York.

Waves

BADGETT, H. H. (1967): "A Radar Wave Height Sensor," *Proceedings of Offshore Exploration Conference* (OECON), pp. 129–149.

CARTWRIGHT, D. E., and N. D. SMITH (1964): "Buoy Techniques for Obtaining Directional Wave Spectra," *Transactions of the 1964 Buoy Technology Symposium*, Marine Technology Society (March), Washington, D.C., pp. 173–182.

DARBYSHIRE, J. (1970): "Wave Measurements with a Radar Altimeter over the Irish Sea," *Deep-Sea Research*, Vol. 17, No. 5 (October), pp. 893–901.

DRAPER, L. (1967): "Instruments for Measurement of Wave Height and Direction in and Around Harbours," *Proceedings of the Institution of Civil Engineers*, Vol. 37 (May), pp. 213–219.

FARMER, H. G. (1963): "A Data Acquisition and Reduction System for Wave Measurement," in *Ocean Wave Spectra*, pp. 227–233, Prentice-Hall, Englewood Cliffs, N.J.

HALPERN, D., et al. (1975): "Surface Wave Height Measurements Made Near the Oregon Coast during August 1972, and July and August 1973," U.S. Dept. of Commerce, National Oceanic and Atmospheric Administration, Environmental Research Laboratories, Boulder, Colorado, *NOAA Technical Report ERL 324-PMEL 22* (April).

HARRIS, D. L. (1974): "Finite Spectrum Analysis of Wave Records," *Proceedings of the International Symposium on Ocean Wave Measurement and Analysis*, New Orleans, La. (September), ASCE, pp. 107–124.

McGRATH, B. L., and D. C. PATTERSON (1973): "Wave Climate at Gold Coast, Queensland," *Preprints*, First Australian Conference on Coastal Engineering, Sydney (May), pp. 8–15.

PIERSON, W. J., JR., G. NEUMANN, and R. W. JAMES (1955): *Practical Methods for Observing and Forecasting Ocean Waves by Means of Wave Spectra and Statistics*, U.S. Navy, Hydrographic Office, *Publication No. 603*.

PLOEG, J. (1971): "Wave Climate Study—Great Lakes and Gulf of St. Lawrence," The Society of Naval Architects and Marine Engineers, *Technical and Research Bulletin No. 2-17* (May).

WATERS, C. B. (1975): "Experiences in the Operation of Waverider Buoys," International Association for Hydraulic Research, *Proceedings*, 16th Congress, São Paulo, Brazil (July–August), Vol. 1, p. 384.

ZOPF, D. O., H. C. CREECH, and W. H. QUINN (1976): "The Wavemeter: A Land-based System for Measuring Nearshore Ocean Waves," *Marine Technology Society Journal*, Vol. 10, No. 4 (May), pp. 19–25.

Currents

BARTOLINI, C., and E. PRANZINI (1977): "Tracing Nearshore Bottom Currents with Sea-bed Drifters," *Marine Geology*, Vol. 23, pp. 275–284.

BELL, W. H., D. M. FARMER, and G. R. KAMITAKAHARA (1976): "A Field Translation System for Aanderaa Data Tapes," Environment Canada, Institute of Ocean Sciences, Patricia Bay, Victoria, B.C., Canada, *Pacific Marine Science Report 76-7* (unpublished manuscript).

CONOMOS T. J., et al. (1970): "Movement of Seabed Drifters in the San Francisco Bay Estuary and the Adjacent Pacific Ocean: A Preliminary Report," U.S. Geological Survey, *Circular 637-B*.

CRAIG, R. E. (1962): "Water Movements over the Sea Floor," *Scottish Fisheries Bulletin*, No. 17 (June), pp. 14–15.

FLEMING, R. H., and D. HEGGARTY (1962): "Recovery of Drift Bottles Released in the

Southeastern Chukchi Sea and North Bering Sea," University of Washington, Dept. of Oceanography, *Technical Report No. 70* (February).

GOULD, W. J., and E. SAMBUCO (1975): "The Effect of Mooring Type on Measured Values of Ocean Currents," *Deep-Sea Research*, Vol. 22, pp. 55–62.

HORRER, P. L. (1967): "Methods and Devices for Measuring Currents," *Estuaries*, edited by G. H. Lauff, American Association for the Advancement of Science, Washington, D.C., *Publication No. 83.*

LAEVASTU, T., D. E. AVERY, and D. C. COX (1964): "Coastal Currents and Sewage Disposal in the Hawaiian Islands," University of Hawaii, Hawaii Institute of Geophysics, *Report No. HIG-64-1* (June).

LANDRY, L. P. (1976): "Radar Tracking of Drift Drogues in Pendrell Sound and Port Mellon, June and September 1974," Environment Canada, Institute of Ocean Sciences, Patricia Bay, Victoria, B.C., Canada, *Pacific Marine Science Report 76-8* (March) (unpublished manuscript).

LAUZIER, L. M. (1967): "Bottom Residual Drift in the Continental Shelf Area of the Canadian Atlantic Coast," *Journal of the Fisheries Research Board of Canada*, Vol. 24, No. 9, pp. 1845–1859.

LEE, A. J., D. F. BUMPUS, and L. M. LAUZIER (1965): "The Sea-Bed Drifter," International Commission for the Northwest Atlantic Fisheries, *Research Bulletin No. 2*, pp. 42–47.

MAYHUE, R. J., and R. W. LOVELADY (1976): "Acoustic Tracking of Woodhead Drifters," *Proceedings of Oceans '76*, Washington, D. C. (September), Marine Technology Society, *Paper 15B.*

MONAHAN, E. C., B. J. HIGGINS, and G. T. KAYE (1975): "A Comparison of Vertical Drift-Envelopes to Conventional Drift-Bottles," *Limnology and Oceanography*, Vol. 20, pp. 141–147.

MORSE, B. A., M. G. GROSS, and C. A. BARNES (1968): "Movement of Seabed Drifters near the Columbia River," ASCE, *Journal of the Waterways and Harbors Division*, Vol. 94, No. WW1 (February), pp. 93–103.

SCHUMACHER, J. D., and B. J. KORGEN (1976): "A Seabed Drifter Study of Near-Bottom Circulation in North Carolina Shelf Waters," *Estuarine and Coastal Marine Science*, Vol. 4, pp. 207–214.

SQUIRE, J. L., JR. (1969): "Observations on Cumulative Bottom Drift in Monterey Bay using Seabed Drifters," *Limnology and Oceanography*, Vol. 14, pp. 163–167.

WIEGEL, R. L., and J. W. JOHNSON (1960): "Ocean Currents, Measurement of Currents and Analysis of Data," *Waste Disposal in the Marine Environment*, edited by E. A. Pearson, pp. 175–245, Pergamon Press, New York.

WINANT, C. D., and J. R. OLSON (1976): "The Vertical Structure of Coastal Currents," *Deep-Sea Research*, Vol. 23, pp. 925–936.

Diffusion and Dispersion

BAUMGARTNER, D. J., M. H. FELDMAN, and C. L. GIBBONS (1971): "A Procedure for Tracing of Kraft Mill Effluent from an Ocean Outfall by Constituent Fluorescence," *Water Research*, Vol. 5, pp. 533–544.

BOURRET, R., and S. BROIDA (1960): "Turbulent Diffusion in the Sea," *Bulletin of Marine Science of the Gulf and Caribbean*, Vol. 10, pp. 354–366.

BRUUN, P. (1970): "Use of Tracers in Harbor, Coastal and Ocean Engineering," *Engineering Geology*, Vol. 4, No. 1 (January), pp. 73–88.

CEDERWALL, K. (1971): "A Float Diffusion Study," *Water Research*, Vol. 5, pp. 889–907.

CSANADY, G. T. (1963): "Turbulent Diffusion in Lake Huron," *Journal of Fluid Mechanics*, Vol. 17, pp. 360–384.

DENNER, W. W., T. GREEN, and W. H. SNYDER (1968): "Large Scale Oceanic Drogue Diffusion," *Nature*, Vol. 219, pp. 361–362.

DIACHISHIN, A. N. (1963): "Dye Dispersion Studies," ASCE, *Journal of the Sanitary Engineering Division*, Vol. 89, No. SA1 (January), pp. 29–49.

HARRIS, T. F., et al. (1964): "Mixing in the Surf Zone," *Advances in Water Pollution Research*, Vol. 3, Pergamon Press, New York.

ICHIYE, T. (1967): "Upper Ocean Boundary-Layer Flow Determined by Dye Diffusion," *Boundary Layers and Turbulence, The Physics of Fluids*, Supplement, pp. S270–S277.

KATZ, B., R. GERARD, and M. COSTIN (1965): "Response of Dye Tracers to Sea Surface Conditions," *Journal of Geophysical Research*, Vol. 70, pp. 5505–5513.

PRITCHARD, D. W., and J. H. CARPENTER (1960): "Measurements of Turbulent Diffusion in Estuaries and Inshore Waters," *Bulletin of the International Association of Scientific Hydrology*, No. 20 (December), pp. 37–50.

VERWEY, C. J., and W. R. MCMURRAY (n.d.): "Tracers for the Study of Mixing in the Surf," Council for Scientific and Industrial Research, Pretoria, South Africa, Research Report 222, pp. 45–49.

VISSER, M. P. (1966): "Note on the Estimation of Eddy Diffusivity from Salinity and Current Observations," *Netherlands Journal of Sea Research*, Vol. 3, pp. 21–27.

Oceanographic Studies for Proposed Outfalls

BOLEY, S. L., and L. S. SLOTTA (1974): "Relevant Data Concerning Proposed Discharges of Domestic Wastes into Netarts Bay, Oregon," Oregon State University, Corvallis, Oregon, Engineering Expt. Station, *Bulletin No. 50* (December) [*NTIS Publication COM-75-10309*].

MURTHY, C. R., and G. T. CSANADY (1971): "Outfall Simulation Experiment in Lake Ontario," *Water Research*, Vol. 5, pp. 813–822.

OAKLEY, H. R., and E. A. DYER (1966): "Investigation of Sea Outfalls for Tyneside

Sewage Disposal," *Proceedings*, The Institution of Civil Engineers, Vol. 33, pp. 201–230.

"Oceanographic Study Specifications for the Humboldt Bay Wastewater Authority Clean Water Grant Project" (1976): State of California, State Water Resources Control Board (January 27).

"Oceanographic Study Specifications for the Monterey Peninsula Water Pollution Control Agency Clean Water Grant Project" (1976): State of California, State Water Resources Control Board (January 27).

RAMBOW, C. A., and P. V. HENNESSY (1965): "Oceanographic Studies for a Small Wastewater Outfall," *Journal of the Water Pollution Control Federation*, Vol. 37, No. 11 (November), pp. 1471–1480.

RANCE, P. J. (1966): "Investigation of Wind-Induced Currents and Their Effects on the Performance of Sea Outfalls," *Proceedings*, The Institution of Civil Engineers, Vol. 33, pp. 231–260.

STEVENSON, R. E. (1962): "Sewage Disposal and the Sea in Southern California," *Water and Sewage Works*, Vol. 109, No. 12 (December), pp. 452–456 and Vol. 110, No. 1 (January, 1963), pp. 49–51.

6

Collecting Other

Pertinent Marine Data

6-1 GEOPHYSICAL INVESTIGATIONS

Bathymetry

Introduction: The most fundamental type of information required for a coastal region concerns the distribution of water depths. This is referred to as the _bathymetry_ of the area. For numerous coastal areas around the world, considerable detail on water depths exists and is shown on appropriate charts discussed in Appendix A.

Although charts provide considerable detail on bathymetry, they seldom provide enough detail for proper design of an ocean outfall system, and it is common that additional bathymetric surveys be run in order to check existing data and to provide the more detailed information. I am familiar, for example, with a region in nearshore waters off Waikiki, Hawaii, where published charts gave no indication of a trench paralleling the shoreline, and it was apparent from watching waves passing through the area that such a trench should exist. I verified my suspicion by actually surveying the area by lead-line. The _lead-line_ consists of a heavy object on the end of a line marked off in a convenient system of units. The heavy object, when thrown from the vessel, falls to the bottom, taking the line with it. The depth is then inferred from the reading on the line at the water surface. There are practical problems with this technique, such as the angle of the line with the vertical (owing to relative boat movement) and wave action interfering with the definition of the water surface.

Lead-line surveys of the surf zone are sometimes done using helicopters. The cable to the lead contains a weak link so that it will break readily if the lead becomes snagged on an underwater obstruction. The cable is suitably marked, and a surveyor on shore with a level can read off the amount of line below his elevation. He takes a reading when a horn sounds on the helicopter—indicating that the lead has touched bottom. At the same instant, two angles from a known base line to the cable are set by two operators with transits on shore and the position of the survey point can then be geometrically determined. The chart water depths at the various survey points are obtained by relating the level operator's elevation to chart datum.

Depths by Echo-Sounding: Although chart data have been obtained using the lead-line approach, by far the majority have been obtained using depth-sounding techniques. The *fathometer* has been in use for several decades as a means of quickly determining water depths from the water surface. The fathometer uses ultrasonic pulses emitted by a source that is most often mounted on the hull of a boat. These pulses, involving frequencies of approximately 7–200 kHz depending upon the commercial model, are directed vertically downward in a narrow beam (e.g., 3°) and reflected off the bottom. The pulse echoes are then received by a *hydrophone** mounted on the boat. The time between transmission of a particular pulse and its reception back at the boat is translated into water depth, knowing or assuming the velocity of sound in the water, and displayed on a suitable chart recorder. The fathometer can be used to obtain water depth while a boat is underway, so that a vessel can obtain the depth profile along a desired line for which the end points are known using one of the positioning methods indicated in Appendix A. Insufficient depth data on a published chart for an area of interest can quite readily be overcome by using a boat equipped with suitable fathometric instrumentation and an accurate means of positioning. Allowance for the stage in the tide when the survey was carried out must be made when transferring depths to a chart, although in many fathometer systems it is possible to make allowance for tidal stage by using an internal control.

Bigger boats and ships usually have their own permanent fathometer systems with the *transducer* (source) hull-mounted and the display in the wheel house. Many bathymetric surveys are run from small boats, however, that do not have such integral systems. For this reason, portable systems, which can easily be shipped, if need be, mounted, and then taken down, have great utility. There are many such systems commercially available.

There are various problems with fathometric depth sounding systems that require mentioning. The translation of the time of ultrasonic pulse round-trip into the depth, for display on the fathometer recorder, is done using a preset constant value of the velocity of sound in sea water, typically 4800 fps (1463 m/s). If the local oceanographic conditions are such that this value, or whatever else has been set, is in error, then the depths displayed will also be in error. The papers by Wilson (1960) and Del Grosso (1972, 1974), for example, present data

---

*An electroacoustic transducer that responds to water-borne sound waves and delivers essentially equivalent electric waves.

showing how the acoustic velocity varies with temperature, salinity, and pressure. Albers (1960) has given an empirical expression for the acoustic velocity of water in cm/s as

$$c = 141{,}000 + 421\,T - 3.7\,T^2 + 110\,S + 0.018\,d \qquad (6\text{-}1)$$

where T is the temperature (°C)
$\quad S$ is the salinity ($^0/_{00}$),
$\quad d$ is the depth (cm).

Fathometer depth sounding systems work best when the bottom is firm. Since ooze and slit are poor reflectors, a "bottom" trace in such cases may be fuzzy. On the other hand, however, the fuzziness of the bottom trace may allow a skilled operator to judge the nature of the seabed.

When the outboard motor on a boat is reversed, or when the propeller on a bigger vessel is reversed, aerated water passes under the vessel. The result, when a fathometric survey is being run, is to lessen the sound intensity generated by the transducer, with a consequent poorer definition of the bottom. Heavy wave action can also pump unusually large amounts of air into the surface water with equivalent results.

Heavy wave or swell action can also cause another effect. Under such conditions, a single pass over an area will not permit the operator to completely determine which undulations shown for the sea floor are real and which actually stem from wave-induced heaving or rolling of the boat. However, this problem can largely be eliminated, as done in a resurvey for the Mokapu outfall (#47 in Table 10-1), by making several passes (in opposite directions) along the same track.

A bathymetric survey produces a large amount of data, and for this reason thought should be given to using the computer both in data processing and then in driving a plotter to spell out the position-depth information. Contour lines can be drawn by machine or by hand.

One of various approaches in the field, when the survey is being run, is to have time and two Autotape ranges (Appendix A) printed on paper tape every 10 s; at the same instant a marker line is placed across the sounding trace. Later at the office, a keypunch operator can place the double range and time information on cards or other suitable medium. The depth information can be read off the fathometric trace at the appropriate positions with a digitizer and placed directly on computer cards or tape.

The density of the fathometer lines run in the field should be a function of the complexity of the bottom conditions. Before a particular outfall alignment is tentatively decided upon, a preliminary survey can be run using a wide, e.g., 100-yard (90-m) spacing. But once a proper diffuser location is provisionally established and a route to shore for the pipe is indicated, final survey grids should be arranged such that excellent detail is provided.

Bottom and Subbottom Conditions

Introduction: The general concept of remote sensing of the bottom (from the water surface) has been extended beyond the depth-sounder concept to two very important means of obtaining necessary geological data for a marine outfall project. The first approach, the *side-scanning system*, provides information pertaining to the configuration or microbathymetry of the bottom on either side of the track followed by a survey vessel. The second approach, the *seismic profiling system*, describes the subbottom layering of materials along a particular route followed by the boat. Both approaches fall under the general heading of *geophysical surveys*.

Side-Scanning Systems: The side scan sonar system is used to map the topographical features of the ocean bottom without having to run vast numbers of fathometer runs at closely spaced intervals, in itself scarcely achievable out on the water due to both positioning errors and the difficulties of steering precise tracks.

In this side-scanning technique, two short bursts (e.g., 0.1 ms) of very high frequency (e.g., 100 kHz) ultrasound are transmitted, one to either side of the survey vessel, in fanlike beams. Seen from the stern of the vessel, the upper edges of the beams are 10° below horizontal, and each beam extends through 20–60° of angle. The beam angle seen from above is typically 2°. The ultrasonic waves are reflected off objects such as wrecks, boulders, rock outcroppings, sandwaves, submarine pipelines, as well as the bottom itself plus objects above the bottom such as buoys and fish. Some of the reflected energy returns to the vicinity of the boat where it is picked up. The very high frequency used in the side-scanning system means virtually complete reflection (i.e., no penetration) from bottom features.

It is customary that the side-scan sonar system be used in conjunction with a torpedo-shaped "fish" towed behind the survey vessel by a cable containing electrical circuits. Within this fish are housed both the ultrasound transmitter and the pickups that sense the returning echoes. The resulting signals pass through appropriate electronic circuitry and are displayed on a suitable recorder, with sea floor features on one side of the boat lying along one side of the chart and those on the other along the other side. The result on the chart is an excellent graphical representation of sea floor features flanking the boat's track as well as the acoustic shadows created by the reflecting targets. The objects are shown dark on the chart with the shadows (nonreflecting areas) light. The length of the shadows, used in conjunction with input on the acoustic velocity in the water, the water depth, and the distance of the object from the boat's track, provides a means of estimating the height of that object above the sea floor (whether connected to the bottom or not). Also, a skilled interpreter can translate the recorded reflectivities of the bottom features into meaningful statements on bottom surfaces, bathymetry, and composition.

Commercially available side-scan sonar systems typically have several range scale settings permitting features up to various distances to either side of the vessel to be charted. One manufacturer, for example, has range scales of 50, 100, 125, 200, 250, and 500 m.

The towing of a side-scanning sonar system over an area through which an outfall is to be laid can quickly indicate features that should be avoided, such as wrecks, rock outcroppings, or holes, or on the other hand natural paths for such a line, such as openings through ledges. Use of a side-scanning sonar system during preliminary work for the Barbers Point outfall on Oahu, Hawaii, in 1973, for example, turned up three possible routes for the pipe through two submarine ledges indicated by the bathymetric survey. These were at depths of 50–75 ft (15–23 m) and 120–150 ft (37–46 m).

The Barbers Point side-scan survey involved a mile-square (2.6-km²) area between the 40-ft (12-m) and 300-ft (92-m) depth contours. Survey transects approximately 300 ft (100 m) apart were used, and for the most part the 500-ft (150-m) setting on the side-scan system was set. This work involved 18 sonar traverses and was completed in 2 days. Positioning was done using the Autotape system, and time lines were generated across the side-scan record when fixes were obtained.

The fish is always towed some distance behind the boat, and allowance must be made for this time lag in interpreting the chart records. This distance may differ in different survey depths. In the Barbers Point work, for example, the transducer was towed approximately 100 ft (30 m) behind the boat for water depths less than 150 ft (46 m) with a towing distance of approximately 200 ft (60 m) for depths above 150 ft (46 m). Tow speeds are typically 4 to 5 knots.

Seismic Reflection Profiling: The object of seismic reflection profiling is to determine the vertical extent and nature of materials composing and underlying the sea bed. A sudden electrical discharge, or equivalent, at the end of a cable towed behind a survey vessel acts as a sound source. This acoustic energy radiates out from the energy source; some is reflected from the sea surface, some from the sea floor, and some from *horizons* (lines of demarcation between zones of different reflective properties) beneath the sea floor. Of all of this energy, only a portion is reflected back to a hydrophone array also towed behind the boat.

Acoustic energy from commercial seismic profiling energy sources is directed vertically downward in a beam of total angle 6.5–55°. It then "sees" only what is beneath it, unlike the side-scanning system. The relative amounts of acoustic energy reflected from the sea floor or else initially transmitted through the bottom depend upon the frequency of the acoustic signal. High-frequency pulses, as used with side-scan systems and fathometers, are essentially completely reflected at the bottom surface.

The *resolution* of a seismic profiling system refers to its ability to indicate the position of a sharp interface to a certain accuracy. One problem in seismic profiling is that whereas low frequencies (<1 kHz) mean good penetration, they also mean poor resolution. A typical commercial seismic profiling system for example, uses frequencies in the 3.5–7.0 kHz range. One means of achieving both deep penetration and good resolution near the sea bed is to make concurrent use of two different energy sources, one of low frequency and high power, the other of higher frequency and low power (Alpine Geophysical Associates, Inc., 1971).

Once energy enters the bed, it is reflected from different interfaces between rock strata or from surfaces within one formation where there is a change in a characteristic such as density. The ratio of the amount of reflected energy (E_r) to the amount of incident energy (E_i) for an acoustic wave approaching at right angles to the interface is

$$\frac{E_r}{E_i} = \left(\frac{\rho_2 V_2 - \rho_1 V_1}{\rho_2 V_2 + \rho_1 V_1} \right)^2 \tag{6-2}$$

where ρ represents the density of a medium, V is the acoustic speed through that medium, and the subscripts refer to two different media.

The generation of an acoustic pulse is followed by the arrival at the towed hydrophone array of a series of echoes, first from the water surface, then from the bottom, and finally from a host of reflecting horizons within the bottom. The echoes, as they are received, are amplified and filtered and then sent to a suitable recorder where they influence the writing of a pen on a strip chart recorder as it traverses the chart. Thus any pass across the chart leaves a series of imprints representing echo reception. A record of these passes over a length of time leaves a picture of the sea surface location, the bottom position, and the location of the various reflecting surfaces.

The across-the-chart representation, printed in terms of the time of echo reception and of known total duration, is not valid for showing all the distances involved since the speed of acoustic propagation through water and solid material is not the same. A sea water acoustic velocity considered to be suitable for the local ocean waters is used to convert the space between the water surface and the bottom to length scale. A determination of the velocity of sound in the sediments or rock, independent of the seismic profiling system, should ideally be carried out. Where great precision in knowing exactly where bottom and subbottom horizons are located is not necessary, average acoustic velocity figures may be sufficient.

An average figure for the sonic velocity in sea water, for example, is 4800 fps (about 1450 m/s) although this figure varies with salinity, temperature, and pressure, as outlined earlier. A value of the acoustic velocity of about 5400 fps (1650 m/s) might be considered representative for unconsolidated sediments such as sand with 6500–7500 fps (approximately 2000–2300 m/s) in consolidated sediments. Representative values for limestone are 11,500–21,300 fps (3500–6500 m/s); for granite, values are 15,000–23,000 fps (4600–7000 m/s).

Multiple reflections can sometimes cause confusion in a seismic profiling record. Such effects result from the reverberating effect of the sound between the bottom or subbottom layers and the water surface.

A marker pen can be triggered on seismic profiling recorders to show where position fixes were obtained. Such fixes can be labeled for correlation later with position information. The position desired is that of the hydrophone array towed perhaps 150 ft (46 m) behind the survey vessel, rather than the boat itself. Allowance must be made in the final analysis for the stage in the tide at which the information was obtained.

The EG & G (Waltham, Mass.) Unibom Seismic Profiling System is a system with a stated resolution of 0.5 ft (0.15 m) and possible penetrations up to 200 ft (61 m). The energy source consists of a vibrating plate, and this is mounted on a small catamaran towed behind the survey vessel. The hydrophone array is towed separately.

A preliminary survey of the sea floor area near the Barbers Point outfall in Hawaii, referred to earlier, was made using the Uniboom system for 2 days and 16 profiles in 1973, at the same time as the side scanning survey was run. Later, the Uniboom system was again employed for 1 day to obtain detail along the proposed outfall alignment. This was done by having the vessel run three tracks, one along the intended line and two at a distance of 100 ft (30 m) to each side of that line. This particular survey was run out to a water depth of 90 ft (27 m), from a 10-ft (3-m) depth, and was prompted primarily by a pocket of deep sand found during coring operations.

In ocean outfall work, the full penetration capabilities of the Uniboom system or equivalent are unnecessary because of the limited vertical extent of the work. A distance of 15–20 ft (4.5–6 m) would often mark the limit of interest in subbottom conditions. But there are at least two types of situations wherein information to perhaps 60–80 ft (18–24 m) might be required. First, if a trestle appears likely for laying the outfall in shallow water (Chapter 10), information will be required for the design of the associated piles (e.g., Whitaker, 1970). Secondly, because it is hydraulically desirable to have the outfall always sloping downward in a seaward direction, deep cuts may be necessary in benches or ridges on the sea floor. An example of the latter occurred in preliminary plans for an outfall for Broward County, Florida, where 40–60 ft (12–18 m) of coral were to be blasted out in certain places.

6-2 OBTAINING ENGINEERING PROPERTIES OF SEA FLOOR MATERIALS

Introduction

Surveying of the bottom and subbottom conditions over an extensive area can be rapidly and effectively carried out with geophysical methods as just described. Once a feasible route for the outfall has been selected, however, it would be unusual if further sea floor studies along the route were not carried out, using other means, to assure its complete suitability. Geophysical surveys leave largely unanswered, for example, the particular form of some sea bed layers,* and, in addition, such surveys give no information on the dynamics of materials on the sea floor. Furthermore, tolerances in the geophysical work are such that it may be impossible to determine whether bedrock is, say, 8–12 ft (2.5–4 m) below the bottom at a particular station. This can be a critical piece of information in nearshore waters where an outfall is customarily buried. This is because considerably larger construction costs are involved when rock must be blasted out to form a burial trench for a submarine pipe or removed in locations where piles for a construction trestle are placed. The depth to rock may also be important from the standpoint of anchoring vessels during construction (Bemben and Kalajian, 1969; Koster, 1974).

Geophysical investigations are impractical if not impossible in truly nearshore waters. In such regions, at the least, another form of sea floor investigation has to be employed. The importance of carrying out a thorough sea floor and immediately subbottom investigation for the whole outfall length cannot be overemphasized. I am aware of one outfall where an area within the surf zone was only casually observed by a diver before design began. It turned out that a raised mound in the path of the outfall consisted of a dense volcanic rock rather than the softer materials imagined. Construction times and costs were considerably increased through this area. It can be remarked, in this case, that a permanent raised bench in the surf zone should be suspected to consist of unusually resistant material.

Potential pipe failure due to the nature of bottom material must be considered. This material must have adequate bearing capacity for the pipe, and the possibilities of liquefaction under wave or seismic action should be appraised. Knowledge of the size and nature of the bottom material can aid in assessing its potential instability under storm waves. This same knowledge, coupled with data on sea floor slopes, can assist in estimating the potential of sea floor slumping. Submarine pipes have failed for each of the above reasons as will be outlined in Chapter 8.

The spacing of sampling stations along the intended route of the outfall should be carefully considered as a function of the complexity of the ocean bottom along the alignment. Samples are particularly important in locations where there is confusion in the interpretation of the seismic data, and sampling is warranted in sand pockets, to check their depth, and in areas where bedrock is close to the surface as remarked earlier.

---

*In fact, geophysically "established" conditions can sometimes be completely untrue. As an example, for a specific outfall in New Zealand the geophysical consultant predicted 50 ft (15 m) of sand at and below the bottom. It turned out that the sea floor was solid mudstone.

Sampling from the Surface

Drilling and Sampling: Subbottom investigations may be performed using a standard terrestrial rotary drilling rig mounted on the side of a suitable ship or barge in order to make required borings. The drilling equipment required may consist of a gas- or diesel-powered rotary drill with hydraulic pressuring system, a pipe derrick or tripod, a pump, casing, drill rods and bits, special tools, and a drive hammer for sampling. The casing is steel pipe [e.g., 4- or 5-in. (102- or 127-mm)], which is used outside the drill rods on many, but not all, marine drilling jobs. Sufficient casing is run down into the water to reach the sea floor. The casing is then turned by the drill head and, due to a cutting bit at its lower end and the hydraulic pressuring device, penetrates into the bottom material. This drilling-in of the casing is normally done in 5- or 10-ft (1.6- or 3.3-m) steps since that is the length of the replacement pipe sections and is often suitable as a sampling or coring interval. When the casing has been drilled in, its upper end is disconnected from the drill head and it becomes a freely standing pile contained in the well of the drilling platform unless there is only limited casing penetration into the bottom. Then the upper end of the casing is clamped. A drill bit, attached to a drill rod, is run into the hole. Rotation of the drill string and pressure applied to the upper end forces the drill bit down through the bottom material. A continuous flow of water or "drilling mud" brings the bottom material cuttings to the top of the casing where they and the water or mud flow over the top. At that location, it is possible for a geologist on the drill vessel to examine the cuttings and to have some idea of the material being drilled.

For engineering purposes it is important to obtain samples through different levels of the bottom materials and to make the results available in a boring log. It is also important to note the rate of drilling progress. If the drilled material is adequately soft, sampling is done by drilling to the bottom of the casing, cleaning out the debris with the wash water, and then withdrawing the drill stem.* A split-spoon sampler or equivalent (Hough, 1957) is lowered by wire to the bottom of the hole, seated, and then pounded 18–24 in. (0.46–0.61 m) into the material using a standard hammer falling a specified distance, since the so-called *blow count* can be correlated with a bearing strength of the material. After driving, the sampler is raised to the surface and the sample carefully removed, packed in plastic bags and capped cylinders and stored for later soil tests. If the drilled material is hard, coral rock for example, a core sample may be taken by a special, internally open drill into which the core passes as the drill moves downward with the help of the wash water.

The U.S. Coast Guard, or equivalent organization in other countries, should be advised of marine drilling plans. Notification to mariners then follows in the

---

*An exception to this occurs when the bore of the drill pipe and hole in the bit will allow the sampling system to pass (Emrich, 1971).

"Notice to Mariners." The following, for example, appeared in the August 3, 1976 issue from the Fourteenth Coast Guard District, Honolulu, Hawaii.

HAWAIIAN ISLANDS—HAWAII—BARBERS POINT

Commencing 26 July and continuing through August, a small barge and scaffolding will be drilling test moorings in the Barge Channel. Mariners are requested to give a wide berth when transiting this area.

Sea Floor Samplers for Non-drilling Applications: A *bottom sampler* should satisfy at least six criteria (Hopkins, 1964): 1. The device should have a minimum number of moving parts and these should be corrosion-resistant. 2. The sampler should be sturdy enough to stand up under rough usage on deck and impacts on the bottom. 3. The system's size, weight and closing arrangement should be such that handling of it on deck is not difficult or dangerous. 4. The sampler should orient properly before contacting the bottom. 5. The system should have sufficient self-weight or power to assure adequate penetration into the bottom material. 6. The device should not permit any sample loss during retrieval but easy sample removal on deck. Many different forms of sampler, falling in several separate classes, have found use in sampling bottom material and resident organisms in the sea. Not all such devices have satisfactorily fulfilled the preceding list of six requirements, however.

The three main classes of marine bottom samplers deployed from ships are for our purposes:

1. Grab-type samplers, devices with jaws that are forced shut by weights, lever arms, springs, or cords
2. Gravity coring sampler-tubes, with trap valves on top and core retainers at the cutting end, that are driven into the sediment by an encircling weight
3. Piston corers, tubes that are forced pneumatically by weight or vibrations into the bottom around an enclosed piston that remains stationary at the surface of the sediment

The grab sampler is lowered by lines from a surface vessel. Figure 6-1 shows a small grab sampler in the process of closing. When the sampler has been actuated, the jaws are closed and the arm shown in the figure extends to the left roughly perpendicular to the line of the device. To cock the sampler, prior to lowering, the jaws are forced apart against the spring, and they catch in the open position under the collar shown in the diagram just above the jaws. The arm, meanwhile, has oriented itself downward, and when the sampler impacts the bottom the arm is struck and the jaws are released by the collar. The height of the brass sampler shown in Figure 6-1 is 12.5 in. (318 mm).

The grab samplers are not generally of great use for all engineering purposes since they customarily obtain a considerably disturbed sample plus

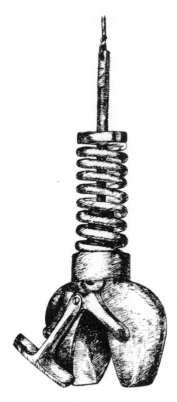

Figure 6-1 Small grab sampler in process of closing

give no indication of truly subbottom materials. They are also only useful in sandy/muddy bottoms. The reason the various types have been included above is that they find considerable application in the sampling of benthic organisms. Thus details on these samplers and their operation are provided in Section 6-5.

Using Gravity and Piston Corers: Gravity and piston samplers are useful for gathering engineering information on the sea floor. Hopkins (1964) has described 22 different types of gravity corers and 3 different types of piston corers. Smith (1962), Richards and Parker (1967), and Noorany (1972) detail the use of gravity and piston corers. The gravity corer is allowed to free fall from some distance above the sea bed. It has a hardened steel coring head attached to the lower end of the tubing. A thin plastic liner within the corer is used to enclose the sample obtained. These types of corers are useful for supplementing geophysical data in waters too deep for diver work or for small-scale sea floor drilling, both of which will be considered presently. One such corer was used, for example, in deeper waters along the proposed routes of the Sand Island No. 2 outfall and the Barbers Point outfall in Hawaii. Table 10-1 contains details on these outfalls.

Corers can be used in conjunction with an *accelerometer* (acceleration-measuring device) to provide additional insights into the degree of sample

Figure 6-2 40-foot Vibracore being lowered by crane along alignment of Rochester, N.Y., outfall, Lake Ontario (Courtesy, Alpine Geophysical Associates, Inc.)

disturbance caused by the coring operation (Scott, 1967b, 1970; Preslan, 1969) among other things. The accelerometer can also be used with a *penetrometer*, a cone-shaped device whose resistance to penetration into a medium can be correlated with soil density or consistency.

Vibrocorers: The *Vibracore* of Alpine Geophysical Associates, Inc. (Norwood, N.J.), a sampling system for waters to 300-ft (90-m) depth, has been described by Tirey (1972) and has been used in ocean outfall situations to gather information on bottom conditions (see Figure 6-2). The Vibracore consists basically of a pneumatic vibrator at the top of a 40-ft-long* (12-m), 4-inch-diameter (102-mm) steel pipe within which there is a plastic liner for a 3.5-in. (89-mm) core of bottom material. Air for the vibrator is supplied by a compressor on the support vessel. The pipe is supported by an "H" beam mast held in vertical orientation by a frame, consisting of four legs, set on the sea bed. The sampler is designed to be driven into the bottom and pulled back into the frame

---

*Shorter lengths are possible if a shorter core is satisfactory. These smaller systems are more easily handled by a small vessel.

before being lifted off the bottom. The base of the mast has a door that seals the end of the core barrel after it has been pulled back into the frame. Full-length cores of sand, gravel, silt, and clay, alone or in combinations, are possible with the Vibracore. Short cores from approximately 6–18 in. (150–450 mm) in length are possible in soft sedimentary rock.

The sequence of materials making up a sea bottom profile can of course be obtained from the core sample. Koutsoftas et al. (1976) report that natural water content, Atterberg limits, sieve analyses, and maximum and minimum densities can be accurately determined from Vibracore samples. However, the samples, being somewhat disturbed, yield low estimates of strength, at least in the case of clays. The volume by Hough (1957) is representative of the many soil mechanics books that explain the terms used above.

Work Involving Divers

In the previous chapter, the use of divers in the taking of physical oceano-graphic data pertinent to ocean outfall-related marine investigations was casu-ally outlined. During the obtaining of other physical data, as well as chemical, biological and geological information, the topics covered in this chapter, the diver again can play a major role. The word "diver" here is intended to mean "scientist-diver." It is my strong belief that the underwater tasks required in oceanographical studies should at the least be under the direct supervision of a qualified scientist who has also received thorough training in SCUBA diving. It is further suggested that virtually all members of the diving party have the requisite scientific training, as well as that pertaining only to diving. Somers (1975), for example, has outlined a rigorous diving training program for poten-tial scientist divers. The *NOAA Diving Manual* (1975) is an excellent reference book on the many aspects of diving, and additional material has been included in Appendix B of this volume.

When water depths exceed 100–150 ft (30–46 m), the use of divers for sea floor investigations becomes questionable.* In such depths, to complement the results of geophysical studies and/or those outlined in the previous section, a submersible is a possibility. Such a vehicle was used, for example, in surveying the planned alignment for the Sand Island No. 2 outfall, offshore from Honolulu, Hawaii.† Another survey possibility, but not an easy one, involves a television camera suspended from a boat and towed along the route. A TV or movie camera can also be carried by a submersible.

---

*The diving survey and probing for the Barbers Point outfall, however, extended to a depth of 185 ft (56 m).

†A major difficulty in using submersibles for such work is keeping them on line. Limited success was achieved in the Sand Island case by dragging an anchor along the planned alignment and by placing numbered, submerged buoys along the line.

But many outfalls do not extend into waters too deep for comfortable diving. Even for "deep" outfalls such as the Sand Island No. 2 mentioned above discharging into 235 ft (72 m) of water, 64% of the 12,500 ft (3800 m) of line was shoreward of the 150-ft (46-m) contour. Thus divers can be used extensively to study soils in place and to obtain samples along planned outfall alignments.

In situ investigations can vary from mere observations to extensive probing and sampling. The technique of *probing* can be used to back up the results of geophysical surveying. In this case, lengths of pipe can be gradually assembled on the sea bed as the down pipe is forced into the sea floor using both the hand pressure of diving attendants, or perhaps a pneumatic hammer, and the washing action of a water jet out the bottom of the pipe. This flow is provided by a pump mounted on an attending work boat, and this type of probing is more properly called *jetting*. As the probe is lowered in the bottom material, the diver attendant can feel the resistance to penetration, and thus has a good idea of the nature of the material being penetrated. If the sediment cover is not too deep, then probing proceeds to refusal, an indication of horizons hopefully duplicated by the geophysical survey. Care should be taken to identify locations where temporary refusal may indicate only a harder crust over softer material or where refusal means a boulder rather than solid rock. Samplers may be used in some cases at the end of a particular jetting run.

Diver sampling can vary from crude scooping up of bottom material (into a container) to the use of somewhat more sophisticated corers (Hopkins, 1964). In the latter case a thin-walled cylinder with an adequate cutting tip may be pushed into the bottom material if it is soft. Otherwise the sampler can be driven into the sea bed by means of a slide hammer (with the blow count noted). In either case, it is possible to withdraw the sample by coning out a region around the sampler through the use of a water jet and then placing one's hand or a cutting tool over the lower end of the sampler before withdrawing. Turner et al. (1968) and Lukin et al. (1971) describe the use of plastic coring tubes in sand deposits.

Diver observations on the bottom materials can be of great value to the design engineer. Extensive vegetative growth over pockets of sand, for example, may indicate relative stability; on the other hand, recent coral fragments buried under several feet of sand may indicate an unstable region which the chosen outfall route should probably avoid. The nature of materials right on the bottom may be important in planning construction techniques. Silt is easily kicked up by divers and causes impairment of water visibility.

Drilling/coring rigs that sit on the bottom and are controlled by divers have been used on some pre-ocean-outfall-design surveys, e.g., Barbers Point in Hawaii in 1973 and Wanganui, New Zealand, in 1976. In both cases, the rig was used in conjunction with a surface support vessel. Submersible rig power, pneumatic in the Hawaii case and hydraulic in the New Zealand one, came from the boat. The rigs were both of rotary drilling type. Water was used for cooling and bringing up cuttings and washings in the Hawaii case, drilling mud in the

New Zealand work. In both cases, the debris was collected and later examined, and in Hawaii core recovery was achieved in the denser material. The compactness of the various strata encountered was judged from the speed of drilling advance.

Wave surge and currents pose real problems for divers drilling on the sea floor, pulling on hoses and the diver himself and tending to overturn the rig. This was the experience during the sea floor drilling work for the Wanganui outfall. Visibility is also a problem. Whereas the operator divers had good visibility in the Hawaiian work, visibility during the New Zealand investigation was virtually zero.

The bearing capacity of sea floor materials can be determined directly by means of a special underwater loading frame, operated by divers, such as described by Harrison and Richardson (1967). Shear strength of the bottom material can be obtained *in situ* using a diver-held vane-shear device such as described by Dill and Moore (1965). This involves a small vane attached to the shaft of a calibrated torque screwdriver. Scott (1967a) reviews this technique as well as others used to gather *in situ* sediment strength data.

The sea floor investigations carried out in May 1976 for the Doctor's Gully No. 2 outfall at Darwin, N.T., Australia made use of a diver-operated 33-mm vane tester for *in situ* strength measurements and a 150-mm hand auger for samples. Probing was done with pipe using water flow from an attending barge.

Complete Field Studies for Outfalls

The Orange County No. 2 outfall extends 27,400 ft (8400 m) to a depth of 195 ft (59 m) in the Pacific Ocean off Newport Beach, California. Borings were drilled at 500-ft (150-m) intervals from outside the surf zone to an approximate water depth of 40 ft (12 m) using a rotary wash drill rig. This was mounted on a platform that could be towed in floating mode from one station to the next and then raised above the water surface on four jack-up legs that extended to the sea floor.* Casing was installed between the drilling platform and the bottom, and drilling mud was used to prevent caving of the borings. Borings were drilled to depths ranging from 20–35 ft (approximately 6–11 m) below the sea floor, and samples were taken at frequent intervals for laboratory inspection and testing.

Bottom soil conditions for depths greater than 40 ft (12 m) were obtained by diver observations and shallow sampling. Over part of the alignment jetting was done by divers. Jetting was performed with water, using a $^3/_4$-in. (19-mm) pipe and portable pump delivering 25 gal/min (1.6 ℓ/s) at approximately 50 psi (34 N/cm²). Jetting was done in those areas where planned trench excavation might encounter rock.

There were eight marine borings made to investigate bottom and subbottom conditions along part of the alignment for the Sand Island No. 2 outfall. Three

---

*This eliminated the very troublesome problem of barge or ship movement in waves.

borings, out to a depth of 18 ft (5.5 m), were done from a scaffold mounted on the sea floor. The remaining five borings, in depths of 25–70 ft (7.5–21.5 m) were done using divers and a raft on the water surface.

For the drilling done from the scaffold, casing was extended above the water surface from the boring and advanced, where necessary, to prevent caving. Drilling was carried out using a hand-operated chipping drill in conjunction with wash techniques, and this proceeded to a depth of approximately 20 ft (6 m) below bottom. Relatively undisturbed samples were taken at selected depths by driving a sampler having inside diameter 2.4 in. (61 mm) with a 140-lb (623-N) weight falling through 30 in. (0.76 m). The number of hammer blows required to drive the sampler into the undisturbed subsurface was recorded at 6-in. (150-mm) intervals.

For the deeper borings, a pontoon-raft on the surface provided diver support plus a power cathead with which to drive the casing and sampler. At each boring, divers set up a casing guide on the sea floor and casing was driven to the required sampling depth. In cases where materials were hard, predrilling was used to pave the way for the casing. The technique used in the predrilling, and also in clearing inside the casing, involved a hand-operated chipping drill lifted and rammed into the hole by the divers. Flushing of the hole was accomplished by high-pressure water supplied to the top of the drill by a hose connected to a pump on the raft. Driving of both casing and a sampler was in part done on the bottom by the divers and in part by the power cathead on the raft. Blow count data were recorded and during drilling operations the cuttings were observed. Waves caused this operation considerable problems, first in terms of raft movement and second in terms of bottom surge, and operations had to be terminated on several occasions.

Exploratory work for two proposed ocean outfalls at San Francisco, California, in 1977 involved the use of a walking platform, Spider II described in Chapter 10, in the surf zone and an exploratory drill ship offshore. However, this approach is expensive. The three holes drilled from Spider II cost $250,000; the drilling ship cost $7000 per day.

Laboratory Analyses

The first step in the size gradation of a soil sample is to dry the sample in an oven typically set at a temperature of 105–110°C. The sample is then placed on a nest of sieves on a shaker, the nest composed of the following sieve sizes: 4 (4.76-mm); 10 (2.00); 20 (0.840); 40 (0.420); 70 (0.210); 140 (0.105); and 270 (0.053). The amount of the sample remaining on each sieve is weighed and the results plotted in cumulative probability fashion such as shown in Figure 6-3, where d represents particle size. For the very small size fraction that passes the No. 270 sieve, use of a hydrometer is called for if a size breakdown is required.

Hydrometer analysis is detailed in, for example, Lambe (1951). A *hydrometer* is usually a glass item consisting of a lower bulb and an upper graduated rod.

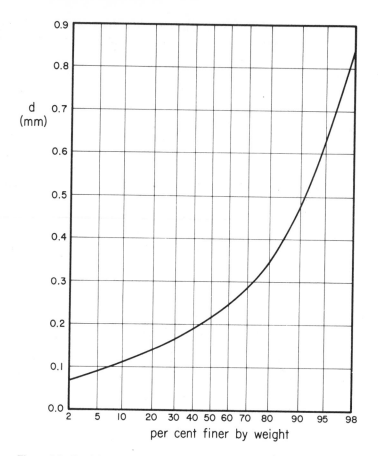

Figure 6-3   Particle size gradation of sea floor sand sample

The hydrometer floats in a vertical attitude, floating lower and lower in the water as the specific gravity of the liquid in which it is floating decreases. The intersection of the liquid surface and the graduated neck of the hydrometer is read as the specific gravity of the liquid. The idea involved in the determination of the gradation among finer soil fractions is that the specific gravity of an initially well-mixed soil-water combination will decrease with time as soil settles to the bottom. The time taken for a soil particle to settle out depends primarily on its diameter, although there are other factors (Vanoni, 1975). The time record of hydrometer reading can then be translated into the distribution of particle size for the sample.

There are special counters commercially available (Coulter) that can be used to obtain size fractions of sediments. Granular materials are characterized by the titles given in Table 2-4, e.g., gravel, sand, silt, and clay. The percent by weight of a soil falling within any such category can be determined from the cumulative size distribution plot for the granular material. The example plot in

Figure 6-3, clearly represents a sand (0.063–2 mm). In the sample, the proportion of medium sand (0.25–0.5 mm) is (91–60)% or 31% by weight. The median size is 0.22 mm.

Consider a sea floor sample contained in a two-part container. The *shear strength* of the sediment refers to the ability of the sample to withstand the sliding of one part of the sample with respect to the other under a laterally applied force. In general, this shear strength contains a portion (*solid friction*) that is dependent on the magnitude of an applied (normal) force pressing together the potential two halves of the sample and a second part independent of that force. The second type of resistance to shear deformation is called *cohesion* and is typical of clays (very fine-grained sediments) that have sizes less than 0.004 mm. On the other hand, the ratio of the increase in normal force that goes into an increase in shear resistance is taken as the tangent of an angle called the *friction angle*. Coarse granular materials such as sands and gravels have their shear resistance completely in this form.

The amount of load that can be applied to a given area of marine sediment, such as by a submarine pipe, depends critically upon the cohesive and solid friction characteristics of that material. Thus the sample of marine sediment should permit determination of the cohesion and friction angle. The correct friction angle is not difficult to establish with standardized soil mechanics laboratory procedures (e.g., Lambe, 1951; Bowles, 1970). However, the very act of sampling tends to disturb the natural structure of the sediment on the sea bed, such that what is tested in the laboratory no longer retains the full measure of cohesive strength possessed by the sea floor sample. One might question the use of "undisturbed" cohesive values in any event, since, to put the pipe in place on the sea bed, some disturbance of the material supporting it must be imagined.

The *undrained shear strength* of a soil sample is its strength to shearing failure when whatever pore water was present in the sample is imprisoned, causing pore water pressure changes as the testing is carried out. Ladd and Lambe (1963) have presented methods of adjusting laboratory values of undrained shear strength of clays to take account of the disturbance that occurred during sampling. The more recent report by Lee (1973) provides additional insights into this problem.

6-3 SHORELINE AND SEA FLOOR DYNAMICS

Introduction

A great many outfalls proceed into the receiving waters through a beach area. A beach is a notoriously variable zone, and the cycles of erosion and deposition as materials are moved onshore and offshore are of various time scales. A big storm or tsunami inundation can extensively erode a beach system in a matter of hours; seasonal changes in wave characteristics generally cause cut in the winter and fill in the summer; many beaches undergo cycles of

buildup and erosion stretching over multiyear periods. These are natural alterations to a beach system. It is well-known that the construction of *groins* (barriers
perpendicular to the beach) or breakwaters result in the buildup of materials
moving along the coast on the upcurrent side with a cutting back of the beach
on the downcurrent side. Even man's activities inland, such as the damming of
rivers and the resulting trapping of river-borne sediments, can lead to extensive
beach regression since there is no replenishment of eroded materials.

Ideally, a survey should be carried out early in the planning for an outfall
to check on possible man-caused alterations to the entire littoral zone in the
vicinity of the outfall's penetration into the sea. Such work as dredging, groin
and breakwater construction, harbor creation, and river damming or diversions
should be pinpointed and carefully considered.

There is no substitute for an extensive, multiyear study of shoreline and
nearshore bottom dynamics as a prelude to ocean outfall design. As is commonly the case, however, the need for an outfall is virtually immediate, and
there is insufficient time to mount the type of research investigation really
necessary. Usually all that could be possible for the outfall planners is a
single-year study of shoreline and nearshore processes. The results of such work
would then be supplemented by whatever data had been obtained in the past,
during unrelated projects, on the stretch of coastline concerned or others near or
similar to it. Past air photographs of a coastal area can be extremely useful in
assessing the variable nature of a particular area.

As will be discussed in some detail in later chapters, an outfall is customarily buried through the surf zone, so that it is not directly acted upon by
pronounced wave action. But a pipe buried today under 10 ft (3 m) of naturally
occurring bottom material will not necessarily be so buried a month or two from
today if the bottom material is readily erodable and wave action is such that it is
eroded. This is not just a supposition. Nearshore bottom surveys for the
Humboldt Bay Wastewater Authority, California, outfall in October and December, 1975, then March and August, 1976, provided graphic evidence of this
phenomenon. Another example concerns a set of three pipelines proceeding
from a refinery on shore to an offshore tanker mooring facility off Barbers
Point, Hawaii. This installation undergoes considerable cycles of burial and
exposure as the unconsolidated sediments in which it is buried suffer erosion
and deposition. To ensure outfall stability through such a dynamic region, it is
prudent to bury the outfall sufficiently deeply that it is very unlikely to be
exposed during the planned duration of the installation. As will be remarked
later, backfilling of the pipe, after it is laid in a trench, need not be done with
parent material. Stability is better assured by backfilling with more stable
material of larger size.

A real problem with respect to the burial of outfalls, and submarine
pipelines in general, is that a trench dug in the nearshore zone can become
refilled with natural material in a short time, before the pipe is placed in it.
Ideally this should not occur, since preferred construction practice is for the

excavating and pipe laying to proceed in tandem, such that the pipe is laid in a freshly completed trench of the depth and width required in the specifications. But, practically, this is not always the case. Also, there is a construction technique, which will be detailed later, wherein the pipe is pulled out from shore into the prepared trench. For a long pipeline, it is virtually impossible for the contractor to ensure that the trench is not somewhat refilled throughout its length, and it is very possible that the distance of the pipe below the sea bed will be less than that specified over certain lengths of the pipe.

Thus it is important in ocean outfall studies that the volume rates of sediment transport past the proposed construction area be at least approximately known. Large transport rates mean rapid filling-in of prepared excavations. If such transport rates are to be estimated without any field data, a precarious undertaking, the paper by Komar (1975) may be useful.

Sediment Transport Rates and Beach Profiles

There are two types of sand tracer commonly used in assessing sediment transport rates: radioactive and fluorescent tracers. The advantages and disadvantages of these two types of tracer, as well as details on their usage, are spelled out by Ward and Sorensen (1970) and Noda (1971). Additional detailed information is available in the *Shore Protection Manual* (1973). Space does not permit us a thorough discussion of this form of field study; the reader is referred to these references for guidance and orientation in cases where sediment transport rates are required for outfall projects.

Beach profiles are established by using a level, rod, and tape, or equivalent, in the measurements. This surveying should extend as far seaward of the water edge as is feasible. Plotting of the resultant profiles provides a littorally graphic illustration of maximum and minimum beach levels.

Sea Floor Dynamics

Not only material on the beach and material immediately fronting the beach suffer movement under wave action but also material to the edge of the continental shelf (Komar et al., 1972). Not only waves are involved in effecting sea floor changes but also quasisteady currents such as those of tidal origin.

In coastal areas where the sea floor offshore is unstable, diving studies on scour such as those carried out by Palmer (1969) may be useful in assessing the degree to which wave-induced erosion might take place in parent bottom material. At the least, such studies would complement predictions made with techniques to be explained in Chapter 8.

Stride (1970) has described what must be a particularly severe form of unstable bottom, that in the North Sea. He has investigated migrating sand waves in depths between 20 and 110 m, with an average of 30 m. Movement of material in these sand waves was due to tidal currents, and some sand waves with trough-to-crest height of 11 m were measured. Crest separations of these waves were from approximately 200–500 m. It is obvious simply from these data that the design of pipelines for the North Sea is no trivial matter.

6-4   MARINE FLORA AND FAUNA

Introduction

A marine area planned as the recipient of a wastewater effluent is customarily intensively studied from physical, chemical, geological, biological, and bacteriological viewpoints. In Section 6-5 attention will be focused on the securing of data having to do with the marine biota indigenous to the area concerned and the analysis of such data. Complete data would thoroughly survey the phytoplankton and the zooplankton, the benthic plants and animals, and the nekton. The present section provides background for Section 6-5.

There are two ways of naming members of the marine *flora* (plants) and *fauna* (animals); first by referring to their common names, e.g., great white shark, second by making use of their scientific names, in this case *Carcharodon carcharias*. The scientific name is a part of an extensive classification system. The identification of marine plants and animals should make use of *both* name tags for every type of specimen. First of all, a common name is often not very precise; it includes a variety of organisms more or less but not exactly the same. Secondly, the common name for a particular organism in one place may bear little or no resemblance to that employed in another location, whether or not there is a difference in language, with the result that a listing by common name would be at least in part unfathomable to someone outside the immediate geographic area. Although not a particularly flagrant example, the great white shark is otherwise known in English as a white pointer or white shark.

Identification by purely scientific name is largely ineffectual as it shuts out a sizable group of persons interested in marine outfalls who are not marine biologists. Some comprehensive cross-reference lists for marine creatures are available in the literature. An example is "A List of Common and Scientific Names of Fishes from the United States and Canada" (1970).

The two subsections to follow will present short summaries of the general classification of marine plants and animals. The division of plants and animals into a classification system (based on evolutionary relationships) is called *taxonomy*.

Plants

There is considerable variability in schemes for classifying members of the plant kingdom (Plantae), and the one used below (Keeton, 1969) is simply an example. In this arrangement, there are 10 divisions of which six are of marine interest.

One of these divisions, Traecheophyta (vascular plants), includes the Angiosperms. One of the major representatives of this class is so-called "eel grass," *Zostera*, having leaves, rhizome (runner or creeping stem) and true roots, and growing in extensive beds. Another division is Pyrrophyta, the dinoflagellates, part of the phytoplankton. The dinoflagellates are typically unicellular, some armored by plates of cellulose, a carbohydrate; all have flagella for locomotion; reproduction is by simple cell division; and they are found in all seas but chiefly in warmer waters. The four remaining divisions of the kingdom of Plantae of marine interest involve different forms of algae (singular alga).

The pigments of the green algae (Chlorophyta) include the two types of chlorophyll, *a* and *b*. The cell walls of these plants are largely cellulose. A common form of green algae is *Ulva*, or sea lettuce, that can grow abundantly in protected places in the intertidal and upper subtidal zones. Green algae are found most commonly in warmer seas and from the upper littoral zone to a depth of about 10 m, especially in the lower half of the tidal zone.

Large brown seaweeds, particularly kelp, are included within the brown algae (Phaeophyta). Kelp grow in marine forests or kelp beds. A particularly long-stalked (35 m or more) giant kelp is *Nereocystis*. It anchors itself to a hard layer by means of a *holdfast*; there are no true roots. The stalk, or *stipe*, is hollow for most of its length and extends up to a hollow bulb that gives the plant buoyancy. Fronds extend from the bulb near the surface for maximum light usage.

The red algae (Rhodophyta) are usually small in size. They are widely distributed geographically but most abundant in the temperate seas. They prefer subdued light. Coralline algae, that serve to cement coral reef structures and/or to form reefs themselves, are representatives of the red algae. A limestone coating that is secreted beneath the living portion of the plant remains after the plant matter has died and decomposed and forms the hard substrate.

The division Chrysophyta, contrary to the other algal forms discussed above, primarily comprises floating forms. This division includes three classes: yellow-green algae, golden-brown algae, and diatoms. The diatoms are basically unicellular, although they may occur in chains or other groups, and have shells composed of translucent silica. The most common form of diatom propagation is by simple cell division.

A fifth form of algae, blue-green algae, is actually a division (Cyanophyta) of another kingdom, Monera. These are primitive algal types. Unicellular forms reproduce by simple cell division. They are widely distributed in fresh and brackish waters, and they may cause the phenomenon of sliming.

Animals

It is convenient to organize animals into groups and then subgroups according to similarities in structure. The system of classification that has evolved has led to the following headings from most general to absolutely specific: phylum, class, order, family, genus, and species. Even this list is further subdivided in some instances, subdivisions extending for example to subphyla, subclasses, and suborders.

It is common when studying (marine) animals to divide such creatures into two distinct groups: *vertebrates* and *invertebrates*. This is in a sense peculiar since neither grouping corresponds to any phylum or group of phyla. In fact the term vertebrate designates only part of one phylum, and the remaining members of this phylum as well as all other phyla then fall under the classification of invertebrate. The phylum that contains both vertebrates and invertebrates is Chordata, customarily divided into three subphyla: Urochordata, Cephalochordata, and Vertebrata. The first two are invertebrate, i.e. they have no backbone.

The marine vertebrates are summarized in Table 6-1, the various categories of marine invertebrates in Table 6-2. There is not complete agreement on the organization and naming of the scientific classification system, but the systems shown are common.

Table 6-1: Marine Vertebrates (Phylum Chordata, Subphylum Vertebrata)

Class	Examples
Agnatha (jawless fishes)	lampreys, hagfishes
Chondrichthys (cartilaginous fishes)	sharks, rays, skates
Osteichthys (bony fishes) [Pisces (true fishes)]	herring, smelt, halibut, cod, sole, perch, pipefish, salmon
Reptilia (reptiles)	turtles, sea snakes
Mammalia (mammals)	whales, porpoises

Diversity

A complete description of the marine biota associated with a particular bottom area and the water column overlying it would involve an enumeration of all of the species of benthic plants and animals living therein, of resident nekton, and of pseudoresident phytoplankton and zooplankton and of the numbers of all such organisms. It is certain, unless the area concerned were particularly sterile, that the mass of information resulting from this thorough assessment would be overwhelming.

A parameter of some utility in illustrating the wealth of species and organisms in a particular area without detailing it is the *diversity index*. A mass of data on species and the number of organisms within any such species can be translated into a single number.

The diversity index, referred to above, as a number usually incorporates two measures: *richness* or *variety* (i.e., number of species), and *equitability* or

Table 6-2:　Selected Marine Invertebrates

Phylum	Subphylum	Class	Subclass	Order	Examples
Chordata	Tunicata (Urochordata)	—	—	—	sea squirts
Chaetognatha	—	—	—	—	(arrow-worms)
Echinodermata	—	Holothuroidea	—	—	sea cucumbers
		Asteroidea	—	—	starfish
		Ophiuroidea	—	—	brittle stars
		Echinoidea	—	—	sea urchins, sand dollars
		Crinoidea	—	—	sea lilies
Arthropoda	—	Crustacea	Ostracoda	—	(ostracods)
			Cirripedia	—	barnacles
			Copepoda	—	(copepods)
			Malacostraca	Euphausiacea	krill
				Amphipoda	(amphipods)
				Isopoda	(isopods)
				Decapoda	crabs, lobsters, shrimp
Annelida	—	Polychaeta	—	—	(polychaetes)
Mollusca (shellfish)	—	Gastropoda (univalve)	—	—	abalone, snails, cone shells, cowries, conches, limpets
		Pelecypoda (bivalve)	—	—	clams, oysters, mussels, scallops
		Cephalopoda (valveless)	—	—	squid, octopus
Brachiopoda	—	—	—	—	(lamp shells)
Nematoda	—	—	—	—	(round worms)
Nemertina	—	—	—	—	(ribbon worms)
Platyhelminthes	—	—	—	—	(flatworms)
Ctenophora	—	—	—	—	(comb jellies)
Cnidaria (Coelenterata)	—	Hydrozoa	—	—	Portuguese man-of-war, small medusae
		Scyphozoa	—	—	jellyfish, large medusae
		Anthozoa	—	—	corals, sea fans, sea anemones
Porifera	—	—	—	—	(sponges)
Protozoa	—	Sarcodina	—	Foraminifera	—
				Radiolaria	—
		Ciliata	—	—	—

evenness (i.e., the distribution of individuals among the species) (e.g., Odum, 1969). Various equations have been developed to quantify biological diversity, with the following one receiving great usage.

$$DI = \sum_{i=1}^{n} p_i \log_2(1/p_i) \qquad (6\text{-}3)$$

In Equation (6-3), DI is the (Shannon or Shannon-Weaver) diversity index, n is the total number of species in a biological sample, i is a (summation) index referring to any particular species, and p_i is the proportion (fraction) of the total number of individuals in the sample representing the species i.

The maximum value taken on by the diversity index (DI_{max}) occurs for equal numbers of individuals making up each of the n species present. It can be readily shown that $DI_{max}=\log_2 n$. The minimum value of DI occurs for one individual in each species except one, all remaining individuals in that particular species. This value, DI_{min}, depends on the total number of individuals present, N. Table 6-3 presents information on extreme and intermediate values of DI for $n=4$ and $N=64$. The number of individuals in species i is represented by n_i.

Table 6-3: Sample Values of Diversity Index

n_1	n_2	n_3	n_4	DI
16	16	16	16	2.00
24	24	8	8	1.81
32	16	8	8	1.75
30	30	2	2	1.34
60	2	1	1	0.43
61	1	1	1	0.35

Fish Characteristics

Fish surveys by divers, to be described in Section 6-5, yield data on the species, numbers, and sizes of the fish encountered. To estimate fish biomass, a correlation between the size and weight of various fish species is required. There can be considerable variability in the length-to-weight ratio for individuals within only one fish species (e.g., Mearns and Harris, 1975). However, this is the nature of statistics, and for a single species a line of best fit to length-versus-weight data can be drawn. This line has the form

$$W' = \beta L^b \tag{6-4}$$

where L and W' are the length and predicted weight, respectively, and β and b are empirical constants. These constants can be chosen by simple linear regression statistical means (e.g., Hoel, 1976) by expressing Equation (6-4) in the form

$$\log W' = \log \beta + b \log L \tag{6-5}$$

Brown and Caldwell, Consulting Engineers (1975b) carried out 28 different regressions according to species and subdivided according to adult or juvenile, male or female, or different time. The values obtained for b varied between 2.33 and 3.26, averaging 2.85. That this value is so close to 3 is hardly surprising since the volume of a specific fish individual is proportional to the cube of its length. Quast (1968), in a study of 35 species of kelp-bed fishes had b range from 2.72 to 4.51 with an average of 3.28.

As a simplification, and recognizing the inherent variability in the data, it is generally standard to assume that the weight of a certain species of fish (leaving out age or sex) is truly proportional to the cube of its length. From Equation (6-4), this leaves only one coefficient to be determined, and paired length-weight data can be used to establish that value.

The values used for six different types of fish are shown in Table 6-4. The smallness of the coefficient for the trumpet fish, for example, reflects the fact that its volume is far less than the cube of its length.

Table 6-4:   Length-Weight Relationships for Certain Fish

Common Name	Scientific Name	Shape	$\beta\,(lb/in.^3)$
Trumpet fish	*Aulostomus chinensis*	long, thin	0.000060
Goatfish	*Parupencus chryserydros*	elongated	0.00053
Squirrel fish	*Myripristis berndti*	flat, oval	0.00076
Damsel fish	*Pomacentrus jenkinsi*	flat, round	0.00101
Parrot fish	*Scarus dubius*	thick-bodied	0.00075
Grouper	*Cephalopholis argus*	thick-bodied	0.00050

6-5   SAMPLING THE BENTHOS AND NEKTON

Introduction

Several methods of securing information on the nekton and on benthic plants and animals will be presented in this section. Two of these methods are used from surface vessels: trawls and the grab samplers mentioned in Section 6-2. A third approach involves the gathering of related information by divers.

Trawls and Trawling

Bottom Trawls: Trawls are extensively used across the world in commercial fishing operations. They also find great utility in collecting, as a part of biological studies, organisms living on or in close proximity to the bottom. Although it is this latter category that is of interest in this section, material will be presented below on the types of trawls used in commercial fishing operations. The reason for inclusion of this material is that large commercial fishing trawls pose problems for pipes lying on the ocean floor and some background material on such systems is useful before the problem of outfall design against trawls is considered briefly in Chapter 8.

The *trawl* (Figure 6-4) is a tapered bag of netting that can be towed over the sea bed to capture *demersal fish* or *groundfish* (dominantly bottom feeders)— species such as cod, haddock, and sole, and invertebrates such as shrimp. The trailing end of the trawl, where fish or other creatures collect after entering the opening, is known as the *cod end*. This is usually a narrow sleeve of net that is both stronger and of finer mesh than the net in the main part of the trawl. Once the trawl is taken back on board the towing vessel, the accumulated fish are removed from the back of the cod end by undoing a securing rope, the *cod line*. To assure against chafing, the underside of the cod end is covered with nylon. At the junction of the main bag and the cod end there may be a *flopper*, a flap of netting hanging down to act as a valve preventing escape of fish from the cod end if the trawl should stop moving.

A *beam trawl* is one wherein a long beam holds the mouth of the trawl open. The upper leading edge of the bag is attached to a strong *headrope* lashed to the beam. The lower edge of the bag is attached to a longer *groundrope* that drags along the bottom somewhat behind the headrope so that fish disturbed by the groundrope are already "in the bag." The beam is maintained approximately 2 ft (0.5 m) above the sea floor by two strong curved runners, attached to the ends of the beam, that make contact with the bottom. A bridle connected to the runners is used for towing.

Beam trawls tend to be small, and in modern fishing have been replaced with the larger and more versatile *otter trawl*. However, beam trawls are still used in small boat trawling and in biological work.

The after part of the otter trawl is similar to that of the beam trawl. In this case, however, the bag is kept open by two *otterboards* (Fig. 6-4), connected to two mesh wings extending out to the sides of the bag, usually by means of cables 180 ft (55 m) or so in length. The idea involved is that two otterboards are not towed head on, but by means of two strong cables, or *warps*, pulling from the insides of the two otterboards, if they are flat, such that hydrodynamic lift forces them apart, thereby holding the net open. V-shaped otter boards are also used.

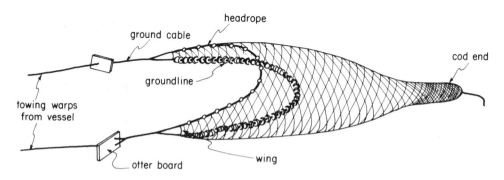

Figure 6-4 Otter trawl

The under edges of the otterboards, or *doors*, slide over the sea bed and are shod with steel for protection. The common otterboard is a rectangular, wood and/or steel structure. These can weigh up to 3500 lb (about 15,500 N) apiece (Guerry, 1976).

The upper lip of the trawl net is connected to a headrope, typically some 100–130 ft (30–40 m) in length in commercial use. This headrope is buoyed up by numerous metal floats that keep it several feet above the bottom. In some cases, elevator boards (*kites*) assist in providing headrope lift. The groundrope, typically 140–180 ft (approximately 40–60 m) long, is a steel wire rope to which the lower part of the net is connected. Rubber discs toward the ends of the groundrope and steel *bobbins* toward the center assist in the trawl's overriding sea floor obstructions.

A typical towing speed for otter trawls is 3 knots, but speeds to 5 knots for fast-swimming fish are possible. The length of warps is usually about three times the water depth, although this varies (Pereyra, 1963). Further details, such as on the letting out and taking in of such trawls, are available in Tait and DeSanto (1972), Sainsbury (1971), and Hjul (1972).

Biological Sampling by Trawling: The bottom trawl gear used in the Brown and Caldwell (1975a) biological sampling offshore from San Francisco was towed at between 2 and 3 knots by a 65-ft (20-m) commercial trawler. The tow cable length was such as to give a 10:1 scope. The two cables were attached to 40×72-in. (1.02×1.83-m) steel otter boards, and the "mud lines" (ground cables) extended another 450 ft (137 m) back to the trawl itself. This had a 42-ft (13-m) headrope, 1.25- (stretch measurement) and 0.5-inch (32 and 13-mm) mesh in the body of the net and cod end, respectively, and a foot rope weighed with steel chain. The optimum tow time in the San Francisco work was found to be 5 min.

For trawl sampling inside San Francisco Bay, a 21-ft (6.5-m) Boston Whaler was used as the towing vessel. Typical speeds were 1 to 2 knots and times from 10 to 30 min. A 24-ft (7.3-m) otter board trawl was used, the doors in this case being 16×30 inches (0.41×0.76 m) in size connected directly to the net and with metal skids. Tow warps were 75-ft (23-m) lengths of 0.5-in. (13-mm) nylon line. The body of the net was fabricated of 1.5-in. (38-mm) mesh, the cod end 0.5-in. (13-mm). A safety line, with float, was attached to the cod end in order to assist in freeing it in the event it became snagged on a bottom obstruction.

The Marine-Estuarine Technical Committee of the California State Water Resources Control Board has attempted to standardize trawling procedures because of the inherent variability in the size and nature of catches with different sizes and types of trawls. It was stipulated that towing should be along an *isobath* (constant-depth contour line) with paired night and day tows separated in time by a maximum of 48 h. This is done because different results, sometimes radically so, are obtained in day and night trawls in the same area (e.g., Roessler, 1965; SCCWRP, 1973).*

The characteristics of the otter trawl (*try net*) are spelled out. It should have a minimum of 24-ft (7.3-m) headrope and a single warp bridle three times the length of the headrope. Furthermore, a semiballoon net should be used [with 1.25–1.5-in. (32–38-mm) stretch mesh in the body with 0.5-in. (13-mm) maximum liner in the cod end—0.25-in. (6-mm) preferred]; the otter boards should be 18×30-in. (0.46×0.76-m) minimum and fitted with steel or iron shoes.

Treatment of Organisms and Data Presentation: When organisms are brought back on deck after a trawling session they can be placed in numbered sacks. After the cruise, these organisms are promptly refrigerated at the laboratory or else preserved in 10% formalin, as an example. Later during sorting they are separated into species by trawl station and identified. The numbers of organisms for each species and station are recorded and standard length measurements for fish (from the tip of the snout to the base of the vertebral column) are taken. Weights are also taken for fish unless a length-weight correlation has been established for the individual species involved.

The resulting data can be presented in a table containing the following columns: scientific name of organisms, common name, trawling station number, date of survey, time of day, number of individuals of the particular species, average or total length, and average or total weight. The concentration of organisms per unit bottom area can be given. Pooled data in a table permit the reader to examine seasonal differences in the species composition of groundfish and benthic animals at a single station or differences among stations for the same time of year. The latter might involve stations closer to or more remote from a wastewater effluent source.

Trawling and Other Approaches: The sampling of benthic animals living on the bottom and fish living in close proximity to it can be done using approaches other than trawling. Fish can be taken by laying out long lines with baited hooks at discrete intervals. Crabs can be taken with crab nets. Another possibility, which is the topic of the following subsection, is for biologist-divers to survey the bottom. Fager et al. (1966), for example, found that trawl estimates for the density of a particular flatfish were only 12–29% (median 17%) of that found using a diver "raking" technique.

The same results cannot be anticipated from trawl data on demersal organisms as from other approaches. There are several reasons for this (e.g.,

---

*For one thing, otter trawls by day are biased toward smaller and less mobile species. However, capture of juveniles for study is important.

Brown and Caldwell, Consulting Engineers, 1975a):

1. The size of the organisms vis-à-vis the net mesh size
2. Their ability to avoid the net
3. Their schooling behavior
4. Their tendency to congregate by size or age class or by sex

    Additional discussion of the efficiency of trawling operations is available in Roessler (1965), Pereyra et al. (1967) and SCCWRP (1973), as well as in Beamish (1967) and other papers delivered at the FAO Conference on Fish Behavior (Bergen, Norway, 1967).

Surveys by Divers

    The Transect: The *transect*, in the context herein, is basically a study path along the sea floor. It is customarily marked in some way, such as by a stainless steel cable or nylon line strung between small concrete blocks or tied to the bottom. Observations are made by divers as they move along the line. There are essentially two different types of transects. One type is carefully chosen—to be typical of a particular part of the ocean floor, such as from the beach to a 100-ft (30-m) depth or along a specified path such as a proposed outfall line. Permanent markings are often used with such transects since they will usually be revisited from time to time. The other type is of a random nature. An area of interest is occupied by a boat and a marker is dropped to the bottom. The transect, of predetermined length (say 20 m), is then set out from that marker in a direction indicated by a pointer or other direction-giving item on the marker (e.g., Brock, 1954). A transect should be of limited length or else a range of habitats can be involved.

    It is a moderately standard procedure for two divers to swim along the transect line together, one observing on one side of the line with the other diver on the other. Organisms out to a distance of 10 or 20 ft (3 or 6 m) to the side of the line are normally counted. When fish are involved a typical vertical distance of 10 ft (3 m) is used. In a region of relative abundance of marine flora and fauna, a pair of divers on one pass along the line can only deal with one category of benthic plants, benthic animals, or nekton. Data on these three separate groups of organisms can either be collected by the one pair of divers in repeated passes along the transect or by other divers, each pair having its own group responsibility. The latter is more usual because of differences in specialties. Duplicate passes along the line should be made with the pair of divers having changed sides. This can be done easily on the way back to the starting point.

Estimation by a diver of lateral distance may not be very accurate. One way of helping the diver in this regard is to tie small floats on short lines to the transect line at 5-m intervals. A diver then moves down the transect at 5-m intervals. Stationed at one float, he notes the distance to the next float and uses this visual aid to estimate the lateral extent of his study area. This approach has the additional advantage that notes can be made sequentially in an orderly fashion and that the fish are not swept down the transect in front of the divers but have a chance to settle down.

The Data: Fager et al. (1966) have described an underwater tape recorder system for recording information on the organisms encountered. The more usual case, however, is to take notes by pencil on specially prepared sheets of underwater-writing paper, mounted on a clipboard, which list the various types of organisms anticipated and have columns for entries about the organisms. However, doing this quickly and efficiently underwater, especially when there is wave surge, is no easy task.

Fish, for example, are counted by species, and the estimated sizes are entered in the table. This length information can be used later, with a coefficient appropriate to that species, to estimate the total weight of that species along the transect. From the transect length and the lateral distance being used for inclusive observation, the weight of that species per unit bottom area can be derived. Combining all such data yields the total weight of all fishes per unit of bottom area.

The nekton can be divided into resident species, wandering schooling species, and solitary wanderers. The results of a study on nekton can be considerably altered in the event of encounter or nonencounter of a large wandering school. To eliminate this problem some investigators ignore such a school in their calculations, but they may list the encounter in their report because the fish observed are a part of the overall biological community.

The Density of Organisms: The *quadrat* is a sampling plot used to study and analyze plants and/or animals. For our purposes the quadrat is usually a square area, bounded by a frame, used to estimate the density of benthic organisms. The frame is composed typically of four lengths of aluminum tubing that can be easily carried in a small boat and then assembled either in the boat or on the bottom using elbows and bolts. The most typical quadrat area is 1 m^2, but quadrats of area 0.25 m^2 are also used.

In some cases, the quadrat frame is simply dropped from the boat to adopt some random position on the sea floor. In others, a diver, out of sight of the bottom, releases the quadrat for another random position. In still other cases, the quadrat is placed where it would appear to be truly representative of the area being studied. Whatever the method of assigning the quadrat position, the diver's task once it is placed is in part to estimate, using a measuring tape as necessary, the proportion of the total area taken up by different species of sessile benthic organisms. It is also possible to estimate the concentration of infrequent species by flipping the quadrat over and over in order to sample a larger area.

If estimates of benthic algae biomass are desired, all such plant material within the quadrat can be picked and placed in plastic bags for transport to the boat and thence to the laboratory.* Picking is best done from four or five quadrats for representativeness. The algae are then separated into species and the wet weights for each such species measured. Drying for 24 h at 100°F then permits the dry weight of that particular species to be determined.

When estimates are desired of the areal concentration of infrequent members of the benthic plants and animals, one approach followed is that given by Batcheler (1971). This technique has been used to measure density and distribution of sea urchins and coral heads, for example. A random point along the transect line is selected, and the distance from that point to the nearest member of the species concerned is measured. In addition, the distance separating that member and its nearest neighbor is also determined. This is done a number of times, typically 25. Mathematical treatment of the data provides an estimate of the density of the organism.

A technique used to determine percent cover of, say, corals is to measure the length of a transect line, randomly located and oriented, that falls above the species involved. Another involves the subdivision of a 1 m^2 quadrat into square-decimeter areas through use of wire, and then the estimation of the percent coral cover for each small square.

Bottom Materials and Notification: Samples of bottom materials can also be obtained by divers using techniques and equipment such as described by Fager et al. (1966). Basically, a box sampler (e.g., Rosfelder and Marshall, 1967) is pushed into the bottom material, a cover is pushed through the bottom material to close the open end of the sampler, and the sampler and contents are returned to the surface. Where the bottom material is harder, say coral rock, divers can use hammers and chisels to remove samples.

It is probably a good idea to advise the U.S. Coast Guard, or equivalent organizations in other countries, of planned marine biological surveys. The following, for example, appeared in the "Local Notice to Mariners" issued by the Fourteenth Coast Guard District, Honolulu, on 3 August 1976.

HAWAIIAN ISLANDS—HAWAII—HONOKAHAU HARBOR

A marine biological survey will be taken in Honokahau Harbor from 20 thru 22 August 1976. Divers and related underwater equipment will be utilized. Mariners are urged to use extreme caution while transiting this area on the above dates.

---

*This technique is not workable with coralline algae cemented to the reef.

Sampling the Benthos with the Grab Sampler

Introduction: Hopkins (1964) has described 19 different types of grab samplers. Numerous others are listed in McIntyre (1970). There are five particular types that have seen considerable use in marine surveys related to outfalls:

1. Petersen Grab
2. van Veen Grab
3. Ekman Grab
4. Smith-McIntyre Grab
5. Orange Peel Grab

Drawings of 1, 3, 4, and 5 are contained in *Standard Methods*...(1976). Photographs of the van Veen appear in Ingham (1975) and in Word (1975).

The Petersen Grab has two weighted jaws that snap together when the shock of the sampler contacting the bottom releases a catch. The sampler takes 0.1 m² out of the bottom, biting deeper the heavier the grab and the softer the sediment. The van Veen Grab has the same jaw construction and sampling area as the Petersen, but long arms are used to provide a stronger bite. This sampler will take more material on hard bottoms than the Petersen, but the arms cause handling problems in small boats and can lead to premature grab closure when wave action is too severe.

The Ekman Grab is a box-shaped sampler that has taken various forms. It operates best on soft sediments and obtains samples with moderately undisturbed stratification. Closure is automatic as the sampler strikes the bottom.

The Smith-McIntyre Grab samples 0.1 m² and is a somewhat complex and heavy device designed for rough weather and hard sand bottoms, more effective in both respects than the van Veen Grab (Smith and McIntyre, 1954). The jaws are semi-cylindrical, and two springs acting on the axis of the jaws aid in forcing the jaws into the bottom. As the sampler is retrieved the hoisting line acts on levers to close the jaws. The Orange Peel Grab has four sectors that, when closed, take a roughly hemispherical bit out of the bottom material. The closing mechanism can be operated by messenger or by a second cable.

Performance: Comparative tests on seven different types of grab samplers have been carried out by the Southern California Coastal Water Research Project (SCCWRP) in an effort to establish which is the best sampler from engineering and biological viewpoints. The tests have been summarized by Word (1975, 1976). The samplers tested were examined in terms of their abilities to satisfy the following five criteria:

1. Operate from a variety of small marine research vessels
2. Function properly in a range of from shallow, sandy bottoms to deep, clay bottoms

3. Function in rough weather without being a hazard

4. Not leak or cause an extreme pressure wave (blowing surface sediment away when sampling)

5. Return to the vessel with a relatively intact sample of the surface sediment

The tests by SCCWRP (pronounced "squirp") involved bottoms composed of coarse sand, fine clay, and sludge, and were undertaken in depths of 13–280 m. An underwater television camera assisted in evaluating performance.

The original list of seven samplers was pared down to three: the van Veen with standard tripping bar mechanism, a modified ("chain-rigged") version of the van Veen, and the Smith-McIntyre. This group was considered comparable; all sampled 0.1 m² of surface area with relatively little disturbance of the surface sediment and leakage. The standard van Veen was eliminated because it was found to be extremely dangerous to handle at the surface. The modified van Veen was chosen over the Smith-McIntyre because it was the simpler machine with fewer moving parts. Both were very similar in their physical sampling characteristics on the types of bottoms tested (i.e., not hard).

Examples of the use of grab samplers in marine work related to wastewater discharges are as follows. The Petersen Grab was used to gather benthic data for the Humboldt Bay outfall in northern California in 1975 but rejected as unsuitable in work offshore from San Francisco because it was ineffectual in the compact sediments encountered (Brown and Caldwell, Consulting Engineers, 1975b). The Petersen, as well as an Ekman Grab, were used in earlier San Francisco area work (Filice, 1958). SCCWRP (1973) used the modified van Veen Grab in its work whereas Turner et al. (1968) employed the Orange Peel Grab in their sampling.

Sample Handling and Data Presentation: A benthic sample containing sand and silt or organic material should be emptied into a tub or large pail, diluted with water, and then swirled for mixing purposes. Ultimately, the resulting slurry should be passed through a U.S. Standard No. 30 sieve (0.590-mm), but coarse screening to remove larger material, such as sticks and rocks, may be necessary beforehand. Such material should be checked for attached and resident organisms before discarding.

The No. 30 sieve can be rotated to ease in the screening process, which should continue until fine material is washed through, retaining only the organisms. At this point some investigators prefer to carry out a general sorting of the organisms, when alive, on a by-phylum or by-class basis. Whether or not this step is included, the resulting organisms are washed into containers (typically wide-mouthed and tapered pint freeze jars) half full of 10% formalin or some other preserving medium. In time, the organisms in the containers are counted and identified to the species level.

Details concerning the field work as well as information derived from the follow-up laboratory analysis can be combined on one form. Typically, pertinent field data should include the following: station number, sequence number, date and time, collected by, capture method, water depth, type of substrate, and volume sampled. Heading the data reporting sheet should be the following information covering the laboratory tests: laboratory identification, analyzed by, aliquot (ml), number of species, number of organisms/ℓ, predominant organisms, diversity index.

The main portion of the data-reporting pertains to the various types of organisms and organic debris (such as snail shells or shell debris) found in the bottom sample. Five columns describing such information can be the following: common name, scientific name, number of organisms per aliquot, number of organisms/m^3 of bottom material and/or per m^2 of bottom surface area, remarks.

In addition, information on biomass per species and per major taxonomic group can be tabulated, and these can be expressed in terms of weight per unit bottom surface area or per unit volume of bottom material. Since it is likely that several different grabs will be made at the same station, a statistical representation for the data is useful. This can include mean and median as measures of central tendency and the standard deviation plus maximum and minimum values (from which one can determine the range) as measures of dispersion.

Sampling by Camera

The percent substrate cover does not have to be evaluated *in situ* by a diver. A technique that has been used with success involves the placing on the sea floor of a rectangular metal frame supporting an underwater camera and a suitable electronic flash unit within an appropriate underwater housing. The frame positions the camera and flash unit a fixed distance from the bottom so that a uniformly illuminated area is photographed. A typical plan dimension for the frame is $1 \times 2/3$ m^2, and the camera height (approximately, 1.2 m) is adjusted so that the frame is just visible around the border of the resulting transparency. The percent of substrate cover is determined by projecting the slide on a graduated screen (e.g., a piece of cardboard) and then estimating the percent cover within each unit. These estimates are then summed and averaged. An advantage of this method over that of *in situ* diver estimation, in addition to in-water time and accuracy, is that a permanent record of the bottom is obtained. Disadvantages, however, include the necessity of having water of very good visibility and the fact that three-dimensional surface areas are difficult to estimate. A further disadvantage concerns the fact that only organisms that are visible can be counted and then only identified if larger than about 0.5 to 1 in (12–25 mm) in size. The percent algal cover on the bottom cannot be accurately appraised using the camera technique because many types tend to sway back and forth with the wave surge. The algae are better cropped, placed in a plastic bag, and taken to the laboratory for biomass determinations as outlined earlier.

Carrothers (1967) has described a photographic system for benthic organisms and nekton. McIntyre (1956) has compared the benthic sampling efficiency of a trawl, a grab sampler, and a camera. Tests that were run made use of a 0.1 m^2 van Veen Grab and a 6-ft (1.8-m) beam trawl. It was found that the trawl always gave the lowest estimates of organism abundance.

However, McIntyre felt that the trawl was good for widely dispersed organisms, plus had the advantage that it would take large and active animals. The trawl, of course, gives no information on the fine structure of aerial distribution of organisms. The grab gives aerial distributions, but since the sample area is small, at least several repeats are necessary at the same station to minimize the chance of sampling errors leading to spurious conclusions. It was found by McIntyre that the camera system was good on hard substrates. The camera approach gives an indication of aerial variability between estimates of the trawl and the grab. McIntyre listed the following representative sampling areas for a 0.5-h period: trawl, 2000 m^2; camera, 50 m^2; grab sampler, 2 m^2.

6-6 GATHERING DATA ON PLANKTON

Sampling in the Field

The gathering of data on plankton can be illustrated by going on a hypothetical, coastal waters outing with two marine biologists, starting at the dock. The boat is a 16-ft-long (5-m) fiberglass Boston Whaler equipped with a 40-hp outboard motor. A small winch, powered by a gasoline engine, is located amidship, and the light steel cable from the winch runs over a pulley at the end of a stanchion opposite it on the starboard side (see Figure 6-5). The cable supports a pair of plankton nets* held in a dumbell-shaped, open frame.

In the bow of the boat is a pump/filter system with inlet and discharge hoses of tygon tubing. An anchor and line are stowed forward. On the floor of the boat are the following: an ice chest, with several inches of ice covering the bottom; a box of labeled, plastic bottles; and a box of assorted items such as a notebook, pencils, screwdrivers, stainless steel hose clamps, wash bottles, and plastic cod ends. An additional plankton net and its polypropylene towline are stowed along the port side of the boat.

The boat is taken to the first sampling station and securely anchored. The water depth is measured with a lead line, and a Secchi disc is put over the side and the Secchi depth obtained as explained in Section 6-7. Covered plastic cod ends are inserted into the after ends of the washed nets. These are secured in place inside the nets with the stainless-steel hose clamps. The reading of a flowmeter in the mouth of one of the nets is taken, and the nets and their frame

---

*Such nets are customarily made of nylon or silk, materials that are stiff yet flexible, durable, and resistant to swelling.

Figure 6-5   Boat set up for sampling plankton

are then lowered to a position near the seabed (as judged from a wire payout meter near the pulley). The winch is then engaged, and the net assembly is brought vertically upward to the boat at the desired speed, between 0.5 and 1.0 m/s. The time of upward transit is recorded as a check. Filtered sea water is then played over the nets to wash organisms down into the cod end. The cod end, with its contents, is then removed, and a wash bottle with nozzle is used to wash organisms on the walls of the low end of the net into the cod end. The contents from the cod end are then dumped into a sample bottle, which is placed on ice.

The final reading on the flowmeter is taken, and this figure is added to other tabulated data for the station such as the station number, water depth, Secchi depth, sample bottle numbers, initial reading of flowmeter, weather conditions, date, time, and persons involved. Another vertical run is made at the same station for checking purposes. This is called *replication*.

The two nets have different mesh sizes. One of the nets is of 333-μ size (see Figure 6-6). The other "net" is really a pair of nets, one inside the other. On the inside is another 333-μ net, but of smaller overall size than the first one, and on the outside is a 35-μ net. The catch of the inner net is discarded; the outer net then traps members of the plankton lying in the 35–333-μ size range.

Upon completion of the second run, the anchor is brought up, the boat put underway, and the single plankton net (also of 333-μ net size) towed in a circular path around the station where the vertical haul was made. This horizontal haul in the surface to sample the *neuston*, or surface plankton, lasts for a timed interval of 1–3 min. The net is brought back aboard (Figure 6-7), and the same procedure as for the other nets is followed until the filled sample jar is in the ice chest.

Figure 6-6   Plankton net and frame

Figure 6-7   Net being retrieved after neuston tow

Laboratory Work

When the boat returns to the biological station, the ice chest with its store of samples is taken to the laboratory where other workers carry on with the remaining steps. If the number of planktonic organisms is small, they can be concentrated in various ways (*Standard Methods...*, 1976). Often, however, there is an abundance of planktonic organisms; then the sample is mixed with filtered sea water and "split." This involves pouring the "soup" into a rotatable, circular device (*Folsom splitter*) with a middle wall. The contents are then dumped out into two separate containers. The sample is repeatedly split until the concentration of organisms remaining is satisfactory for analysis. The number of dilutions of the original sample is recorded, and the result is known, for example, as a one-quarter *aliquot*.

The two parts of the finally split sample are treated differently. One of these is treated with formalin as a preserving step and filed away on a shelf in a labeled bottle until the time-consuming and tedious counting and identification of the organisms can be undertaken. The strength of formalin used in such preserving varies from laboratory to laboratory and person to person. Figures quoted vary from approximately 1–10%. A preferred intermediate value is 4%.

The aqueous part of the other half of the final sample is separated from the organisms by passing the sample through a special glass fiber filter. The organisms remaining on the filter are then dried in an oven, and the final dry weight of organic (and whatever inorganic) matter is obtained. The filter and associated organic and inorganic matter are then combusted after being ground up together. The glass filter remains behind as a solid ball, whereas the organic matter burns off. The C, H, and N constituents are given off as gases, and these gases move away through a column. The amounts of each gas are measured, and from this information a C/N ratio can be established. A high value is associated with plant matter and a low value with animal matter. The ratio of organic to inorganic material can also be obtained.

Another use for the nonformalin part of the sample is to study, under a fluorescence microscope, for example, the amount of chlorophyll in the sample. With the amount of chlorophyll *a* known, algal biomass can be approximated by multiplying by 67 (*Standard Methods...*, 1976) since approximately 1.5% of algal biomass is chlorophyll *a*.

Some planktonic samples consist primarily of a single type of organism. This, however, is a rarity, and planktonic samples usually contain a number of different kinds of planktonic organisms. The identification of such organisms to a species level is no trivial matter. Often, one person will be well versed in one phylum or class of organisms with another individual specializing in another. Thus before identification to species level is attempted, it is not unusual that an initial coarse separation of the organisms into by-phylum or by-class groups be carried out. The individual samples resulting from this preliminary step are then passed on to those equipped to fathom their secrets. One coarse screening I have seen for zooplankton was as follows: fish eggs and larvae, watery forms (e.g., jellyfish, tunicates), chaetognaths, euphausids, halobates (part of the neuston), mollusks, polychaetes, and higher crustacea (e.g., isopods, decapods).

The findings of the various biologists involved for the different aspects of any single sample are put together in order to specify the number of organisms and the number of species for that sample. Correction must of course be made for the dilution of the original sample. The number of organisms per unit volume in the sea can be obtained by dividing the number of organisms by the water volume sampled. This volume is obtained as follows. The difference between the final and initial readings of the flowmeter can be related through calibration factors to the distance traveled by the net. This distance multiplied by the large open area of the net gives the volume required. If the meter is just inside the net openings, no correction need be made for the "efficiency" of the net. This involves the ratio of volume of water that actually moves into the net to that approaching the net directly in front of the opening. The effective distance of "fishing" of the net is not necessarily the real distance travelled by the boat because of this efficiency effect and because of the possible existence of currents.

Other Approaches

The methods illustrated in the previous example are by no means the only ways of obtaining plankton samples. The Clarke-Bumpus sampler is a small net with flowmeter and a closing disc that can be opened by messenger. The Isaacs-Kidd midwater trawl is a large net with a V-shaped vane, or depressor, that tends to keep the trawl below the surface (Newell and Newell, 1963). Aron (1959), for example, describes the use of the Isaacs-Kidd trawl; Lukin et al. (1971) used a Clarke-Bumpus sampler.

The "bongo net" has many adherents. Two nets trail backward from a yoke consisting of two circular frames to hold the nets open and a joining bar to or through which a haul line from the ship is connected. A canvas cover over the mouth of the net prevents it from "fishing" until the time and depth are right. Then, a messenger dispatched from the surface releases one side of the cover and it flops back into the entrance to the net. At the same moment, a flowmeter in the mouth is actuated. Upon completion of the run, another messenger from the surface releases a wire around the outside of the net, and the net slides backward. As this happens the flowmeter rotation is arrested and a slip line is pulled taut inside the mouth of the net, sealing it closed.

Several bongo nets can be deployed at one time at different depths along the same tow line. A depth recorder can be used in such a string, below the lowest net, to give a better indication of net submergence than would be possible from judging from the length of line paid out and the line angle.

Investigations concerning plankton are not limited to the use of nets and the analysis of organisms obtained in this way. In-the-water studies are possible.

For example, primary productivity, the rate at which phytoplankton convert inorganic carbon to the organic form, can be assessed using the light-dark bottle technique. Two pairs (for replication) of 1-ℓ bottles, one clear and one painted black, are filled with water at a desired depth, sealed, and attached to anchors and floats to suspend them about 1 m above the bottom. A third bottle is also filled with water at the same depth, but in this case the bottle and contents are brought to the surface where dissolved oxygen is fixed to be measured later by the Winkler technique detailed in *Standard Methods*... (1976). After several hours, the light and dark bottles are retrieved and similarly treated. Primary productivity can be determined by measuring the changes in the oxygen (and CO_2) concentrations in the samples. The loss in oxygen in the dark bottle is used as an estimate of respiration, that in the clear bottle in an estimate of net production.

There are commercially available,[*] self-contained submersible fluorometers designed to obtain an *in vivo* record of chlorophyll *a* at a depth which the instrument measures. Such systems can either be lowered from a stationary vessel or towed.

An Integrated Marine Biological Study

The work of Turner et al. (1968) illustrates a complete biologically oriented marine study. Site for the work was just north of San Diego, California, along four transects perpendicular to the shoreline. Rather than swim the entire transects and gather only token information, these investigators closely studied the flora and fauna at five stations along each transect. These were at approximate depths of 20, 40, 60, 80, and 100 ft (about 6, 12, 18, 24, and 30 m) as determined from an on-board fathometer. A float and flag were dropped at each station so that the exact position could be determined through the use of transits from shore.

At each station the following types of operations were involved:

1. vertical plankton tow such as described earlier
2. a diver-obtained core sample
3. a scoop sample of the bottom material for use in analyzing for polychaetes and foraminiferans[†]
4. water depth
5. bottom water temperature
6. water clarity, judged by the divers
7. general description of the bottom area

[*]Zone Research Inc., Washington, D.C.

[†]Benthic foraminifera have been used as tracers of sediment movement and indicators of sediment sources.

8. an enumeration, by estimate, of the larger plants and animals (including fish)

9. the quantitative enumeration of the larger plants and animals in a specified circular study area

10. quantitative sampling, by actual removal, of growths within a quadrat of area $1/4$ m^2

11. photographic records of the bottom and of each quadrat before sample removal

The circular study area was assigned and positioned by using the spot where the anchor "set" as its center. A line of specified length was then rotated around this spot to mark out a desired area. In sandy areas, a 3.1-m line was used to yield a study region of area 30 m^2. On harder bottoms, with the visible biota more diverse, a 2.2-m radius line was used to give an area of 15 m^2. An appendix in the report of Turner et al. (1968) provided a cross-reference list between the common and scientific names of the organisms encountered and listed in the main body of the text. Another integrated marine biological survey is described by Smith et al. (1973).

6-7 OBTAINING WATER-COLUMN INFORMATION

Orientation

The discussion of Chapter 5 and this chapter to this point have largely ignored a multitude of water-column parameters that should be established for a reasonably complete characterization of that water column. For example, an important physical parameter not yet presented involves the clarity/turbidity of the water, and this matter will be considered later in this section. The following types of chemical information are possibilities: concentrations of heavy metals; concentrations of plant nutrients; levels of man-introduced pollutants like radioactive materials, synthetic organics, and hydrocarbons; DO and BOD; and pH. Information of bacteriological importance concerns fecal and total coliform counts as well as levels of pathogenic bacteria and viruses. By and large, treatment of the above topics in this section will be extremely brief or nonexistent, chiefly because they are covered in detail in *Standard Methods...* (1976) and, for sea water, by Strickland and Parsons (1972). A good part of the information referred to above should perhaps more properly be included in Chapter 11 where monitoring of wastewater discharges will be discussed, but has been included here to avoid fragmentation of the treatment.

Collecting Samples

The collecting of samples of the water column at a particular station and water depth was described in Section 5-7. Since metallic sampling bottles can contaminate the samples and invalidate values of chemical parameters (e.g., pH and Fe concentration) obtained from them, plastic samplers are preferred (e.g., Betzer and Pilson, 1970; Takahashi et al., 1970). In particularly sensitive cases, a once-only sampling container is employed. Niskin (1962) has developed a particularly useful ("sterile bag") sampler of this type that is marketed by General Oceanics, Inc. (Miami, Florida).

A water sample to be used for chemical (or other) analysis purposes does not have to come from a discrete sampler lowered over the side of a boat or ship or taken down by a diver and activated. A technique that can be used to advantage involves the continuous pumping of water from a specific depth up to the boat. Sample bottles are then filled from the discharge hose from the pump, which should be of the positive displacement type so that dissolved air concentrations are not altered as the water passes through it.

The Ultramar Chemical Water Laboratory (1969), for example, used this approach. The pump ran from a 12-V battery. Heavy rubber hose (75 feet or 23 m) served as the suction line. At the low end of this line, there was a strainer, and near this point a 5-ft (1.5-m) length of leader was connected, hanging down to a 8-lb (36-N) weight. The hose was lowered until the weight contacted the bottom, and the water about 5 ft (1.5 m) off the bottom then entered the line. It was felt that this distance up from the sea bed largely reduced chances of stirred-up bottom material entering the system. The Ultramar researchers also did some sampling, difficult in an unruly seaway, with 3-ℓ Van Dorn bottles handled by heavy fishing tackle—viz. rod and reel.

On a large, well-equipped oceanographic vessel, it is often possible to carry out a wide range of chemical and bacteriological tests on board. But in most oceanographic work associated with proposed outfalls considerably smaller boats are used, and they do not have space for an elaborate testing setup. Most of the analysis then must be done on shore, and samples obtained at sea must be properly handled so that any tendency for the water samples to change before they reach the laboratory can be minimized. As an example, samples to be used in bacteriological tests need to be refrigerated; those to be used for dissolved oxygen levels need to be "fixed." However, even on small boats there are often several types of analysis that can be run. These will be mentioned later.

Chemical Parameters

Analytical Methods: In trace metal analysis, great care must be taken to avoid changes in the sample not only during collection but also during handling. Contamination can come at the least from samplers, laboratory equipment, glassware, and impurities in reagents. Losses can occur due to absorption on sampling and handling vessels. Procedures for analyzing water samples for specific metallic ions are detailed in *Standard Methods...* (1976), as are techniques for evaluating concentrations of radioactive substances as well as foreign organic material such as grease and oil, pesticides, phenols, and surfactants.

Plant nutrients have been discussed several times already in this volume, viz. in Chapters 2, 3, and 4. The primary pair of these consists of nitrogen and phosphorus. The results of analyses for nitrogen normally contain several components. Total *Kjeldahl nitrogen* refers to the sum of organic and ammonia nitrogen. *Total nitrogen* includes organic and ammonia nitrogen, as well as nitrate and nitrite nitrogen. Contrary to the situation for nitrogen, phosphorus in natural waters exists almost solely in the phosphate form. The total amount of phosphate or phosphorus present is sometimes divided into reactive and nonreactive forms. Again, tests for all of the above are in *Standard Methods...* (1976).

The dissolved oxygen in a water column (Figure 6-8) is a particularly important parameter to evaluate. Oxygen is of course required to enable aquatic animals to carry on life functions and for aquatic plants in respiration when they are not photosynthesizing. Low levels of DO can lead, for example, to delayed hatching of fish eggs and to reduced size and vigor of embryos.

Standard Methods... (1976) discusses the collection and preservation of samples for laboratory determinations of DO, and both *Standard Methods...* and Strickland and Parsons (1972) detail the procedures involved in the determination of DO by the Winkler (iodometric) method. The former volume describes procedures for determining the pH of a water sample brought to the laboratory.

Instruments and Probes: It is by no means necessary to employ analytical methods in the laboratory to determine the DO or pH of water at a particular coastal waters station and at a specific depth. There are two other approaches, one of which involves a measurement system that has a probe at the depth desired and a readout on the boat; the other employs a special instrument to determine the value of that parameter on a water sample. The type of instrument used with a water sample can either be used on the boat, immediately after the

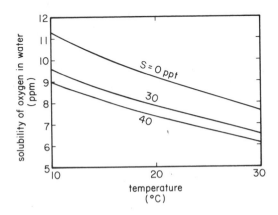

Figure 6-8   Solubility of oxygen in water: wet air at atmospheric pressure

samples are brought aboard, or later in the laboratory.

In the work by Ultramar Chemical Water Laboratory referred to earlier, portable DO and pH meters were used on the boat. As a calibration check for the DO meter, some samples were fixed and later taken to the laboratory for DO determination by the Winkler method.

The Rexnord (Malvern, Pa.) Model 350 Dissolved Oxygen Analyzer is an example of an instrument that can determine chemical parameters directly. It has a range of 0–20 mg/ℓ; its response is 99% complete in 15 s. The probe in the instrument is a galvanic cell with platinum and lead as the electrodes, and a tiny agitator provides stirring near the probe tip. The probe assembly is inserted by an operator into a flask or beaker containing the sample. A direct digital readout of DO is provided.

Numerous manufacturers have setups with multiple probes that can be lowered into the ocean from a boat or ship to obtain data on conductivity, salinity, temperature, depth, DO, pH, and sound velocity. Various of these systems have the readout for any of the various quantities on a panel meter with data recording done by an operator with a pad and pencil. In some cases the same instrument can be used for either sample analysis or *in situ* measurements.

Martek Instruments, Inc. (Newport Beach, Calif.) has several different models that provide *in situ* measurement and deck readout capability for up to six water quality parameters as follows: temperature, conductivity, salinity, depth, DO, and pH. The Mark IV design, for example, provides for measurement and digital readout, if desired, of all six parameters for depths down to 300 m. Martek manufactures companion recording instrumentation for which one option is a digital magnetic tape cassette recorder. Similar systems are available from various manufacturers.

Data Presentation: A useful (computer) technique for presenting both the time and depth variation of any particular water-quality parameter for various stations is as follows. The range of possible values of the parameter is placed along the abscissa, the range of possible sampling depths along the ordinate. Different symbols (e.g., letters) are then used to show the depth variation of the parameter for a given station and time. A considerable amount of data can be combined in this way, and temporal and spatial (horizontal and vertical) trends are readily apparent.

Water Clarity

A technique that has been used for many years to determine the clarity (or conversely, the *turbidity*) of a water column involves a circular disc, called a *Secchi disc*, that is lowered into the water (see Figure 6-9). These discs are usually 30 cm diam, although ones with diameters of 20 and 25 cm have also been used. It has been virtually standard that the Secchi disc be painted white,

Figure 6-9 Oceanographer retrieving Secchi disc

but again there are variations. A black and white-striped version is one such variant, one with alternating black and white quadrants another.

The Secchi disc is customarily connected to a lowering line through a three-part bridle that maintains the disc in the horizontal plane. A weight hangs under the disc, again normally through a three-part bridle. The "Secchi depth" is the average of the depth at which the disc disappears from view on the way down and the depth at which it reappears when being retrieved from some distance below the first estimate of the Secchi depth. It is necessary that the lowering line be marked off in convenient length intervals so that the length of line in the water can be determined.

The Secchi depth is admittedly crude. Its estimate must at the least depend upon the following factors: cloud cover, angle of the sun, shadow, surface ripple, and whether or not some form of glass-bottomed viewing box is used. It is probably best for the sake of consistency that observers do not wear sun glasses and that they do use a viewing box. In work described by McAllister (1968), for instance, such a viewing box was used with the Secchi disc.

Although an approximate measure of water clarity, the Secchi depth has been correlated with various water column characteristics. These include the concentrations of suspended inorganic or organic matter, chlorophyll, the attenuation coefficient for light, and the compensation depth (e.g., Clarke, 1940; Atkins et al., 1954; Jones and Wills, 1956; Postma, 1961; Williams, 1968a, b; Qasim et al., 1968).

Because of the crude aspects of the estimation of water clarity by the Secchi disc many researchers lean toward electrical instruments with appreciable accuracy. These are the transmissometer and the photometer. However, before these two instruments are very briefly described, one might counter the dismissal of the Secchi disc as a viable oceanographic tool by simply making the following statement. The natural small-scale temporal and spatial variation in water clarity is often sufficiently large that using a complicated instrument in such a case is paramount to measuring very accurately a sample of one. The accuracy of the measurement may not be worth a great deal.

The *transmissometer* measures water turbidity by determining the percentage transmission of a columnated light beam through a given path length (between source and pickup) in the water. The transmissometer supplies its own light in contrast to the *photometer* that indicates the illumination falling on it from the sun. Thus the photometer gives a measure of the average light attenuation through a water column of certain depth, whereas the transmissometer gives such an indication for a specific depth. An instrument that indicates the spectral makeup of light at a specific depth is known as a *spectrophotometer*.

6-8   REFERENCES

"A List of Common and Scientific Names of Fishes from the United States and Canada" (1970): American Fisheries Society, Washington, D.C., *Special Publication No. 6*, 3rd ed.

ALBERS, V. M. (1960): *Underwater Acoustics Handbook*, The Pennsylvania State University Press, State College, Pa.

ALPINE GEOPHYSICAL ASSOCIATES, INC. (1971): "Geophysical Survey Offshore Davenport, California," Report prepared for Pacific Gas and Electric Co., San Francisco, Calif., Norwood, N.J. (February).

ARON, W. (1959): "Midwater Trawling Studies in the North Pacific," *Limnology and Oceanography*, Vol. 4, pp. 409–418.

ATKINS, W. R. G., P. G. JENKINS, and F. J. WARREN (1954): "The Suspended Matter in Sea Water and Its Seasonal Changes as Affecting the Visual Range of the Secchi Disc," *Journal of the Marine Biological Association of the U.K.*, Vol. 33, pp. 497–509.

BATCHELER, C. L. (1971): "Estimation of Density from a Sample of Joint Point and Nearest-Neighbor Distances," *Ecology*, Vol. 52, pp. 703–709.

BEAMISH, F. W. H. (1967): "Photographic Observations on Reactions of Fish ahead of Otter Trawls," in *Proceedings of the FAO Conference on Fish Behavior in Relation to Fishing Techniques and Tactics*, Bergen, Norway (October), Food and Agriculture Organization of the United Nations, Rome, *Fisheries Reports No. 62*, Vol. 3, pp. 511–521.

BEMBEN, S. M., and E. H. KALAJIAN (1969): "Vertical Holding Capacity of Marine Anchors in Sand," ASCE, *Proceedings*, Civil Engineering in the Oceans II, Miami Beach, Fla. (December), pp. 117–136.

BETZER, P. R., and M. E. Q. PILSON (1970): "The Nansen Bottle—A Major Contributor to Reported Concentrations of Particulate Iron in Sea Water," *Deep-Sea Research*, Vol. 17, pp. 671–674.

BOWLES, J. E. (1970): *Engineering Properties of Soils and Their Measurement*, McGraw-Hill, New York.

BROCK, V. E. (1954): "A Preliminary Report on a Method of Estimating Reef Fish Populations," *Journal of Wildlife Management*, Vol. 18, pp. 297–308.

BROWN and CALDWELL, Consulting Engineers (1975a): "A Predesign Report on Marine Waste Disposal," Vol. 4, prepared for City and County of San Francisco, at Walnut Creek, Calif. (October).

BROWN and CALDWELL, Consulting Engineers (1975b): "A Predesign Report on Marine Waste Disposal," Vol. 5, prepared for City and County of San Francisco, at Walnut Creek, Calif. (October).

CARROTHERS, P. J. G. (1967): "Automatic Underwater Photographic Equipment for Fisheries Research," Fisheries Research Board of Canada, Ottawa, Ontario, *Bulletin 159*.

CLARKE, G. L. (1940): "Observations on Transparency in the South-western Sector of the North Atlantic Ocean," *Journal of Marine Research*, Vol. 4, pp. 221–230.

DEL GROSSO, V. A. (1974): "New Equations for the Speed of Sound in Natural Waters (with Comparisons to Other Equations)," *Journal of the Acoustical Society of America*, Vol. 56, pp. 1084–1091.

DEL GROSSO, V. A., and C. W. MADER (1972): "Speed of Sound in Sea Water Samples," *Journal of the Acoustical Society of America*, Vol. 52, pp. 961–974.

DILL, R. F., and D. G. MOORE (1965): "A Diver Held Vane-Shear Apparatus," *Marine Geology*, Vol. 3, pp. 323–327.

EMRICH, W. J. (1971): "Performance Study of Soil Sampler for Deep-Penetration Marine Borings," in *Sampling of Soil and Rock*, American Society for Testing and Materials, Philadelphia, *Special Technical Publication 483*, pp. 30–50.

FAGER, E. W., et al. (1966): "Equipment for Use in Ecological Studies Using Scuba," *Limnology and Oceanography*, Vol. 11, pp. 503–509.

FILICE, F. P. (1958): "Invertebrates from the Estuarine Portion of San Francisco Bay and Some Factors Influencing Their Distribution," *The Wasman Journal of Biology*, Vol. 16, pp. 159–211.

GUERRY, T. L. (1976): "Highlights of 1975 North Sea Pipe Line Operations," *Ocean Industry*, Vol. 11, No. 2 (March), pp. 23–28.

HARRISON, W., and A. M. RICHARDSON, JR. (1967): "Plate-Load Tests on Sandy Marine Sediments, Lower Chesapeake Bay," in *Marine Geotechnique*, edited by A. F. Richards, University of Illinois Press, Urbana, Illinois, pp. 274–290.

HJUL, P., Ed. (1972): *The Stern Trawler*, Fishing News (Books) Ltd., London.

HOEL, P. G. (1976): *Elementary Statistics*, 4th ed., Wiley, New York.

HOPKINS, T. L. (1964): "A Survey of Marine Bottom Samplers," in *Progress in Oceanography*, Vol. 2, edited by M. Sears, pp. 215–256, Pergamon Press, New York.

HOUGH, B. K. (1957): *Basic Soil Engineering*, Ronald Press, New York.

INGHAM, A. E., Ed. (1975): *Sea Surveying*, 2 vols., Wiley, New York.

JONES, D., and M. S. WILLS (1956): "The Attenuation of Light in Sea and Estuarine Waters in Relation to the Concentration of Suspended Solid Matter," *Journal of the Marine Biological Association of the U.K.*, Vol. 35, pp. 431–444.

KEETON, W. T. (1969): *Elements of Biological Science*, W. W. Norton, New York.

KOMAR, P. D. (1975): "Longshore Currents and Sand Transport on Beaches," *Proceedings*, Civil Engineering in the Oceans III, Univ. of Delaware, Newark, Del. (June), pp. 333–354.

KOMAR, P. D., R. H. NEUDECK, and L. D. KULM (1972): "Observations and Significance of Deep-Water Oscillatory Ripple Marks on the Oregon Continental Shelf," in *Shelf Sediment Transport: Process and Pattern*, edited by D. J. P. Swift, D. B. Duane, and O. H. Pilkey, pp. 601–619, Dowden, Hutchinson and Ross, Inc., Stroudsburg, Pa.

KOSTER, J. (1974): "Digging in of Anchors into the Bottom of the North Sea," Delft Hydraulics Laboratory, The Netherlands, *Publication No. 129* (June).

KOUTSOFTAS, D. C., et al. (1976): "Evaluation of the Vibracorer as a Tool for Underwater Geotechnical Explorations," *Proceedings*, Eighth Annual Offshore Technology Conference, Houston, Texas (May), Vol. 3, pp. 107–121.

LADD, C. C., and T. W. LAMBE (1963): "The Strength of 'Undisturbed' Clay Determined from Undrained Tests," in *Laboratory Shear Testing of Soils*, American Society for Testing and Materials, Philadelphia, *Standard Technical Publication 361*, pp. 342–371.

LAMBE, T. W. (1951): *Soil Testing for Engineers*, Wiley, New York.

LEE, H. J. (1973): "In-Situ Strength of Seafloor Soil Determined from Tests on Partially Disturbed Cores," U.S. Navy, Civil Engineering Laboratory, Port Hueneme, Calif., *Technical Note N-1295* (August).

LUKIN, L. D., et al. (1971): "Limitation and Effects of Waste Disposal on an Ocean Shelf," Florida Ocean Science Institute, Deerfield Beach, Fla., Report (December) (*NTIS Publication PB-226-727*).

MCALLISTER, R. F. (1968): "Demonstrations of the Limitations and Effects of Waste Disposal on an Ocean Shelf," Florida Atlantic Ocean Sciences Institute, Boca Raton, Florida, Report (March) (*NTIS Publication PB-215-585*).

MCINTYRE, A. D. (1956): "The Use of Trawl, Grab and Camera in Estimating Marine Benthos," *Journal of the Marine Biological Association of the U.K.*, Vol. 35, pp. 419–429.

McIntyre, A. D., Ed. (1970): "Bibliography on Methods of Studying the Marine Benthos," Food and Agriculture Organization of the United Nations, Rome, *Fisheries Technical Paper No. 98.*

Mearns, A. J., and L. Harris (1975): "Age, Length, and Weight Relationships in Southern California Populations of Dover Sole," Southern California Coastal Water Research Project, El Segundo, Calif., *TM 219* (June).

Newell, G. E., and R. C. Newell (1963): *Marine Plankton: A Practical Guide,* Hutchison Educational Ltd., London.

Niskin, S. J. (1962): "A Water Sampler for Microbiological Studies," *Deep-Sea Research,* Vol. 9, pp. 501–503.

Noda, E. K. (1971): "State-of-the-Art of Littoral Drift Measurements," *Shore and Beach,* Vol. 39, No. 1 (April), pp. 35–41.

Noorany, I. (1972): "Underwater Soil Sampling and Testing—A State-of-the-Art Review," in *Underwater Soil Sampling, Testing, and Construction Control,* American Society for Testing and Materials, Philadelphia, *Special Technical Publication 501,* pp. 3–41.

Odum, E. P. (1969): "The Strategy of Ecosystem Development," *Science,* Vol. 164, No. 3877 (18 April), pp. 262–270.

Palmer, H. D. (1969): "Wave-Induced Scour on the Sea Floor," ASCE, *Proceedings,* Civil Engineering in the Oceans II, Miami Beach, Fla. (December), pp. 703–716.

Pereyra, W. T. (1963): "Scope Ratio—Depth Relationships for Beam Trawl, Shrimp Trawl, and Otter Trawl," *Commercial Fisheries Review,* Vol. 25, No. 12 (December), pp. 7–10.

Pereyra, W. T., H. Heyamoto, and R. R. Simpson (1967): "Relative Catching Efficiency of a 70-foot Semiballoon Shrimp Trawl and a 94-foot Eastern Fish Trawl," *Fishery Industrial Research,* Vol. 4, No. 1 (December), pp. 49–71.

Postma, H. (1961): "Suspended Matter and Secchi Disc Visibility in Coastal Waters," *Netherlands Journal of Sea Research,* Vol. 1, No. 3, pp. 359–390.

Preslan, W. C. (1969): "Accelerometer-Monitored Coring," ASCE, *Proceedings,* Civil Engineering in the Oceans II, Miami Beach, Fla. (December), pp. 655–678.

Qasim, S. E., P. M. A. Bhattathiri, and S. A. H. Abidi (1968): "Solar Radiation and Its Penetration in a Tropical Estuary," *Journal of Experimental Marine Biology and Ecology,* Vol. 2, pp. 87–103.

Quast, J. C. (1968): "Estimates of the Populations and the Standing Crop of Fishes," State of California, Dept. of Fish and Game, *Fish Bulletin 139,* pp. 57–79.

Richards, A. F., and H. W. Parker (1967): "Surface Coring for Shear Strength Measurements," *Proceedings of the Conference on Civil Engineering in the Oceans,* ASCE, San Francisco, Calif. (September), pp. 445–488.

Roessler, M. (1965): "An Analysis of the Variability of Fish Populations Taken by Otter

Trawl in Biscayne Bay, Florida," *Transactions*, The American Fisheries Society, Vol. 94, pp. 311–318.

ROSFELDER, A. M., and N. F. MARSHALL (1967): "Obtaining Large, Undisturbed, and Oriented Samples in Deep Water," in *Marine Geotechnique*, edited by A. F. Richards, University of Illinois Press, Urbana, Ill., pp. 243–263.

SAINSBURY, J. C. (1971): *Commercial Fishing Methods—An Introduction to Vessels and Gear*, Fishing News (Books) Ltd., London.

SCOTT, R. F. (1967a): "In-Place Soil Mechanics Measurements," in *Marine Geotechnique*, edited by A. F. Richards, University of Illinois Press, Urbana, Illinois, pp. 264–271.

SCOTT, R. F. (1967b): "In-Place Measurement of the Strength of Ocean Floor Soils by Accelerometer," *Proceedings of the Conference on Civil Engineering in the Oceans*, ASCE, San Francisco, Calif. (September), pp. 419–444.

SCOTT, R. F. (1970): "In-Place Ocean Soil Strength by Accelerometer," ASCE, *Journal of the Soil Mechanics and Foundations Division*, Vol. 96, No. SM1 (January), pp. 199–211.

Shore Protection Manual (1973): U.S. Army, Corps of Engineers, Coastal Engineering Research Center, Fort Belvoir, Virginia, Vol. 1.

SMITH, R. J. (1962): "Engineering Properties of Ocean Floor Soils," in *Field Testing of Soils*, American Society for Testing and Materials, Philadelphia, *Special Technical Publication 322*, pp. 280–302.

SMITH, S. V., et al. (1973): "Atlas of Kaneohe Bay, A Reef Ecosystem Under Stress," University of Hawaii, Honolulu, Hawaii, Sea Grant Programs, *Publication TR-72-01* (February).

SMITH, W., and A. D. MCINTYRE (1954): "A Spring-Loaded Bottom-Sampler," *Journal of the Marine Biological Association of the U.K.*, Vol. 33, pp. 257–264.

SOMERS, L. H. (1975): "Research Diver Training Program," The University of Michigan, Ann Arbor, Sea Grant Program, Report (June).

Southern California Coastal Water Research Project (1973): "The Ecology of the Southern California Bight: Implications for Water Quality Management," *Technical Report No. 104* (March).

Standard Methods for the Examination of Water and Wastewater, 14th ed. (1976): American Public Health Association, American Water Works Association and Water Pollution Control Federation, Washington, D.C.

STRICKLAND, J. D. H., and T. R. PARSONS (1972): *A Practical Handbook of Seawater Analysis*, Fisheries Research Board of Canada, Ottawa, Ontario, *Bulletin 167*, 2nd ed.

STRIDE, A. H. (1970): "Shape and Size Trends for Sand Waves in a Depositional Zone of the North Sea," *Geological Magazine*, Vol. 107, pp. 469–477.

TAIT, R. V., and R. S. DeSANTO (1972): *Elements of Marine Ecology*, 2nd ed., Springer-Verlag, New York.

TAKAHASHI, T., et al. (1970): "A Carbonate Chemistry Profile at the 1966 Geosecs Intercalibration Station in the Eastern Pacific Ocean," *Journal of Geophysical Research*, Vol. 75, No. 36, pp. 7648–7666.

The NOAA Diving Manual (1975): U.S. Dept. of Commerce, National Oceanic and Atmospheric Administration, Manned Undersea Science and Technology Office, Washington, D.C.

TIREY, G. B. (1972): "Recent Trends in Underwater Soil Sampling Methods," in *Underwater Soil Sampling, Testing, and Construction Control*, American Society for Testing and Materials, Philadelphia, *Special Technical Publication 501*, pp. 42–54.

TURNER, C. H., E. E. EBERT, and R. R. GIVEN (1968): "The Marine Environment Offshore from Point Loma, San Diego County," State of California, Dept. of Fish and Game, *Fish Bulletin 140*.

Ultramar Chemical Water Laboratory (1969): "Water Quality Study: Nearshore Waters of the Island of Kauai," prepared for State of Hawaii, Dept. of Health, at Honolulu, Hawaii (July).

VANONI, V., Ed. (1975): *Sedimentation Engineering*, ASCE, New York.

WARD, M., and R. M. SORENSEN (1970): "A Method of Tracing Sediment Movement on the Texas Gulf Coast," Texas A and M University, College Station, Texas, Sea Grant Program, *Publication No. TAMU-SG-71-204* (December).

WHITAKER, THOMAS (1970): *The Design of Piled Foundations*, Pergamon Press, London.

WILLIAMS, J. (1968a): "A Mathematical Model for the Description of the Optical Properties of Turbid Water in Terms of Suspended Particle Size and Concentration," Chesapeake Bay Institute, *Technical Report 47* (November).

WILLIAMS, J. (1968b): "Determination of Particle Size and Concentration from Photometer and Secchi Disc Measurements," Chesapeake Bay Institute, *Technical Report 48* (November).

WILSON, W. (1960): "Speed of Sound in Sea Water as a Function of Temperature, Pressure, and Salinity," *Journal of the Acoustical Society of America*, Vol. 32, pp. 641–644, 1357.

WORD, J. Q. (1975): "A Comparison of Grab Samplers," Southern California Coastal Water Research Project, El Segundo, Calif., *Annual Report*, pp. 63–66.

WORD, J. Q. (1976): "Biological Comparison of Grab Sampling Devices," Southern California Coastal Water Research Project, El Segundo, Calif., *Annual Report*, pp. 189–194.

6-9 BIBLIOGRAPHY

Biological

"An Oceanographic and Biological Survey of the Southern California Mainland Shelf" (1965): State of California, State Water Quality Control Board, *Publication No. 27*.

BARNES, R. D. (1974): *Invertebrate Zoology*, W. B. Saunders, Philadelphia.

BROWN, M. E., Ed. (1957): *The Physiology of Fishes*, Vol. 1: *Metabolism*, Academic Press, New York.

BROWN, M. E., Ed. (1957): *The Physiology of Fishes*, Vol. 2: *Behavior*, Academic Press, New York.

FAGER, E. W. (1963): "Communities of Organisms," in *The Sea*, edited by M. N. Hill, Vol. 2, pp. 415–437, Wiley (Interscience), New York.

Food and Agriculture Organization of the United Nations (1972): *FAO Catalogue of Fishing Gear Designs*, Fishing News (Books) Ltd., London.

GARNER, J. (1967): *Modern Deep Sea Trawling Gear*, Fishing News (Books) Ltd., London.

GOSLINE, W. A., and V. E. BROCK (1960): *Handbook of Hawaiian Fishes*, University of Hawaii Press, Honolulu, Hawaii.

HART, J. L. (1973): *Pacific Fishes of Canada*, Fisheries Research Board of Canada, Ottawa, Canada, *Bulletin 180*.

HAYES, F. R. (1964): "The Mud-Water Interface," in *Oceanography and Marine Biology: An Annual Review*, edited by H. Barnes, Vol. 2, pp. 121–145.

JOHNSON, M. E., and H. J. SNOOK (1927): *Seashore Animals of the Pacific Coast*, Macmillan, New York.

LEIM, A. H., and W. B. SCOTT (1966): *Fishes of the Atlantic Coast of Canada*, Fisheries Research Board of Canada, Ottawa, Ontario, *Bulletin 155*.

LIGHT, S. F., et al. (1970): *Intertidal Invertebrates of the Central California Coast*, University of California Press, Berkeley and Los Angeles, Calif.

LOESCH, H., et al. (1977): "Technique for Estimating Trawl Efficiency in Catching Brown Shrimp (Penaeus Aztecus), Atlantic Croaker (Micropogon Undulatus) and Spot (Leiostomus Xanthurus)," *Gulf Research Reports*, Vol. 5, No. 2, pp. 29–33.

LORENZEN, C. J. (1966): "A Method for the Continuous Measurement of *In Vivo* Chlorophyll Concentration," *Deep-Sea Research*, Vol. 13, pp. 223–227.

MARAGOS, J. E. (1974): "Coral Communities on a Seaward Reef Slope, Fanning Island," *Pacific Science*, Vol. 28, pp. 257–278.

MARSHALL, T. C. (1966): *Tropical Fishes of the Great Barrier Reef*, American Elsevier, New York.

MARTIN, J. H., and G. A. KNAUER (1973): "The Elemental Composition of Plankton," *Geochimica et Cosmochimica Acta*, Vol. 37, pp. 1639–1653.

MEGLITSCH, P. A. (1972): *Invertebrate Zoology*, 2nd ed., Oxford University Press, New York.

MIEDECKE, J. G., and W. STEPHENSON (1977): "Environmental Impact and Baseline Studies of the Soft Bottom Marine Environment in the Vicinity of Ocean Sewer Outfalls on the New South Wales Central Coast," Institution of Engineers,

Australia, *Proceedings of the Third Australian Conference on Coastal and Ocean Engineering*, Melbourne (April), pp. 157–162.

MILLER, D. J., and R. N. LEA (1972): "Guide to the Coastal Marine Fishes of California," State of California Dept. of Fish and Game, *Fisheries Bulletin 157*.

National Research Council (1969): *Recommended Procedures for Measuring the Productivity of Plankton Standing Stock and Related Oceanic Properties*, Committee on Oceanography, Panel on Biological Methods, National Academy of Sciences, Washington, D.C.

NICHOLS, F. H. (1973): "A Review of Benthic Faunal Surveys in San Francisco Bay," U.S. Geological Survey, *Circular 677*.

OWRE, H. B., and J. K. LOW (1976): "A Mechanically Operated Opening-Closing Macroplankton Net," *Marine Technology Society Journal*, Vol. 10, No. 6 (July–August), pp. 3–7.

POPE, J. A., et al. (1975): "Manual of Methods for Fish Stock Assessment: Part III—Selectivity of Fishing Gear," Food and Agriculture Organization of the United Nations, Rome, *Fisheries Technical Paper No. 41*, Revision 1.

REISH, D. J. (1959): "A Discussion of the Importance of the Screen Size in Washing Quantitative Marine Bottom Samples," *Ecology*, Vol. 40, pp. 307–309.

RICKETTS, E. F., and J. CALVIN (1962): *Between Pacific Tides*, 3rd ed., revised by Joel W. Hedgpeth, Stanford University Press, Stanford, Calif.

SCAGEL, R. F. (1966): "Marine Algae of British Columbia and Northern Washington, Part I: Chlorophyceae (Green Algae)," National Museum of Canada, Ottawa, Ontario, *Bulletin 207*.

SCOTT, W. B., and E. J. CROSSMAN (1973): *Freshwater Fishes of Canada*, Fisheries Research Board of Canada, Ottawa, Ontario, *Bulletin 184*.

SHAPIRO, S., Ed. (1971): *Our Changing Fisheries*, U.S. Dept. of Commerce, National Oceanic and Atmospheric Administration, National Marine Fisheries Service, Washington, D.C.

SHIH, C-T., A. J. G. FIGUERIA, and E. H. GRAINGER (1971): *A Synopsis of Canadian Marine Zooplankton*, Fisheries Research Board of Canada, Ottawa, Ontario, *Bulletin 176*.

SMITH, K. L., JR., and J. D. HOWARD (1972): "Comparison of a Grab Sampler and Large Volume Corer," *Limnology and Oceanography*, Vol. 17, pp. 142–145.

SMITH, S. V., and R. S. HENDERSON (1976): "An Environmental Survey of Canton Atoll Lagoon 1973," Naval Undersea Research and Development Center, San Diego, Calif., NUC TP 395 (June).

STEELE, J. H., and I. E. BAIRD (1961): "Relations between Primary Production, Chlorophyll and Particulate Carbon," *Limnology and Oceanography*, Vol. 6, pp. 68–78.

STRICKLAND, J. D. H. (1960): "Measuring the Production of Marine Phytoplankton," The Fisheries Research Board of Canada, Ottawa, Ontario, *Bulletin No. 122*.

TURNER, C. H., E. E. EBERT, and R. R. GIVEN (1964): "An Ecological Survey of a Marine Environment Prior to Installation of a Submarine Outfall," *California Fish and Game*, Vol. 50, No. 3, pp. 176–188.

VERNBERG, W. B., and F. J. VERNBERG (1972): *Environmental Physiology of Marine Animals*, Springer-Verlag, New York.

WALLS, J. G. (1975): *Fishes of the Northern Gulf of Mexico*, T. F. H. Publications, Neptune City, N.J.

WATERMAN, T. H., Ed. (1960a): *The Physiology of Crustacea*, Vol. 1, *Metabolism and Growth*, Academic Press, New York.

WATERMAN, T. H., Ed. (1960b): *The Physiology of Crustacea*, Vol. 2, *Sense Organs, Integration, and Behavior*, Academic Press, New York.

WESTREE, B. L. (1975): "Biological Baseline Studies for a Proposed Marine Terminal," The Third International Ocean Development Conference, Tokyo, Japan (August), *Preprints*, Vol. 4, pp. 131–137.

WILBUR, K. M., and C. M. YONGE, Eds. (1964): *Physiology of Mollusca*, 2 vols., Academic Press, New York.

YENTSCH, C. S., and J. H. RYTHER (1957): "Short-term Variations in Phytoplankton Chlorophyll and Their Significance," *Limnology and Oceanography*, Vol. 2, pp. 121–130.

Quantitative Approaches in Biology

ANDERBERG, M. (1973): *Cluster Analysis for Application*, Academic Press, New York.

BEALS, E. (1960): "Forest Bird Communities in the Apostle Islands of Wisconsin," *The Wilson Bulletin*, Vol. 72, No. 2 (June), pp. 156–181.

BURLINGTON, R. F. (1962): "Quantitative Biological Assessments of Pollution," *Journal of the Water Pollution Control Federation*, Vol. 34, pp. 179–183.

CAIRNS, J., JR., and K. L. DICKSON (1971): "A Simple Method for the Biological Assessment of the Effects of Waste Discharges on Aquatic Bottom-Dwelling Organisms," *Journal of the Water Pollution Control Federation*, Vol. 43, pp. 755–772.

CLARK, P. J., and F. C. EVANS (1954): "Distance to Nearest Neighbor as a Measure of Spatial Relationships in Populations," *Ecology*, Vol. 35, pp. 445–453.

COTTOM, G., and J. T. CURTIS (1956): "The Use of Distance Measurements in Phytosociological Sampling," *Ecology*, Vol. 37, pp. 451–460.

CURTIS, J. T., and R. P. McINTOSH (1950): "The Inter-Relations of Certain Analytic and Synthetic Phytosociological Characters," *Ecology*, Vol. 31, pp. 434–455.

DICKMAN, M. (1968): "Some Indices of Diversity," *Ecology*, Vol. 49, pp. 1191–1193.

EDDEN, A. C. (1971): "A Measure of Species Diversity Related to the Lognormal Distribution of Individuals among Species," *Journal of Experimental Marine Biology and Ecology*, Vol. 6, No. 3 (April), pp. 199–209.

FAGER, E. W. (1972): "Determination and Analysis of Recurrent Groups," *Ecology*, Vol. 38, pp. 586–595.

FAGER, E. W. (1972): "Diversity: A Sampling Study," *The American Naturalist*, Vol. 106, pp. 293–310.

FISHER, R. A., A. S. CORBET, and C. B. WILLIAMS (1943): "The Relationship between the Number of Species and the Number of Individuals in a Random Sample of an Animal Population," *Journal of Animal Ecology*, Vol. 12, pp. 42–58.

HEIP, C., and P. ENGELS (1974): "Comparing Species Diversity and Evenness Indices," *Journal of the Marine Biological Association of the U.K.*, Vol. 54, pp. 559–563.

HURLBURT, S. H. (1971): "The Nonconcept of Species Diversity: A Critique and Alternative Parameters," *Ecology*, Vol. 52, pp. 577–586.

JOHNSON, R. G. (1970): "Variations in Diversity Within Benthic Marine Communities," *The American Naturalist*, Vol. 104, pp. 285–300.

MARGALEF, R. (1968): *Perspectives in Ecological Theory*, University of Chicago Press, Chicago, Illinois.

McCAMMON, R. B. (1968): "The Dendrograph: A New Tool for Correlation," *Geological Society of America Bulletin*, Vol. 79, pp. 1663–1670.

PEET, R. K. (1975): "Relative Diversity Indices," *Ecology*, Vol. 56, pp. 496–498.

PIELOU, E. C. (1966): "The Measurement of Diversity in Different Types of Biological Collections," *Journal of Theoretical Biology*, Vol. 13, pp. 131–144.

PIELOU, E. C. (1969): *An Introduction to Mathematical Ecology*, Wiley (Interscience), New York.

POOLE, R. (1974): *An Introduction to Quantitative Ecology*, McGraw-Hill, New York.

RICKER, W. E. (1975): *Computation and Interpretation of Biological Statistics of Fish Populations*, Fisheries Research Board of Canada, Ottawa, Ontario, *Bulletin 191*.

SANDERS, H. L. (1968): "Marine Benthic Diversity: A Comparative Study," *The American Naturalist*, Vol. 102, No. 925 (May–June), pp. 243–282.

SIMPSON, G. G. (1960): "Notes on the Measurement of Faunal Resemblance," *American Journal of Science*, Bradley Volume, Vol. 258A, pp. 300–311.

SOKAL, R. R., and F. J. ROHLF (1969): *Biometry, the Principles and Practice of Statistics in Biological Research*, W. H. Freeman, San Francisco, Calif.

SORENSEN, T. (1948): "A Method of Establishing Groups of Equal Amplitude in Plant Sociology Based on Similarity of Species Content," *Biologiske Sknifter*, Vol. 5, No. 4, pp. 2–47.

WHITFORD, P. B. (1949): "Distribution of Woodland Plants in Relation to Succession and Clonal Growth," *Ecology*, Vol. 30, pp. 199–208.

WILHM, J. L., and T. C. DORRIS (1968): "Biological Parameters for Water Quality Criteria," *Bioscience*, Vol. 18, pp. 477–481.

Geological

AISIKS, E. G., and I. W. TARSHANSKY (1969): "Soil Studies for Seismic Design of San Francisco Transbay Tube," in *Vibration Effects of Earthquakes on Soils and Foundations*, American Society for Testing and Materials, Philadelphia, *Special Technical Publication 450*, pp. 138–166.

ANGEMEER, J. (1972): "Foundation Investigations for Drilling Platforms in Cook Inlet," in *Underwater Soil Sampling, Testing, and Construction Control*, American Society for Testing and Materials, Philadelphia, *Special Technical Publication 501*, pp. 232–241.

BIJKER, E. W. (1974): "Coastal Engineering and Offshore Loading Facilities," *Proceedings*, Fourteenth Coastal Engineering Conference, Copenhagen, Denmark (June), pp. 45–65.

BEARD, R. M. (1976): "Expendable Doppler Penetrometer," Interim Report, U.S. Navy, Civil Engineering Laboratory, Port Hueneme, Calif., *Technical Note N-1435*, (April).

BLIDBERG, D. R., and D. W. PORTA (1976): "An Integrated, Acoustic, Seabed Survey System for Water Depths to 2000 Feet," *Proceedings*, Eighth Annual Offshore Technology Conference, Houston, Texas (May), Vol. 3, pp. 447–464.

BOUMA, A. H., et al. (1972): "Comparison of Geological and Engineering Parameters of Marine Sediments," *Preprints*, Fourth Annual Offshore Technology Conference, Houston, Texas (May), Vol. 1, pp. 21–34.

"British Firm Takes a Rig Down to Seafloor Drilling Shot Holes in Shallow Water" (1975): *Offshore*, Vol. 35, No. 12 (November), pp. 54–56.

BROWNFIELD, W. G. (1973): "Using Geology to Lay Subsea Pipelines," *Offshore*, Vol. 33, No. 10 (September), pp. 52–57, 60.

CIANI, J. B., and R. J. MALLOY (1975): "Seafloor Surveying in the Nearshore Zone," *Preprints*, Seventh Annual Offshore Technology Conference, Houston, Texas (May), pp. 743–751.

COLP, J. L., W. N. CANDLE, and C. L. SCHUSTER (1975): "Penetrometer System for Measuring In Situ Properties of Marine Sediment," Marine Technology Society, *Ocean 75 Record*, pp. 405–411.

COOK, D. O., and D. S. GORSLINE (1972): "Field Observations of Sand Transport by Shoaling Waves," *Marine Geology*, Vol. 13, No. 1 (June), pp. 31–55.

CROSS, R. H. (1974): "Hydrographic Surveys Offshore—Error Sources," ASCE, *Journal of the Surveying and Mapping Division*, Vol. 100, No. SU2 (November), pp. 83–93.

DEAN, R. G. (1976): "Beach Erosion: Causes, Processes and Remedial Measures," C.R.C., *Critical Reviews in Environmental Control*, Vol. 6, Issue 3 (September), pp. 259–296.

DEMARS, K. R., and D. G. ANDERSON (1971): "Environmental Factors Affecting the

Emplacement of Seafloor Installations," Naval Civil Engineering Laboratory, Port Hueneme, Calif., *Technical Report R 744* (October).

DEMARS, K. R., and R. J. TAYLOR (1971): "Naval Seafloor Soil Sampling and In-Place Test Equipment: A Performance Evaluation," U.S. Naval Civil Engineering Laboratory, Port Hueneme, Calif., *Technical Report R 730* (June).

DIXON, S. J., and K. L. WILSON (1974): "Geological Planning for Ocean Outfalls," ASCE, National Meeting on Water Resources Engineering, Los Angeles, Calif. (January), *Preprint 2181*.

DUANE, D. B., and C. W. JUDGE (1969): "Radioisotopic Sand Tracer Study, Point Conception, California—Preliminary Report On Accomplishments July 1966–June 1968," U.S. Army, Coastal Engineering Research Center, *Miscellaneous Paper No. 2-69* (May).

EWING, J. I. (1963): "Elementary Theory of Seismic Refraction and Reflection Measurements," *The Earth Beneath the Sea*, Vol. 3, Chapter 1, pp. 3–19, Wiley (Interscience), New York.

FLEMMING, B. W. (1976): "Side-Scan Sonar: A Practical Guide," *International Hydrographic Review*, Vol. 53, No. 1, pp. 65–92.

FLETCHER, G. F. A. (1965): "Standard Penetration Test: Its Uses and Abuses," ASCE, *Journal of the Soil Mechanics and Foundations Division*, Vol. 91, No. SM4 (July), pp. 67–75.

FLETCHER, G. F. A. (1969): "Marine-Site Investigations," in *Handbook of Ocean and Underwater Engineering*, edited by J. J. Myers, C. H. Holm, and R. F. McAllister, pp. 8–18 to 8–27, McGraw-Hill, New York.

GARCIA, W. J., JR. (1971): "Literature Survey and Bibliography of Engineering Properties of Marine Sediments," University of California, Berkeley, Hydraulic Engineering Laboratory, *Technical Report HEL 2-27* (February).

GORDON, A. D., and P. S. ROY (1977): "Sand Movements in Newcastle Bight," Institution of Engineers, Australia, *Proceedings of the Third Australian Conference on Coastal and Ocean Engineering*, Melbourne (April), pp. 64–69.

GRANGE, A., and A. M. M. WOOD (1970): "The Site Investigations for a Channel Tunnel 1964–65," *Proceedings*, The Institution of Civil Engineers, Vol. 45, pp. 103–123.

HERBICH, J. B. (1975): *Coastal and Deep Ocean Dredging*, Gulf Publishing Co., Houston, Texas.

HERBICH, J. B., and Z. L. HALES (1971): "Remote Sensing Techniques Used in Determining Changes in Coastlines," *Preprints*, Third Annual Offshore Technology Conference, Houston, Texas (April), Vol. 2, pp. 319–334.

HERSEY, J. B. (1963): "Continuous Reflection Profiling," in *The Sea*, Vol. 3, *The Earth Beneath the Sea*, Chapter 4, pp. 47–72, Wiley (Interscience), New York.

HETHERINGTON, H. A. (1963): "Drilling Muds for Mineral Drilling and Water-well

Construction," in *Grouts and Drilling Muds in Engineering Practice*, pp. 206–210, Butterworths, London, England.

HIRONAKA, M. C., and R. J. SMITH (1967): "Foundation Study for Materials Test Structure," *Proceedings of the Conference on Civil Engineering in the Oceans*, ASCE (September), pp. 489–530.

INMAN, D. L. (1957): "Wave-Generated Ripples in Nearshore Sands," U.S. Army Corps of Engineers, Beach Erosion Board, *Technical Memorandum No. 100* (October).

JOHNSON, S. J., J. R. COMPTON, and S. C. LING (1972): "Control for Underwater Construction," in *Underwater Soil Sampling, Testing and Construction Control*, American Society for Testing and Materials, Philadelphia, Pa., *Special Technical Publication 501*, pp. 122–181.

JONASSON, A., and E. OLAUSSON (1966): "New Devices for Sediment Sampling," *Marine Geology*, Vol. 4, pp. 365–372.

KNEBEL, H. J., and D. W. FOLGER (1976): "Large Sand Waves on the Atlantic Outer Continental Shelf around Wilmington Canyon off Eastern United States," *Marine Geology*, Vol. 22, No. 1 (September), pp. M7–M15.

KOMAR, P. (1976): *Beach Processes and Sedimentation*, Prentice-Hall, Inc., Englewood Cliffs, N.J.

LAY, E. P. (1969): "Core-Sampling Techniques," *Oceans*, Vol. 1, No. 3 (March), pp. 27–29.

LEONARDS, G. A., Ed. (1962): *Foundation Engineering*, McGraw-Hill, New York.

LOWELL, F. C., JR., and W. L. DALTON (1971): "Development and Test of a State-of-the-Art Sub-Bottom Profiler for Offshore Use," *Preprints*, Third Annual Offshore Technology Conference, Houston, Texas (April), Vol. 1, pp. 143–158.

MACKENZIE, K. V. (1960): "Reflection of Sound from Coastal Bottoms," *Journal of the Acoustical Society of America*, Vol. 32, pp. 221–231.

Manual on Test Sieving Methods (1972): American Society for Testing and Materials, Philadelphia, Pa., *Special Technical Publication 447A*.

MEYER, R. E. (1972): *Waves on Beaches and Resulting Sediment Transport*, Academic Press, New York.

MILLER, D. R. (1965): "Marine Studies for the Design and Construction of Offshore Pipelines," ASCE, *Coastal Engineering*, Santa Barbara, Calif., Specialty Conference (October), pp. 991–1006.

MOBERLY, R., JR., and T. CHAMBERLAIN (1964): "Hawaiian Beach Systems," University of Hawaii, Hawaii Institute of Geophysics, *Publication No. HIG-64-2* (May).

MOORE, D. G., and H. P. PALMER (1967): "Offshore Seismic Reflection Surveys," *Proceedings of the Conference on Civil Engineering in the Oceans*, ASCE, San Francisco, Calif. (September), pp. 789–806.

MURRAY, D. A. (1976): "A Light Weight Corer for Sampling Soft Subaqueous Deposits," *Limnology and Oceanography*, Vol. 21, pp. 341–344.

NOORANY, I., and S. F. GIZIENSKI (1970): "Engineering Properties of Submarine Soils: State-of-the-Art Review," ASCE, *Journal of the Soil Mechanics and Foundations Division*, Vol. 96, No. SM5 (September), pp. 1735–1762.

PAGE, G. L. (1974): "Modification of a Pneumatic Track Drill for Underwater Use by Divers," U.S. Navy, Civil Engineering Laboratory, Port Hueneme, Calif., *Technical Note N-1339* (April).

PAWLOWICZ, E. F. (1971): "Ocean Engineering Significance of Marine Seismic Reflection Profiling Technology," U.S. Naval Civil Engineering Laboratory, Port Hueneme, Calif., *Technical Note N-1157* (May).

PECK, R. B., W. E. HANSON, and T. H. THORNBURN (1974): *Foundation Engineering*, 2nd ed., Wiley, New York.

Proceedings of the Conference on In Situ Measurement of Soil Properties (1975): North Carolina State University, Raleigh, N.C., American Society of Civil Engineers, New York, N.Y. (June).

RASHID, M. A., and J. D. BROWN (1975): "Influence of Marine Organic Compounds on the Engineering Properties of a Remoulded Sediment," *Engineering Geology*, Vol. 9, pp. 141–154.

RITCHIE, W. (1974): "Environmental Problems Associated with a Pipeline Landfall in Coastal Dunes at Cruden Bay, Aberdeenshire, Scotland," *Proceedings*, Fourteenth Coastal Engineering Conference, Copenhagen, Denmark (June), pp. 2568–2580.

ROBINSON, L. A. (1977a): "Marine Erosive Processes at the Cliff Foot," *Marine Geology*, Vol. 23, pp. 257–271.

ROBINSON, L. A. (1977b): "Erosive Processes on the Shore Platform of Northeast Yorkshire, England," *Marine Geology*, Vol. 23, pp. 339–361.

SMITH, D. T., and W. N. LI (1966): "Echo-Sounding and Seafloor Sediments," *Marine Geology*, Vol. 4, pp. 353–364.

STAFFORD, D. B. (1971): "An Aerial Photographic Technique for Beach Erosion Surveys in North Carolina," U.S. Army, Corps of Engineers, Coastal Engineering Research Center, Washington, D.C., *Technical Memorandum No. 36* (October).

STEERS, J. A., and D. B. SMITH (1958): "Detection of Movement of Pebbles on the Sea Floor by Radioactive Methods," *Geographical Journal*, Vol. 122, pp. 343–345.

SWART, D. H. (1974): "Offshore Sediment Transport and Equilibrium Beach Profiles," Delft Hydraulics Laboratory, The Netherlands, *Publication No. 131* (December).

TAYLOR, R. J., and K. R. DEMARS (1970): "Naval In-Place Seafloor Soil Test Equipment: A Performance Evaluation," Naval Civil Engineering Laboratory, Port Hueneme, Calif., *Technical Note N-1135* (October).

THORNTON, E. B. (1968): "A Field Investigation of Sand Transport in the Surf Zone," *Proceedings of the Eleventh Coastal Engineering Conference*, London, England (September), Vol. 1, pp. 335–351.

TRASK, P. D., Ed. (1968): *Recent Marine Sediments*, Dover, New York.

VAN REENAN, E. D. (1964): "Subsurface Exploration by Sonic Seismic Systems," in *Soil Exploration*, American Society for Testing and Materials, Philadelphia, *Special Technical Publication No. 351*, pp. 60–73.

VAN REENAN, E. D. (1970): "Geophysical Approach to Submarine Pipeline Surveys," *Preprints*, Third Annual Offshore Technology Conference, Houston, Texas (April), Vol. 1, pp. 367–378.

YOST, E. (1975): "Satellite Measurements of Turbidity of Coastal Waters," *Proceedings*, Civil Engineering in the Oceans III, University of Delaware, Newark, Delaware (June), pp. 1170–1180.

Other

GAUDETT, H. W., et al. (1974): "An Inexpensive Titration Method for Determination of Organic Carbon in Recent Sediments," *Journal of Sedimentary Petrology*, Vol. 44, pp. 249–253.

GILBERT, W. E., W. M. PAWLEY, and K. PARK (1967): "Carpenter's Oxygen Solubility Table and Nomograph for Seawater as a Function of Temperature and Salinity," *Journal of the Oceanographic Society of Japan*, Vol. 23, No. 5, pp. 252–255.

GOLDBERG, E. K. (1965): "Minor Elements in Seawater," in *Chemical Oceanography*, edited by J. P. Riley and G. Skirrow, pp. 163–196, Academic Press, New York.

GREEN, E. J., and D. E. CARRITT (1967): "New Tables for Oxygen Saturation in Seawater," *Journal of Marine Research*, Vol. 25, pp. 140–147.

HOGDAHL, O. T. (1963): "The Trace Elements in the Ocean: A Bibliographic Compilation," The Central Institute for Industrial Research, Blindern, Oslo, Norway (December).

"Humboldt Bay Wastewater Authority—Oceanographic Study" (1976): Winzler and Kelly, Consulting Engineers, Eureka, Calif., Report (January).

MAJOR, G. A., et al. (1972): "A Manual of Laboratory Techniques in Marine Chemistry," C.S.I.R.O., Division of Fisheries and Oceanography, Cronulla, N.S.W., Australia, *Report 51*.

"Manual of Methods in Aquatic Environment Research; Part 1—Methods for Detection, Measurement and Monitoring of Water Pollution" (1975): Food and Agriculture Organization of the United Nations, Rome, *Fisheries Technical Paper No. 137*.

REYNOLDS, J. R. (1969): "Comparison Studies of Winkler vs. Oxygen Sensor," *Journal of the Water Pollution Control Federation*, Vol. 41, pp. 2002–2009.

"Tables for Rapid Computation of Density and Electrical Conductivity of Sea Water" (1956): U.S. Navy Hydrographic Office, Washington, D.C., *Special Publication 11*, (May).

TRUESDALE, G. A., A. L. DOWNING, and G. F. LOWDEN (1955): "The Solubility of Oxygen in Pure Water and Sea-Water," *Journal of Applied Chemistry*, Vol. 5, Part 2 (February), pp. 53–62, with discussion, Vol. 5, Part 9 (September), p. 502.

U.S. Environmental Protection Agency (1971): *Methods for Chemical Analysis of Water and Wastes*, Analytical Quality Control Laboratory, Cincinnati, Ohio.

WILLIAMS, J. (1968): "The Meaningful Use of the Secchi Disc," Chesapeake Bay Institute, *Technical Report 45* (November).

The Hydraulics of
Outfalls, Diffusers, and Effluents

7-1 MARINE DISPOSAL OF WASTEWATERS

Wastewater Outflow and Mixing

The fundamental intent of wastewater disposal is to mix the effluent thoroughly with large volumes of ambient water. The type of pipe terminus shown in Figures 1-1 and 8-7 does not promote adequate mixing of the effluent, and it has been shown by many investigators that the most effective way of effecting high wastewater dilutions is through use of a *diffuser*.

A *manifold* (e.g., McNown, 1954) is a pipe, closed off at one end and having a supply of water at the other, with holes of constant size along its side or sides permitting water to discharge. The pipe diameter and hole sizes are such that a fairly uniform internal pressure is maintained throughout the length of the manifold with resulting equal flows from the holes. The diffuser is an elongated manifold, and the holes through which the wastewater passes into the marine environment are known as *ports*. A very large number of outfalls have the ports in the wall, as in Figure 7-1. Others have the wastewater pass out of the diffuser through smaller pipes, or *risers*.

The wastewater flow that passes out through the ports in the diffuser is lighter than the surrounding seawater, and it tends to rise because of this net buoyancy. The term *plume* is used to describe the form of the effluent if these buoyancy effects are of primary importance, whereas the term *jet* is preferred when the momentum (velocity) effects of the exiting flow are of more importance in determining the degree of mixing. A buoyant jet becomes a plume after a certain length of travel.

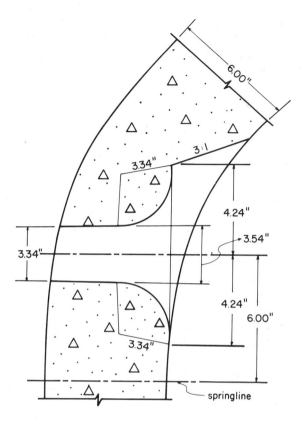

Figure 7-1 Port for Sand Island No. 2 outfall diffuser, 66-inch-diameter segment (Courtesy, City and County of Honolulu, Sewers Division)

As the plume or jet rises toward the surface of a stagnant water body, the flow mixes with the ambient liquid. The term *dilution* is used for the effluent as follows. If 1000 cc of clean seawater are added to 10 cc of a miscible liquid (e.g., wastewater) in a large beaker, then the dilution of that liquid is said to be 101 : 1 or, following the conventional approach in the dilution of plumes and jets, simply 101. The wastewater in the core of a plume or jet has less opportunity to mix with the ambient stagnant liquid than does the effluent around the periphery, and at a given station along the plume or jet axis the dilution at the centerline of the jet will always be the minimum. The dilution used to describe mixing in the marine disposal of wastewaters is taken to be that at the centerline as a conservative procedure.

The average dilution within a buoyant jet is theoretically 1.74 times the centerline value for a round jet and $\sqrt{2}$ times the centerline value for a two-dimensional (*line* or *slot*) jet (Brooks, 1972). It may not at first be obvious why a slot jet should be considered when the diffuser consists of a series of round ports, but jets from different ports remain separate entities for a given depth, only if their spacing exceeds a certain value. Otherwise the jets along the diffuser tend to merge, approximating the efflux from long slots along both sides of the outfall pipe.

The ambient sea water into which a jet or plume flows has a two-fold classification: (1) stagnant or moving; (2) constant density or a density increasing with depth. Theoretical solutions, with experimental verification, exist for the mixing of buoyant three- and two-dimensional (inclined) jets in stagnant receiving waters, either of homogeneous or of linearly density-stratified form. Some results have been obtained for moving receiving waters and some theoretical work has been done for arbitrary-density-stratified receiving waters. The methods to be used in the various cases will be summarized in following sections; these will be limited to horizontal jets for the most part.

The outstanding feature of a density-stratified receiving water with regard to wastewater disposal is that, with proper design and site conditions, the effluent may never reach the water surface. A submerged effluent field develops in such cases, because through dilution with deeper, more dense receiving waters the mixed effluent reaches a density equal to that of the receiving water at some intermediate depth. The rising jet or plume overshoots this level somewhat, then sinks because of its greater relative density and ultimately spreads out at this depth.

The reason why a diffuser is used for the marine disposal of wastewaters rather than an open-ended pipe is that higher *initial dilution*, as the buoyant jet rises toward the surface, can be realized. In an open coast situation, it is desirable to obtain high initial dilutions of the effluent before the sewage cloud diffuses and/or is advected by currents to other areas. In an estuary with poor flushing, however, a high level of initial dilution may not be of primary importance, since sooner or later the effluent will be rather uniformly spread through the estuary whatever the initial dilution. Here, however, we are considering the open coastline problem.

Wastewater Discharge through Outfalls

The history of a wastewater flow when it enters a marine disposal system may be conveniently divided into six parts:

1. Flow inside the pipe and diffuser
2. Jet flow in the *zone of flow establishment* that exists for a short distance beyond the port (about six times the port diameter)

3. Flow in the *zone of established flow* as the diluted wastewater rises toward the surface of the receiving water

4. Development of a homogeneous sewage field, either submerged or at or near the ocean surface

5. Overflow of diluted effluent on the ambient water surface owing to the effluent's residual buoyancy (buoyant spreading)

6. Dispersion of the sewage field due to oceanic processes such as diffusion and advection

In this chapter, methods to calculate the initial dilution of the effluent from a diffuser (items 2 and 3) are described. In this analysis it is assumed that equal flows issue from all ports. But equal flow from all ports is not automatic; it is a consequence of proper hydraulic design of the diffuser system. This topic is part of item 1 and will be considered in Section 7-3. The possible use of pumps will be sketched in Section 7-4.

It is important to be able to assess the degree of dilution of the effluent after the initial dilution process (items 4 and 5), and later as the diluted effluent near the surface spreads laterally and is advected by existing current systems (item 6). The matter of initial dilution will be dealt with in Sections 7-5 and 7-6.* A method has been developed to estimate dilutions in the *secondary dispersion* stages, which are described toward the end of the chapter. The model will permit the designer to predict what concentration of wastewater exists at any point in the receiving water body. Such computations are important in assuring that water-quality standards and other objectives are met.

I believe it is appropriate to enter here a word of caution before beginning the formal topics that compose this chapter. The equations (and coefficients) that are introduced to predict effluent dilution and subsequent dispersion have been obtained through refined mathematical analyses and painstaking laboratory experiments. But the ocean never was and never will be susceptible to easy quantification even at one time and place, let alone over a period of time and over a surface area. Furthermore, wastewater discharges are never steady for any sustained time period, nor are the constituent concentrations within the wastewater constant values. For these reasons, i.e., variability in both the marine environment and in the wastewater flows, I believe it is folly to pretend that the computations carried out to predict the dilution of an effluent at various stages, which will be introduced as this chapter unfolds, have great precision in *the marine environment*.

---

*The standard semi-empirical models implicitly assume that an effluent is all liquid phase. No attention is given to the fate of associated particulates, involving matters such as flocculation, breakup, settling, and resuspension.

7-2 HYDRAULICS BACKGROUND*

Outflow

The specific weight and density of liquids are represented, respectively, by the symbols γ and ρ. The subscript d is used throughout the chapter to indicate the wastewater (discharge). When the density of the receiving (ambient) water is constant throughout the water column, the condition now assumed, the subscript a will be employed.

An important term in the marine disposal of wastewaters is

$$g_0' = \left(\frac{\gamma_a - \gamma_d}{\gamma_a}\right)g \equiv \left(\frac{\rho_a - \rho_d}{\rho_a}\right)g \tag{7-1}$$

where g is the acceleration due to gravity. The concept of the acceleration g is familiar; a body dropped in a vacuum (no air resistance) accelerates at the rate g. Similarly, the quantity g_0' relates to the upward acceleration a buoyant body would undergo in an inviscid fluid.

The effluent passes through a port of minimum diameter d_0 (area a_0) into the receiving water. It is a common concept in fluid mechanics that the minimum area occupied by a liquid issuing from a port a_e is less than a_0, the ratio being the *contraction coefficient* C_c, where $0 < C_c \leqslant 1$. Thus

$$a_e = C_c a_0 \tag{7-2}$$

Expressing the two areas in Equation (7-2) in terms of diameters

$$d_e = \sqrt{C_c}\, d_0 \tag{7-3}$$

The discharge through the port is

$$q_0 = u_e a_e \tag{7-4}$$

where u_e is the mean jet velocity at the *vena contracta*, the location of the minimum jet area. The *effective densimetric Froude number* is defined to be

$$\mathbf{F}_e = \frac{u_e}{\sqrt{g_0' d_e}} \tag{7-5}$$

The *densimetric Froude number* is defined as

$$\mathbf{F}_0 = \frac{u_0}{\sqrt{g_0' d_0}} \tag{7-6}$$

---

*All symbols used in this chapter are listed and defined in Appendix C as well as at their points of introduction.

where

$$u_0 = q_0 / a_0 \tag{7-7}$$

Define the parameter

$$\mathbf{B}'_e = \frac{h/d_e}{\mathbf{F}_e^2} \tag{7-8}$$

where h is the depth of water. Then

$$\mathbf{B}'_e = \frac{h}{d_e} \frac{g'_0 d_e}{u_e^2} \equiv \frac{g'_0 h}{u_e^2} \tag{7-9}$$

Consider a mass Δm of buoyant liquid. Then

$$\mathbf{B}'_e = \frac{1}{2} \left\{ \frac{\Delta m g'_0 h}{\frac{1}{2} \Delta m u_e^2} \right\} \tag{7-10}$$

Written in the form (7-10), it is clear that $\mathbf{B}'_e$ represents a ratio of potential to kinetic energy of the buoyant mass. Following the earlier discussion of jets and plumes, $\mathbf{B}'_e$ then appears as a useful parameter in determining whether an efflux is more jetlike or plumelike. Since a plume has been defined as being dominated by buoyancy effects, it is seen that a plume occurs for relatively large values of $\mathbf{B}'_e$.

It is customary in the marine disposal of wastewaters to use a parameter slightly different from $\mathbf{B}'_e$, namely

$$\mathbf{B}_e = \frac{h/d_e}{\mathbf{F}_e} \tag{7-11}$$

As for $\mathbf{B}'_e$, large values of $\mathbf{B}_e$ are associated with plumes and small values with jets.

It is convenient to define a parameter similar to $\mathbf{B}_e$ as

$$\mathbf{B}_0 = \frac{h/d_0}{\mathbf{F}_0} \tag{7-12}$$

Liseth (1970) has reported that $\mathbf{F}_0$ for major U.S. Pacific coast outfalls is usually in the range 15–30, with peak values in the vicinity of 40–50. Let the average between-port spacing be l. Then Liseth states that, for the same Pacific coast outfalls, the parameter h/d_0 falls in the general range 100–700, whereas h/l lies between 2 and 75 with the latter a rare high number. The parameter $\mathbf{B}_0$

was reportedly remarkably stable for these outfalls, usually lying between 15 and 18, having a maximum value of 20.

Pipe Flow

The *flow capacity* of a pipe refers to its ability to carry volumes of water. This flow capacity is heavily dependent on the roughness of the pipe, as well as on such other factors as pipe length, area, and slope.

There are several ways of taking account of roughness in pipe discharge computations; two will be considered here. The first measure involves the average height of pipe wall protuberances and is represented by ϵ_*. The second measure, called *Manning's roughness factor n*, has only a loose connection with ϵ_*. The factor *n* is not obtained directly from protuberance height measurements but indirectly from pipe flow capacity measurements. Such determinations then give *n* values for different pipe materials.

Manning's *n* values for various pipe (and channel) materials are given in Table 7-1.

Table 7-1:   Manning's Roughness Factors for Various Materials (English Units)

Wall Material	*Pipe*	*Channel*	*n* Min	*n* Max
Plastic	yes	no	0.009	0.011
Wood-stave	yes	yes	0.010	0.013
Vitrified sewer pipe	yes	no	0.010	0.017
Concrete (precast)	yes	yes	0.011	0.013
Cast iron	yes	no	0.013	0.017
Steel (welded)	yes	no	0.010	0.014
Sand	no	yes	0.018	0.025
Gravel	no	yes	0.022	0.030

7-3   INTERNAL HYDRAULICS OF DIFFUSERS

Introduction

The hydraulic design of the diffuser can be set up as a sequence of calculations that can be carried out expeditiously on the computer. In this section, the pertinent equations are developed, constraints are discussed, and methods of solution are outlined. The treatment follows the same general pattern as that of Brooks (1970b).

The design of the diffuser should be such that the total effluent flow is essentially equally distributed across the various ports of the diffuser. This is by no means a trivial exercise because of the variation in flows sent through the diffuser (see Figure 3-1) and the variability of water depths, and thus ambient pressure, possible at any opening. There is little practical opportunity to even out wastewater flow fluctuations, but the diffuser can be designed to lie along a constant-depth contour in order to equalize external pressures at all ports.

The constraint of assuring an even distribution of flows across all the ports is important but by no means the only consideration. There are at least three other hydraulic requirements (Koh and Brooks, 1975).

1. There should be adequate flow speeds in the diffuser to prevent deposition of solids carried with the flow. It appears as if, practically speaking, this is very difficult to achieve for low flows and that minimum flow speeds in the 2 to 3 fps (0.6 to 0.9 m/s) range should be achieved for peak flows in order to scour any material that settled during low flows. The scouring flow speed is strictly speaking a function of the level of wastewater treatment, which helps determine particle size. Deposition is a particular problem at the outermost end of a diffuser, and often special ports are placed in the end of the diffuser to assure that adequate flow speeds are maintained

2. The overall head losses should be kept as low as possible to minimize the level of pressure head at the upstream end of the line and the amount of pumping required. Pumps involve use of energy and operation and maintenance worries

3. All the ports should be fully occupied by discharging wastewater; i.e., no seawater intrusion should occur while the diffuser is in operation. Brooks (1970b) has suggested that this can be obtained by assuring, for all ports, that $\mathbf{F}_e > 1$

Some *practical* considerations concerning diffusers (e.g., manholes, cleaning, provision for flushing) will be deferred until Chapter 8. However, one practical note is appropriate at this point about the design of discharge ports. These should be functional, but simple in design to assure ease of construction as well as minimal maintenance problems. The design shown in Fig. 7-1 has worked out well on all counts.

Experimental research has uncovered two important considerations pertaining to satisfaction of the equal-port-flow requirement and the three requirements above. First, it is often convenient to reduce the pipe diameter, in one or more steps, with increasing distance out along the diffuser (see Figure 8-2). Second, the total area of ports downstream of a pipe section should not exceed $\frac{1}{2}-\frac{2}{3}$ the area at that section.

Finally, if it is not possible, because of particular circumstances, to have equal flow from all ports for the range of design discharges, then it is better to have the higher flows from the outermost ports to satisfy minimum velocity considerations (requirement 1 above).

Basic Equations

The diffuser ports are numbered consecutively from the offshore end; the index j is used for this purpose. The (smallest) diameter of the jth port is d_j, and the pipe diameter (which may change along the diffuser) at this point is D_j. The associated areas are represented by a_j and A_j, respectively. The distance between ports is l.

The pipe discharge just upstream of the jth port is Q_j, and the flow rate through that port is q_j. Let the associated mean flow velocities be V_j and u_j, respectively. The mean pressure within the pipe just upstream of the jth port is p_j. These and other variables are represented in Figure 7-2.

The water column height over the jth port is h_j. With the horizontal datum plane fixed in the sea surface, the elevation of any port is given by $-h_j$. The ambient pressure at that depth, assuming a constant sea water specific weight over the depth, is $\gamma_a h_j$, but to handle the general case this pressure will be represented by p_{a_j}.

Flow Analysis

The Energy Equation: The types of *useful* or *mechanical energy* in fluid mechanics are potential, kinetic, and pressure energies. However, energy is not considered as such in fluid mechanics; what is considered is energy per unit weight, or *head*. Potential, kinetic, and pressure energies are then manifested as elevation head, velocity head, and pressure head, respectively. The sum of these three terms is known as the *total head*.

The *energy equation* in fluid mechanics sets an initial total head equal to a final total head plus a *head loss*, the amount of useful energy converted to heat as a single unit weight of the liquid travelled from the initial to final station.

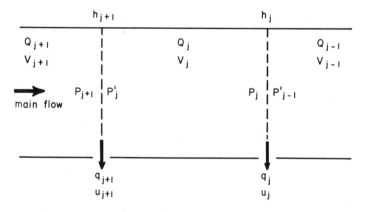

Figure 7-2   Schematic diagram of diffuser

Flow through Ports: Consider the flow from the inside of the manifold out the jth port into the marine environment. The energy equation can be written

$$-h_j + \frac{V_j^2}{2g} + \frac{p_j}{\gamma_d} = -h_j + \frac{u_j^2}{2g} + \frac{p_{a_j}}{\gamma_d} + k_1 \frac{u_j^2}{2g} \tag{7-13}$$

where g is the acceleration due to gravity and k_1 is a dimensionless local or *minor head loss coefficient.* The last term in Equation (7-13) then represents a local head loss. Equation (7-13) can be rewritten

$$u_j = \sqrt{1/(1+k_1)} \ \sqrt{2g} \ \sqrt{E_j} \tag{7-14}$$

where

$$E_j = \frac{p_j - p_{a_j}}{\gamma_d} + V_j^2/2g \tag{7-15}$$

a measure of available (net) head. Let

$$q_j = C_j' a_j u_j \tag{7-16}$$

where C_j' is a contraction coefficient. Combining Equations (7-14) and (7-16), and writing a discharge coefficient

$$C_j = C_j'/\sqrt{1+k_1} \tag{7-17}$$

$$q_j = C_j a_j \sqrt{2g} \ \sqrt{E_j} \tag{7-18}$$

An empirical equation for C_j for a bell-mouth port (Figure 7-1) flowing full is (Brooks, 1970b)

$$C_j = 0.975 \left\{ 1 - \frac{V_j^2/2g}{E_j} \right\}^{3/8} \equiv 0.975 \{1 - J\}^{3/8} \tag{7-19}$$

This equation is applicable only for a port (nozzle) contraction ratio of $4:1$ or greater and only for $d_j < 0.1 D_j$. Since V_j and E_j will be different for different ports, it is apparent that C_j will be variable along the length of the diffuser.

In analyzing the flow from diffusers, it is simplest to work upstream from port to port rather than proceeding downstream. Thus, to obtain conditions at an upstream point $(j+1)$, it is assumed that conditions immediately downstream are known. In Figure 7-2, for example, the following are assumed known: the flows Q_{j-1}, q_j, Q_j; the velocities V_{j-1}, u_j, and V_j; the pressures p_{j-1}' and p_j; the port centerline elevations h_j and h_{j+1}.

The energy equation for the jth pipe segment can be written

$$h_{j+1} + \frac{V_j^2}{2g} + \frac{p_j'}{\gamma_d} = h_j + \frac{V_j^2}{2g} + \frac{p_j}{\gamma_d} + h_{lj} \tag{7-20}$$

which reduces to

$$h_{j+1} + \frac{p_j'}{\gamma_d} = h_j + \frac{p_j}{\gamma_d} + h_{lj} \tag{7-21}$$

The last term on the right-hand side of Equation (7-21) is the *head loss*, initially unknown. Immediately-known terms in Equation (7-21) are h_{j+1}, h_j, and p_j/γ_d, and the head loss can be computed* according to the following equation

$$h_{l_j} = f_j \frac{l}{D_j} \frac{V_j^2}{2g} \tag{7-22}$$

The f_j in Equation (7-22) is the *Darcy-Weisbach friction factor*, a function of the relative roughness of the pipe and Reynolds number. The *relative roughness* is the ratio ϵ_*/D_j and the Reynolds number is given by

$$\mathbf{R}_j = V_j D_j / \nu \tag{7-23}$$

where ν is the kinematic viscosity of the flowing liquid. The determination of f_j from known ϵ_*/D_j and $\mathbf{R}_j$ is done using the *Moody diagram* (Moody, 1944) given in standard fluid mechanics textbooks (e.g., Streeter, 1971; Daugherty and Franzini, 1965). With f_j known, the head loss can be determined, and then p_j'/γ_d is obtained from Equation (7-21). It is admissible, in analyzing flow from a diffuser, to assume that the total head of the main flow remains constant as it passes a port. Considering Figure 7-2 it is possible to write that

$$\frac{V_{j+1}^2}{2g} + \frac{p_{j+1}}{\gamma_d} = \frac{V_j^2}{2g} + \frac{p_j'}{\gamma_d} \tag{7-24}$$

where the right-hand side is known. Comparing the right-hand side of Equation (7-15) and the left-hand side of Equation (7-24), it can be seen that for assumed known ambient pressure $p_{a_{j+1}}$ outside the $(j+1)$st port, E_{j+1} is known. However, $V_{j+1}^2/2g$ itself is not known, and this term is required in Equation (7-19) for C_{j+1}. As a first trial, $V_j^2/2g$ can be used in place of $V_{j+1}^2/2g$ in Equation (7-19) to determine a first approximation to C_{j+1}. Then a first approximation to q_{j+1}

*Losses beyond those due to pipe friction may have to be included in certain pipe (e.g., manhole) sections.

can be determined using Equation (7-18) and, after computing

$$Q_{j+1} = Q_j + q_{j+1} \tag{7-25}$$

a first approximation to V_{j+1} can be found from the continuity equation

$$V_{j+1} = Q_{j+1} / \left(\frac{\pi}{4} D_{j+1}^2 \right) \tag{7-26}$$

A second approximation to $V_{j+1}^2/2g$ is then available and the process can be repeated. A sufficient number of iterations can be employed until a desired small difference between successive values of V_{j+1} is obtained. In many cases, the port discharge is only a small part of the pipe discharge and the first trial using the actual downstream velocity head in Equation (7-19) is sufficient.

The procedure described here can be carried on and on upstream along the pipe. The natural starting place for the entire sequence is the outermost port where, to start the calculations, an assumption must be made about E_1. As was the case for the general port, some iteration will be necessary before the velocity head assumed for use in Equation (7-19) conforms to that computed from the pipe flow velocity obtained via the continuity equation from

$$Q_1 = q_1 \tag{7-27}$$

Calculations proceed upstream step by step until the most upstream (the Nth) port is reached. The total discharge through the diffuser can then be computed as the sum of all the port discharges or simply $Q = Q_N$ which is equivalent. The pressure p_N is also an important quantity.

At each port, checks against the various constraints given earlier should be carried out. For example, if the sum of all the port areas downstream of a particular pipe station exceeds one-half the area at that section, it may be time to increase the pipe size.

Computations by Computer

The step-by-step computational procedure outlined in previous paragraphs is ideally suited to computer solution, and various consultants have developed such programs. With such an approach, the effects on flow distribution through the ports of variations in total flow to the diffuser can be assessed for a specified diffuser design. Also, the influence of port sizing on the distribution of port flows can be quickly appraised, or the effects of various-size pipes can be explored.

The distribution of port discharges along the Sand Island No. 2 diffuser is shown for two different values of Manning's roughness n in Figure 7-3. The corresponding flow velocities in the pipe are shown in Figure 7-4. The Sand Island diffuser is shown in Figure 8-2. Port discharge and pipe velocity plots for the Barbers Point outfall are shown in Figures 7-5 and 7-6.

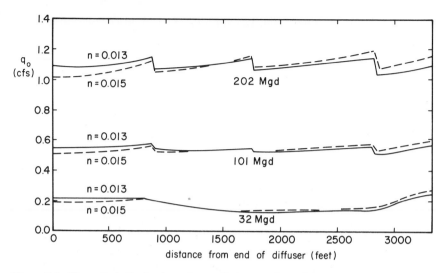

Figure 7-3 Theoretical distribution of port discharges along diffuser for Sand Island No. 2 outfall (Courtesy, City and County of Honolulu, Sewers Division)

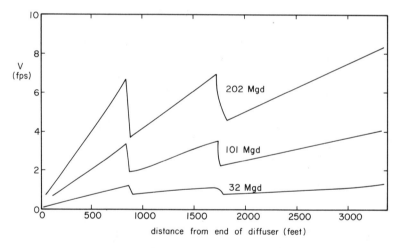

Figure 7-4 Theoretical variation of mean flow velocity in the diffuser for the Sand Island No. 2 outfall (Courtesy, City and County of Honolulu, Sewers Division)

In the design of the diffuser for the Sand Island No. 2 outfall, four different values of Manning's roughness *n* were used to study the effect of variation in that parameter on port discharge and flow in the diffuser itself. These values were 0.013, 0.014, 0.015, and 0.016. Changes were so small for flow velocities

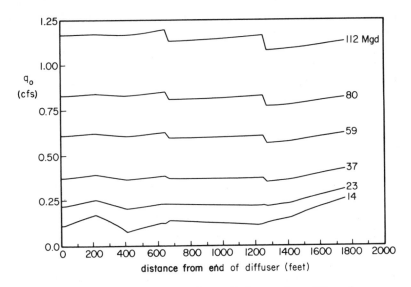

Figure 7-5 Theoretical distribution of port discharges along diffuser for Barbers Point outfall (Courtesy, City and County of Honolulu, Sewers Division)

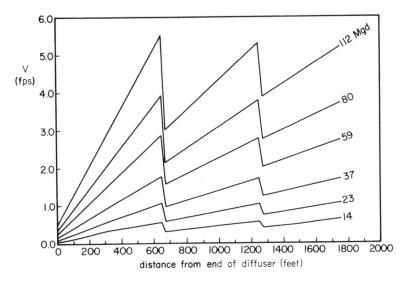

Figure 7-6 Theoretical variation of mean flow velocity in the diffuser for the Barbers Point outfall (Courtesy, City and County of Honolulu, Sewers Division)

that curves for the four roughnesses fell on one line as shown in Figure 7-4. The three discharges used in Figures 7-3 and 7-4 had the following significance: 32 Mgd (1.4 m^3/s), 1972 minimum flow; 101 Mgd (4.4 m^3/s), predicted 1990 peak dry weather flow; 202 Mgd (8.8 m^3/s), predicted 2020 peak wet weather flow. Other flows were included in the actual design computations for the Sand Island No. 2 outfall. All design flows for the Barbers Point outfall are shown in Figures 7-5 and 7-6.

Risers

In many modern outfall diffusers, the ports are bell-mouthed and cast integrally with the pipe wall. The openings are located slightly above the *springline* (median line) of the diffuser pipe as shown in Figure 7-1. This type of design requires that the pipe be laid on the sea floor with ballast rock kept below the springline level.

It is not uncommon, however, that the desire is to bury the diffuser part of the outfall. This could be possible, for example, in shallow waters where wave effects would pose stability problems for an unburied line (see Chapter 8). When the diffuser is buried, the technique of discharging the effluent is through use of vertical pipes (*risers*) leading from the outfall up through the covering material to a termination point above the seabed. Risers are used sometimes even in unburied diffusers to keep the outflow above the zone of extensive sand transport and deposition. Examples of outfalls using risers are the following: Sand Island No. 1 (#7 in Table 10-1) in Hawaii; two other Hawaiian outfalls, one at Hawaii Kai and the other at Waianae; New York City area outfalls at Brooklyn, Coney Island, and Bowery Bay; Lion's Gate (#36) and Annacis Island at Vancouver, B.C.; Agaña and Agat in Guam; the Hastings outfall (#27) in England; the Hammermill Paper Co. line into Lake Erie at Erie, Pa.; the City of Newport and the Georgia-Pacific Paper Co. outfalls at Newport, Oregon; the Menasha Corporation outfall at Coos Bay, Oregon; the Grimsby-Lincoln District, Ontario, outfall into Lake Ontario; and the ITT Rayonier Inc. outfall at Port Angeles, Washington.

Riser pipes (without upper terminations) are shown on the diffuser in Figure 7-7. A very real problem for riser-type diffusers is that the risers can be broken off, e.g., by dragged anchors and by dredges. Or corrosion and wave action can effect the same result—as in the City of Newport line referred to above. If the break should be below seabed level, backfill materials can enter the outfall, with the potential for blockage. Layton (1976) has discussed the advantages, in this regard, of a rubber, weak-link flange in the riser, above the seabed level, where the riser could fail if hooked by an anchor or dredge. The riser can be repaired by using the flange arrangement.

The Annacis Island outfall at Vancouver, B.C. (Figure 7-7), with design capacity 360 cfs (10.2 m^3/s), consisted of three parallel steel pipes; two 66 in.

Figure 7-7 Diffuser for Annacis outfall before pulling (Courtesy, Greater Vancouver Sewerage and Drainage District)

diam (1676 mm) and one 48 in. diam (1219 mm). Almost full-size riser pipes at the ends of the three lines allowed for flushing the system. Risers were also used to discharge the effluent under normal conditions. For both outfall sizes the discharge risers (six for each pipe) discharged horizontally about 25 ft (7.5 m) above the buried outfall crown, about 2.5 ft (0.8 m) above the bottom. Elbow "heads" turned the flow at the summit. The riser pipe size was 18 in. (457 mm) for the 48-in.-diam (1219-mm) outfall and 24 in. (610 mm) for the 66-in.-diam (1676 mm) outfalls.

Tests for possible Annacis Island riser discharge pipe configurations were carried out by the Western Canada Hydraulic Laboratories, Ltd. (1973). Both elbows and tees capping the risers were tested. Discharge coefficients were derived for different external flow conditions (no change over test range) and for different proportions of velocity to total net head immediately downstream of the riser pipe exit opening. Let this latter ratio be represented by $\mathbf{J}'$; it is little different from $\mathbf{J}$ in Eq. (7-19). For the tee-shaped discharge nozzle the discharge coefficient was found to be 0.73 for $\mathbf{J}' \leqslant 0.013$. C_j then decreased, reaching the value of 0.66 at the limit of the tests, $\mathbf{J}' = 0.15$. At this same point C_j for the elbow nozzle was 0.60. For $\mathbf{J}' \leqslant 0.005$, C_j for this nozzle was 0.70.

Koh (1973) has also done experimental work on discharge coefficients for risers. Extensive tests were made on models of various prototype risers of height 10.0 ft (3.05 m) and diameters of 2.5 (chiefly) and 2.0 ft (0.76 or 0.61 m). Riser flow was turned through an angle slightly less than a right angle at the top of the riser and then discharged. Koh's discharge coefficients for $\mathbf{J} = 0$ varied from 0.648 to 0.864. In all test cases the discharge coefficients decreased with increasing $\mathbf{J}$ as in Equation (7-19).

The flow discharge characteristics of risers are inferior to those of pipe wall ports as could be anticipated because of the lead losses through the entry to the riser, along the riser, and then in the flow-turning system at the top of the riser. Despite this, it can be imagined that some form of equation like Equation (7-19) could be developed for any particular riser. Then the diffuser hydraulics calculations would be no different in plan than those used earlier for wall ports.

7-4 PUMPING AND HYDRAULIC TRANSIENTS

Driving Head

The discharge $Q = Q_N$ and the pressure at the upstream end of the diffuser p_N must be provided or the particular port flow analysis carried out for the diffuser is invalid. The pipe size found at the most upstream port is normally the size continued back to shore to the sewage treatment plant and/or pumping station. The analyst has at his disposal all of the variables required to determine the head at the shore installation to provide the pressure p_N at the upstream end of the diffuser. The flow and pipe diameter are known; the elevation of the pipe at the shore end and at the upstream end of the diffuser are known; presumably the pipe material has also been established. Thus the energy equation can be used to compute the required shoreline head.

If it is feasible to naturally provide the head required at the shoreline, then gravity flow conditions can be fully used for the outfall. If, however, the head called for exceeds that available, which it will tend to do particularly for larger flow rates, then pumping is required. In rare instances, e.g., for the Point Loma outfall at San Diego (#18 in Table 10-1) and the Barbers Point outfall in Hawaii (#49), there is so much head available at the plant on the shoreline that energy-dissipating techniques (e.g., hydraulic jumps, vortex structures) must be used to obtain the required flow conditions. However, to realize this condition, it is more than likely that pumping was used somewhere in the wastewater collection system before it reached the beginning of the outfall.

Hydraulic Transients

A surge tank is normally a cylindrical tank mounted over a pipe and connected into it. The surge tank serves a useful function in situations wherein rapid changes in flow conditions either must be effected or in cases where such changes are caused by some effect, such as a power failure or blockage, beyond the control of the pumping station operator. Basically, the surge tank reduces the over- or under-pressures that would result in the pipe system from such change(s).

The large over-pressures that can result from (rapid) valve closure can be explained as follows. A pipe, full of water moving at speed V, is closed off at the downstream end in a time Δt. If the pipe is long and the closure time short, very large pressures must be built up toward the downstream end of the pipe in order to provide the required deceleration force for this mass of water. If there is some way in which less water can be decelerated, the force buildup decreases accordingly. This is one of the actions of the surge tank. The side access from the main pipe into the surge tank allows some of the water to flow into the tank.

Although normally considered to be incompressible, the small compressibility of water gives rise, under conditions of rapid flow changes, to a system of pressure fluctuations known as *waterhammer*. The presence of a surge tank in a pipe system also helps to reduce the levels of internal waterhammer pressures. Two outfalls with surge tanks were Orange County No. 1 (#9 in Table 10-1) and Orange County No. 2 (#35). Both surge tanks were at the upstream ends of the lines.

Surface Wave Effects on Diffuser Operation

A diffuser is designed to deliver a range of discharges over a range of tidal stands of the receiving water surface. The tide changes so gradually (relatively speaking) that any tidal condition can be considered as a static situation, independent of other tidal ones. But when large waves pass over the diffuser, a static analysis may well not be adequate to assure that the desired flow can be obtained. The passing waves cause the ambient pressure at the ports to continuously cycle back and forth, and the outfall may not carry the flow discharge desired.* In addition, temporary storage on shore may be necessary for the rejected flow volumes.

Henderson (1967) has carried out an analytical study of this problem. His analysis is based on several simplifying assumptions as follows: the densities of effluent and ambient liquid are the same; the wave-induced (ambient) pressure variation is sinusoidal (with amplitude $\Delta h/2$); the head in a storage tank on shore is a constant; the change with time in the sum of the exit velocity head (through the ports) and head loss through the pipe is negligible; the (ambient) pressure variation is in phase at all ports; and the total pipe head loss is a constant. Henderson found that the storage required to be handled at the

---

*That wave action can perturb the continuous operation of a multiport diffuser has been confirmed at the least by diver observations of two outfalls in Hawaiian waters. Both diffusers had riser-type outlets. At one outfall (Hawaii Kai) a slug of dye released opposite a port as a crest passed moved directly into the port to reappear with a burst of effluent and seawater on the next trough. There was some mixing within the diffuser since dye would appear at other ports in later cycles. At another outfall (Waianae) the port flow was slowed under wave crests but was not actually reversed. Such situations as related above not only cause hydraulic perturbations but can lead to sand in suspension entering lines and settling inside them. Adequate flow must issue from all ports at all times. This can be achieved early in the life of an outfall, when there may be relatively low flows, by keeping caps on some of the ports.

upstream end of the pipe during half a wave cycle is

$$\Delta\Psi = \frac{A'g(\Delta h)T^2}{8\pi L_1} \tag{7-28}$$

where the variables are as follows: A' is the (constant) cross-sectional area of the pipe, g is the acceleration due to gravity, T is the wave period, and L_1 is the length of the pipe. It is convenient to normalize in Equation (7-28) by the volume of water in the pipe at any time. Thus

$$\frac{\Delta\Psi}{A'L_1} = \frac{g(\Delta h)T^2}{8\pi L_1^2} \tag{7-29}$$

Equation (7-29) basically shows a reaction equivalent to that when a high-frequency force is applied to a mechanical system with a low natural frequency; there is very little response. The smallness of $\Delta\Psi/A'L_1$ can be illustrated by assuming reasonable valuables for the terms involved. Consider, as an example, the effect of a wave with pressure double-amplitude 10 ft (3 m)* and period 15 s on a fairly short outfall of length 5000 ft (1500 m). From Equation (7-29), $\Delta\Psi/A'L_1 \approx 1 \times 10^{-4}$. In general, from this analysis, unless the outfall is very short, and probably gravity flow is involved, it can be concluded that the effect of wave action on the delivery of the required flow is secondary. Realistically, for modern long outfalls and ports distributed over hundreds of feet of diffuser such that small (because of the depth) wave-induced pressure variations would not be in phase, the problem can be ignored.

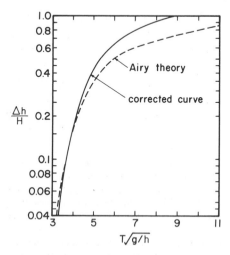

Figure 7-8   Theoretical and corrected curves for pressure variation under waves

---

*The linear theory discussed in Chapter 8 can be used, as a first approximation, to relate surface wave height and pressure "height." Information on the accuracy of this prediction is available in Figure 7-8 derived from data in Grace (1970).

7-5 INITIAL DILUTION OF EFFLUENTS IN HOMOGENEOUS RECEIVING WATER

Introduction

A *homogeneous receiving water* means one in which there is negligible variation in sea water density with depth in the vicinity of the planned diffuser. This of course implies that the mixed layer extends throughout the depth concerned and implies further that the diffuser is probably in relatively shallow water. It is impossible to give a single depth in this case. Even in a water depth as comparatively shallow (for modern outfalls) as 90 ft (27 m), summer conditions can show marked differences between the temperatures, and hence densities, between the surface and the bottom (e.g., Stewart et al., 1971).

In the case considered herein, as the buoyant effluent rises toward the surface it entrains ambient water as is always the case, but because the buoyancy of the mixture never reaches that of the ambient, the mixed effluent always rises to the water surface. When depth and seasonal considerations conspire to render an ocean receiving water of essentially constant density from bottom to surface, the results of this section may be applied.

The problem of wastewater discharge into an *oscillating* current such as that caused by tidal effects has not yet been adequately considered either theoretically or experimentally. The problem that has been solved concerns *steady* currents flowing at any angle to a diffuser, and this is the situation that will be considered in this section. Some material on the oscillating case is included in the Bibliography.

Six quantities determine the initial dilution of effluents discharged into steadily moving receiving waters of constant density. The first of these is the *unit buoyancy flux*, given by the equation

$$b = g_0' q \qquad (7\text{-}30)$$

where q is the discharge per unit length of the diffuser and g_0' was defined in Equation (7-1). The second quantity of interest, and the only important dynamic parameter, is a Froude number defined by the equation

$$\mathbf{F} = u^3 / b \qquad (7\text{-}31)$$

where u is the current speed. The angle of that current with the line of the pipe, represented by θ, is the third quantity of prime concern.

Let the fourth quantity be

$$\bar{S} = uh / q \qquad (7\text{-}32)$$

It is apparent that $\bar{S}$ indicates a type of overall average, two-dimensional dilution for the effluent, since the ratio of unit current and effluent discharges is involved. The fifth quantity M represents the ratio of the (vertical) momentum of the source to (over) that of the ambient current. The final quantity is L_2/h where L_2 is the length of the diffuser.

It has been found (Roberts, 1977a) that there are three flow regimes for a plume of *infinite length* in a perpendicular current ($\theta = 90°$). These separate regimes are characterized by the range of **F** values as shown in Figure 7-9. However, any real wastewater discharge situation involves a diffuser of *finite length*. The overall flow patterns in the two- and three-dimensional cases are conceptually quite different for $\theta = 90°$. In the three-dimensional case, the effluent can spread laterally whereas this is not possible in the two-dimensional situation.

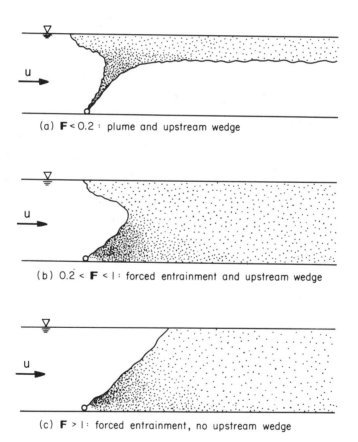

(a) **F** < 0.2 : plume and upstream wedge

(b) 0.2 < **F** < 1 : forced entrainment and upstream wedge

(c) **F** > 1 : forced entrainment, no upstream wedge

Figure 7-9 Two-dimensional flow regimes for discharge of plume into perpendicular current [Roberts (1977a), with permission]

Surface Dilution of Slot Plumes

In the marine disposal of wastewaters, interest centers on the *minimum dilution* S_m of the effluent at the water surface. For the assumption of fully turbulent flow conditions (i.e., high Reynolds numbers), Roberts (1977a) has shown that the following general functional relationship exists among the various variables

$$\frac{S_m}{\overline{S}} \equiv \frac{S_m q}{uh} = f\{\mathbf{F}, L_2/h, \theta, M\} \tag{7-33}$$

All terms in this equation have been defined. If the assumption of a pure plume is made, the parameter M disappears and $S_m q/uh$ is a function of only the three remaining terms.

The pure plume experiments of Roberts (1977a) were designed to establish the relationship between $S_m/\overline{S}$ and $\mathbf{F}$, L_2/h, and θ. Three different diffuser angles were considered: 90°, 45°, and 0°. Mean lines drawn through the data obtained are shown in Figure 7-10. This plot clearly shows the dependence of $S_m q/uh$ on $\mathbf{F}$ and θ, but does not demonstrate any effect of L_2/h. At least within the range $3.7 \leqslant L_2/h \leqslant 30$ used by Roberts there were no clear differences in normalized dilution results as L_2/h varied. The asymptotic solution for small $\mathbf{F}$ is

$$S_m q/uh = 0.27/\mathbf{F}^{1/3} \tag{7-34}$$

Equation (7-34) can be rewritten as

$$S_m = 0.27\frac{b^{1/3}h}{q} \tag{7-35}$$

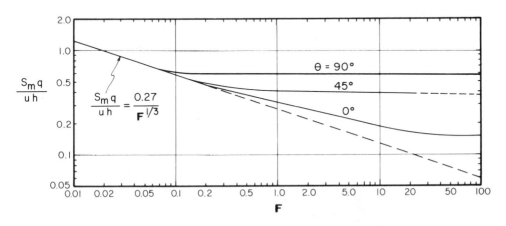

Figure 7-10 Minimum surface dilutions of line plumes in steady currents [Roberts (1977a), with permission]

As would be expected the current speed disappears. It is clear from Figure 7-10 that use of Equation (7-35) [or equivalently Equation (7-34)] greatly underestimates the initial dilution of the effluent if there is an appreciable current. This is an important conclusion, as initial dilution models used before Roberts' work ignored the effect of current (and three-dimensionality) on initial dilution. It appears from Figure 7-10 that Equation (7-35) applies for $\mathbf{F} \lesssim 0.1$.

The surface configuration of effluent for $\mathbf{F} \approx 0.1$ (and $L_2/h \approx 15$) for the current perpendicular to the diffuser is shown in Figure 7-11. The completely different effluent plume shape for $\mathbf{F} \approx 10$ (and $L_2/h \approx 15$) is shown in Figure 7-12. It is noteworthy that the downstream plume shapes for the two source orientations in Figure 7-12 are virtually indistinguishable, whereas at the same time, from Figure 7-10, the $S_m q/uh$ values vary considerably.

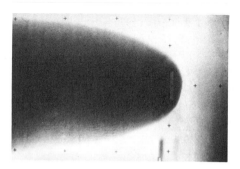

Figure 7-11 Surface plume configuration for $\mathbf{F} \approx 0.1$ and current perpendicular to diffuser [Roberts (1977a), with permission]

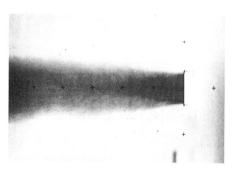

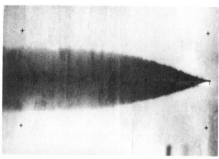

Figure 7-12 Surface plume configurations for $\mathbf{F} \approx 10$ and two source orientations [Roberts (1977b), with permission]

Roberts compared his own results to those of other investigators who had carried out similar (but chiefly two-dimensional) experiments but with nonnegligible M. The effect of increasing M was to increase S_m, but the relative increase decreased with increasing $\mathbf{F}$.

Dilution of Round Jets and Plumes

Initial dilutions of effluents when momentum effects are appreciable can be derived on the basis of results of Liseth (1970, 1973, 1976) for stagnant or weakly moving receiving waters. Liseth's two-dimensional data have been replotted by Roberts (1975) and the result is shown in Figure 7-13. The variable ξ in this figure is a vertical coordinate, positive upward, with origin at the center of the port. The upper bound on this variable occurs at the water surface, where $\xi = h$ if the relatively small distance between the port and the seafloor is ignored.

Toward the left-hand side of Figure 7-13 the centerline dilutions S_c are independent of h/l. This is the region within which the individual jets issuing from adjacent diffuser ports do not intermingle. For $h/l > 5$, however, there is

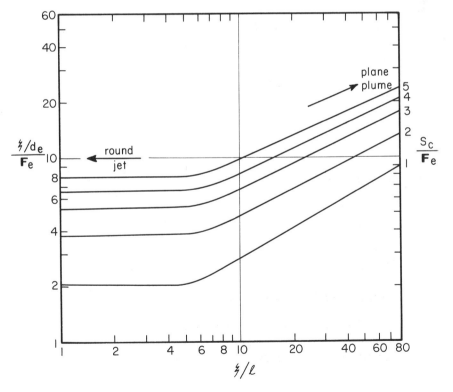

Figure 7-13   Contours of constant values of centerline dilution for merging, round buoyant jets into uniform, stagnant water [Roberts (1975), with permission]

interaction of adjacent jets. Toward the right-hand side of the plot the flow paths are so long that buoyancy effects have taken over in mixing the effluent. Thus the right-hand side of Figure 7-13 corresponds to a plume. It also corresponds to a *slot* plume case because of the extensive intermingling of the jets from the various ports. Thus the lines shown in Figure 7-13 should approach values predicted for slot plumes.

The variation in centerline dilution for a two-dimensional plume in stagnant waters is

$$S_c = 0.42 b^{1/3} \xi / q \tag{7-36}$$

Equation (7-36) is an excellent fit to the data lines in Figure (7-13) near the right border of the plot. The minimum dilution of the effluent cloud at and near the surface, however, is (Roberts, 1977a)

$$S_m = 0.29 b^{1/3} h / q \tag{7-37}$$

It is clear that setting $\xi = h$ in Equation (7-36) overestimates S_m. This discrepancy brings up the problem of *blocking*. Since there is a cloud of diluted effluent occupying the upper part of the water column,* the latter stages of plume rise are not associated with influx of pure dilution water. Because of blockage it is not correct to consider S_c and S_m as equivalent. The ratio S_m/S_c for $\xi = h$ is 0.69.

The centerline dilution of a round plume in a stagnant or weakly moving receiving water is (Roberts, 1977a)

$$\frac{S_c}{F_e} = 0.107 \left[\frac{\xi / d_e}{F_e} \right]^{5/3} \tag{7-38}$$

A round jet or plume remains distinct only for $\xi / l < 5$, so that constraint applies here. Also, for a plume to be approximated,

$$\frac{\xi / d_e}{F_e} \gtrsim 20 \tag{7-39}$$

If the two constraints above are satisfied, Equation (7-38) can be used to extend the results along the left margin in Figure 7-13.

Effluent with the Same Density as Ambient Liquid

When the effluent and the ambient liquid are of the same density, as with wastewater flow into a lake, the densimetric Froude number (7-6) becomes infinite. Such a condition is not represented in Figure 7-13 but has been

---

*A rounded figure of 30% is normally considered appropriate for this proportion. This has the support of theoretical and laboratory results (e.g., Koh, 1976) as well as that of various field studies (e.g., Sunn, Low, Tom and Hara, 1973).

represented in Figure 7-14. In this plot for stagnant conditions, both the round (x/d_0) and the slot (x/B_0) jets have been included. x is the distance of jet travel, d_0 the initial round jet diameter, and B_0 the initial total slot jet width. Since slots are not used for discharge of effluents, B_0 is obtained according to the following equation

$$B_0 = \left(\frac{\pi}{4} d_0^2 \right) / l \qquad (7\text{-}40)$$

Papers and reports in the Bibliography provide further information on discharges of wastewater into fresh water bodies.

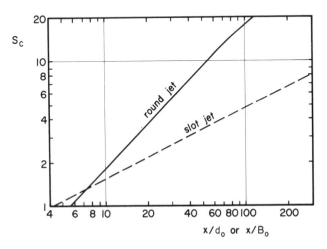

Figure 7-14 Centerline dilutions of round and slot jets into water of same density [Abraham (1963), with permission]

7-6 INITIAL DILUTION OF EFFLUENTS: STAGNANT RECEIVING WATER OF VARIABLE DENSITY

Introduction

Adequate effluent dilution is normally the primary design objective for ocean outfall systems. A desirable secondary objective, at least on aesthetic grounds, is to create a submerged sewage field.* A submerged field requires, of course, that the effluent be discharged in a body of water that possesses the requisite density stratification (see Figures 7-15 and 7-16). This is not normally possible for sustained periods of time in depths of 100 ft (30 m) and less, since a fairly homogeneous water mass probably exists in such regions for at least part

---

*Deep submergence of the field may possibly be undesirable since it could reach a shoreline during a period of upwelling, dilutions achieved are smaller than those obtained on the sea surface, and bacterial die-off rates are lessened.

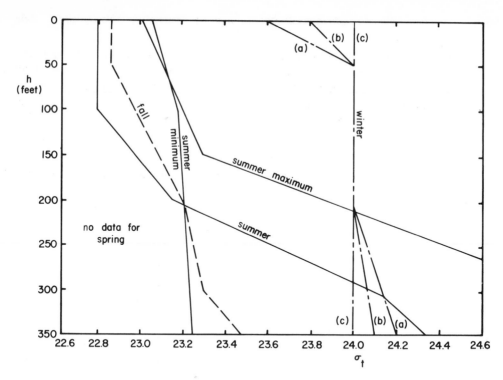

Figure 7-15 Design density profiles for Sand Island No. 2 outfall (Courtesy, City and County of Honolulu, Sewers Division)

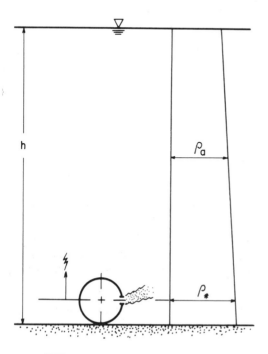

Figure 7-16 Density stratified receiving water

of the year. It is more usual that density-stratified design conditions would be obtained in depths approaching at least 200 ft (60 m). As an example, the Sand Island No. 2 outfall, which made use of a design density stratification structure illustrated in Figure 7-15, has a diffuser depth of approximately 235 ft (see Figure 8-2).

In the density stratified case, let ρ_a represent the density of the receiving water at any depth. If ξ represents the vertical distance upward from the port, then it will be assumed that $d\rho_a/d\xi$ is a (negative) constant. ρ_* will represent the receiving water density at the level of the port (see Figure 7-16). Although only the case of linearly varying density will be considered, a methodology for dealing with a more realistic density variation has been discussed by Brooks (1972). However, the density structure of 200-ft-deep (60-m) waters may change so quickly and vary so considerably (see Figure 7-17) that it is doubtful that anything beyond the model to be used here, possibly in conjunction with an assumed mixed zone of constant density near the surface, need be applied.

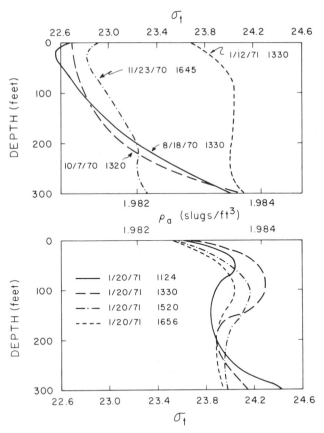

Figure 7-17 Measured density profiles offshore from Sand Island, Hawaii (Courtesy, City and County of Honolulu, Sewers Division)

Round Buoyant Jet Discharged Horizontally

In this case it is convenient to define a *stratification parameter*

$$T_0 = \frac{\rho_* - \rho_d}{d_0(-d\rho_a/d\xi)} \tag{7-41}$$

This typically has values in the range of 5×10^3 to 25×10^3. *Momentum* and *volume flux parameters* are defined, respectively, as

$$m_0 = 0.426 F_0^2 T_0^{-1} \tag{7-42}$$

and

$$\mu_0 = 2.64 F_0^{1/4} T_0^{-5/8} \tag{7-43}$$

F_0 in Equations (7-42) and (7-43) makes use of the receiving water density ρ_*. Typical values for these two parameters for real outfalls are as follows: $0.005 \lesssim m_0 \lesssim 0.1$; $0.01 \lesssim \mu_0 \lesssim 0.03$. The maximum height of rise of the jet (ξ_{max}) and the centerline dilution relative to the discharge at that level (S_t) will be computed from the respective equations

$$\frac{\xi_{max}}{d_0} = 1.32\phi_t F_0^{1/4} T_0^{3/8} \equiv \delta\phi_t \tag{7-44}$$

and

$$S_t = 1.15\mu_t/\mu_0 \tag{7-45}$$

where

$$\delta = 1.32 F_0^{1/4} T_0^{3/8} \tag{7-46}$$

Values for ϕ_t and μ_t are obtained from Figures 7-18 and 7-19.

In the limiting plume case, when $m_0 \approx 0$ and $\mu_0 \approx 0$, it turns out that

$$S_t = 0.75 F_0^{-1/4} T_0^{5/8} \tag{7-47}$$

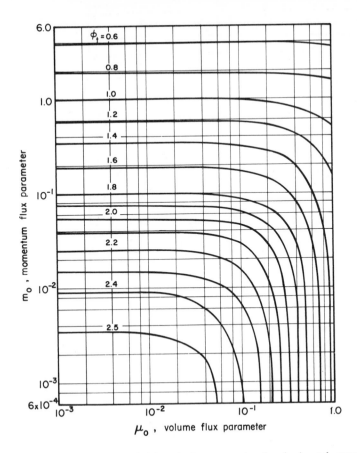

Figure 7-18 Terminal height of rise parameter for horizontal round buoyant jets in a linearly density-stratified environment [Fan and Brooks (1969), with permission]

and

$$\frac{\xi_{max}}{d_0} = 3.67 F_0^{1/4} T_0^{3/8} \qquad (7\text{-}48)$$

For intermediate cases between jets and plumes Figure 7-20 may be useful. The total width of the jet or plume at the point of maximum rise is

$$2w_t = 8d_0 \left[0.464 \mu_t / m_0^{1/4} \right] \qquad (7\text{-}49)$$

A value of $2w_t$ can be determined to see if there is any interaction among jets.

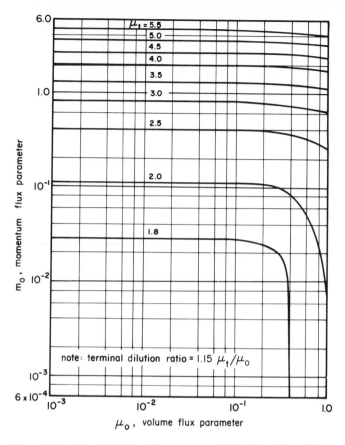

Figure 7-19 Terminal volume flux parameter for horizontal round buoyant jets in a linearly density-stratified environment [Fan and Brooks (1969), with permission]

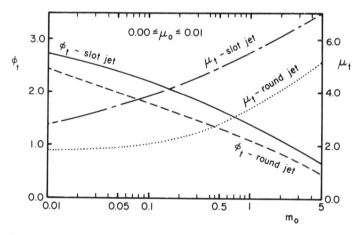

Figure 7-20 Terminal parameters for horizontal buoyant jets in a linearly density-stratified environment

Slot Buoyant Jet Discharged Horizontally

In this case the assumed slot jet width B_0 is obtained from Equation (7-40). Let

$$F_0' = \frac{u_0}{\sqrt{g_0'B_0}} \tag{7-50}$$

and

$$T_0' = \frac{(\rho_* - \rho_d)}{B_0(-d\rho_a/d\xi)} \tag{7-51}$$

Also let

$$m_0 = 0.500(F_0')^2(T_0')^{-1} \tag{7-52}$$

and

$$\mu_0 = 1.85(F_0')^{2/3}(T_0')^{-1} \tag{7-53}$$

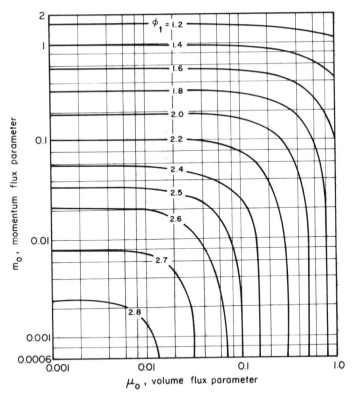

Figure 7-21 Terminal height of rise parameter for horizontal slot buoyant jets in linearly density-stratified environment [Fan and Brooks (1969), with permission]

315

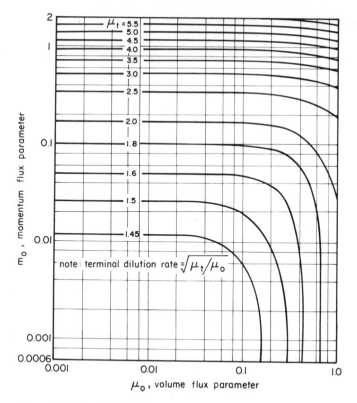

Figure 7-22   Terminal volume flux parameter for horizontal slot buoyant jets in linearly density-stratified environment [Fan and Brooks (1969), with permission]

Then the height of rise and dilution are given by

$$\frac{\xi_{max}}{B_0} = 0.96\phi_t(F_0')^{1/3}(T_0')^{1/2} \tag{7-54}$$

and

$$S_t = \sqrt{\mu_t/\mu_0} \tag{7-55}$$

The values of ϕ_t and μ_t are read from Figures 7-21 and 7-22.

In the limiting case of a plume ($m_0 \approx \mu_0 \approx 0$) the following is obtained

$$\frac{\xi_{max}}{B_0} = 2.84(F_0')^{1/3}(T_0')^{1/2} \tag{7-56}$$

and

$$S_t = 0.87(F_0')^{-1/3}(T_0')^{1/2} \tag{7-57}$$

7-7 SECONDARY DISPERSION PROCESS IN COASTAL WATERS

Introduction

I have already discussed in some detail the dilution of a sewage effluent wherein that effluent either rises to the surface or to a position of vertical stability below the surface. Clearly the computation of dilutions for the wastewater cannot end there, for the effluent will tend to migrate due to local current systems and spread due to diffusive mechanisms and may move into certain areas of coastal waters where the concentration of a marker substance or organism must be below a prescribed level. A model should be available for predicting dilutions of the effluent during this secondary dispersive process. In addition, for nonconservative indicators,* means should be available for assessing the die-off factor as the secondary process unfolds.

For effluents rising to the surface, another cause of spread concerns gravitational forces. An effluent reaching the water surface may, for density-stratified conditions, have the same density as the surface water, but in most cases the density of the mixed effluent is somewhat less than that of the ambient surface waters. The effluent can then spread by flowing over the surface in a process called *buoyant* or *surface spreading*. Buoyant spreading does not occur for a sufficiently small difference in mixed effluent and surface water densities, and it has been suggested (McBride, 1975) that the phenomenon can be ignored if $S_m > 50$, a level commonly surpassed in the field. It will be assumed herein that initial dilutions are adequate to eliminate concern over surface spreading and that the densities of mixed effluent and ambient surface water are the same. Details on the surface spreading process are provided by Koh (1976) and Roberts (1977a).

Dilutions during the rise of the effluent from the diffuser port to its surface or subsurface position for modern outfalls are typically two orders of magnitude. For example, it is dictated by law, in waters off California, that dilutions of 100 should be obtained for at least 50% of the time. The conservative-substance dilution typically involved in the secondary dispersive process is one order of magnitude, so that the lion's share of conservative-substance dilution takes place in the initial stages. For nonconservative substances, however, the concentration decrease during the secondary dispersion process may exceed that in the initial dilution process because initial dilution of a given mass of effluent takes place only during a few minutes whereas subsequent secondary dispersion occupies hours. Dieaway of nonconservative substances is discussed in Chapter 4.

---

*The concentration of a conservative substance can change only through the physical dilution process. Nonconservative substances can undergo concentration changes through additional mechanisms.

Because of its importance in determining the secondary dispersion of an effluent, the diffusion of a marker substance in marine waters will be discussed first in this section. Presentation will then follow of two mathematical models that have been developed to map the spread of a cloud of marker substance.

Diffusion in the Ocean

Introduction: If a thimbleful of dye is dropped into a pail of water, after a certain length of time, say t_1, the dye can be regarded as being fully and evenly dispersed throughout the pail. The physical mechanism by which the dye is spread throughout the pail is called *diffusion*, and the type of diffusion involved in spreading the dye throughout the pail is *molecular diffusion*. Consider a similar experiment, but where a system of fine screens is oscillated up and down in the pail. This activity causes small eddies, or rotating regions of liquid, to form. In this case the time taken for the dye to evenly spread throughout the pail t_2 would be substantially less than t_1, because the small eddies provide an additional dispersive agent for the dye. The latter form of diffusion is of interest herein. It is important to note that implicit within it is the assumption that the size of eddies involved is small with respect to the physical scale of things involved (i.e., the size of the pail).

The mathematical formulations that apply to diffusion processes are in the form of partial differential equations, and many practical situations have been represented by such mathematical models. Solutions have been obtained for the rate at which a dye, released either continuously or in one discrete entity (*slug*), disperses to the surrounding medium. The variation with time and distance of the concentration of dye (either on a per unit volume or per unit weight basis) in the medium is charted. This concentration is represented by c.

A major component in diffusion determinations is the *diffusion coefficient* ε. In many mathematical analyses involving diffusion, it is assumed that ε is a constant for all time and space and the same in all directions. In this case, ε basically represents the time rate of transport of the dye across a unit area of the receiving medium divided by the gradient of dye concentration, in the direction of dye movement being considered, at that station. Solutions to more or less standard diffusion situations with constant ε are available in the literature (e.g., Dobbins, 1963).

Richardson's Law: Consider a patch of marker substance being transported by a large current system. Clearly the patch is being *advected*. But the particular current system involved could well be part of a very large eddy whose scale is orders of magnitude larger than that of the diffusing patch. But if the patch is being advected by the local part of the larger system, it is apparent that the larger eddy, as such, plays no role in the speed of spread of that patch about its center. Thus it is possible to make the generalization that the eddies that influence the spread of a marker substance about its center are only those that have a scale smaller than the instantaneous size of the patch.

But as the size of a patch of marker substance expands, it is apparent that eddies that were at one time considerably larger than the size of the patch (and would have been considered as an advective mechanism) become of influence in terms of causing spread of the patch about its center. As the patch increases in size, therefore, there are more dispersive eddies to consider, for one can imagine that coastal waters contain large numbers of eddies of various scales.

It was generalized earlier that the more eddying in the ambient liquid the faster the spread of the patch, i.e., the higher the diffusion coefficient. Thus one can reason that the diffusion coefficient ε in the ocean should increase as the scale of the marker substance patch grows. Because the size of a patch of marker substance can be more easily visualized in terms of area than of a pure length scale, it is reasonable to assume that the diffusion coefficient must increase faster than the length scale of the patch L. It is not clear on the basis of the reasoning above, however, if the diffusion coefficient should increase as fast as L^2, the scale of the patch in terms of its area. Experience suggests that a good relationship to use for diffusion in the open ocean relates the diffusion coefficient to the four-thirds power of the length scale of the patch. This is expressed as *Richardson's Law* (Richardson, 1926; Stommel, 1949):

$$\varepsilon = \alpha L^{4/3} \tag{7-58}$$

There can be and has been considerable debate about whether or not the constant α in Equation (7-58) is really a constant in the sea. In truth it cannot be expected to be so; for one thing, it appears clear that the same size of patch in the same coastal water area will spread at different rates on 2 days, one of which features calm water and the other a wave environment churned up by a strong local wind. The motion of water particles caused by waves is a legitimate example of a type of eddying. The marker substance is spread quicker in the presence of waves because the wave-induced water motion carries the marker more quickly into formerly untainted areas. It is not surprising, then, that the "constant" in Equation (7-58) is not really so. According to Koh and Brooks (1975), data indicate that the approximate range for α is 1.5×10^{-4} to 5×10^{-3} ft$^{2/3}$/s (0.0015–0.049 cm$^{2/3}$/s). Pearson (1961) suggested $\alpha = 0.001$ ft$^{2/3}$/s (0.01 cm$^{2/3}$/s).

Another reason why α in Equation (7-58) is not a constant even though Equation (7-58) perhaps remains the best one for predicting the increase in the diffusion coefficient, is that the presence or absence of shore boundaries must exert an influence on the rate of spread of a patch of marker. Consider, for example, the spread of a patch against a coast. Obviously the spread can only take place along the coast and offshore. Richardson's Law is basically only for situations wherein there is unobstructed spread of the marker substance. In cases where the spread of the marker is hampered by the presence of a coast, it is unreasonable to assume that the diffusion coefficient would increase as fast as the four-thirds power of the length scale of the patch. It is more reasonable to consider that the spread in such a situation is due to a diffusion coefficient that increases as the first power of the length scale of the patch. Thus if Richardson's Law is to be force-fitted to data for dispersion along the coast, the coefficient α must be adjusted, and this could lead to the range of values given earlier.

All of the foregoing discussion leads up to the following general conclusions related to the marine disposal of wastewaters. The diffusion coefficient to use in assessing the rate of spread of a body of pollutant cannot be a constant, independent of the scale of the process. The coefficient should increase as the scale of the process increases, and it appears that Richardson's Law approximately describes the average manner in which this diffusion coefficient grows to the point where the cloud has moved against a coast. From that point on, it is better to assume that the process continues to spread, but through a diffusion coefficient that increases less fast, say linearly, with the length scale of the process.

The actual coefficient value to be used in a particular design situation (with Richardson's Law) should take into account the particular situation especially in terms of the wave climate. A location with a moderate wave climate means a basic constant in Equation (7-58) that is larger than that for another location where there is very little wave action. A case in point in Hawaiian waters is the Hawaii Kai outfall that lies in a moderate to severe wave environment and where the best estimate for α was twice the value obtained for quieter waters (Sunn, Low, Tom and Hara, 1973).

Coastal Currents

Water masses adjacent to coasts come under the influence of various disturbing influences causing them to change position. Perhaps most important for our purposes are the winds and the tide-producing forces. One should also include the influence of very large-scale current systems moving past offshore such as off the east coast of Florida (Stewart et al., 1971). The prediction of coastal currents from even fairly well set out causative factors remains a very inexact practice.

Even if exhaustively studied, I believe that ocean currents will always include unexplained random components. For the purposes of marine wastewater disposal, the coastal current regime can best be established by a lengthy (usually covering all seasons) current-measuring program at stations considered to be representative for the outfall both in terms of horizontal position and depth of submergence. The data can be classified, for each station, in terms of a joint frequency distribution of current speed and direction (see Chapter 5). One feature lacking from such a representation, however, is the duration of currents of approximately constant speed and direction; another lack concerns an indication of how sinusoidal the current variation is over certain spans of time. Although it is doubtful that coastal currents can be considered as unidirectional for appreciable lengths of time, the mathematical models for secondary dispersion to be considered have used precisely this assumption. *Workable* models based on more realistic assumptions are not available.

Brooks' Model

Introduction: Figure 7-23 shows schematically the line source model considered by Brooks (1959). A liquid with a concentration c_0 of some conservative contaminant enters the current of speed u setting perpendicular to the line of the source. As the effluent is swept downstream it spreads laterally through diffusion; in so doing, the peak concentration along the centerline of the spreading cloud decreases.

Beyond the assumption of a constant current speed normal to the line source, Brooks made the following assumptions in solving for the concentration of contaminant at any station downstream of the source.

1. Vertical mixing is negligible. Thus, if we consider what is truly a line source, the entering substance must stay in the surface. Looked at in another manner, the model is a two-dimensional one. This should be a reasonable assumption in light of the fact that the diffusion coefficient in the vertical direction in the ocean is usually several orders of magnitude less than that in the horizontal direction because of density changes with depth, but opinions vary (e.g., Foxworthy et al., 1966). In addition, the depth (the limit on vertical spread of the field) is of course smaller than the horizontal distances over which the contaminant can range.

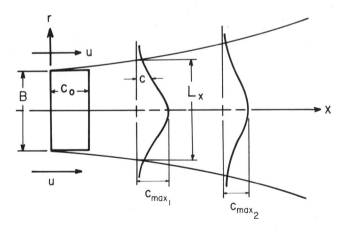

Figure 7-23 Schematic diagram for Brooks' surface plume model [Brooks (1959), with permission]

2. Mixing in the direction of the current is negligible. This is a tenable assumption as long as the speed at which the contaminant front advances is substantially greater than the speed of advance due to turbulent diffusion in the absence of advection.

3. The effluent moves with the current system. This implies that the effluent has the same density as the ambient water body. The effluent, initially lighter than the receiving medium, must be thoroughly mixed with the surrounding water in order to achieve a density approximately equal to its density. This implies large initial dilutions of the effluent as discussed earlier.

4. Mixing in the lateral direction can be described by the diffusion process with variable ε. The diffusion coefficient depends on the nominal width of the plume L_x, and is thus a function solely of the x coordinate of position, not of r, the lateral coordinate of position.

Brooks made the assumption that the following equation governs the variation of the diffusion coefficient

$$\varepsilon = \alpha L_x^{n_0} \tag{7-59}$$

and he chose three different values of the exponent n_0: 0, 1, and 4/3. The first of these corresponds to an assumed constant diffusion coefficient, the second is consistent with a coastline situation discussed earlier, and the latter conforms to Richardson's Law also outlined earlier.

Equations and Results: Brooks focused on the concentration of the contaminant along the centerline of the plume. This is a conservative measure, in terms of predicting contaminant concentrations at a downstream station, since the centerline value is the maximum for any fixed value of x. The distribution of c across any station x was taken to be bell-shaped (normal or Gaussian distribution), and the nominal width of the plume at that point was taken to be

$$L_x = 2\sqrt{3}\,\sigma \tag{7-60}$$

where σ is the standard deviation of the concentration profile at x. Since the standard deviation of a rectangular distribution of width B is

$$\sigma = B/2\sqrt{3}\,, \tag{7-61}$$

the use of a width defined by Equation (7-60) meant that the width of the plume at the beginning was as it should be, i.e., B.* Since only 8.3% of a Gaussian

---

*B can be greater than L_2, the diffuser length, as a result of jet/plume spread and early buoyant spreading.

distribution lies outside the limits $\pm \sqrt{3}\,\sigma$, the use of Equation (7-60) is admissible in terms of including the bulk of the contaminant at a particular x station. Let the diffusion coefficient for $x=0$ be ε_0. Then, from Equation (7-59)

$$\varepsilon_0 = \alpha B^{n_0} \tag{7-62}$$

As far as application of Brooks' model to the marine disposal of wastewaters is concerned, it is interesting to study the rate of spread of the surface plume and the rate of decrease of the centerline concentration c_{max}. The results are shown in Table 7-2 where

$$\beta = 12\varepsilon_0 / uB. \tag{7-63}$$

Table 7-2: Results of Brooks' Model

Values of n_0 in Equation (7-59)	L_x / B	c_{max} / c_0	Equation
0	$\left[1 + 2\beta \dfrac{x}{B}\right]^{1/2}$	$\mathrm{erf}\left\{\left[3 / \left(4\beta \dfrac{x}{B}\right)\right]^{1/2}\right\}$	(7-64)
1	$\left[1 + \beta \dfrac{x}{B}\right]$	$\mathrm{erf}\left\{\left[\dfrac{3}{2} / \left[\left(1 + \beta \dfrac{x}{B}\right)^2 - 1\right]\right]^{1/2}\right\}$	(7-65)
4/3	$\left[1 + \dfrac{2}{3}\beta \dfrac{x}{B}\right]^{3/2}$	$\mathrm{erf}\left\{\left[\dfrac{3}{2} / \left[\left(1 + \dfrac{2}{3}\beta \dfrac{x}{B}\right)^3 - 1\right]\right]^{1/2}\right\}$	(7-66)

The *error function* erf{ } given in Table 7-2 is defined by the equation

$$\mathrm{erf}\{a\} = \frac{2}{\sqrt{\pi}} \int_0^a e^{-t^2} dt \tag{7-67}$$

Tables for erf$\{a\}$ are available, such as in Abramowitz and Stegun (1964). However, the error function can be evaluated from more common tables of normal distribution probabilities (e.g., Hoel, 1976) by a change in dummy variable. Figures 7-24 and 7-25 are plots of the c_{max} / c_0 and L_x / B information in Table 7-2. Figure 7-26 depicts the variation of concentration across the plume.

It has been shown in earlier sections that a long properly designed diffuser means high initial dilutions of the effluent. The advantage of this greater length in terms of initial mixing is somewhat offset by the fact, in secondary dispersion, that a longer diffuser (wider initial sewage field in Brooks' model) means a larger value of c_{max} / c_0 at any particular station x. Since c_0 is in itself lower for the longer diffuser, however, the real changes are not obvious until one runs out the appropriate calculations.

The relative centerline concentration is a function solely of the term $\beta x / B$ as shown in Figure 7-24. Using Richardson's Law for the initial diffusion

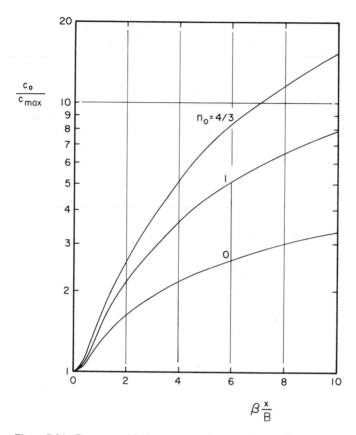

Figure 7-24 Decrease with downstream distance of centerline concentration of conservative substance [Brooks (1959), with permission]

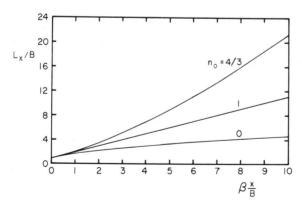

Figure 7-25 Increase in width of surface plume with downstream distance [Brooks (1959), with permission]

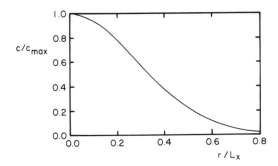

Figure 7-26 Gaussian distribution for concentration distribution across surface plume

coefficient, viz.

$$\varepsilon_0 = \alpha B^{4/3} \tag{7-68}$$

and using Equation (7-63),

$$\beta x / B = 12 \alpha x / u B^{2/3} \tag{7-69}$$

If one fixes on a specific ratio of c_{max}/c_0, a specific value of $\beta x/B$ is required (see Figure 7-24). But, from Equation (7-69), to obtain a fixed value of $\beta x/B$ for a greater value of B requires that one consider a greater distance x. Thus equivalent values of c_{max}/c_0 are pushed further and further away from the source as its width increases. However, as implied earlier, if the initial contaminant concentration c_0 can be related to the length of the diffuser, the total dilution of the effluent at any station x can be computed.

Brooks' basic model ignores nonconservative substances. Brooks showed, however, that such substances can easily be accommodated. Assuming a die-away equation of the form (Chapter 4)

$$c = c_0 e^{-k_0 t} \tag{7-70}$$

then the concentration of nonconservative substance can be computed from c_{max} in Figure 7-24 by multiplying by

$$e^{-k_0 x/u}$$

This is the same as $e^{-k_0 t}$ since the time taken for the effluent to travel a distance x with steady speed u is

$$t = x/u \tag{7-71}$$

If coliform bacteria are being considered as the nonconservative markers, then a design value of k_0 can be obtained from t_{90} data (Chapter 4).

Another Unidirectional Current Model

Foxworthy et al. (1966) have referenced several mathematical models proposed for atmospheric and oceanic diffusion in *isotropic* (homogeneous) and *anisotropic* (directionally variable) turbulent conditions. The model they concentrated on for representing the dispersion of a surface field was called a *volume source model*. In it, it was assumed that a flow of effluent Q passed into a current of speed u setting in the positive x direction and constant over all r (horizontal) and z (vertical) space. The authors concentrated on anisotropic conditions, and the spread of the field as it proceeded downstream was given in terms of the standard deviations of the concentration profiles for any station x. Equations were given for the concentration at any (x,r,z) point in the plume, but again, as in the case of Brooks' model, primary attention was directed to the concentration along the centerline of the plume lying in the ocean surface and assumed (as a simplification) to be nonmeandering. Let the local standard deviations in the horizontal and vertical directions be σ_r and σ_z, respectively.

The concentration along the central, surface axis is

$$c_{max}(x) = \frac{c_0}{\left\{\left[\dfrac{2\sigma_r^2(x)}{\sigma_r^2(0)}+1\right]^{1/2}\left[\dfrac{2\sigma_z^2(x)}{\sigma_z^2(0)}+1\right]^{1/2}\right\}} \tag{7-72}$$

where

$$c_0 = \frac{2Q}{\pi u \sigma_r(0)\sigma_z(0)} \tag{7-73}$$

and is the initial concentration of material.

The approximate equations for only horizontal diffusion are

$$c_{max}(x) = \frac{c_0}{\left[\dfrac{2\sigma_r^2(x)}{\sigma_r^2(0)}+1\right]^{1/2}} \tag{7-74}$$

where

$$c_0 = \frac{Q/H_*}{u\sqrt{\pi}\,\sigma_r(0)} \tag{7-75}$$

where H_* is the average depth of the waste field assumed to have a uniform concentration distribution.

An estimate of the concentration across the plume at any x station can be obtained by combining the result of Equation (7-74) with Figure 7-26. The assumption of Gaussian profiles was made in the development of the Foxworthy et al. model.

Computer Programs

The Department of Civil Engineering at the Massachusetts Institute of Technology (Cambridge, Mass.), for example, has available for purchase two computer programs that can be used in charting the spread of pollutant discharges. Both are two-dimensional (horizontally), one-layer (depth-averaged), finite element models. Each model package consists of a program tape and a user's manual.

The first such model, CAFE-1, predicts the circulation patterns of wind-driven or tide-driven currents within a body of water. DISPER-1 then predicts the dispersion patterns of dissolved or suspended substances within the body of water. Both computer models can handle large-scale processes involving complicated geometries.

7-8　REFERENCES

ABRAMOWITZ, M., and I. A. STEGUN (1964): *Handbook of Mathematical Functions with Formulas, Graphs, and Mathematical Tables*, National Bureau of Standards, Washington, D.C.

ABRAHAM, G. (1963): "Jet Diffusion in Stagnant Ambient Fluid," Delft Hydraulics Laboratory, *Publication No. 29* (July).

BROOKS, N. H. (1959): "Diffusion of Sewage Effluent in an Ocean Current," in *Waste Disposal in the Marine Environment*, edited by E. A. Pearson, pp. 246–267, Pergamon Press, New York.

BROOKS, N. H. (1970b): "Conceptual Design of Submarine Outfalls—II: Hydraulic Design of Diffusers," California Institute of Technology, W. M. Keck Laboratory of Hydraulics and Water Resources, *Technical Memorandum 70-2* (January).

BROOKS, N. H. (1972): "Dispersion in Hydrologic and Coastal Environments," California Institute of Technology, W. M. Keck Laboratory of Hydraulics and Water Resources, *Report No. KH-R-29* (December); also U.S. E.P.A. *Publication 660/3-73-010* (1973).

DAUGHERTY, R. L., and J. B. FRANZINI (1965): *Fluid Mechanics with Engineering Applications*, 6th ed., McGraw-Hill, New York.

DOBBINS, W. E. (1965): "Diffusion and Mixing," *Journal*, Boston Society of Civil Engineers, Vol. 52, No. 2 (April), pp. 108–128.

FAN, L-N., and N. H. BROOKS (1969): "Numerical Solutions of Turbulent Buoyant Jet Problems," California Institute of Technology, W. M. Keck Laboratory of Hydraulics and Water Resources, *Report No. KH-R-18* (January).

FOXWORTHY, J. E., R. B. TIBBY, and G. M. BARSOM (1966): "Dispersion of a Surface Waste Field in the Sea," *Journal of the Water Pollution Control Federation*, Vol. 38, No. 7 (July), pp. 1170–1193.

GRACE, R. A. (1970): "How to Measure Waves," *Ocean Industry*, Vol. 5, No. 2 (February), pp. 65–69.

HENDERSON, F. M. (1967): "The Effect of Wave Action on the Flow in the Proposed Ocean Outfall of a Sewage Line, Napier," Report to Steven and Fitzmaurice Consulting Engineers, Christchurch, New Zealand.

HOEL, P. G. (1976): *Elementary Statistics*, 4th ed., John Wiley, New York.

KOH, R. C. Y. (1973): "Hydraulic Tests of Discharge Ports—Hydraulic Investigations of Thermal Outfalls for the San Onofre Nuclear Power Plant," California Institute of Technology, W. M. Keck Laboratory of Hydraulics and Water Resources, Progress Report No. 3 (*Technical Memorandum No. 73-4*) to Southern California Edison Co. (March 30).

KOH, R. C. Y. (1976): "Buoyancy-driven Gravitational Spreading," *Proceedings*, Fifteenth International Conference of Coastal Engineering, Honolulu, Hawaii (July), Vol. 4, pp. 2956–2975.

KOH, R. C. Y., and N. H. BROOKS (1975): "Fluid Mechanics of Waste-Water Disposal in the Ocean," *Annual Review of Fluid Mechanics*, Vol. 7, Annual Reviews, Inc., Palo Alto, Calif., pp. 187–211.

LAYTON, J. A. (1976): "Design Procedures for Ocean Outfalls," *Proceedings*, Fifteenth International Coastal Engineering Conference, Honolulu, Hawaii (July), Vol. 4, pp. 2919–2940.

LISETH, P. (1970): "Mixing of Merging Buoyant Jets from a Manifold in Stagnant Receiving Water of Uniform Density," University of California, Berkeley, Hydraulic Engineering Laboratory, *Technical Report HEL 23-1* (November).

LISETH, P. (1973): "Mixing of Merging Buoyant Jets from a Manifold in Stagnant Receiving Water of Uniform Density," *Advances in Water Pollution Research*, edited by S. H. Jenkins, pp. 921–934, Pergamon Press, New York.

LISETH, P. (1976): "Wastewater Disposal by Submerged Manifolds," ASCE, *Journal of the Hydraulics Division*, Vol. 102, No. HY1 (January), pp. 1–14.

MCBRIDE, G. B. (1975): personal communication.

MCNOWN, J. S. (1954): "Mechanics of Manifold Flow," ASCE, *Transactions*, Vol. 119, pp. 1103–1142, with discussion.

MOODY, L. F. (1944): "Friction Factors for Pipe Flow," ASME, *Transactions*, Vol. 66, pp. 671–684, with discussion.

PEARSON, E. A. (1961): "Marine Waste Disposal," paper presented at the 75th Annual General Meeting of the Engineering Institute of Canada, Vancouver, B.C. (May).

RICHARDSON, L. F. (1926): "Atmospheric Diffusion Shown on a Distance-Neighbour Graph," *Proceedings of the Royal Society of London*, Series A, Vol. 110, pp. 709–737.

ROBERTS, P. J. W. (1975): "The Diffusion of Buoyant Effluent from Outfall Diffusers of Finite Length—Progress Report," California Institute of Technology, W. M. Keck Laboratory of Hydraulics and Water Resources, *Technical Memorandum 75-1* (August).

ROBERTS, P. J. W. (1977a): "Dispersion of Buoyant Waste Water Discharged from Outfall Diffusers of Finite Length," California Institute of Technology, W. M. Keck Laboratory of Hydraulics and Water Resources, *Report No. KH-R-35* (March).

ROBERTS, P. J. W. (1977b): personal communication.

STEWART, R. E., et al. (1971): "Diffusion of Sewage Effluent from an Ocean Outfall," ASCE, *Journal of the Sanitary Engineering Division*, Vol. 97, pp. 485–503.

STOMMEL, H. (1949): "Horizontal Diffusion due to Oceanic Turbulence," *Journal of Marine Research*, Vol. 8, pp. 199–225.

STREETER, V. L. (1971): *Fluid Mechanics*, 5th ed., McGraw-Hill, New York.

Sunn, Low, Tom, and Hara, Inc. (1973): "Investigation of the Effects of Hawaii Kai Treatment Plant Effluent on Water Quality and Ecosystems off Sandy Beach," prepared for Kaiser-Aetna, Honolulu, Hawaii (June).

Western Canada Hydraulic Laboratories Ltd. (1973): "Diffuser Study for Annacis Island Sewage Treatment Plant," prepared for Greater Vancouver Sewerage and Drainage District, B.C., at Port Coquitlam, B.C., Canada (August).

7-9 BIBLIOGRAPHY

General

BACHE, D. H. (1976): "Density Current Surges I: Their Rate in the Initial Dispersion of a Surface Field," *Journal of Hydraulic Research*, Vol. 14, No. 1, pp. 1–7.

BUMPUS, D. F., W. R. WRIGHT, and R. F. VACCARO (1969): "Considerations on a Sewer Outfall off Nobska Point," Woods Hole Oceanographic Institution, Woods Hole, Mass., Reference No. 69-87 (unpublished manuscript) (December) [*NTIS Publication AD-700-895*].

Design and Construction of Sanitary and Storm Sewers (1969): American Society of Civil Engineers, New York.

FISCHER, H. B. (1970): "Prediction of Coastal Waste Concentrations," University of California Berkeley Extension, Pollution of Coastal and Estuarine Waters (January), Section 9B.

HARREMOËS, P. (1966): "Prediction of Pollution from Planned Wastewater Outfalls," *Journal of the Water Pollution Control Federation*, Vol. 38 (August), pp. 1323–1333.

HEANEY, F. L. (1961): "Hydraulic Problems of Ocean Disposal," Boston Society of Civil Engineers, Seminar Papers on Waste Water Treatment and Disposal (April).

Metcalf and Eddy, Inc. (1972): *Wastewater Engineering: Collection, Treatment, Disposal*, McGraw-Hill, New York.

Parker, Homer W. (1975): *Wastewater Systems Engineering*, Prentice-Hall, Englewood Cliffs, N.J.

Rambow, C. A. (1969): "Submarine Disposal of Industrial Waste," *Proceedings*, 24th Industrial Waste Conference, Purdue University, pp. 1486–1493.

Thomann, R. V. (1972): *Systems Analysis and Water Quality Management*, McGraw-Hill, New York.

Flow in Pipes, Losses, Pumping, and Hydraulic Transients

Benami, A. (1968): "New Head-Loss Tables for Sprinkler Laterals," ASCE, *Journal of the Irrigation and Drainage Division*, Vol. 94, No. IR2 (June), pp. 185–197.

Fox, J. A. (1977): *Hydraulic Analysis of Unsteady Flow in Pipe Networks*, Wiley, New York.

"General Principles of Pumping Station Design and Layout" (1962): U.S. Army Corps of Engineers, *Engineering Manual 1110-2-3102* (December).

Hicks, T. G., and T. W. Edwards (1971): *Pump Application Engineering*, McGraw-Hill, New York.

"Mechanical and Electrical Design of Pumping Stations" (1962): U.S. Army Corps of Engineers, *Engineering Manual 1110-2-3105* (December).

Rich, G. R. (1963): *Hydraulic Transients*, Dover, New York.

Stepanoff, A. J. (1957): *Centrifugal and Axial Flow Pumps*, 2nd ed., Wiley, New York.

Streeter, V. L., and E. B. Wylie (1967): *Hydraulic Transients*, McGraw-Hill, New York.

Internal Hydraulics of Diffusers

Camp, T. R., and S. D. Graber (1968): "Dispersion Conduits," ASCE, *Journal of the Sanitary Engineering Division*, Vol. 94, No. SA1 (February), pp. 31–39; with Closure, Vol. 95, No. SA5 (October, 1969), pp. 943–947.

Camp, T. R., and S. D. Graber (1970): Discussion of "Internal Hydraulics of Thermal Discharge Diffusers," by S. Vigander, R. A. Elder, and N. H. Brooks, ASCE, *Journal of the Hydraulics Division*, Vol. 96, pp. 2631–2635.

French, J. A. (1972): "Internal Hydraulics of Multiport Diffusers," *Journal of the Water Pollution Control Federation*, Vol. 44, No. 5 (May), pp. 782–795.

Diffuser Analysis and Design

Argaman, Y., M. Vajda, and N. Galil (1975): "Use of Jet Pumps in Marine Waste Disposal," ASCE, *Journal of the Environmental Engineering Division*, Vol. 101, No. EE5 (October), pp. 703–711.

BAUMGARTNER, D. J., and D. S. TRENT (1970): "Ocean Outfall Design; Part I—Literature Review and Theoretical Development," U.S. Federal Water Quality Administration, Northwest Region, Corvallis, Oregon (April).

BAUMGARTNER, D. J., D. S. TRENT, and K. V. BYRAM (1971): "User's Guide and Documentation for Outfall Plume Model," U.S.E.P.A., Pacific Northwest Water Laboratory, Corvallis, Oregon, *Working Paper No. 80* (May).

BROOKS, N. H. (1970a): "Conceptual Design of Submarine Outfalls—I: Jet Diffusion," California Institute of Technology, W. M. Keck Laboratory of Hydraulics and Water Resources, *Technical Memorandum 70-1* (January).

BURCHETT, M. E., G. TCHOBANOGLOUS, and A. J. BURDOIN (1967): "A Practical Approach to Submarine Outfall Calculations," *Public Works*, Vol. 98, No. 5 (May), pp. 95–101.

CARTER, H. H., D. W. PRITCHARD, and J. H. CARPENTER (1966): "The Design and Location of a Diffuser Outfall for a Municipal Waste Discharge at Ocean City, Maryland," The John Hopkins University, Chesapeake Bay Institute, *Special Report 10* (January).

CHAO, T. L., and C. M. CAMPUZANO (1972): "Simplified Method of Ocean Outfall Diffuser Analysis," *Journal of the Water Pollution Control Federation*, Vol. 44, No. 5 (May), pp. 806–812.

"Conceptual Designs of Outfall Systems for Desalination Plants" (1970): Report prepared by the Dow Chemical Company for the U.S. Dept. of the Interior, The Office of Saline Water (April).

HUANG, W. H. (1975): "Determination of Length and Direction of Ocean Outfall Diffuser at Wantagh Sewage Treatment Plant, Nassau County, New York," *Proceedings*, Civil Engineering in the Oceans III, University of Delaware, Newark, Del. (June), ASCE, New York, pp. 1294–1313.

SHUMAS, F. J. (1975): "Diffuser Design," *Preprints*, 1975 Environment Improvement Conference, Vancouver, B.C. (October), Technical Session, Canadian Pulp and Paper Assn., Montreal, Quebec, pp. 17–24.

SILVESTER, R. (1967): "Jet Mixers in Sewage Outfalls," *The Journal of the Institution of Engineers*, Australia, Vol. 39, No. 3 (March), pp. 33–37.

Jets and Plumes: Theoretical, Numerical, and Laboratory Studies

ABRAHAM, G. (1965a): "Entrainment Principle and Its Restrictions to Solve Problems of Jets," Delft Hydraulics Laboratory, Delft, The Netherlands, *Publication No. 55* (July).

ABRAHAM, G. (1965b): "Horizontal Jets in a Stagnant Fluid of Other Density," ASCE, *Journal of the Hydraulics Division*, Vol. 91, No. HY4 (July), pp. 139–154.

ABRAHAM, G. (1967): "Jets with Negative Buoyancy in Homogeneous Fluid," Delft Hydraulics Laboratory, Delft, The Netherlands, *Publication No. 56* (December).

ABRAHAM, G. (1970): "The Flow of Round Buoyant Jets Escaping Vertically into Ambient Fluid Flowing in a Horizontal Direction," *Advances in Water Pollution Research*, Vol. 2, Paper III-15.

ABRAHAM, G. (1971): "The Flow of Round Buoyant Jets Issuing Vertically into Ambient Fluid Flowing in a Horizontal Direction," Delft Hydraulics Laboratory, Delft, The Netherlands, *Publication No. 81* (March).

ABRAHAM, G. (1974): "Jets and Plumes Issuing into Stratified Fluid," Delft Hydraulics Laboratory, Delft, The Netherlands, *Publication No. 141* (December).

ABRAHAM, G., and W. D. EYSINK (1969): "Jets Issuing into Fluid with a Density Gradient," Delft Hydraulics Laboratory, Delft, The Netherlands, *Publication No. 66* (October).

CHAN, T-L, and J. F. KENNEDY (1972): "Turbulent Nonbuoyant and Buoyant Jets Discharged into Flowing or Quiescent Fluids," The University of Iowa, Iowa Institute of Hydraulic Research, Iowa City, Iowa, *IIHR Report No. 140* (August).

CHAN, T-L, and J. F. KENNEDY (1975): Discussion of "Buoyant Forced-Plumes in Cross Flow," by V. H. Chu and M. B. Goldberg, ASCE, *Journal of the Hydraulics Division*, Vol. 101, No. HY8 (August), pp. 1105–1107.

CHU, V. H., and M. B. GOLDBERG (1974): "Buoyant Forced-Plumes in Cross Flow," ASCE, *Journal of the Hydraulics Division*, Vol. 100, No. HY9 (September), pp. 1203–1214.

CREW, H. (1970): "A Numerical Model of the Dispersion of a Dense Effluent in a Stream," Texas A & M University, Dept. of Oceanography, *Technical Report 70-10T*, Project 716 (June).

FAN, L-N. (1967): "Turbulent Buoyant Jets into Stratified or Flowing Ambient Fluids," California Institute of Technology, W. M. Keck Laboratory of Hydraulics and Water Resources, *Report No. KH-R-15* (June).

FAN, L-N., and N. H. BROOKS (1966): Discussion of "Horizontal Jets in a Stagnant Fluid of Other Density" by G. Abraham, ASCE, *Journal of the Hydraulics Division*, Vol. 92, No. HY2, pp. 423–429.

HARLEMAN, D. R. F., and G. H. JIRKA (1974): "Buoyant Discharges from Submerged Multiport Diffusers," *Proceedings*, Fourteenth Coastal Engineering Conference, Copenhagen, Denmark (June), pp. 2180–2198.

HAYASHI, T. (1971): "Turbulent Buoyant Jets of Effluent Discharged Vertically Upward from an Orifice in a Cross-Current in the Ocean," *Proceedings*, Fourteenth Congress of the International Association for Hydraulic Research, Paris, France (August), Vol. 1, pp. 157–165.

KOH, R. C. Y. (1971): "On Buoyant Jets," *Proceedings*, Fourteenth Congress of the International Association for Hydraulic Research, Paris, France (August), Paper A-18.

KOH, R. C. Y., et al. (1974): "Hydraulic Modeling of Thermal Outfall Diffusers for the

San Onofre Nuclear Power Plant," California Institute of Technology, W. M. Keck Laboratory of Hydraulics and Water Resources, *Report No. KH-R-30* (January).

LIST, E. J., and J. IMBERGER (1973): "Turbulent Entrainment in Buoyant Jets and Plumes," ASCE, *Journal of the Hydraulics Division*, Vol. 99, No. HY9 (September), pp. 1461–1474.

LIST, E. J., and R. C. Y. KOH (1975): "Hydraulic Modeling of Thermal Outfall Diffusers: Interpretation of Results," International Association for Hydraulic Research, Sixteenth Congress, São Paulo, Brazil (July).

LIU, S-L. (1976): "Mixing of Submerged Two-Dimensional Buoyant Jets in Uniform Bodies of Water in the Absence and Presence of Wind Action," University of California, Berkeley, Hydraulic Engineering Laboratory, *Technical Report HEL 23-5* (June).

MAXWORTHY, T. (1972): "Experimental and Theoretical Studies of Horizontal Jets in a Stratified Fluid," International Association for Hydraulic Research, International Symposium on Stratified Flows, Novosibirsk, U.S.S.R., *Contribution 17*.

McBRIDE, G. B. (1972): "Mathematical Models of the Flow Phenomena Associated with the Discharge of Sewage or Waste Heat into the Marine Environment," Ministry of Works, Water and Soil Division, Wellington, New Zealand (December).

McBRIDE, G. B. (1973): "Turbulent Buoyant (sic) Jets into Stagnant Non-Linearly Stratified Environments," International Association for Hydraulic Research, *Proceedings of the Fifteenth Congress*, Istanbul, Turkey, Vol. 2, pp. 87–95.

NOSPAL, A., and J-C. TATINCLAUX (1976): "Design of Alternating Diffuser Pipes," ASCE, *Journal of the Hydraulics Division*, Vol. 102, No. HY4 (April), pp. 553–558.

PEÑA, J. M., and S. C. JAIN (1974): "Numerical Analysis of Warm, Turbulent Sinking Jets Discharged into Quiescent Water of Low Temperature," The University of Iowa, Iowa Institute of Hydraulic Research, *Report No. 154* (February).

ROBIDEAU, R. F. (1972): "The Discharge of Submerged Buoyant Jets into Water of Finite Depth," General Dynamics Corp., Electric Boat Division, Groton, Conn., *Report No. U440-72-121* (November).

SHUTO, N. (1971): "Buoyant Plume in a Cross Stream," *Coastal Engineering in Japan*, Vol. 14, pp. 163–173.

SHUTO, N., and L. H. TI (1974): "Wave Effects on Buoyant Plumes," *Proceedings*, Fourteenth Coastal Engineering Conference, Copenhagen, Denmark (June), pp. 2199–2208.

WALLIS, I. G., and D. A. REINSCH (1977): "Relative Influence of Initial Momentum and Buoyancy on the Performance of Deepwater Ocean Outfalls," Institution of Engineers, Australia, *Proceedings of the Third Australian Conference on Coastal and Ocean Engineering*, Melbourne (April), pp. 96–102.

WRIGHT, S. J. (1977): "Mean Behavior of Buoyant Jets in a Crossflow," ASCE, *Journal of the Hydraulics Division*, Vol. 103, No. HY5 (May), pp. 499–513.

Buoyant Spreading

CHAO, J. L. (1975): "Horizontal Spread of Wastewater Field over Calm Ocean Surface," *Journal of the Water Pollution Control Federation*, Vol. 47, No. 10 (October), pp. 2504–2510.

HYDÉN, H., and I. LARSEN (1975): "Surface Spreading," in *Discharge of Sewage from Sea Outfalls*, edited by A. L. H. Gameson, pp. 277–283, Pergamon Press, Oxford, England.

LARSEN, I., and T. SÖRENSEN (1968): "Buoyancy Spread of Waste Water in Coastal Regions," *Proceedings*, Eleventh Conference on Coastal Engineering, London, England (September), pp. 1397–1402.

SHARP, J. J. (1969): "Spread of Buoyant Jets at the Free Surface," ASCE, *Journal of the Hydraulics Division*, Vol. 95, No. HY3 (May), pp. 811–825.

SHARP, J. J. (1971): "Unsteady Spread of Buoyant Surface Discharge," ASCE, *Journal of the Hydraulics Division*, Vol. 97, No. HY9 (September), pp. 1471–1492.

Diffusion

BATCHELOR, G. K. (1950): "The Application of the Similarity Theory of Turbulence to Atmospheric Diffusion," *Quarterly Journal of the Royal Meteorological Society*, Vol. 76, pp. 133–146.

BATCHELOR, G. K. (1952): "Diffusion in a Field of Homogeneous Turbulence. II—The Relative Motion of Particles," *Proceedings* of the Cambridge Philosophical Society, Vol. 48, pp. 345–362.

CARSLAW, H. S., and J. C. JAEGER (1959): *Conduction of Heat in Solids*, 2nd ed., Oxford University Press, Oxford.

GIFFORD, F., JR. (1959): "Statistical Properties of a Fluctuating Plume Dispersion Model," *Advances in Geophysics*, Vol. 6, pp. 117–137.

HOLLEY, E. R. (1969): "Unified View of Diffusion and Dispersion," ASCE, *Journal of the Hydraulics Division*, Vol. 95, No. HY2 (March), pp. 621–631.

Oceanic Diffusion

Allan Hancock Foundation, University of Southern California (1965): "An Investigation on the Fate of Organic and Inorganic Wastes Discharged into The Marine Environment and Their Effects on Biological Productivity," California State Water Quality Control Board, Sacramento, Calif., *Publication No. 29*.

BOWDEN, K. F. (1965): "Horizontal Mixing in the Sea due to a Shearing Current," *Journal of Fluid Mechanics*, Vol. 21, pp. 83–95.

FOXWORTHY, J. E. (1968): "Eddy Diffusivity and the Four-Thirds Law in Near-Shore (Coastal Waters)," University of Southern California, *Report 68-1* [*NTIS Publication PB 217-157*].

GRAY, E., and T. E. POCHAPSKY (1964): "Surface Dispersion Experiments and Richardson's Diffusion Equation," *Journal of Geophysical Research*, Vol. 69, pp. 5155–5159.

JOSEPH, J., and H. SENDNER (1962): "On the Spectrum of the Mean Diffusion Velocities in the Ocean," *Journal of Geophysical Research*, Vol. 67, pp. 3201–3205.

MEERBURG, A. J. (1972): "An Experimental Study of the Turbulent Diffusion in the Upper Few Metres of the Sea," *Netherlands Journal of Sea Research*, Vol. 5, No. 4 (May), pp. 492–509.

OKUBO, A. (1962): "A Review of Theoretical Models for Turbulent Diffusion in the Sea," *Journal of the Oceanographical Society of Japan*, pp. 286–320.

OKUBO, A., and D. W. PRITCHARD (1969): "Summary of Our Present Knowledge of the Physical Processes of Mixing in the Ocean and Coastal Waters...," The John Hopkins University, Chesapeake Bay Institute, *Report No. NY0-3109-40* (September).

SULLIVAN, P. J. (1971): "Some Data on the Distance-Neighbour Function for Relative Diffusion," *Journal of Fluid Mechanics*, Vol. 47, Part 3, pp. 601–607.

TAMAI, N. (1972): "Unified View of Diffusion and Dispersion in Coastline Waters," *Journal of the Faculty of Engineering*, The University of Tokyo, Vol. 31, No. 4, pp. 531–692.

Secondary Dispersion

CALLAWAY, R. J. (1975): "Subsurface Horizontal Dispersion of Pollutants in Open Coastal Waters," *Discharge of Sewage from Sea Outfalls*, edited by A. L. H. Gameson, pp. 297–307, Pergamon Press, Oxford.

CEDERWALL, K. (1970): "Dispersion Phenomena in Coastal Waters," *Journal*, Boston Society of Civil Engineers, Vol. 57, No. 1, pp. 34–70.

LAU, Y. L., and G. B. KRISHNAPPAN (1975): Discussion of "Transverse Dispersion in Oscillatory Channel Flow," by P. R. B. Ward, ASCE, *Journal of the Hydraulics Division*, Vol. 101, No. HY3, pp. 560–561.

SUMER, S. M. (1976): "Transverse Dispersion in Partially Stratified Tidal Flow," University of California, Berkeley, Hydraulic Engineering Laboratory, *Report WHM-20* (May).

WARD, P. R. B. (1974): "Transverse Dispersion in Oscillatory Channel Flow," ASCE, *Journal of the Hydraulics Division*, Vol. 100, No. HY6 (June), pp. 755–772.

Field Studies of Outfalls

AGG, A. R., and A. C. WAKEFORD (1972): "Field Studies of Jet Dilution of Sewage at Sea Outfalls," *Journal*, Institution of Public Health Engineers, London, England, Vol. 71, No. 2 (April), pp. 126–153.

GAMESON, A. L. H. (1968): "Studies of Sewage Dispersion from Two Sea Outfalls," *Chemistry and Industry*, No. 46 (November 16), pp. 1582–1589.

RAWN, A. M., F. R. BOWERMAN, and N. H. BROOKS (1960): "Diffusers for Disposal of Sewage in Sea Water," ASCE, *Journal of the Sanitary Engineering Division*, Vol. 86, No. SA2, pp. 65–105.

STEWART, R. E. (1973): "Unusual Plume Behavior from an Ocean Outfall off the East Coast of Florida," *Journal of Physical Oceanography*, Vol. 3, No. 2 (April), pp. 241–243.

STEWART, R. E., et al. (1969): "Diffusion of Sewage Effluent from an Ocean Outfall," ASCE, *Proceedings*, Civil Engineering in the Oceans III, Miami Beach, Fla. (December), pp. 1151–1185.

Thermal Effluents

HARLEMAN, D. R. F. (1975): "Heat Disposal in Water Environment," ASCE, *Journal of the Hydraulics Division*, Vol. 101, No. HY9 (September), pp. 1120–1138.

HARLEMAN, D. R. F., and K. D. STOLZENBACH (1972): "Fluid Mechanics of Heat Disposal from Power Generation," *Annual Review of Fluid Mechanics*, Vol. 4, pp. 7–32.

"Heat Disposal from Power Generation in the Water Environment: Literature from 1970" (1976): Delft Hydraulics Laboratory, Delft, The Netherlands, *Bibliography B284* (September).

HOOPES, J. A., R. W. ZELLER, and G. A. ROHLICH (1968): "Heat Dissipation and Induced Circulations from Condenser Cooling Water Discharges into Lake Monona," University of Wisconsin, Engineering Experiment Station, *Report No. 35* (February).

KOH, R. C. Y., and L-N. FAN (1970): "Mathematical Models for the Prediction of Temperature Distributions Resulting from the Discharge of Heated Water into Large Bodies of Water," report prepared for the U.S. Environmental Protection Agency, Water Quality Office, Water Pollution Control Research Series, 16130 DW 010/70, (October).

SAYRE, W. W. (1975): "Investigation of Surface-Jet Thermal Outfall for Iatan Steam Electric Generating Station," The University of Iowa, Iowa Institute of Hydraulic Research, *Report No. 167* (April).

SHIRAZI, M. A., and L. R. DAVIS (1972): "Workbook of Thermal Plume Prediction: Vol. 1—Submerged Discharge," U.S. E.P.A., National Environmental Research Center, Corvallis, Oregon, *Report EPA-R2-72-005a* (August).

SHIRAZI, M. A., and L. R. DAVIS (1974): "Workbook of Thermal Plume Prediction: Vol. 2—Surface Discharge," U.S. E.P.A., National Environmental Research Center, Corvallis, Oregon, *Report EPA-R2-72-005b* (May).

STEFAN, H., L. BERGSTADT, and E. MROSLA (1975): "Flow Establishment and Initial Entrainment of Heated Water Surface Jets," U.S. E.P.A., National Environmental Research Center, Corvallis, Oregon, *Report EPA-660/3-75-104* (May).

STOLZENBACH, K. D., and D. R. F. HARLEMAN (1971): "An Analytical and Experimental

Investigation of Surface Discharges of Heated Water," U.S. E.P.A., Water Quality Office, Report (February).

VIGANDER, S., R. A. ELDER, and N. H. BROOKS (1970): "Internal Hydraulics of Thermal Discharge Diffusers," ASCE, *Journal of the Hydraulics Division*, Vol. 96, No. HY2 (February), pp. 509–527.

WILLIAMS, J. M. (n.d.): "The Spread of Warm Water from an Outfall in a Tidal Flow," *Mathematical and Hydraulic Modelling of Estuarine Pollution*, McCorquodale Printers, Ltd., pp. 191–199, London.

Wastewater Disposal in Estuaries

DAILEY, J. E., and D. R. F. HARLEMAN (1972): "Numerical Model for the Prediction of Transient Water Quality in Estuary Networks," M.I.T., Ralph M. Parsons Laboratory of Water Resources and Hydrodynamics, *Report No. 158* (October).

FISCHER, H. B. (1976): "Mixing and Dispersion in Estuaries," *Annual Review of Fluid Mechanics*, edited by M. Van Dyke, W. G. Vincenti, and J. V. Wehausen, Palo Alto, California, Annual Reviews, Inc., Vol. 8, pp. 107–133.

HARLEMAN, D. R. F. (1964): "The Significance of Longitudinal Dispersion in the Analysis of Pollution in Estuaries," *Advances in Water Pollution Research, Proceedings*, Second International Conference, Tokyo (August), Vol. 1, pp. 279–290 (with discussion).

HARLEMAN, D. R. F. (1966): "Pollution in Estuaries," in *Estuary and Coastline Hydrodynamics*, edited by A. T. Ippen, pp. 630–647, McGraw-Hill, New York.

HINWOOD, J. B., and I. G. WALLIS (1973): "Modelling the Movement of Conservative Substances in Tidal Estuaries," *Preprints*, First Australian Conference on Coastal Engineering, Sydney (May), pp. 159–166.

HINWOOD, J. B., and I. G. WALLIS (1975a): "Classification of Models of Tidal Waters," ASCE, *Journal of the Hydraulics Division*, Vol. 101, No. HY10 (October), pp. 1315–1331.

HINWOOD, J. B., and I. G. WALLIS (1975b): "Review of Models of Tidal Waters," ASCE, *Journal of the Hydraulics Division*, Vol. 101, No. HY11 (November), pp. 1405–1421.

HOLLEY, E. R., D. R. F. HARLEMAN, and H. B. FISCHER (1970): "Dispersion in Homogeneous Estuary Flow," ASCE, *Journal of the Hydraulics Division*, Vol. 96, No. HY8 (August), pp. 1691–1709.

IPPEN, A. T. (1966): "Salinity Intrusion in Estuaries," in *Estuary and Coastline Hydrodynamics*, edited by A. T. Ippen, pp. 598–629, McGraw-Hill, New York.

LEENDERTSE, J. J., and E. C. GRITTON (1970): "A Water Quality Simulation Model for Well Mixed Estuaries and Coastal Seas," 3 Volumes, The New York City Rand Institute, *Publications RM-6230-RC, R-708-NYC*, and *R-709-NYC* (February, and July 1971).

PINCINCE, A. B., and E. J. LIST (1973): "Disposal of Brine into an Estuary," *Journal of the Water Pollution Control Federation*, Vol. 45, No. 11 (November), pp. 2335–2344.

PRITCHARD, D. W. (1969): "Dispersion and Flushing of Pollutants in Estuaries," ASCE, *Journal of the Hydraulics Division*, Vol. 95, No. HY1 (January), pp. 115–124.

SELLECK, R. E., and B. GLENNE (1966): "A Model of Mixing and Diffusion in San Francisco Bay," Vol. 7 of Final Report, "A Comprehensive Study of San Francisco Bay," University of California, Sanitary Engineering Research Laboratory, *Report 67-1* (June).

Tracor, Inc. (1971): "Estuarine Modeling: an Assessment," report prepared at Austin, Texas, for the U.S. Environmental Protection Agency, Water Quality Office (February).

Wave Effects on Effluent Dispersion

JOHNSON, J. W. (1959): "The Effect of Wind and Wave Action on the Mixing and Dispersion of Wastes," in *Waste Disposal in the Marine Environment*, edited by E. A. Pearson, Pergamon Press, New York.

MASCH, F. D. (1961): "Mixing and Dispersive Action of Wind Waves," University of California, Berkeley, *I.E.R. Technical Report 138-6* (November).

RICE, E. K., and J. W. JOHNSON (1954): "Waves as a Factor in Effluent Disposal," *Water and Sewage Works*, Vol. 101, No. 4 (April), pp. 172–175.

Wave Effects on Pipe Flow

ISAACS, J. D., and R. L. WIEGEL (1949): "The Measurement of Wave Heights by Means of a Float in an Open-End Pipe," *Transactions of the American Geophysical Union*, Vol. 30, pp. 501–506.

8

The Technical Aspects
of Outfall Design

8-1 INTRODUCTION

Outfall Pipes

An ocean outfall is merely a submarine pipeline with a specific purpose. Of the four terms used in conjunction with pipes (see Figure 8-1), the *springline* and the *invert* are particularly important. When elevations are specified for an outfall, these apply to the invert rather than the outside bottom of the pipe.

As a means of putting the contents of this chapter into context, it is convenient at this stage to sketch the general design features of a specific outfall. The outfall to be considered for illustrative purposes is the 12,500-ft-long (3,800-m) Sand Island No. 2 outfall constructed offshore Honolulu, Hawaii, in

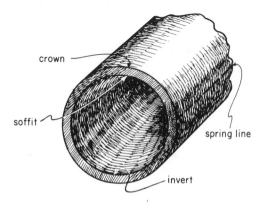

Figure 8-1 Descriptive terms for pipe

1974 and 1975. This specific outfall is considered because of frequent reference to it in Chapter 10.

Sand Island No. 2 Outfall

The straight nondiffuser part of this outfall cuts across depth contours at about a 45° angle, then turns into the straight diffuser that generally lies along the 235-ft (72-m) depth contour as shown in Figure 8-2. There are 282 ports along the overall 3,398-ft (1,036-m) diffuser length. Two 7.00-in. (178-mm) ports are located in the flapgate structure that closes off the end of the pipe.

The main part of the outfall is 84-in. diam (2,134-mm) reinforced concrete pipe. This pipe is completely buried in the bottom out to a depth of 90 ft (27 m) where it gradually emerges to lie along the sea bed. Figures 8-3 and 8-4 show cross-sectional views of the buried pipe in two different depth ranges. The arrangement with the sheet piling extends out 2,500 ft (760 m) from shore to a depth of about 19 ft (6 m). Rock was not only used to cover over the pipe in its trench in nearshore waters, but was also used along the sides of the pipe for support in deeper water. From the point of its emergence from the bottom out to

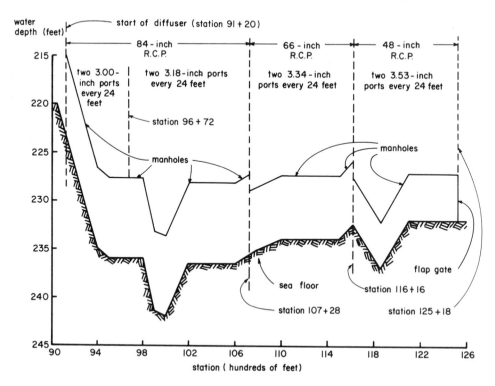

Figure 8-2   Diffuser of Sand Island No. 2 outfall (Courtesy, City and County of Honolulu, Sewers Division)

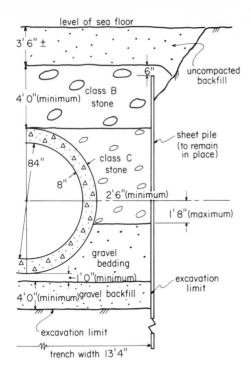

level of sea floor

3' 6" ±

6"

uncompacted
backfill

class B
stone

4' 0"(minimum)

sheet pile
(to remain
in place)

84"

class C
stone

8"

2' 6"(minimum)

1' 8"(maximum)

gravel
bedding

1' 0"(minimum)

excavation
limit

4' 0"(minimum) gravel backfill

excavation limit

trench width 13' 4"

Figure 8-3 Cross-section of Sand Island No. 2 outfall through inshore zone (Courtesy, City and County of Honolulu, Sewers Division)

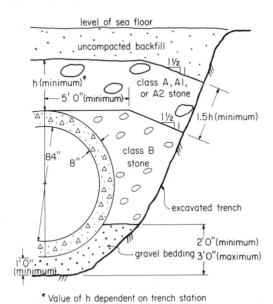

level of sea floor

uncompacted backfill

1½

h (minimum)*

class A, A1,
or A2 stone

5' 0"(minimum)

1½

1.5h (minimum)

84"

8"

class B
stone

excavated trench

2' 0"(minimum)

gravel bedding 3' 0"(maximum)

1' 0"
(minimum)

* Value of h dependent on trench station
and stone class.

Figure 8-4 Cross-section of Sand Island No. 2 outfall buried in offshore zone (Courtesy, City and County of Honolulu, Sewers Division)

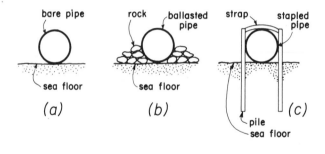

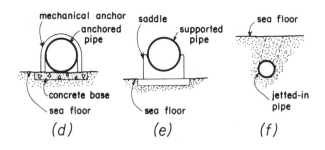

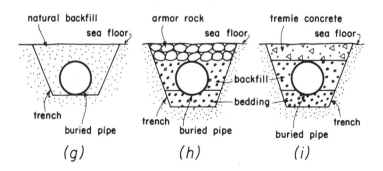

Figure 8-5   Examples of possible configurations for submarine pipelines

a depth of about 198 ft (61 m) the outfall had stone built up level with the crown. The pipe was *completely ballasted*. Toward the end of the diffuser the ballasting extended only up to the springline as shown in Figure 8-5(b).

The rock used for ballasting or for covering the trench varied in size. Five different well-graded classes were employed as follows: A2, 30-in.-diam nominal (762-mm); A1, 24-in. (610-mm); A, 18-in. (457-mm); B, 12-in. (305-mm); and C, 6-in. (152-mm). The B size, for example, was used as the top rock layer in the surf zone out to a depth of approximately 19 ft (6 m). See Figure 8-3.

Orientation

An attempt will be made in this chapter to deal systematically with the various considerations involved in determining the final design of an outfall such as the Sand Island No. 2. (The design of a diffuser was presented in Chapter 7 and will not be covered here.)

Not all ocean outfalls are made of reinforced concrete; Section 8-2 gives information on the various types of materials that can be used, the problems peculiar to each type, and the most efficient applications. One reason for writing this book was to acquaint ocean outfall designers, present and future, with potential problem areas. Too many outfalls have already failed and dumped poorly diluted wastewater into nearshore marine waters because inadequate attention was given to critical details. Section 8-3 describes the sources of potential trouble for marine outfalls, both man-caused and natural, and gives particular instances of outfall damage that have occurred.

Section 8-4 discusses the route for the outfall. Sections 8-5 and 8-7 deal with physical environmental conditions important in ocean outfall design, including quasi-steady water levels, currents, and waves. Outfall design with respect to currents is considered in Section 8-6. Generally speaking, waves provide the sternest natural test for marine outfalls, and much of the remainder of the chapter will consider them. Models for translating the customary wave descriptions of height, period, and water depth into near-bottom water motion that adequately simulates real ocean waves are discussed in Section 8-8. This information and related data are used in Section 8-9 to formulate models for predicting wave-induced forces on unburied outfalls lying on the sea bed. Much of the material for Sections 8-8 and 8-9 was obtained in the ocean during the summers of 1975 and 1976 by a research group under my direction.

Even outfalls laid on the sea floor in 100 ft (30 m) or more of water can be broken by strong water motion near the bottom under large waves. This is a more serious problem in the *surf zone*, that area shoreward of the largest breaking waves. The standard means of protection is to excavate a trench in the bottom, place the pipe in it, and *backfill* with either naturally occurring material or rock. The latter is the more usual since wave action sufficient to cause problems for a pipe lying on the sea bed can probably also erode the naturally occurring bottom material. Outfall burial and protection by means of rock are discussed in Section 8-10. Section 8-11 primarily discusses outfall design against earthquakes. The chapter is rounded off in Section 8-12 by considering particular features of outfalls, such as the end flapgate structure mentioned earlier for the Sand Island No. 2 outfall.

The initial output of the complete design effort is the Preliminary Design Report. The final document, incorporating changes suggested by reviewers such as the owner and funding agencies, is the Final Design Report.

8-2 OUTFALL PIPE

Introduction

As part of his thorough study of the marine disposal of sewage and sludge through outfalls, Pearson (1956) obtained information on 145 such lines. The materials used were as follows: cast iron (65 outfalls); reinforced concrete, including concrete–steel cylinder pipe (39); wrought iron (19); steel (15); wood stave (4); corrugated iron (2); and vitrified clay (1).

Entries in Table 10-1 since 1956 do not include any of cast iron, wrought iron, wood stave, corrugated iron, or vitrified clay pipe, and such materials have apparently seen infrequent use. Reinforced concrete pipe (RCP) has been very popular, and some steel outfalls have been built. Also, the materials used now include three types of plastic: high-density polyethylene, polyvinyl chloride (PVC), and fiberglass-reinforced pipe (FRP). In the following sections brief summaries of RCP, steel pipe, high-density polyethylene pipe, PVC pipe, and FRP will be presented. Because of its historical significance in outfall construction, cast iron will be added to this list and considered first.

Cast Iron Pipe

Gray cast iron is an alloy of iron with a high proportion of carbon in the graphite form. Although quite resistant to attack by many liquids, certain waters and other liquids can cause growths (or *tubercles*) to develop on the inside of such pipes. For this reason cast iron pipes have often been centrifugally lined with an interior coat of cement mortar. Such a layer also increases the flow-carrying capacity of the pipe.

Ductile (cast) iron pipe was developed around the time of World War II; its strength approaches that of steel pipe and it has corrosion resistance superior to that of standard cast iron pipe. It also has good flexibility and excellent impact resistance. The combination of ductile iron pipe and a tightly bonded polyethylene lining is highly regarded as material for sewer force mains. The smoothness of the liner provides excellent initial flow capacity. The stability of polyethylene against attack by hydrogen sulfide and other components in sewage ensures maintenance of the high flow capacity. A ductile iron outfall was installed at Otter Rock, Oregon, in early 1973.

Cast iron is commercially available in various diameters through 54-in. (1,372-mm). The standard lengths vary depending upon the process used to make the pipe and the end conditions specified, i.e., flanged or otherwise; however, 20-ft (6.1-m) lengths are most common. Lengths of pipe are linked at *joints*. The design of these joints is of critical importance to the stability, efficiency, and lifetime of the pipe. They must provide a tight seal; they should prevent pullout; and in many instances they should be flexible.

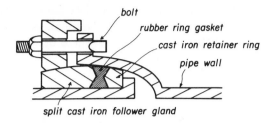

bolt

rubber ring gasket

cast iron retainer ring

pipe wall

split cast iron follower gland

Figure 8-6   Ball and socket joint for cast iron pipe

Cast iron pipe is usually made such that one end (the *bell*) is flared, while the other end (the *spigot*) is not. The spigot of one length of pipe is inserted into the bell of the other throughout the length of the line. In the past, the rigid joints built into cast iron pipe were a real disadvantage, and numerous outfall installations failed because settlement or lateral loads caused the joints to open out (Pearson, 1956). Today, a subaqueous ball-and-socket joint is preferred for cast iron pipe; the bolted version of such a joint is shown in Figure 8-6. This type of joint permits up to 15° changes in direction between any two adjacent lengths of pipe; it is tight at any deflection; and in the bolted version there is no shear stress on the bolts. The flexibility is an excellent feature, since settling of a line on the sea bed is not uncommon. Ball joints were specified for some joints in the surf area for the ductile iron outfall of the Humboldt Bay Wastewater Authority, California, in 1976.

Figure 8-7 shows the end of a 48-in.-diam (1,219-mm) cast iron outfall constructed in 1926 just outside Kewalo Basin, Honolulu, Hawaii; however, it has not been used, except for emergencies (Fig. 3-1), since 1949 when the Sand Island No. 1 outfall was built. This particular unlined outfall was built with rigid joints and supported on concrete pads as in Figure 8-5(d). I have inspected the exterior of this line on numerous occasions from its appearance on the sea bed in about 15 ft (4.5 m) of water to its terminus in 60 ft (18 m). It is still in excellent external condition. It can be inferred from an experiment made on the line in early 1976 that the inside of the pipe is in good condition as well. Using

Figure 8-7   Terminus of Kewalo Basin outfall, Hawaii

various assumptions, I processed flow test data obtained by the City and County of Honolulu, Department of Public Works, Sewers Division, on the line and found Manning's roughness n (Table 7-1) to be somewhere between 0.017 and 0.019.

Steel Pipe

A general idea concerning the sizes and weights of available steel pipe is provided by the information in Table 8-1. Although the size limit in Table 8-1 is 92 in. (2,337 mm), larger-diameter spirally welded steel pipe can be obtained. Standard lengths of the steel pipe sizes shown in Table 8-1 vary. The Dutch firm of Van Leeuwen Buizen, for example, advertises that 8–14-m lengths are available.

Table 8-1:   Selected Standard Seamless or Longitudinally Welded Steel Pipe Sizes

Nominal Diameter*	Outside Diameter		Wall Thickness†		Approximate Weight		Approximate Buoyancy‡ Sea Water Outside, Air Inside		Approximate Buoyancy‡ Sea Water Outside, Water / Sewage Inside	
(in.)	(in.)	(mm)	(in.)	(mm)	(lb/ft)	(N/m)	(lb/ft)	(N/m)	(lb/ft)	(N/m)
16	16.000	406.4	0.250	6.4	42	620	+47	+690	−35	−510
16	16.000	406.4	1.438	36.5	224	3,270	−135	−1,970	−193	−2,820
20	20.000	508.0	0.250	6.4	53	770	+86	+1,260	−43	−630
20	20.000	508.0	1.031	26.2	211	3,080	−72	−1,050	−182	−2,660
24	24.000	609.6	0.250	6.4	63	920	+137	+2,000	−51	−740
24	24.000	609.6	1.000	25.4	246	3,590	−46	−670	−211	−3,080
30	30.000	762.0	0.312	7.9	99	1,440	+214	+3,120	−80	−1,170
30	30.000	762.0	1.000	25.4	310	4,520	+3	+40	−264	−3,850
36	36.000	914.4	0.375	9.5	142	2,070	+309	+4,510	−114	−1,660
36	36.000	914.4	1.250	31.7	465	6,790	−14	−200	−396	−5,780
42	42.000	1,066.8	0.375	9.5	167	2,440	+447	+6,520	−132	−1,930
42	42.000	1,066.8	0.500	12.7	222	3,240	+392	+5,720	−180	−2,630
48	48.000	1,219.2	0.375	9.5	190	2,770	+612	+8,930	−148	−2,160
48	48.000	1,219.2	0.500	12.7	254	3,710	+548	+8,000	−204	−2,980
60	60.000	1,524.0	0.354	9.0	226	3,290	+1,027	+14,990	−169	−2,470
60	60.000	1,524.0	0.750	19.0	474	6,910	+779	+11,370	−386	−5,630
72	72.000	1,828.8	0.472	12.0	361	5,280	+1,443	+21,060	−275	−4,010
72	72.000	1,828.8	0.750	19.0	570	8,320	+1,234	+18,010	−458	−6,680
80	80.000	2,032.0	0.472	12.0	400	5,830	+1,827	+26,660	−300	−4,380
80	80.000	2,032.0	0.750	19.0	634	9,250	+1,593	+23,250	−504	−7,360
92	92.000	2,337.0	1.000	25.4	973	14,200	+1,972	+28,770	−785	−11,450
92	92.000	2,337.0	1.250	31.7	1,211	17,670	+1,734	+25,300	−992	−14,470

*Not all possible diameters are given.

†The thicknesses given for a particular pipe outside diameter indicate the range from low- to high-pressure service.

‡A positively buoyant section would rise to the surface; a negatively buoyant one would sink.

   Steel pipe is widely used in the offshore oil and gas industry, and it has been used as well in numerous outfalls, some of which are given in Table 10-1. Steel is stronger than cast iron, and since the specific gravities of these two materials are virtually the same, pipes for the same use in steel are lighter. Generally speaking, such pipes are also cheaper, easier to construct, and easier to transport than cast iron pipe.

   A major problem with steel is that it corrodes—particularly in the marine environment. For this reason a protective anticorrosion coating is normally applied to the outside of the pipe (e.g., Saroyan, 1969; Hartley, 1971). One form of protective coating for such marine lines has consisted of an underlayer of coal tar enamel and then two or three wraps of fiberglass cloth. An example of an outside coating for a steel pipe is provided in "Submarine Pipeline Wins 1967 Award" (1967) as follows: "After steam curing, each pipe was prime coated, given a coating of coal tar enamel into which a fiberglass wrapping was drawn, then two more coats of enamel, followed by an impregnated asbestos felt wrapping applied while the last coat was still hot."

   The inner wall of steel pipe to be used in the ocean is generally also coated as (partly) an anticorrosive measure. This may take several forms, but materials that have found frequent use for such internal coatings in submarine pipes are cement mortar, coal tar enamel, and tar epoxy. The modern trend has been to move away from coal tar and asphalt—chiefly to epoxies and polyethylene, but also to other materials such as PVC (Thompson, 1976). However, coal tar still finds extensive use (O'Donnell, 1977). The use of cement mortar has largely been discontinued because, being rigid, it will not bend without fracturing, and it is not entirely impervious.

   The last two columns in Table 8-1 show that steel pipe filled with fresh water or sewage and placed in the marine environment is negatively buoyant. However, there is insufficient weight in most of these lines to assure stability in the event of current- or wave-induced forces on the pipe or if excess buoyancy results from liquefaction of the supporting sediments. These topics are discussed later in this chapter.

   Although there are potentially various ways of supplying extra weight to a submarine pipe (e.g., Small and Serpas, 1972) the method customarily used for steel lines is to apply a concrete weight-coating to the outside of the pipe. There are companies with offices worldwide that will supply concrete-jacketed steel pipe. Price Coating Companies (Bartlesville, Oklahoma), for example, supplies concrete thicknesses of 1–8 in. (25–203 mm). Price has three basic densities of concrete jacket—nominally 140, 165, and 190 pcf* (22,000, 25,900, and 29,800 N/m^3). Coatings are applied at a plant, directly over corrosion coatings, such as coal tar enamel, and over a steel-reinforcing cage. Controlled-speed wire brushes throw the concrete mixture, at approximately 100 fps (30 m/s), onto the

---

*Iron ore aggregate can be used to achieve 190 pcf.

prepared pipe surfaces. Galvanized wire reinforcement is machine-fed onto the rotating pipe during application.

Steel pipes for marine use are usually joined by welding, although flanged connections are sometimes used especially near the sea end of an outfall, presumably to permit extension of the line at a later date. When the pipes are joined by welding, however, the weld is checked by radiography. Then the protective coatings are applied to the outside, followed by steel reinforcing and a concrete coat. If cement mortar is being used on the inside of the pipe, then that coat must be made continuous as well. See Figures 10-10 and 10-12.

A joint that will permit angular adjustments of the pipe is desirable in areas of appreciable sea floor dynamics. Some surf zone joints in the concrete-covered steel outfall of the Menasha Corporation at Coos Bay, Oregon, were ball joints designed to provide this flexibility.

Concrete Pipe

Concrete pipe is frequently used for marine outfall installations. Weights and dimensions of commercially available RCP are shown in Table 8-2, and details on constituent materials, fabrication, fittings, and other related matters are contained in American Water Works Association (1965).* All pipe in Table 8-2 is available in standard 16-ft (4.88-m) lengths. Pipe 34 in. (864 mm) diam and greater also has 32-ft (9.75-m) lengths standard. For concrete outfalls, where the number of sections required is very large, virtually any desired section length under 32 ft (9.75 m) can be called for.

For high-pressure situations, a steel cylinder provides the major reinforcement for concrete pipe. In some cases high-tensile steel wire is wound around the outside of the steel cylinders, and the result is referred to as prestressed concrete pipe. In steel cylinder pipe, additional reinforcement is provided by circumferential and longitudinal steel reinforcing bars; such bars are used to provide necessary reinforcing in lower-pressure RCP.

Concrete is very resistant to attack by sea water or marine organisms. However, it is susceptible to attack by acids, and the sulfur content of wastewaters can have a large bearing on the stability of the material. Some protection can be provided by using cements that are particularly acid-resistant or by using vinyl or stainless steel (Type 316, not 304) liners on the inside walls of the pipe in areas where sulfate attack is likely to be most pronounced. This is usually in the upstream reach of an outfall.

Both vinyl and stainless steel liners were used for protective purposes at the Point Loma outfall in San Diego, California, but portions of these became

---

*It is perhaps appropriate to note here that there is much to be said for using AWWA or ASTM (American Society for Testing and Materials) standards for everything possible—e.g., steel reinforcing, aggregate, cement, dimensions. Practically the only exception regarding RCP for marine outfalls would concern extra concrete cover over the reinforcing steel.

Table 8-2: Weights and Dimensions of Standard Reinforced Concrete Subaqueous Pipe

Nominal Inside Diameter		Wall Thickness		Approximate Weight	
(in.)	*(mm)*	*(in.)*	*(mm)*	*(lb/ft)*	*(N/m)*
24	610	4	102	341	4,980
30	762	4	102	416	6,070
34	864	4	102	535	7,810
36	914	4	102	563	8,220
39	991	4	102	604	8,810
42	1,067	$4\frac{1}{2}$	114	731	10,670
46	1,168	$4\frac{1}{2}$	114	794	11,590
48	1,219	5	127	920	13,430
54	1,372	$5\frac{1}{2}$	140	1,130	16,490
60	1,524	6	152	1,360	19,850
66	1,676	$6\frac{1}{2}$	165	1,615	23,570
72	1,829	7	178	1,885	27,510
78	1,981	$7\frac{1}{2}$	191	2,180	31,810
84	2,134	8	203	2,500	36,480
90	2,286	8	203	2,660	38,820
96	2,438	$8\frac{1}{2}$	216	3,010	43,930
102	2,591	$8\frac{1}{2}$	216	3,185	46,480
108	2,743	9	229	3,565	52,030
120	3,048	10	254	4,385	63,990
132	3,353	11	279	5,300	77,350
144	3,658	12	305	6,300	91,940
156	3,962	13	330	7,380	107,700

detached and were lost out the outfall. Up to 8 in. (203 mm) of concrete on the 14-in.-thick (356-mm) pipe walls was eaten away by sulfate attack (coupled with erosion due to high velocities).

Figure 8-8 shows a joint used to connect sections of steel cylinder concrete pipe. The steel in the joints assures strength, and the wall does not have to be built up as is the case when the only steel extending into a joint lip area consists of reinforcing bars. The situation shown in Figure 8-8 is not an ideal solution in the marine environment, however, owing to the necessity of providing corrosion protection for the exposed steel. A preferred joint form, consisting solely of concrete surfaces, is shown in Figure 8-9. This type of pipe is called *extended-bell pipe*.

The joint shown in Figure 8-9 allows for some flexibility. Since by itself it does not provide ample protection from pullout, there is an adaptation of this joint in which steel lugs for draw bolts are located on the springline of the pipe near the ends of the sections. The double-gasket joint-sealing arrangement

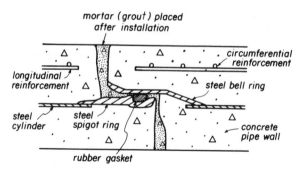

Figure 8-8 Joint for steel cylinder concrete pipe

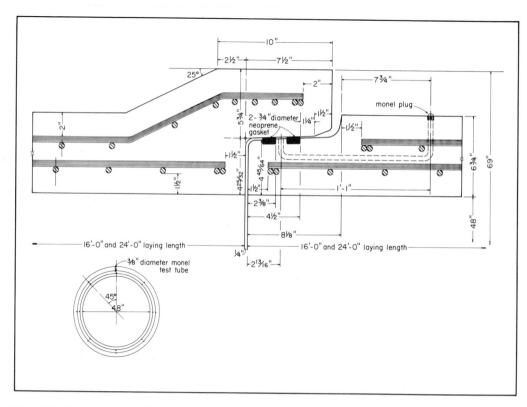

Figure 8-9 Joint for extended-bell reinforced concrete pipe used for Mokapu outfall (Courtesy, City and County of Honolulu, Sewers Division)

shown in Figure 8-9 is an important feature in terms of testing the tightness of a joint underwater as is described in Chapter 10.

Longitudinally-post-tensioned concrete pipe has been used for ocean outfalls (e.g., Williams, 1966). Such pipe has the advantage of being able to flex to some extent, unlike the standard concrete pipe. Another advantage of the post-tensioned pipe is that it can be pulled out in the water rather than painstakingly laid section by section as is the case for the standard concrete pipe. Figure 8-10 shows one arrangement for jointing longitudinally-post-tensioned concrete pipe (Ade, 1970).

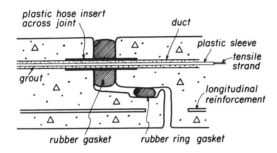

Figure 8-10   Joint for longitudinally-post-tensioned concrete pipe

Plastic Pipe

There are at least three different types of plastics that can be used and have been used (Table 10-1) for ocean outfalls, viz., polyvinyl chloride (PVC), high-density polyethylene, and fiberglass-reinforced pipe (FRP). These materials exhibit excellent resistance to corrosive environments and are light and flexible. Superficially, these appear to be three very desirable attributes for an outfall sewer. Lightness, for example, would mean greater ease in handling during construction; flexibility would mean that the line could adjust to natural cut and fill on the sea bed without failing. Although the latter remains a decidedly positive feature, the lightness can be burdensome. Plastic lines often have weights attached to them in order to assure that they do not surface, and this added weight may completely nullify the inherent lightness of the material ("Polyethylene Pipe Used...," 1972). On the other hand, if weights can be placed over the pipe once in place, or if such pipe can be firmly anchored to the substrate, the lightness can be used to advantage. Plastic outfall pipes are at a disadvantage to the extent that they lack the decades of satisfactory service demonstrated by other materials. Proven materials and techniques are very important in outfalls.

PVC is manufactured only in relatively small sizes [to 24 in. (610 mm)]. Standard pipe lengths are 20 ft (6 m). Although it is not attacked by corrosive liquids, PVC pipe seems to attract marine life. Experience with a PVC outfall at

Rio de Janeiro showed that a marine mollusk drilled through the pipe wall and made an outfall line inoperative within a year. PVC also exhibits low structural strength.

High-density polyethylene pipe ranges up to 63 in. (1,600 mm) in outside diameter in standard form, but up to 80 in. (2,032 mm) in fabricated sizes. Table 8-3 shows sample dimensions and weights. Although pipe lengths of approximately 40 ft (12 m) are standard, lengths up to 60 ft (18 m) have been produced. There have been numerous instances of such pipe being used in protected waters, and, except for heated effluents, its performance is reportedly good.

Table 8-3: Selected Dimensions and Weights of High-Density Polyethylene Pipe*

Nominal Diameter†	Actual Outside Diameter		Wall Thickness‡		Approximate Weight	
(in.)	(in.)	(mm)	(in.)	(mm)	(lb/ft)	(N/m)
16	15.748	400	0.488	12.4	10.3	150
16	15.748	400	1.287	32.7	25.5	372
20	19.685	500	0.612	15.5	16.1	235
20	19.685	500	1.312	33.3	33.1	483
24	24.803	630	0.772	19.6	25.6	374
24	24.803	630	1.653	42.0	52.5	766
28	27.953	710	0.859	21.8	32.1	468
28	27.953	710	1.492	37.9	54.3	792
32	31.496	800	0.980	24.9	41.2	601
32	31.496	800	1.294	32.9	53.8	785
36	35.433	900	1.104	28.0	52.2	762
36	35.433	900	1.458	37.0	68.2	995
40	39.370	1,000	1.225	31.1	64.3	938

*The specific gravity of high-density polyethylene is 0.955.
†Not all available diameters are shown.
‡The extremes of available thicknesses are given.

FRP can be made in very large sizes, to 14 ft (4.3 m) in diameter and to 60 ft (18 m) in length. FRP has a strong physical structure with a much reduced wall thickness when compared to, say, reinforced concrete. Low-pressure FRP of 12-ft (3.7-m) inside diameter, for example, has a wall thickness of only $1\frac{1}{8}$ in. (29 mm). Dimensions and weights of smaller FRP sizes are shown in Table 8-4.

PVC pipe can be joined using flanges linked through a gasket or else slip-on cemented connections. Polyethylene pipe can be butt-fused or connected through flanges. In the latter connection, a pipe section is fitted with a butt-fused stub end and aluminum backup ring. These form the flange assembly. The

Table 8-4: Dimensions and Weights of Standard Filament-Wound FRP*

Inside Diameter†		Wall Thickness‡		Approximate Weight	
(in.)	(mm)	(in.)	(mm)	(lb/ft)	(N/m)
16	406.4	0.180	4.6	7.8	114
16	406.4	0.250	6.4	10.9	159
20	508.0	0.180	4.6	9.7	142
20	508.0	0.313	8.0	17.0	248
24	609.6	0.188	4.8	13.0	330
24	609.6	0.375	9.5	28.5	416
36	914.4	0.188	4.8	18.5	270
36	914.4	0.500	12.7	49.0	715
48	1,219.2	0.188	4.8	25.0	365
48	1,219.2	0.625	15.9	81.6	1,191
60	1,524.0	0.188	4.8	31.0	452
60	1,524.0	0.750	19.1	122.5	1,787
72	1,828.8	0.220	5.6	43.7	638
72	1,828.8	0.875	22.2	171.4	2,501
84	2,133.6	0.220	5.6	51.0	744
84	2,133.6	0.987	25.1	255.6	3,729
96	2,438.4	0.220	5.6	58.4	852
96	2,438.4	1.125	28.6	294.0	4,290

*The specific gravity of FRP is between 1.80 and 2.08.

†Not all available diameters are listed.

‡These sizes reflect the range for a given pipe diameter.

butt-fusion joining process in polyethylene pipe assures strong, leakproof joints and can be done, by machine, at the job site. Some details are provided by Janson (1975) and Campbell (1976).

There are several methods of joining sections of FRP, but perhaps the two major ones for submarine pipe are the welded and flanged joints. In welded joints layers of fiberglass impregnated with catalyzed resin are wrapped over the butted ends of the pipe. In flanged joints the flanges are either formed on the pipe, or they are fused to a pipe section by means of a stub end. Gaskets are used with the flanges.

Plastic pipe has many positive features, but because there are not decades of experience with such materials in outfalls, many designers are loath to try them out. One worry with FRP, for example, is delamination. In addition, experience with FRP pulp and paper outfalls off British Columbia has shown that these pipes can suffer rapid wear from abrasion on the sea bottom, if they are not properly bedded and/or immobilized. Rock backfill has also punctured FRP—as in a line in Yaquina Bay at Toledo, Oregon.

Cathodic Protection

Even with a corrosion-protection coating such as was described earlier for steel pipe, and even in situations where an additional layer of concrete has been applied over that coating, designers are reluctant to leave corrosion prevention to the coating itself. Pinholes ("holidays") can appear in such a layer, and corrosion can then be concentrated at such locations with potentially serious results. The additional corrosion protection of a steel line takes the form of either or both of two systems of cathodic protection. The term *cathodic protection* refers to a technique to control corrosion of a metal surface that involves making that surface the cathode in an electrochemical cell by means of an impressed direct current and/or attachment of sacrificial anodes such as magnesium, aluminum, or zinc.

In *galvanic* (sacrificial) *anode systems*, a low-resistance electrical contact must be maintained between the anode and the metal to be protected throughout the life of the structure. The size and weight of the anode must be sufficient to protect the structure for the desired time. Where anode replacement is difficult, magnesium anodes, even though they are very active and furnish high-current outputs, should not be used because of their relatively short lives. For zinc and aluminum in seawater, the particular composition of the alloy critically influences anode performance. Sacrificial anodes were used in an outfall at Goleta, California. One worrisome aspect of using galvanic anodes on underwater structures such as pipes is that skin divers may visualize such anodes as "salvage" items. This has been the experience in British Columbia waters at any rate.

A standard electrical requirement for cathodic protection is 1 mA/ft^2 (10.8 mA/m^2) of bare pipe surface. For a concrete-coated pipe, "bare surface" can be taken to mean 10% of the surface area of the line. Anode size, weight, and spacing are selected to yield the required current for a nominal anode life of 30 years. The pipe coating is cut away in a small area adjacent to the anode to expose the pipe surface. One end of a cable is then brazed to the pipe at this location with the other end of the cable attached to the anode. A usual anode spacing is 1,000 ft (300 m).

In the *impressed current anode system* an anode (e.g., of lead–silver alloy), or group of anodes, is connected through an insulated conductor to the positive terminal of a direct-current source such as a rectifier or generator. The pipe is then connected to the negative terminal of the direct-current source.

The cathodic protection used for two particular submarine steel lines is described by Keeling (1959) and in "Submarine Pipeline Wins 1967 Award" (1967). The first of these lines is the Hyperion 22-in. (559-mm) sludge line. An example of a steel sewage outfall cathodically protected is the line at Clover Point, British Columbia, Canada. An industrial wastewater outfall, made of steel and similarly protected, is the Georgia-Pacific Paper Co. line at Newport, Oregon. Although the 1959 Hyperion 144-in. (3658-mm) effluent line is of RCP, metal parts were cathodically protected.

Construction Technique Influence on Materials Selection

There are various factors that exert an influence on the type of material, and perhaps also joint details, for an outfall sewer. Both the extent and topography of the land where the outfall enters the water have a distinct bearing on materials selection. A small and/or hilly area, for example, essentially rules out a construction procedure wherein the pipe is assembled in strings on the shore and then pulled to sea on the surface (and later sunk) or along the bottom. Since the "pulling" method usually is associated with steel pipe, occasionally plastic, the absence of a suitable staging area may rule out these materials. However, the oil industry has installed steel lines in such regions by pulling a floating line ashore after bringing it to the site via the sea route. To my knowledge this technique has not been applied to outfalls, and it would be a hazardous operation in exposed waters.

Some outfalls have been designed for two different materials so that the construction contractors have some latitude in bidding for the job. The Hyperion sludge line (#11 in Table 10-1) was designed in both steel (built) and concrete. The Mokapu line (#47) had the same alternatives, but in that case concrete was selected. An outfall at Clover Point, British Columbia, Canada, went to bid with either steel (chosen) or high-density polyethylene as possibilities. The Georgia-Pacific Paper Co. outfall at Newport, Oregon, had three options but actually only two different materials, concrete and steel. The point is that two or more materials may be suitable for a specific outfall, but particular contractor equipment and competence may make construction with one considerably cheaper.

8-3   THE SPECTRUM OF POTENTIAL TROUBLE FOR AN OUTFALL

Introduction

There is a decades-long history of problems with submarine pipelines. There have been numerous instances of damage and breakage from a variety of causes that will be outlined in the following sections. Although most problems have taken place in the offshore oil and gas industry (e.g., Brown, 1967; Grice, 1968; Milz and Broussard, 1972; Demars et al., 1977) a not inconsequential number have concerned outfall sewers (e.g., Herbich, 1977).* Both the Atlantic and Pacific coasts of the United States, as well as the Great Lakes, have been concerned. But this is not a problem limited to the United States; there have been difficulties worldwide. Breakage in an outfall sewer is particularly troublesome because it is virtually inconceivable that the sewage flow can be diverted, and it is certainly not possible to stop it. Thus the result of outfall breakage is usually the flow into nearshore waters of a weakly diluted effluent—with the potential for serious pollution to recreational and shellfish waters. The 1925 Los Angeles, California outfall, for example, reportedly leaked sewage into nearshore waters from the day it was built until it was replaced in 1948. The pollution led to quarantine of 11 miles (18 km) of southern California beaches. However, breakage has also resulted in complete outfall blockage, due to sand entering the pipe through the break. An example concerns the city of Newport, Oregon, outfall which became plugged in 1973; subsequently raw sewage discharged out of manholes and down onto the local beach.

How can such breaks take place? Ships dragging anchors periodically hook onto and break pipelines lying on the sea floor. Radical changes in ocean-bottom configuration due to currents or waves can break pipes, even pipes that were originally buried, or open them at their joints. Currents can impose large loads on unburied pipelines and lead to breakage; the water motion associated with tsunamis (tidal waves) can similarly cause such breaks; and the strong oscillatory water motion associated with large ocean waves can dislocate pipelines. In the latter regard, not only unburied pipelines are in danger but also buried ones, since large waves can cause heavy erosion in the bottom sediments and thus gain access to formerly buried lines. Submarine pipes cannot easily be buried in hard ocean beds and must be anchored to such bottoms if left exposed.

The designer of outfall sewers must be aware of the various ways in which a submarine pipe can fail in order to provide ample insurance against failure. The failure mechanisms for in-place lines will be sketched in the following sections. However, outfalls can fail during construction—as in the case of the first Crown-Simpson industrial outfall near Humboldt Bay, California.

Problems Involving Ships

There have been various cases of fishing trawl systems, usually the doors, damaging submarine oil pipelines (e.g. Guerry, 1976).† However, I know of only a single case where a trawl caught an outfall. Fortunately this resulted simply in the spalling of a small amount of concrete from the pipe. This concerned the Menasha Corporation concrete-coated, mortar-lined steel outfall at Coos Bay,

---

*There is, in truth, little difference between an oil pipeline proceeding from land into the sea and the smaller-size outfall sewers.

†Details on trawling and trawling equipment have been provided in Section 6-5.

Oregon. The trawl itself caught on five of the riser ports in August 1975 and the trawl cables broke when the fisherman tried to pull free. Divers freed the trawl six weeks later.

Theoretical analyses, model tests, and field experiments have been run to determine the force exerted by typical trawl doors impacting an exposed pipe (Gjorsvik et al., 1975; Carstens et al., 1977). In all three of these separate study methods, different tow speeds, different types and weights of doors, and different impact angles were considered. Unfortunately, the data are proprietary and little of direct use in design is available. Brown (1971, 1972, 1973) also discussed the problem of trawl–pipe interaction, but again he left out the vital details. Brown also generally discussed ship anchor–pipe interaction.

Anchors have damaged many submarine oil and gas pipelines (e.g., "Welding 'Hut'...," 1968; O'Donnell, 1971). I know of one case where anchor-inflicted damage to an outfall was very strongly suspected and another where this source was verified. The first concerned the 22-in.-diam (559-mm) sludge line that the City of Los Angeles, California operated at Hyperion. It is theorized that a large oil tanker dropped anchor in a storm in 1959, some two years after the line was installed, and the ship's anchor snagged the pipe. The outfall was pulled up off the bottom and resulting damage required replacement of about 0.25 mile (400 m) of the line at a cost of about $250,000.

The second case concerned an outfall from a pulp and paper mill at Prince Rupert, British Columbia, Canada, that was snagged by a dragged anchor as a ship left port. The pipe was 32-in.-diam (813-mm) plastic and the damage occurred in about 70 ft (21 m) of water.

A standard method for protecting a submarine pipe from trawls or dragged anchors, as well as other effects to be considered later, is to bury it. But the question that naturally arises concerns the depth of burial necessary to be below the depth of potential anchor penetration. It has been reported that ship anchors only 200 lb (900 N) in weight can penetrate sand bottoms over 5 ft (1.5 m) and mud bottoms over 8 ft (2.5 m). A big ship anchor of 6,000 lb (27,000 N) penetrated sand bottoms to almost 7 ft (2 m) and mud bottoms almost 30 ft (9 m). Engineers in Japan, for example, are very wary of ship anchor-caused damage to submarine lines, and there is at least one instance of their burying a line in the vicinity of an offshore loading berth to a depth (in silt) of 20 ft (6 m) (backfilled with sand) (Nippon Steel Corporation, 1973). Koster (1974) has thoroughly discussed and referenced this topic.

Wave-Related Problems

General Mechanisms and Liquefaction: Many unburied submarine pipeline failures have something to do with waves, and it may often be a combination of two or more of the following factors that leads to failure:

1. Liquefaction of the bottom support material

2. Flow-induced force on the pipe
3. (Differential) erosion of the bottom material along the pipe
4. Mass movements of sea floor materials

Liquefaction in a soils context, has been explained as follows (Seed and Lee, 1966).

> If a saturated sand is subjected to ground vibrations, it tends to compact and decrease in volume; if drainage is unable to occur, the tendency to decrease in volume results in an increase in pore-water pressure and if the pore-water pressure builds to the point at which it is equal to the overburden pressure, the effective stress becomes zero, the sand loses its strength completely, and it develops a liquefied state.

The necessary dynamic action to promote liquefaction can come from waves or earthquakes. The danger is that a soil in a liquefied state can exert a very large buoyant force and can cause a pipe formerly buried in it to be moved to its surface—perhaps into direct contact with strong wave-induced water motion. It has been suggested that this phenomenon might have been in part responsible for problems with the large sewer outfall into Lake Ontario at Rochester, New York ("Improper Design…, 1972). The design of a particular power plant outfall in a soils liquefaction context is described by Christian et al. (1974).

Direct Wave Forces: Blumberg (1964, 1966a, b) and Grice (1968) report various failures in submarine gas and oil lines during hurricanes, presumably due to the action of the associated monstrous waves and violent water movement near the sea floor. Some lines in the Gulf of Mexico have been moved *miles*.

Two concrete-clad steel pipelines, one 20-in. (508-mm) and the other 30-in.-diam (762-mm), extend from the Standard Oil refinery, located at Barbers Point, Hawaii, to an offshore discharge anchorage for tankers. Both lines are 10,000 ft (3,000 m) long and terminate in 60 ft (18 m) of water. The lines were laid close together throughout their length and were only buried close inshore. Big swells* in the mid-1960s took 1800 ft (550 m) of the 20-in. (508-mm) line and threw it up and over the bigger line. Although no leakage resulted, the concrete coatings were cracked and spalled off in various places. In addition, as a result of movement of the lines under the influence of wave-induced bottom motion, up to $3\frac{1}{2}$ in. (89 mm) of concrete wear was observed.

The 20-in. (508-mm) line was returned to its former position, concrete was poured between the lines, and they were secured together with monel straps. I swam this line in 1974, and by this time all of the monel straps had parted in the

---

*Waves up to 34 ft (10.4 m) in height were measured.

vicinity of the mass concrete. In addition, sufficient erosion of the coral sea bed in the lee of the 30-in. (762-mm) line had occurred to warrant Standard Oil placing discarded concrete pile tops along the pipe as a protective measure.

In my own ocean research on wave forces on pipes, to be discussed later, I have observed on many occasions the violent, potentially erosive eddy that breaks over the top of a pipe onto the bottom in its lee. Even pipes anchored solidly to wave-cut terraces (in Australia) have been broken apart by large waves breaking on them.

Differential Erosion and Slides: Differential erosion of the bottom material can result in long unsupported pipe lengths.* On the one hand, the static load can pull open joints or overstress the pipe (and/or its jacket). On the other hand, currents or wave-induced water motion can, through the hydrodynamic phenomenon of vortex-shedding, lead to high levels of pipe vibration—with a very great potential for failure. Such failures have in fact occurred in the North Sea (Mes, 1976).

One solution to the problem of erosion along a pipe is to drive pipe-support piles in bottom areas particularly prone to erosion. Alterman (1962), reporting on this procedure, commented that a 60-in.-diam (1524-mm) pipe had been later found 8 ft (2.5 m) "clear" of the bed.

Mass movements of bottom materials in the Gulf of Mexico during storms have caused submarine pipeline failures (e.g., Henkel, 1970; Bea, 1971; Wright and Dunham, 1972; Bea et al., 1975).

General Failure Mechanisms

Examples of Failure: The report by Pearson (1956) listed failures of three United States outfalls as follows:

1. Chula Vista, California, 21-in. (533-mm) reinforced concrete line, laid in trench—joints failed because of settling
2. Ventura No. 1, California, 18-in. (457-mm) steel line with laying details not specified—failed in the surf
3. Wrightsville, North Carolina, 14-in. (356-mm) cast iron outfall on 10-in.-diam (254-mm) piles—failed in the surf

Three major repairs to Honolulu's Sand Island No. 1 outfall (#7 in Table 10-1) took place between its completion in April 1951 and 1975. All three of these problems occurred with the manholes which were 5-ft-high (1.5-m) cast iron cylinders extending up from the outfall pipe. In December 1965 Manhole

---

*One 32-in. (813-mm) line in the North Sea was found to have a 300-ft (90-m) unsupported span length.

No. 1 [water depth 7 ft (2.1 m)] broke off after strong currents removed the overburden, exposing the corroded structure to large wave forces. In May 1966 the bolts holding the cover in place and/or the lugs were broken in Manhole No. 5 in 23 ft (7.0 m) of water. Finally, in February 1971, a flaw in the Manhole No. 2 casting [13 ft (4.0 m) of water] was repaired. Repair of Manhole No. 1 was effected by enclosing the broken manhole with a large-diameter form and filling it with tremie concrete. In the Manhole No. 5 case, the bolts and covers were replaced and covered with concrete. Incidentally, inspections showed that natural cut and fill of the bottom material in about 20 ft (6.1 m) of water accounted for variations of the bottom elevation of up to 4 ft (1.2 m).

There are some details available on failures of outfalls and storm drains in Hawaiian waters other than the Sand Island No. 1 outfall. The tsunami of April 1946 took the exposed storm drain at Hilo, Hawaii, and drove it up on the beach in sections. At the same time the unburied 16-in. (406-mm) outfall at Hilo, held by mechanical anchors, was broken open at many joints. Raw sewage was dumped into Hilo Bay near the beach for many years thereafter until a new line was constructed. In late 1958 storm waves broke open a 60-in. diam (1524-mm) storm drain built near the Kaneohe Marine Corps Air Station. In the mid-1960s waves destroyed the 8-in. (203-mm) unburied outfall at Barking Sands, Kauai, and local beach pollution has continued since that time. At some unknown time the two outfalls at Kahului, Maui, were opened out at various places, possibly owing to wave or tsunami action, and sewage entered the harbor area at that location for many years. The sewage outfall from a hospital at Kapaa, Kauai, broke at the shoreline from unknown causes and caused pollution of the adjacent waters. In early 1969 a joint in a vertical bend in the 54-in. (1372-mm) outfall for Pearl City, Oahu, opened—due to settlement, rotation of the bend, or a dragged ship anchor in Pearl Harbor. Finally, the 48-in.-diam (1219-mm) reinforced concrete outfall at Hilo, Hawaii, suffered a major break in the nearshore portion, in a depth of 25 ft (7.6-m), in 1972. Ten 12-ft-long (3.7-m) sections were displaced. After that break was repaired, divers reported that joints were gradually opening up in the deeper sections as well, primarily in the depths of about 40–50 ft (12–15 m). Repairs in this region were effected in 1977.

Other islands in the Pacific have had their share of bad luck in terms of outfall breakage. At Midway Island early in the 1960s, for example, storm waves broke open two outfalls [8- and 10-in. diam (203- and 254-mm)] in the lagoon. Swimming was thereafter banned in certain areas because of the resulting pollution. In April of 1968 Typhoon Jean caused extensive breakage in two 6-in.-diam (152-mm) outfall lines at Saipan. There has been similar storm-caused damage on Guam. Shortly after Typhoon Pamela struck Guam in May 1976, the three ocean outfalls on the island were inspected by Guam EPA personnel. Damage observed may have been caused by Pamela, an earlier (November 1975) typhoon called June, or perhaps other storms.

The 30-in.-diam (762-mm) cast iron Agaña outfall was still functioning, with the exception of the most offshore right-angled riser nozzle. It discharged raw sewage in a depth of 85 ft (26 m) at the edge of an abrupt drop-off in the sea floor. There was heavy damage to mass concrete poured over the pipe in a depth of about 35 ft (10.5 m), but the outfall was still intact.

In the vicinity of the cast iron Agat outfall there had been several feet of erosion of rocky bottom material, and some sections of the pipe spanned bits of rock. But many sections lay scattered on the sea floor. A secondary effluent of poor quality entered the marine environment through the broken pipe in 6 ft (2 m) of water.

The broken Tipalayao concrete outfall was found to discharge (primary effluent) in the same water depth as the severed Agat line. Two massive boulders on a terrace in about 10 ft (3 m) of water had obviously seen considerable movement during the storm, and there were only a few ground-up remnants of the pipe left within the region of boulder activity.

A large boulder was responsible for breaking open the Otter Rock, Oregon, outfall in the mid 1970's. This was in very shallow water. The boulder fell from a nearby cliff, but it was not clear whether the boulder fell directly on the pipe or whether wave action and resulting motion of the boulder was responsible for the break.

Other Concerns: Pearson (1956) has reported that three California outfalls are known to have been damaged by earthquakes. Other outfalls were extensively damaged in the Alaska earthquake of 1964 (*The Great Alaska Earthquake of 1964: Engineering*, 1973). However, I know of two outfalls whose failure was blamed on natural disasters (one a hurricane and the other an earthquake) in order to secure disaster-relief financing, but in fact the outfalls concerned were in a bad state before the natural calamities.

Even completely unforeseen effects can cause outfall failure. I am aware of one case where the construction contractor left a damaged concrete pipe section some distance off to the side of an outfall and did not remove it after completion of construction. Over a period of years that section was moved by wave action over to the area of the outfall and, acting as a battering ram, succeeded in causing tremendous damage and the complete opening-out of the line.

Concrete-jacketed steel pipes have failed when the concrete separated from the pipe ("Underwater Oil Pipeline...," 1975). The referenced case concerns the same trawl damage situation as mentioned by Guerry (1976), so that the separation may have been given assistance.

There has been no failure in the Goleta, California, outfall, but a potential problem has been observed there. Kelp attaches to the armor stone. Frequently, when the plants are mature and there is sufficient wave-induced water motion, the kelp carries the stone off the pile. Some perspectives on this general phenomenon are provided by Emery and Tschudy (1941) and Kudrass (1974).

Storrie (1947) discussed the breaking of a 72-in. (1829-mm) steel intake pipe by a moving Lake Ontario ice field at spring breakup in 1910 and 1911.

Burial: On the basis of extensive experience with submarine pipelines in the Gulf of Mexico, Krieg (1966) has remarked that the "most successful protection of a pipeline can be obtained by the burial of the pipeline." However, he also has added that "the depths where pipelines need not be buried for stability reasons is [*sic*] ...still unknown." It is now known, in the latter regard, that lines have been destroyed by wave-associated effects in up to 240 ft (73 m) of water. Appreciation of the depth of strong wave effects has taken time in coming. Hyperion Engineers (1957) remarked that for water depths greater than 50 ft (15 m) it was assumed that there were no "serious problems of scour or water forces."

When pipes are simply laid in a trench through the surf zone that is backfilled with native sediment (with or without token, occasional, rock blankets), one may have a dangerous situation. The scouring action of waves and currents can cause shifting zones of cut and fill that bare certain portions of such a pipe and extensively bury other portions. This, for example, is the situation with three lines at the Hawaiian Independent Refinery, Inc., refinery/offshore mooring complex at Barbers Point, Hawaii, 10 miles (16 km) from Honolulu. Experience with a line in New Zealand showed that, in one area within the surf zone, it had actually worked its way down 12 ft (3.6 m) in shifting sand.

All the foregoing serves to indicate that submarine pipes should be protected in zones where wave effects could mean potential pipe instability if it was laid on the bottom or else slightly buried under native material. This protection is normally provided by using large armor rock, which is predicted to be stable under very heavy wave action, to cover over the buried pipe that is laid on a gravel bed [(see Figure 8-5(h)]. Another possibility shown in Figure 8-5(i), is an overlayer of tremie concrete.

Still another option is to place concrete-filled burlap bags over the pipe, as was done at the Otter Rock, Oregon, outfall. Such bags were also placed underneath the Otter Rock line for scour protection. The concrete-clad steel outfall of Georgia-Pacific Paper Co. at Newport, Oregon, had a similar foundation.

8-4 OUTFALL LOCATION AND LENGTH

Introduction

The key decision with regard to outfall location and length is the choice of diffuser position. This location is determined through consideration of a number of factors, some of which are listed below. However, it should be borne in mind that the entire array is not pertinent to every outfall.

1. Area where the required length of diffuser can be fitted to the bottom topography

2. Sufficient depth that the sea water density variation from bottom to surface will mean a submerged sewage field for most of the year

3. A depth less than 200 ft (60 m) to head off large construction diving costs as well as inflated expenses for subsequent maintenance and inspection

4. Location where even strong onshore winds and resulting currents would take sufficiently long to move the wastewater field into nearshore waters that water quality criteria would be met

5. Area of adequate bedding for diffuser and not unduly steep side slopes

6. Position in the way of persistent currents with an offshore component

The point of entry of the outfall into the sea can have a number of determining factors such as the following:

1. Proximity of sewage treatment plant and/or pumping station

2. Ready access to sewer easement or to property owned by municipal body constructing outfall

3. Position adjacent to land area that lends itself as a construction site and staging area for the line

4. Proximity to adequate transportation systems

5. Accessibility to required construction materials

6. Relatively stable area in terms of foreshore morphology

7. Absence of rock outcroppings

8. Ready access to natural submarine channels to deeper water*

The approximate length of an outfall is determined, if one makes the reasonable assumption of a straight line, when the diffuser location and the point of entry of the outfall into the sea are fixed. Thus the factors that influence the choice of diffuser location and orientation largely determine the outfall

---

*Such channels, if they exist, can provide an excellent route for the outfall through the troublesome surf zone. But such channels cannot be found by someone wading. A diver must enter the water and carry out an extensive reconnaissance of the sea floor. Of course, if the area concerned is one of persistent and strong surf, one must bide his time. There are always lulls that can be used for such work, but the data-taking period must allow for this, and the divers must monitor the situation and be ready to enter the water as soon as suitable conditions present themselves. There is no excuse in the critical surf zone area to dismiss obtaining nearshore sea floor information with a remark like "the surf was always too strong for diver work." I am aware of two outfalls, one in Oregon and one in Hawaii, where excellent natural channels, agonizingly close to the chosen alignments, went unused because of purported diver-surf problems.

length. However, there are a few factors that involve *both* items, such as staying within the total head of the system. Other factors will be discussed in subsequent sections.

General Considerations Regarding Pipe and Route

In the preliminary design of an outfall, several pipe sizes are often used, and after overall project cost evaluation for the various pipes, the least costly one is selected for the final design. Briefly, there is an intermediate optimum because although a large pipe can pass larger flow discharges with smaller onshore pumping power expenses, it costs more to purchase, to trench for, and to install. Four pipe sizes were considered for the Orange County No. 2 line until one size (120-in.) (3,048-mm) was shown to be optimum.

The hydraulic design of the pipe/diffuser system was discussed in Section 7-3, and the linkup between pipe size and pumping capacity was implied therein. However, at this point, it will be assumed that the pipe size chosen is not only adequate for carrying the required discharge but also is commercially available. The flow velocities of the chosen pipe must be high enough that the outfall does not act as a settling chamber for whatever suspended solids remain in the flow.

The route of an outfall should, first of all, be along a straight line. There are a number of reasons why this is preferred: the relative ease during construction of keeping on line in terms of both alignment aids and actual pipe laying; all pipe sections are fabricated the same (no beveled or mitered sections); no tricky sea floor anchorages are required at changes in direction. In addition, if there are suspended solids in the effluent, changes in direction can lead to buildup of solids on the pipe wall in the bend.

The line should be routed so that it has a consistent and gentle gradient. Horizontal and adverse slopes should be avoided; buildup of solids can occur at low points and pockets of gas at high points. Regions of outcroppings of rock should be avoided, if possible, because extra expense is involved in blasting such formations and mucking the debris. If cut and fill is necessary to give a consistent line and/or slope, as little filling as possible should be done as this is a highly suspect operation that could lead to pipeline instability under natural environmental influences. Cliffs and singular features such as extensive coral heads, large boulders, and wrecks should be avoided. The chosen route should avoid other submarine pipelines. A peculiarity had the Georgia-Pacific Paper Co. outfall laid by plan right over the tip of the city of Newport, Oregon, sewer outfall.

The line should be buried through the surf zone and in other areas where considerable water movement near the sea floor is possible. Even in quiet waters it may be necessary to bury the pipe to keep it away from ship anchors and trawls. Token burying is not enough to protect a line from either waves or ship anchors, and the depth of burial and the backfill and bedding materials should be carefully chosen. The natural erosion of sea floor material under wave action can bare a submarine pipe if it is not buried deeply enough and if the backfill material is not adequately heavy and/or large. If the outfall is in a ship channel it may be mandatory to bury the line to satisfy regulations (e.g., U.S. Army Corps of Engineers) or to be below the depth of possible future dredging activity.

The Lion's Gate (First Narrows) outfall at Vancouver (#36 in Table 10-1) provides an example concerning future dredging. The line was laid in 1970 in the First Narrows, the entrance to Vancouver Harbor. The design had to take into account future plans for deepening the entrance from 50–60 ft (15–18 m). The surface of the pipe backfill in Figure 8-11 corresponds to the ultimate level of proposed dredging. The trench excavated for the outfall started 10 ft (3 m) above this level, and nominal 1 on 6 wall slopes were specified for the excavation between the bottom level at construction time and the predicted later level. The reinforced concrete cylinders around the two risers and the manhole in Figure 8-11 were to provide some form of protection if a dredge inadvertently strayed over the line—or perhaps also if an anchor was dragged over the line in an emergency. Vancouver has had water lines broken apart in the past by a ship with such a problem in the same general area as the outfall.

Burial of the Lion's Gate diffuser also removed the line from the direct action of currents. Flow speeds in the First Narrows are very large. Values listed in "Canadian Tide and Current Tables: 1976" (1975), for example, show a maximum of 5.7 knots (2.9 m/s).

Finally, submarine pipes that proceed up a steep cliff after leaving the water must have a support pad at the elbow. A 12-in. (305-mm) steel water line at

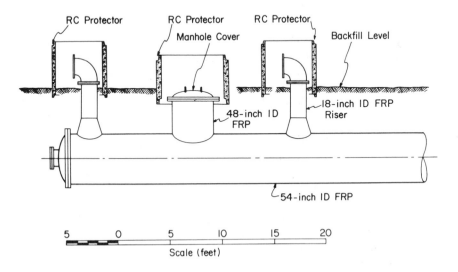

Figure 8-11 End of diffuser for Lion's Gate outfall (Courtesy, Greater Vancouver Sewerage and Drainage District)

South Beach, Yaquina Bay, Oregon, had no such pad and failed at the elbow fitting. The vertical pipe must be well-protected from wave action. The lower part of the Otter Rock, Oregon, vertical section failed in heavy surf even though encased in concrete.

Foundation Stability and Settlement Considerations

A submarine pipe can be regarded as a continuous structure that must not impose more static load on its foundation than can be sustained. In addition, it should be ensured that excessive differential settlements along the line are avoided. Insights into these design issues are provided by Wilson (1970) and Small et al. (1972), and the pipe route should be chosen to satisfy these two criteria.

The ultimate *bearing capacity* for a structure placed on a soil is a subject thoroughly covered in standard soil mechanics texts, e.g., those by Peck et al. (1953) and Hough (1957). The total resistance is comprised of three parts: soil cohesion, solid friction, and *surcharge* (i.e., loading of soil alongside the structure). In order to appraise the foundation stability of a structure, the friction angle, the cohesion, and the specific weight of the soil must be taken into account. These values come from the types of laboratory and *in situ* analyses discussed in Chapter 6.

Structural settlement is a topic also covered in standard soil mechanics textbooks. There can be a complication for outfalls, with environmental loadings leading to increased settlements due to dynamic effects.

For both bearing capacity and settlement considerations, large factors of safety are desirable. This is particularly true for clay bottom materials whose engineering characteristics are never known with precision. Spreading of the outfall load over a wider area, in order to better assure satisfactory conditions, can be accomplished through use of a gravel underlayer. Whatever material supports the pipe should be analyzed from the point of view of liquefaction as well as in terms of standard bearing capacity and settlement considerations (Christian et al., 1974).

8-5 DESIGN WATER LEVELS AND CURRENTS

Water Levels

Introduction: A design water level is required for an outfall in order to determine the peak head to be developed by pumps feeding the outfall. An increased depth over an outfall also permits larger waves to pass over it, and thus information on extreme water levels is important when assessing the stability of the outfall and/or rock cover to wave effects.

The water level without a wave signature is influenced almost purely by the astronomical tides; only at rare times is there an increase in this level due to

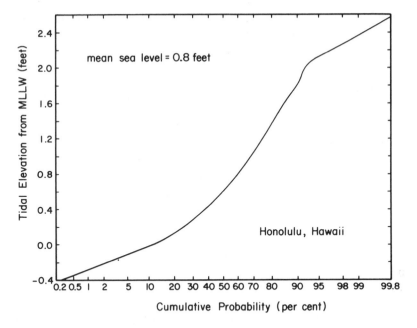

Figure 8-12 Cumulative distribution for astronomical tidal levels at
Honolulu, Hawaii

storms. There are various ways of representing astronomical tidal ordinates for a
given location, and one of these, the cumulative distribution representation, is
shown in Figure 8-12.* Concern over storm-induced increments to the normal
astronomical tidal levels is generally more pronounced in regions where tidal
range is small, such as the Gulf of Mexico, than in areas where the tidal range is
large, such as Darwin, Australia. In either case, however, storm-induced water
superelevations should be considered in the outfall design.

    Large-scale cyclonic storms, involving the pouring of air into a central
low-pressure zone, are often considered in design. Only if the maximum sustained wind speed exceeds 64 knots (119 km/h) is such a cyclone of tropical
origin known as a *hurricane*. It is important to consider such great storms in
coastal engineering design, since they can lead to unusually high superelevations
of the surface, to strong currents, and to monstrous waves. In addition, hurricanes can dump vast amounts of rainfall with attendant risks of flooding to
compound those already possible from the high water levels (*storm tide* or *storm*

---

*Care should be exercised not to confuse coastal chart datum and terrestrial map datum when
assessing outfall discharge depths. Map datum is at approximate mean sea level (MSL); chart datum
is at mean low water in the Atlantic and mean lower low water (MLLW) in the Pacific Ocean. See
Appendix A. Confusion in this regard at Otter Rock, Oregon, where MLLW lies 4.1 feet (1.25 m)
below MSL, resulted in the initial outfall terminating in what was a negligible water depth at very
low tides.

surge) created by a combination of low pressure within the hurricane and strong winds pushing water up against the coast.

A hurricane is a vast storm both in terms of power and areal extent. The pressure builds from the central *eye* out to the fringes of the storm. The difference between the central pressure and the boundary pressure is called the *central pressure reduction* (Δp) and is one of the parameters used to characterize hurricanes. In part it provides an index of peak wind speeds. In all hurricanes, horizontal wind speeds increase from low values in the eye to a maximum just beyond the edge of the eye, then taper off gradually with increasing distance. The average distance from the center of the eye to the circle of peak wind speed is known as the *radius of maximum winds* (R) and has been adopted as a second descriptive parameter for a hurricane—describing its size or lateral extent. The third hurricane descriptor is the velocity of forward movement of the storm, V_f. This, of course, reflects both speed and direction. At any one time an actual or supposed hurricane can be characterized by the above parameters. These can of course vary along the hurricane's path or *track*.

Cyclone Tracy, which demolished Darwin, Australia, on Christmas Day, 1974, was a small hurricane with a central pressure of 940 mb, a radius to maximum winds of only 4 nautical miles (7.5 km), a forward speed of 4 knots (2 m/s), a peak sustained wind of 80 knots (148 km/h), and a maximum estimated gust in excess of 140 knots (260 km/h).

Hurricane Carla that passed through the Gulf of Mexico between September 9 and 12, 1961, is typical of a widespread hurricane. Its central pressure was 931 mb, its radius to maximum winds about 17 nautical miles (31.5 km), and it had a forward speed of 8 knots (4 m/s). Gale force winds [> 34 knots (63 km/h)] extended about 280 miles (450 km) from the eye of the hurricane, and hurricane force winds [> 64 knots (119 km/h)] covered 150 miles (240 km) (Colón, 1966).

Hurricanes take some peculiar paths. Cyclone Tracy followed a general southwesterly course, then turned abruptly through 90° about 80 miles (130 km) outside Darwin and passed directly over that city.

Computing Storm Surge: An accurate way of *forecasting* the spatial and temporal variation in storm surge heights for an approaching storm may be very important, since the water levels determined may indicate that certain areas will be flooded and that the local populace should be moved. Such predictions are carried out for a storm that has been characterized by its central pressure depression, its radius of maximum wind speed, and its velocity (speed and direction) of forward motion. The bathymetric details of the coastal area concerned must be known. Perhaps the best-known of the comprehensive computer models that can be used in such storm surge forecasting is SPLASH (Special Program to List Amplitudes of Surges from Hurricanes), developed by Jelesnianski and others (e.g., Barrientos and Jelesnianski, 1976) and available on magnetic tape from the (U.S.) National Technical Information Service (NTIS,

Order No. COM-75-10180/8WO) for about $300. Jelesnianski and Taylor (1974) have prepared a user's guide for the SPLASH programs.

SPLASH is by no means limited in application to predictions for advancing storms of established characteristics. It can be used as well to compute the spatial and temporal variation in surge heights for a hurricane of assumed characteristics. If a number of storms considered likely to occur over a time of, say, 50 years, is thus considered, the recurrence intervals of storm surge-induced water superelevations can be computed. The U.S. National Oceanic and Atmospheric Administration, National Weather Service (Ho, 1976) has used such computations for the Federal Insurance Administration of the Department of Housing and Urban Development in fixing flood insurance rates. For example, the estimated 50-year storm surges along the coast of the state of Georgia vary from 12.2 ft (3.7 m) at the north boundary to 10.4 ft (3.2 m) at the south, with an intermediate peak of 13.6 ft (4.1 m).

Storm surge-frequency information derived for a particular stretch of coastline where an outfall is to be placed (e.g., Ho and Tracey, 1975) can be used in the design of the outfall. In other cases, storm surge data may have been accumulated for a specific location near a proposed outfall and can be used in the design. In the 1975 design of the Waitara, New Zealand outfall, for example, the report of A. H. Glenn and Associates (n.d.) was used to determine design water levels. The outfall site was 11 miles (17 km) northeast along the coast from the report site.

Rather than use a probabilistic approach to design it is possible simply to use one very powerful hurricane *landfalling** at a particularly bad position in terms of results at the design location. One suggestion for the Gulf of Mexico, for instance, has been to use hurricane Camille in design work. This 1969 hurricane, with central pressure of 908 mb at landfall, was devastating. Peak storm surge was about 25 ft (7.5 m).

The Bathystrophic Storm Tide Model: The bathystrophic storm tide model has been available for some years (Freeman et al., 1957). It does not provide the accuracy of more modern numerical approaches (e.g., Pagenkopf and Pearce, 1975; Pearce and Pagenkopf, 1975) especially in situations of complicated bathymetry and nonideal coastal shapes, and its errors are generally nonconservative.† However, one never knows what really is a proper design hurricane, so that extreme accuracy in assessing the effects of such a storm of specified (but truly error-prone) characteristics may not be warranted.

The bathystrophic model can be applied either by hand calculations or by

---

*This word refers to a hurricane passing from the water to the land. Such a storm produces double or higher the water superelevation of an exiting storm (from land to water) of the same characteristics.

†For example, unpublished computations made for storm surge at Darwin, Australia, showed up to 45% amplification of the open coast surge in coastal inlets in the area.

computer. The report by Bodine (1971) contains a computer program for such determinations, whereas Bretschneider (1967a) details the hand calculation approach. The useful Bretschneider (1967a, c, e) model was entirely two-dimensional. The storm tide was divided into three parts: *pressure* tide, or the barometric effect on the water of the low internal hurricane pressure; *wind tide* due to the wind component directed onshore; and *Coriolis tide* due to currents, probably wind-driven or wind-initiated, along the shoreline. Then the (total) storm tide according to this model is

$$S_{st} = S_{pt} + S_{wt} + S_{Ct} \tag{8-1}$$

where S_{pt} = pressure tide,
  S_{wt} = wind tide, and
  S_{Ct} = Coriolis tide.*
It is convenient for later analysis to write

$$S_* = S_{wt} + S_{Ct} \tag{8-2}$$

There is an effect of water depth and forward speed of the hurricane on the pressure tide, but to a first approximation the pressure tide can be set equal to the height of water that gives the same pressure deficiency that exists within the storm. For example, a hurricane with central pressure depression of 50 mb (~0.05 atm) would lead to an ideal seawater superelevation of about 1.6 ft (0.5 m).

The wind tide occurs because the dragging force of the wind on the water surface must be balanced by a reverse force. This force is derived from water pressure differences, and the water level thus rises in a downwind (assumed onshore) direction.

The Coriolis tide results from the water pressure force differential that must exist across a moving current to give it its actual (global) acceleration.[†] It is assumed that the direction of the current is such as to yield a larger onshore than offshore water surface superelevation.

If one takes Bretschneider's general differential equation for storm surge, assumes steady-state conditions, then integrates using the assumption that the sum of undisturbed water depth (h) plus storm surge increment decreases linearly toward shore from an offshore shelf-break point, it can be shown that

$$S_* = \frac{kW^2 \cos \zeta}{gm} \ln\left(\frac{h_0}{S_*}\right) + \frac{6}{7} f_C \frac{W}{gm}\left(\frac{k}{K}\sin\zeta\right)^{1/2}\left(h_0^{7/6} - S_*^{7/6}\right) \tag{8-3}$$

---

*All symbols used in this chapter are listed and defined in Appendix C as well as at their points of introduction.

†The Coriolis force is described in Chapter 2.

Here S_* [Equation (8-2)] applies where $h=0$,

 h_0 = water depth at the shelf break point (where there is no storm surge)
 k = surface wind stress parameter (suggested value 3.0×10^{-6})
 W = wind speed at 10-m height above the sea surface
 ζ = angle wind makes to the perpendicular to the coast
 g = acceleration due to gravity
 m = sea floor slope
 f_C = Coriolis parameter
 K = bottom friction parameter

A further equation is that

$$f_C = 2\Omega \sin \phi \tag{8-4}$$

where ϕ = latitude
       Ω = angular velocity of the earth (7.29×10^{-5} rad/s).
Also,

$$K = 14.6n^2 \tag{8-5}$$

where n = Manning's roughness factor given in Table 7-1. Suggested values of K are between 0.006 and 0.010 ft$^{1/3}$ (n values of 0.020 and 0.026 ft$^{1/6}$).

Solution for Equation (8-3) is iterative. After values for all known terms are inserted, a value of S_* is assumed on the right-hand side. Calculations are repeated until the value of S_* assumed on the right-hand side emerges as the result of the operation on the left.

I applied Equation (8-3) to predict the peak surge height at Townsville, Australia, during Cyclone Althea. The prediction and the observed maximum were virtually indistinguishable. Others have obtained similar results. When used for simple bathymetries and uncomplicated coastlines, the bathystrophic model of Bretschneider appears to provide reasonable estimates of storm surge.

Currents

Local storms not only generate storm surge, but also large waves and swift currents. The design of submarine pipelines usually takes passage of an assumed large wave as the supreme test. Under such conditions a peak flow speed above 10 fps (3.0 m/s) would not be unusual. The strong currents associated with local storms could amount to, say, 3 fps (0.9 m/s). The design force, dependent on the peak velocity squared, is thus heavily dependent on the additional current speed.

Bretschneider (1967e) has extended his analysis on storm tides—when alongshore and onshore effects are additive—to the associated currents parallel to shore. Let the average alongshore current over the depth h at any time t after process initiation be $\overline{V}$, with $\overline{V}_{st}$ representing the steady-state condition.

Bretschneider's result is

$$\frac{\overline{V}}{\overline{V}_{\text{st}}} = \tanh\left[\frac{Wt}{h}\left(\frac{kK\sin\zeta}{h^{1/3}}\right)^{1/2}\right] \tag{8-6}$$

where

$$\frac{\overline{V}_{\text{st}}}{W} = h^{1/6}\left(\frac{k}{K}\sin\zeta\right)^{1/2} \tag{8-7}$$

Bretschneider (1967e) has also considered more realistic situations than having the wind blow at constant speed W at a fixed angle ζ to a line perpendicular to the coast.

8-6   OUTFALL DESIGN WITH RESPECT TO CURRENTS

Introduction

The strong currents generated during great storms, discussed in the previous section, provide a true test for an outfall installation and should be considered in design. It is likely, however, that such currents would enhance the effects of wave-induced water motion discussed in Section 8-9.

There are many marine areas where strong currents are a daily occurrence. Typically such regions are not along exposed coastlines, where waves could pose problems, and the currents provide the design test for the outfall installation. Examples of such water bodies are Long Island Sound between Long Island, New York, and the state of Connecticut, and the Strait of Georgia and Queen Charlotte Strait between Vancouver Island and the mainland part of the Canadian province of British Columbia.

This section will be divided into two parts. The first part, and most of the material (three subsections), will concern unburied outfalls. The final subsection will cover buried pipes.

Forces on Exposed Outfall: Current Perpendicular to Pipe

The Basic Equations: A flowing liquid exerts a force on an object immersed in it. The component of such a force acting in the line of the velocity vector of the approach flow is called the *drag force*. The equation used to determine this force (F_D) for a liquid of (mass) density ρ and approach flow speed V is

$$F_\text{D} = C_\text{D}\frac{\rho}{2}AV^2 \tag{8-8}$$

where C_D is the *drag coefficient* and A is the projected area of the object as seen by the approaching flow. For a cylinder or pipe of diameter D and length l_* at right angles to the flow,

$$A = Dl_* \tag{8-9}$$

An asymmetrical body or one close to a boundary experiences a steady *lift force* (P_L) perpendicular to the incident velocity vectors. The equation for this force is very similar to (8-8) and is written

$$P_L = C_L \frac{\rho}{2} A V^2 \tag{8-10}$$

where C_L is a *lift coefficient*.

It is convenient to define two dimensionless parameters that will be referred to repeatedly throughout this chapter. The first of these is the *Reynolds number* defined by the equation

$$\mathbf{R} = \frac{VD}{\nu} \tag{8-11}$$

where ν is the kinematic viscosity of the flowing liquid. The second parameter is called the *relative clearance* defined by the equation

$$\mathbf{G} = \frac{\Delta}{D} \tag{8-12}$$

where Δ is the clearance between the edge of the pipe and the boundary.

It is necessary to consider the case of nonzero pipe clearance because not all exposed outfalls are laid along the sea bed. For example, the 144-in.-diam (3658-mm) Los Angeles City, California line built at Hyperion in 1948 and placed on saddles had a 5-ft (1.5-m) clearance between the pipe and sea floor. Other lines with designed clearance between the pipe and the bottom have been located as follows: Gentofte, Denmark; Tipalayao, Guam; the Weyerhaeuser Co. installation at Longview, Washington; and the nuclear power plant at Rainier, Washington. The latter two pipes, both in the tidal reach of the Columbia River, were supported by timber frames.

Values for the Coefficients: It is well established that the steady flow drag coefficient for a circular cylinder perpendicular to the flow far from boundaries depends upon a multiplicity of factors that includes Reynolds number, roughness of the cylinder surface, the turbulence intensity in the flow and the scale of the eddies, and whether or not there is vibration of the cylinder (Schlichting, 1960; Blumberg and Rigg, 1961; Hoerner, 1965; Jones, 1968). For large Reynolds numbers, i.e., values that are encountered in engineering design situations, the drag coefficient for a stationary cylinder becomes independent of Reynolds number and, apparently, the turbulence characteristics of the flow, at a value dictated by the roughness of the surface. For a very rough surface, typical of surfaces in the marine environment, the drag coefficient attains a value of 1.02. This value was obtained by Blumberg and Rigg (1961) for Reynolds numbers up to 6×10^6.

There is substantially less drag coefficient information available for cylinders maintained close to a boundary than there is for cylinders remote from a wall. The pertinent work, in terms of high Reynolds numbers, of Jones (1970, 1971, 1976), Beattie et al. (1971), and Littlejohns (1972) has turned up a C_D value of about 0.6 to 0.9, but these tests did not involve the same high roughness as in the tests of Blumberg and Rigg (1961). Since it turns out that the value of C_D for a pipe appears independent of the relative clearance of the pipe from the wall, I consider a design value of $C_D = 1.0$ valid for pipes perpendicular to a steady current, regardless of the relative clearance.

By contrast, the proximity of the pipe to the boundary makes a sizable difference to the design value of the steady-state lift coefficient. Based on available data, it is proposed that the curve shown in Figure 8-13 be used for pipes subjected to steady flows perpendicular to their lines. As a parenthetical remark, it should be noted that lift coefficients for smooth pipes are higher than those for rough pipes, but Figure 8-13 reflects the more conservative situation.

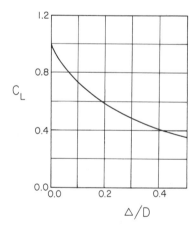

Figure 8-13   Variation of steady flow lift coefficient with relative clearance of pipe perpendicular to flow

Despite the trend shown in Figure 8-13, it is clear from extensive steady, uniform flow tests on lift coefficients for circular cylinders that proximity to a wall makes little difference to the *peak* lift force experienced through the shedding of vortices by the cylinder and the resultant oscillatory cylinder loading (see, for example: Humphreys, 1960; Gerrard, 1961; Jones, 1968; Sarpkaya, 1976). This high-frequency forcing function should not be disregarded in design for pipes some distance from the bottom.

Forces on Exposed Outfall at Angle to Current

When a steady flow approaches a pipe at some angle (θ) other than the perpendicular ($\theta = 90°$), it is convenient to think of the horizontal force on the pipe perpendicular to its line rather than the drag force. Let this force be

$$F_{\mathrm{H}} = C_{\mathrm{H}} \cdot \frac{\rho}{2}(Dl_*)V^2 \tag{8-13}$$

Note that the area being used is still Dl_* and that the flow speed employed is that of the flow rather than the component perpendicular to the pipe. This makes the forms of Equations (8-13) and (8-8) compatible.

The force away from the boundary for the angled pipe can be written

$$P_{\mathrm{V}} = C_{\mathrm{V}} \frac{\rho}{2}(Dl_*)V^2 \tag{8-14}$$

The ratios $C_{\mathrm{H}}/C_{\mathrm{D}}$ and $C_{\mathrm{V}}/C_{\mathrm{L}}$ have been plotted in Figure 8-14. These curves were derived by Grace (1973) using data from several sources. The reference value for C_{D} is 1.0 as outlined earlier. The reference value for C_{L} can be determined using Figure 8-13.

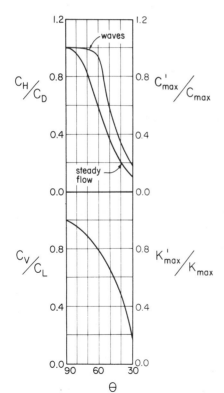

Figure 8-14 Proposed design curves for outfalls of any clearance, angled to currents or wave fronts

Sliding Stability of Pipe

The effective weight of a pipe underwater, the *buoyant weight*, is the weight of pipe wall and contents less the weight of water displaced by the exterior volume of the pipe. Let the buoyant weight of a pipe of length l_* be W_B.

Consider a current of speed V moving perpendicular to a pipe of diameter D resting on the bottom. The flow forces involved are the drag and lift forces (F_D and P_L); other forces are the buoyant weight (W_B) and the *normal force* (F_N) between the bottom and the pipe.

The resistance of a body to sliding under a lateral force cannot exceed the product of the normal force and a *coefficient of static friction* or *friction coefficient*, μ_s. Lyons (1973) has reported on the friction coefficient between a submarine pipe and the bottom for steady applied forces. A range of values was obtained depending upon the finish of the pipe and the nature of the sea floor. Both sand and clay surfaces were used. Coefficient values for sand varied from about 0.6 to 1.4, with those for clay from about 0.2 to 0.7. The friction coefficient was independent of the normal force for sand, but not for clay.

A static force balance for the case of a pipe on the verge of moving under a perpendicular current yields

$$W_B = \frac{F_D}{\mu_s} + P_L \tag{8-15}$$

$$\equiv \left(\frac{C_D}{\mu_s} + C_L\right) \frac{\rho}{2} (Dl_*) V^2 \tag{8-16}$$

The buoyant weight has been placed on the left-hand side of the above expression because in practice this is the quantity that is adjusted to achieve stability. For steel pipes this is done by applying a concrete *weight coating* to the outside of the pipe. Note should be taken, however, of the following two facts: adding concrete to the exterior of a steel pipe will change μ_s; increasing D may change C_D and C_L and will certainly increase the flow forces. If it is not possible to achieve stability through weight, the pipe can be stapled to the seabed (Chapter 10), can be set in an open trench, or can be buried.

Stabilizing an Outfall

Outfalls can be stabilized in an exposed condition through use of the arrangement shown in Figure 8-5(c). The vertical piles can be either screwed into the bottom or grouted into a drilled hole. Engineering calculations for the required pullout resistance of the piles, and their longitudinal spacing, can be effected by summing moments for a force distribution derived for a design wave.

Outfalls can also be stabilized by burying them as in Figure 8-5(f)–(i). In this case the size and specific gravity of the cover material must be adequate to resist erosion by the current that boded ill for the unburied line in the first case.

"Shore Protection, Planning and Design" (1966) contains information on this problem. The design curve for *embedded* particles (i.e., many particles grouped together) of specific gravity 2.65 and a horizontal bottom is

$$V_i = 12.2d^{0.5} \qquad d \geqslant 0.01 \text{ ft} \qquad (8\text{-}17)^*$$

where V_i is the minimum (steady) flow speed (fps) required to erode a particle of size d (ft).

8-7 WAVES

Wave Generation and Decay

The characteristics of wind waves in deep water are dependent on the wind speed, the length of time the wind blows, and the over-water distance the wind traverses to the point in question. The latter is called the *fetch*. Bretschneider (1967d) has published a series of curves that permits the determination of the significant wave height and the associated period from the three independent variables. Slight revisions to the forecasting curves were made later (Bretschneider, 1970).

Wave determinations using the plot referred to above are either *forecasting* or *hindcasting*. Forecasting refers to the prediction of wave conditions in the future from derived meteorological information, whereas hindcasting relates to the determination of supposed wave conditions for a time in the past. Forecasting could be very important in construction operations, as will be discussed in Chapter 10, but for the purposes of design the situation is hindcasting.

A *synoptic* (instantaneous) weather chart (Figure 8-15) provides the basis for wave hindcasting. The configuration of imaginary lines of constant pressure shown on the chart (*isobars*) provide a clue to the winds, and the computed wind speeds are used to determine wave conditions.[†] The synoptic chart in Figure 8-15 represents the surface situation in the northern Pacific at 0000 Z (0 hours, according to Greenwich, England, time) on Nov. 10, 1976. The letters "H" and "L" indicate high- and low-pressure areas. The contour lines represent pressures in millibars (mb), with the initial 9 or 10 omitted. A 4-mb spacing is used for the isobars. The sweeping line with the black triangular-like additions represents a *cold front*—a region of advancing cold air. An *occluded front* (separated from the surface) is shown to the right of the central low. The circles and attached lines represent the wind velocity at measuring points—ships and shore stations. The direction of the radial line indicates the wind direction (toward the circle). The

---

*This becomes $V_i = 0.213d^{0.5}$, $d \geqslant 3$ mm, for V_i in m/s. and d in mm.

[†]Ship reports may provide some information on wind conditions in a wave-generating area, but these do not give comprehensive coverage, especially in storm-prone areas ships would likely avoid. Hindcasters do not normally put too much stock in the ship observations of winds, using them only as an approximate check, and even less in ship observations of wave conditions.

barbs on the line indicate the wind speed. Each full attached line indicates 10 knots (18.5 km/h), each half-line denotes 5 knots (9.3 km/h). Thus the wind at Midway Island was blowing from the southwest at about 15 knots (27.8 km/h). The flag on a wind direction line indicates 50 knots (92.7 km/h).

Pressure gradients can be determined from the spacing of the isobars on the synoptic chart, and the wind resulting from the two factors of pressure gradient and latitude (representing Coriolis force) is known as the *geostrophic wind*. Such an idealized wind blows parallel to isobars. The surface wind is estimated from the geostrophic wind by considering both the radius of curvature of the isobars and atmospheric stability. Owing to surface friction, the open ocean surface wind blows across the isobars at about a 20° angle—toward low pressure.

The first characteristic for a fetch chosen on the synoptic chart is that it must be largely aimed at the location for which wave information is desired. Care should be taken to differentiate between straight lines on a Mercator chart (rhumb lines) and the great circles along which waves propagate. The selection of fetch is one of the most subjective decisions in wave hindcasting. The fetch

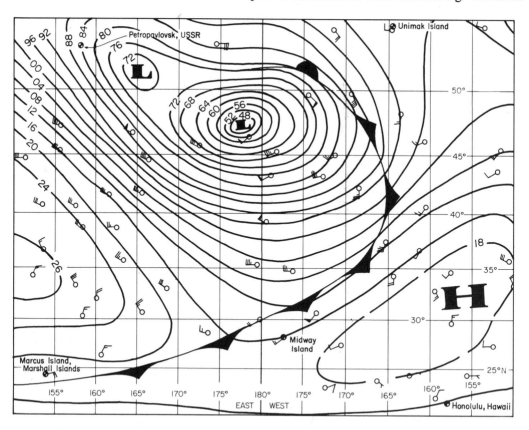

Figure 8-15 Example synoptic chart

length is normally taken as a distance slightly across a moderately straight isobar, terminated at either end when the wind direction exceeds 15° from the average, or when the wind speed is 5 knots (2.5 m/s) from the mean.

The day-to-day (or 6-h to 6-h) history of isobar patterns permits the hindcaster to determine the time variation of wind speed. With information on wind duration, fetch, and wind speed, the significant wave height at the downwind end of the fetch and the associated wave period can be determined. The resulting wave energy spreads both laterally and downwind. Longer waves outrace shorter ones in a process known as *dispersion*. Decay in wave height as the longer waves pass downwind results mainly from spread of the energy rather than frictional effects. For computational purposes the hindcast point should generally lie within a 30° angle from the dominant wind direction in the fetch for it to be considered directly downwind. For big seas, however, appreciable swell energy may propagate at angles of 40°, 50°, or more from the main wind-generating direction. All other things being equal, the waves generated by wide fetches decay more slowly than those due to narrow fetches because of reinforcement.

The above discussion is simplified to the extent that it does not consider preexisting waves, fetches moving, especially at an angle to the fetch (Wilson, 1955), and wind speed changes with time. However, such additions do not pose undue problems and the same curves can generally be used.

The decay of wave heights and changes in effective wave periods as swell propagates can be predicted using curves in Bretschneider (1968). The result involves significant swell heights, and associated wave periods, in deep water offshore from the point at which the hindcast is desired. If the hindcasting point is located at the end of the fetch considered, there is of course no decay involved.

Shoaling, Refraction, and Reflection of Waves

Few coasts display bottom slopes exceeding 1 : 30. Data from Greslou and Mahe (1954) show that, for deep water wave steepnesses in the range $0.005 \leqslant H_0/\lambda_0 \leqslant 0.05$, the wave reflection coefficient is less than 5%. There is not sufficient accuracy in wave prediction, and in shoaling and refraction (to be discussed below), to warrant concern over this small percent, and in most cases wave reflection can be ignored.

As waves approach a coastline their heights change, and they finally break in a water depth theoretically 1.28 times the wave height at breaking *for small foreshore slopes*. The wave height alterations between deep water and breaking can be considered as dependent on three effects. The first effect, called *shoaling*, is due to the decreasing depth. The second is due to bending of the wave front, variations in depth along its length, and resultant focusing and defocusing of the wave's energy; this is called *refraction*. The third influence comes from the bottom itself—energy-dissipating effects such as friction or flow into and out of the bottom material (Bretschneider and Reid, 1954). Although this latter effect can be nonnegligible for very long wave travel distances over shallow depths, such as in the Gulf of Mexico, it can normally be ignored.

The wave height H in any depth h can be computed by the equation

$$H = K_s K_r H_0 \equiv K_H H_0 \qquad (8\text{-}18)$$

where H_0 is the deep water wave height, K_s is a shoaling coefficient, and K_r is a refraction coefficient. Tables in Wiegel (1964), "Shore Protection, Planning and Design" (1966), or other publications can be used to determine K_s. These tables assume the validity of the linear or Airy wave theory (e.g., Ippen, 1966), and K_s is a function solely of the dimensionless parameter $T\sqrt{g/h}$ where T is the (constant) wave period and g is the acceleration due to gravity.

An *orthogonal* is an imaginary line drawn perpendicular to the crest of a wave at any point. Let the shortest distance between two closely spaced orthogonals to a wave in deep water be represented by b_0. As this wave passes toward shore it is possible to visualize the history of the pair of orthogonals. If b_* is the perpendicular distance between them at any depth h, then the refraction coefficient at that depth can be shown theoretically (e.g., Ippen, 1966) to be

$$K_r = \sqrt{b_0/b_*} \qquad (8\text{-}19)$$

The techniques of graphically constructing a refraction diagram for a marine area described by its bathymetry are given by Wiegel (1964) or Silvester (1974). Such approaches give not only the refraction coefficient, by considering two orthogonals, but the direction of wave propagation at any point along a chosen orthogonal. Where large expanses of water are to be studied and/or the bathymetry is complicated, it is more convenient to use computer programs that have been developed to obtain the same information (Griswold, 1963; Goldsmith et al., 1974). Comprehensive programs of this nature combine the effects of both shoaling and refraction, and program input consists basically of a grid point representation of the bathymetry of the area and the direction of deep water wave approach (e.g., Worthington and Herbich, 1971).

Although the approach given above may provide physically reasonable refraction patterns well outside the surf zone, the results for waves nearing their break points may leave much to be desired (Walker, 1974).

Observed Conditions

The swell from the storm shown in Figure 8-15 was measured by a pressure sensor (Section 5-4) in 80 ft (24.4 m) of water on the stretch of Oahu, Hawaii, coastline facing northwest. Using the data obtained, applying the Airy theory wave pressure factor and an empirical coefficient (Figure 7-8), and doing some shoaling calculations yields a maximum breaking wave height of 37 ft (11.3 m). This wave had an 18-s period, and a theoretical 32-ft (9.8-m) deep water wave height.

The Design Wave

A possible but rare wave is usually selected and the whole outfall system designed to be stable under its attack. This *design wave* is often very large. For the Sand Island No. 2 outfall off Honolulu, Hawaii, the period and deep water height of the design wave were 11.7 s and 46 ft (14 m), respectively.

The term *recurrence interval* refers to the average time between occurrences of specific independent events. This could concern, for example, a wave height above 33 ft (10 m). In the design of coastal structures it is not unusual for a 50-year event to be considered. Since wave-measuring efforts at specific stations do not extend over time spans even approaching 50 years, gathered data must be extrapolated. The standard approach. is to plot the annual maximum values on special graph paper derived from a specific ("asymptotic extreme value") probability distribution. One common form of these distributions is the Type II (Fréchet) (e.g., Benjamin and Cornell, 1970).

The Fréchet distribution has been used by Thom (1971) in deriving rare wave information for locations in the Atlantic (nine) and Pacific Oceans (three) occupied by "ocean station vessels" positioned on airlanes. There were 152 total values of observed annual significant wave heights considered, and each station's values were extrapolated to various recurrence intervals, e.g., 50 years, to give the "maximum significant wave height." An "extreme wave height" for each station and recurrence interval was obtained by multiplying the associated maximum significant value by 1.8.

This same factor was used to relate extrapolated maximum significant and extreme wave data by Quayle and Fulbright (1975) who used a form of the Fréchet distribution to derive rare wind and wave predictions for 52 areas, chiefly along the U.S. coast. Recurrence intervals to 100 years were involved. The wave predictions of Quayle and Fulbright were based on local climatological data rather than wave observations. Hindcasting has the advantage that it provides a larger sample to extrapolate than wave observations. Whether or not the hindcasting procedure gives adequate results should be tested by comparing observed and hindcast conditions during periods of actual wave measurement at the site.

Swell can, of course, propagate great distances (e.g., Snodgrass et al., 1966) and reach enormous size (e.g., Rudnick and Hasse, 1971). There are many locations where swell may constitute a more serious design condition than local storm waves.

Short-term local wave data may have their chief use in dictating outfall construction methods as well as anticipated delays in construction work due to unworkable seas. The conditions giving rise to a true design wave may be completely different from those causing the normal run of waves. A rare local hurricane is an example.

Bretschneider (1973) has discussed in detail his derivation of the design wave for the Sand Island No. 2 outfall (mentioned earlier) on the basis of an imagined hurricane located in a critical position about 20 miles (32 km) southwest of the outfall area. The hurricane used, of presumed 50-year recurrence interval, had a central pressure reduction of 34 mb, a radius to maximum winds of 20 nautical miles (37 km), and a northwest velocity of advance of 12 knots (22.2 km/h).

Details on the generation of waves by hurricanes are available in Wilson (1957), Bretschneider (1972a, b), and Cardone et al. (1976). Borgman (1973) has discussed the statistics of such waves.

8-8 WAVE-INDUCED, NEAR-BOTTOM WATER MOTION

The Theoretical Basis

Theory and Reality: Two flagrant idealizations are basic ones underlying what is known as the *Airy* or *linear* wave theory. The first of these is that the two-dimensional water surface is sinusoidal in time and space; the second is that the water is *inviscid*, a term indicating that there is an absence of viscosity (Ippen, 1966; McCormick, 1973). According to the Airy theory the water motion just above the sea floor is parallel to it and takes the form of *simple harmonic motion*, the planar projection of a sine wave. According to the theory the water particles attain a maximum flow speed in the direction of movement of the surface wave feature under its crest, and they reach the (same) peak flow speed in the reverse direction under the trough. This peak flow speed is uniquely determined by specifying the wave height (H), the period (T), and the water depth (h). See Figure 8-16.

The two major assumptions on which the Airy theory is based are patently untrue. With respect to the first of these, waves in coastal waters display neither a constancy of time from wave crest to wave crest nor a constancy in the series of wave heights. In addition, the water surface has long shallow troughs and peaked crests as in Figure 8-17. Further, the real water motion near the sea floor under shoaling waves is parallel to the bottom, but it is by no means simple harmonic motion. Flow speeds under the long troughs develop relatively slowly and peak at a value less (and often considerably less) than the peak flow speed under the crest. This maximum flow speed is very quickly approached, as befits the short time span of the crest.

One would suppose from the discussion of the assumptions underlying the Airy theory that its prediction of the real peak flow speed, near the bottom, under a shoaling wave would be grossly in error. But it is remarkable that this is by no means the case. The Airy theory does a very good job of predicting this maximum speed, even under near-breaking waves (LeMéhauté et al., 1968), with

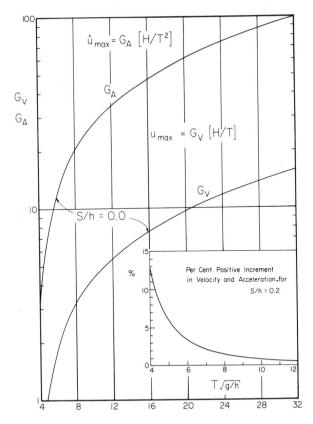

$$\dot{u}_{max} = G_A \left[H/T^2 \right]$$

G_A

$$u_{max} = G_V \left[H/T \right]$$

G_V

S/h = 0.0

Per Cent Positive Increment
in Velocity and Acceleration, for
S/h = 0.2

%

$T \sqrt{g/h}$

Figure 8-16 Airy theory prediction of near-bottom maximum water particle flow speed and acceleration

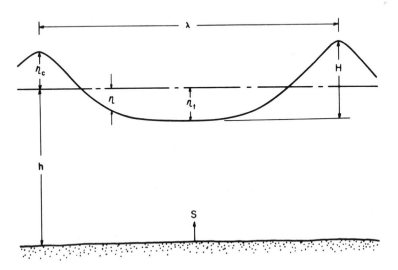

Figure 8-17 Wave profile

the following qualifications. The Airy theory works only in an average sense. It is an experimental fact that a number of waves having the same height, period, and depth do not produce the same maximum speeds. This has been clearly shown both in the laboratory (Goda, 1964) and in the ocean (Grace, 1976). Upon contemplation this is not surprising, since there is an infinity of possible water surfaces that will lie within the vertical constraints imposed by a specified wave height and the horizontal constraints determined by a specified wave period or, equivalently for the Airy theory, a specified wavelength.*

This same observation can be used to predict that peak, near-bottom flow accelerations under waves are not the same just because the waves involved had the same H, T, and h. This is true, and the relative dispersion of the data in this case is more severe than in the case of the flow speeds. The mean behavior of such accelerations is not, however, adequately predicted by the Airy theory (Figure 8-16), the chief reason being the fact, alluded to earlier, that the duration of the wave crest is considerably shorter than half the time between successive wave crests.† Thus the time involved in reaching the real peak flow speed (well-predicted by the Airy theory) is considerably less than that involved in the linear theory. Thus the Airy theory should err on the low side in predicting mean behavior, and it does.

Design Computations: Throughout the discussion on wave-induced water motion to follow, the Airy theory will be used to determine initial estimates of peak velocities and accelerations. Factors will be applied to these kinematical values in order to make them more consistent with extreme values found in the ocean. It will have been noted here by those familiar with more complicated classical wave theories, providing surface profiles closer to those actually observed in the sea (Figure 8-17), that these have been ignored. Such theories are inferior to the Airy on two counts: first, they require much greater expenditure in effort to determine the required kinematical values; second, such predictions are less accurate (and in some cases considerably less accurate) than those of the linear theory (LeMéhauté et al., 1968). There is thus no reason for employing them.

A complicated numerical wave theory that appears to be slightly more accurate than the Airy theory in predicting the average behavior of peak velocities and accelerations under design-type waves is the *stream function theory* (Dean, 1965, 1974; Dalrymple, 1974). Dean (1974) gives tabulated kinematical

---

*Because of all the variability, probabilistic (or *stochastic*) wave models have been developed. In my view such models, used for evaluating wave effects on stationary pipelines and on sea floor materials, are inferior to the *deterministic* (or periodic) type of model.

†The time between crests will be called the "period" in what follows even though it is understood that this time changes somewhat from wave to wave.

data (maximum) for this theory, but only for four ratios of wave height to breaking wave height.

Wave-Induced Kinematics in Context

Introduction: The outfall itself and its protective rock matrix must be designed to be stable under water motion associated with the design wave. Strictly speaking, this statement is true only in water depths seaward of the point at which the design wave breaks. The reformed design wave (after breaking) would be smaller than would another wave, smaller than the design wave in deep water, but just on the verge of breaking in whatever inshore depth considered. It is then standard practice to assume breaking wave conditions shoreward of the break point of the design wave. But there is a peculiarity in this respect. It has always been implicitly assumed that the waves are just about to break rather than in the process of breaking. It can be imagined that the latter can cause stronger water motion and greater forces than the former, and this has been generally observed to be the case (e.g., Grace and Castiel, 1975). However, at this time it will be assumed that nonbreaking waves are involved. Correction factors can be applied later to allow for the fact that incipiently breaking waves are actually breaking waves.

I find it convenient to visualize that a train of supposedly "identical" ocean waves will yield a sample of peak flow speeds of from 0.6 to 1.5 times that value predicted by the Airy theory (Figure 8-16) on the basis of the three wave variables H, T, and h. The range in accelerations is from about 0.6 to 2.5 times the Airy theory peak acceleration (also shown in Figure 8-16).

It is convenient, as will be discussed in the following section, to consider the following dimensionless parameter in the determination of wave forces:

$$\psi = U_{\max}^2 / \dot{U}_{\max} D \qquad (8\text{-}20)$$

In Equation (8-20) $U_{\max}$ and $\dot{U}_{\max}$ correspond, respectively, to estimates of the real peak flow speed and acceleration under the design wave and D is the pipe diameter. It will be shown in Section 8-9 that total maximum pipe force coefficients, both in the horizontal and vertical directions, are functions of ψ, falling off with increasing values of the parameter.

Design Conditions: It is convenient for computations in Section 8-9 to use the following equation for ψ:

$$\psi = \beta_3 \left[\frac{u_{\max_{\text{Airy}}}^2}{\dot{u}_{\max_{\text{Airy}}} D} \right] \qquad (8\text{-}21)$$

where β_3 is an empirical factor.

8-9 UNBURIED OUTFALL DESIGN AGAINST WAVES

Introduction

The offshore portions of many outfalls are exposed in one of the configurations shown in Figure 8-5(a)–(e). The stability of such lines to currents has already been considered in Section 8-6. This section concerns the stability of exposed outfalls to wave effects. Many engineers appear to feel that an exposed submarine pipeline is out of danger with respect to waves if it is in 100 ft (30 m) or more of water. I would like to stress that this is not necessarily the case. Submarine pipelines have failed during storms in depths to at least 240 ft (72 m). The stability of a proposed exposed outfall should be checked in all depths.

Most of the data that have been obtained on flow forces on pipes have been for the current or the waves moving at right angles to the line of the pipe. Although this orientation would not be unusual for a pipe in a current, it is unusual for an outfall. Outfalls are normally oriented so that they largely parallel the average direction of approach of the waves partly because the wave effects on the pipe are lessened. Thus for waves a range of *angles of attack*, or the angles between velocity vectors and the line of the pipe, must be considered.

Some data on wave force coefficients for submarine pipes have been obtained through small-scale laboratory tests. Such work may yield results that are valid in the field when inertia forces overwhelm drag forces [see Equation (8-22)]. But the design of submarine pipelines virtually never conforms to this requirement, and in cases where the drag force component is appreciable, and often dominant, small-scale laboratory modeling of real wave forces is an exercise in futility (see Rance, 1969).

I believe that wave coefficient data should be obtained by experiments in the sea. For one thing, that is where the structure will have to be placed. For two years, starting in September 1974, a team under my direction worked in the ocean to obtain wave force coefficient information for a pipe set parallel to the wave fronts.* We used a 16-in.-diam (406-mm) steel pipe 17.5 ft (5.3 m) long. A 39.5-in. (1-m) segment of the pipe was set up as the force sensor. A current meter set near the pipe measured the wave-induced water motion and a wave staff provided wave height and period information. The site was in a 37 ft (11.3 m) depth in coastal waters offshore from Honolulu. The results of this research will be used as appropriate in the following sections.

We measured waves up to 13 ft (4.0 m) in height and encountered even larger ones that we were unable to measure. Wave periods were typically 12–17 s. We often encountered peak flow speeds of 5 fps (1.5 m/s) and above while working on the bottom. Those who have not experienced flow speeds of this order, with the accompanying strong water accelerations and decelerations,

---

*It was envisioned that a later investigation would involve the pipe angled with respect to the wave fronts.

can be assured that it is all a diver can do to hold on to a firm anchorage point. I mention this because designers may tend to neglect such flow speeds when checking on pipe stability in fairly deep water. Even a flow speed as apparently trivial as 5 fps (1.5 m/s) can exert surprisingly large horizontal and vertical forces on an outfall.

Wave Forces: Pipe Parallel to Wave Fronts

The Morison Equation: Let the total wave-induced horizontal force on a pipe set parallel to the wave fronts (thus perpendicular to the water motion) be represented by F. The Morison equation is written

$$F = F_D + F_I \tag{8-22}$$

where F_D is the drag force encountered earlier and F_I is an *inertia force*. The equation for the drag force in the case of waves is

$$F_D = C_D \frac{\rho}{2} A U |U| \tag{8-23}$$

The product of the flow speed and its absolute value preserves the proper sign of the applied force (that would be lost by squaring the speed).

The peak value of U is U_{max}. For wave loading situations the Reynolds number is defined by

$$\mathbf{R}_w = \frac{U_{max} D}{\nu} \tag{8-24}$$

Whatever the value of $\mathbf{R}_w$, it should not be confused with $\mathbf{R}$ [Equation (8-11)] in picking force coefficients.

The Inertia Force: The equation for the inertia force is

$$F_I = C_I \rho \Psi \dot{U} \tag{8-25}$$

where C_I is an *inertia coefficient*, $\dot{U}$ is the actual flow acceleration, and Ψ is the pipe volume where

$$\Psi = \frac{\pi D^2}{4} l_* \tag{8-26}$$

Thus the inertia force is an acceleration-dependent term. Since accelerations are very much a part of wave-induced water motion, such a term is an indispensable inclusion in the general wave force case. The inertia coefficient can be written as

$$C_I = (1 + C_m) \tag{8-27}$$

where C_m is an *added mass coefficient* (Robertson, 1965; Valentine, 1959). C_I is a function of the relative clearance of a pipe from a boundary. For ideal fluid flow, it has been shown that C_I drops from the value 3.29 for the pipe against the boundary to the asymptotic value 2.00 for the pipe remote from the boundary (Dalton and Helfinstine, 1971) (see Figure 8-18). In practice, real fluid effects mean that the ideal fluid flow values for C_I have to be adjusted.

Maximum Horizontal Forces: The submarine pipeline designer seeks the maximum wave-induced force exerted on his (static) pipe during the passage of a prescribed design wave. Although the Morison equation remains the premier practical approach to general wave force problems, I have found in my own ocean research on wave-induced forces on submarine pipelines that there is an apparently better way to proceed in this case. Let

$$F_{max} = C_{max} \frac{\rho}{2} A U_{max}^2 \qquad (8\text{-}28)$$

The upper envelope curve to our ocean data is shown in Figure 8-19. The parameter ψ [Equations (8-20) and (8-21)] chosen reflects both the velocity and acceleration aspects of the flow, and it is to be emphasized that the kinematical values shown are the designer's best estimates of the real peak kinematical

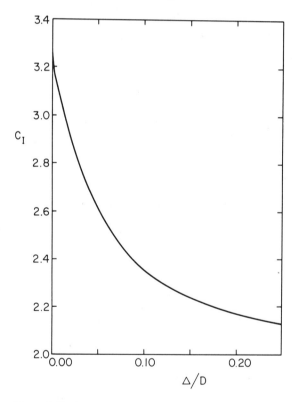

Figure 8-18   Variation of inertia coefficient with relative clearance for an ideal fluid

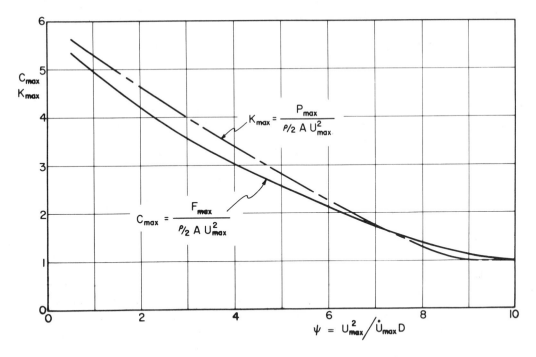

Figure 8-19   Proposed design curves for outfalls at any clearance, parallel to wave fronts

quantities, by no means simply the raw estimates of some pet wave theory. We found in the ocean research referred to (experimental scatter apart) that the C_{max} data were independent of Reynolds numbers and pipe clearance. The latter is not surprising from the discussion of steady flow results in Section 8-6; the former is very reasonable considering the relatively high Reynolds numbers encountered in our research (to 8×10^5) and the high levels of turbulence present. It is believed that this highly turbulent flow field (the pipe is always being subjected to its own turbulent wake) also is responsible for another striking circumstance we noted in our work: the upper envelope for C_{max} shown in Figure 8-19 is independent of the roughness of the pipe. It is proposed that the C_{max} line shown in Figure 8-19 be used in the design of submarine pipes parallel to the wave fronts. Details will be provided later.

For those cases where the value of the parameter ψ is below 0.5, the peak horizontal wave force can be computed simply from the expression

$$F_{max} = C_I \rho \Psi \dot{U}_{max}, \qquad \psi < 0.5 \tag{8-29}$$

where C_I is chosen from Figure 8-18.* For small ψ the inertia force contribution to the total force is dominant. For large ψ, the drag force is all-important.

---

*It is assumed that the pipe size is not so large that diffraction effects become important (e.g., Garrison and Rao, 1971).

Maximum Vertical Forces: Let

$$P_{max} = K_{max} \frac{\rho}{2} A U_{max}^2 \qquad (8\text{-}30)$$

The variation of K_{max} with ψ is shown in Figure 8-19. Again it is suggested that this curve be used in design, along with concepts given in the following section. In our ocean research on wave forces on a pipe parallel to the wave fronts, two different relative clearances were used: 0.03 and 0.19. Although it was clear that the average tendency was for vertical force coefficients at the lower Δ/D to be higher, it was also clear than an upper envelope curve for the two cases was virtually the same. In light of vortex-shedding results cited earlier for steady flows, this result is not totally surprising. Thus the curve suggested for peak vertical wave forces on pipes in the ocean does not reflect the value of Δ/D.

Design Computations: If 10 ocean waves having the height, period, and perpendicular approach direction of the design wave were to move past the pipe in a specific water depth, 10 different F_{max} and P_{max} values would be measured. The object of the design is to pick a reasonable representative of these forces. Since the curves in Fig. 8-19 are upper envelope curves, they already contain a factor of safety. For that reason it is suggested that mean values can be used for kinematic terms used in the force computations.

The design horizontal and vertical pipe forces are given by Equations (8-28) and (8-30), respectively. These can be rewritten as:

$$F_{max} = C_{max} \frac{\rho}{2} (Dl_*) \beta_1 u_{max_{Airy}}^2 \qquad (8\text{-}31)$$

$$P_{max} = K_{max} \frac{\rho}{2} (Dl_*) \beta_1 u_{max_{Airy}}^2 \qquad (8\text{-}32)$$

where β_1 is a correction factor. Based on our ocean research work, it is suggested that β_3 [Equation (8-21)] be taken as 0.86 and that β_1 [Equations (8-31) and (8-32)] be set at 1.08.*

---

*The distribution of (222) β_3 values was normal with expectation 0.86 and standard deviation 0.30. There was no demonstrable correlation of β_3 with the Ursell parameter

$$U = \frac{H/h}{(h/\lambda)^2} \qquad (8\text{-}33)$$

The correlation coefficient between β_3 and β_1 was 0.53, and a regression line

$$\beta_1' = 0.56 + 0.61 \beta_3 \qquad (8\text{-}34)$$

gave a standard error of estimate of 0.29.

The design procedure then begins with the conversion of the pipe and design wave characteristics into

$$\left[\frac{\rho}{2} (Dl_*) u^2_{\text{max}_{\text{Airy}}} \right] \quad \text{and} \quad \left[\frac{u^2_{\text{max}_{\text{Airy}}}}{\dot{u}_{\text{max}_{\text{Airy}}} D} \right]$$

Using $\beta_3 = 0.86$ yields ψ from Equation (8-21) and C_{max} and K_{max} are obtained from Figure 8-19. Use of Equations (8-31) and (8-32), coupled with $\beta_1 = 1.08$, then yields the design values of F_{max} and P_{max}.

Wave Forces: Pipes at Angle of Attack

The general approach followed for steady flow force coefficients in the perpendicular flow and angled flow cases will also be used here. The peak horizontal and vertical forces are computed using the approach outlined in the preceding section, then adjusted using the curves of Figure 8-14. It is to be recalled that the horizontal force considered is that perpendicular to the pipe.

The outfall itself must resist the flexural stresses caused by the alternating directions of wave-induced loading over its length. The following is a suggested approach toward deriving a reasonable loading diagram for the pipe for a specific design wave characterized by its period and deep water wave height. The first step is to derive a refraction diagram for the design wave. At any station along the outfall the depth is known. The variation of wave height (shoaling, refraction) and length can then be plotted against outfall station. When the design wave itself has broken, incipiently breaking wave heights can be used from that outfall station in to the location where the outfall disappears beneath the sea floor.

The Airy theory can be used to give first estimates of the peak velocities and accelerations. Using the factors outlined above allows for the determination of C_{max} and K_{max}, and the angle of the wave orthogonal to the pipe (from the refraction diagram) allows for the determination of θ. A specific value for this angle then allows the designer to enter Figure 8-14 and derive the altered maximum wave force coefficients. The forces can then be computed as

$$F'_{\text{max}} = C'_{\text{max}} \frac{\rho}{2} (Dl_*) \beta_1 u^2_{\text{max}_{\text{Airy}}} \tag{8-35}$$

and

$$P'_{\text{max}} = K'_{\text{max}} \frac{\rho}{2} (Dl_*) \beta_1 u^2_{\text{max}_{\text{Airy}}} \tag{8-36}$$

Here the horizontal force F'_{max} is perpendicular to the pipe. The resulting diagram in no way represents the distribution of wave-induced force along the exposed outfall at any time; it assumes that a crest is located all along the pipe. The design involves taking various pipe stations separated along the pipe by distances $\bar{\lambda}/\sin\bar{\theta}$, where $\bar{\lambda}$ and $\bar{\theta}$ correspond to average wave lengths and angles of attack between the offshore crest point and the onshore crest point.

As an adequate design procedure, the variation of applied force between two adjacent wave crest stations can be represented by roughing in an approximate cosine function. The resulting force distribution all along the pipe can then be used in a flexural stress analysis for the pipe. Other locations for the crests than the first set should be used in order to find that particular situation that gives rise to the most serious stresses.

Pipe Stability Against Wave Forces

Exposed submarine pipelines have frequently failed by sliding, the main examples being in the Gulf of Mexico (Section 8-3). Another pipeline that failed in this way was a water line, linking Magnetic Island and Townsville, Australia, that was battered by Cyclone Althea in December 1971. The potential for sliding can be appraised using the same general method as for the steady flow case.

Although an exposed pipe may itself be stable when subjected to heavy wave surge, the area along the pipe on the side away from the approach of the waves is very vulnerable to attack. When strong wave-induced water motion moves over a pipe, there is a particularly powerful eddy that breaks off the top of the pipe and impinges on the bottom. I have felt the power of these eddies myself when lying in the "shelter" of the test pipe referred to earlier, and have seen the deep erosion caused by such eddies along an exposed submarine pipeline proceeding to an offshore crude oil offloading station at Barbers Point, Hawaii.

Thus the complete design of an exposed outfall should go beyond the consideration of the stability of the outfall *per se*, and should possibly include provision for armoring the sea bed along the line. Partial pipe ballasting may be insufficient, since erosion caused by the eddy could undermine the rock matrix.

Breaking Wave Forces

The expression *depth-limited design criterion* refers to the use in design of a wave on the verge of breaking. For relatively flat sea floor slopes it is usually assumed by coastal engineers that the ratio between the height of a breaking wave and the depth in which it breaks is 0.78.* It was remarked in Section 8-8 that the implicit assumption in computing wave forces under such conditions is that the wave is about to break—not in the process of breaking. This assumption

---

*Breaking wave heights can substantially exceed the water depth on or beyond steep bottom slopes.

underlies all the relevant material in Section 8-9 when the depth-limited design condition was reached.

Grace and Castiel (1975) made model tests on the forces exerted by nonbreaking and breaking waves on a pipe parallel to the wave fronts. The small scale of the work renders invalid the use of their force coefficients in an absolute sense, but the ratio of the breaking to nonbreaking peak forces obtained might still be representative of what would be experienced in the field. The approach was to consider a hypothetical nonbreaking wave with the breaking wave height, then to compute the peak horizontal and vertical forces for that wave, and finally to multiply by a factor to obtain the peak force under the actual (breaking) wave. Let the two factors be represented by α_H and α_V for the horizontal and vertical directions, respectively. It was found that these factors displayed no demonstrable correlation with various wave parameters. The samples of both factors (sample size = 26) displayed normal distributions with the following statistics: horizontal—mean = 1.12 and standard deviation = 0.55; vertical—mean = 0.75 and standard deviation = 0.28. The mean values show that *for these tests* whether a wave is on the point of breaking or in the process of breaking makes little difference. However, numbers larger than unity are possible in the right-hand tails of the distributions, and Grace and Castiel (1975) suggested $\alpha_H = 2.0$ and $\alpha_V = 1.2$ for design use.

8-10   PIPE BURIAL AND BALLASTING

Backfilling with Native Material

It is a possibility that outfalls, after being placed in a trench, can be mechanically or naturally backfilled using naturally occurring bottom material. But this would be unusual, since the fineness of such material means that it can usually be readily eroded during a time of heavy wave action. Such erosion can be particularly severe in the trench-backfilling area where the relative density of the material is lower than in the surrounding region.

The potential instability of bottom materials under design wave action can be predicted using information gathered by the Hydraulics Research Station (HRS) in Wallingford, England. Data were obtained for a range of sediment sizes in a special pulsating flume (Dedow, 1966) that produced simple harmonic motion. The data obtained have been presented in parametric form by Rance and Warren (1968) and in terms of wave conditions in "Threshold of Movement of Shingle Subjected to Wave Action" (1968, 1969). The latter assumed the validity of the Airy wave theory for both peak near-bottom flow speeds and accelerations, an unsafe assumption in terms of real waves, and this manner of presenting the data will not be used herein.

In parametric form the data can be represented by the following double equation:

$$\frac{\dot{U}_{max}d}{U_{max}^2} = \begin{cases} 6.69\left(\dfrac{\dot{U}_{max}}{g'}\right)^{2.278}, & 0.067 \leqslant \dfrac{\dot{U}_{max}}{g'} \leqslant 0.158 & \text{(8-37a)} \\[4mm] 0.533\left(\dfrac{\dot{U}_{max}}{g'}\right)^{1.340}, & 0.005 \leqslant \dfrac{\dot{U}_{max}}{g'} \leqslant 0.067 & \text{(8-37b)} \end{cases}$$

The variables in Equations (8-37) are as follows:

U_{max} = peak wave-induced flow speed
$\dot{U}_{max}$ = peak wave-induced flow acceleration
d = diameter of sediment particle on the verge of moving

$$g' = \left(\frac{\gamma_s - \gamma}{\gamma}\right)g \qquad\qquad\qquad \text{(8-38)}$$

γ_s = specific weight of sediment
γ = specific weight of liquid
g = acceleration due to gravity

The two parameters shown in Equations (8-37a) and (8-37b) are not as given by Rance and Warren, but are equivalent for strictly simple harmonic motion. The forms used herein should be more appropriate for the general coastal engineering case and real waves where there is not a simple relationship between U_{max} and $\dot{U}_{max}$. The unknown in Equations (8-37) is the incipiently unstable sediment size d.

Estimates of $U_{max}^2/\dot{U}_{max}$ and $\dot{U}_{max}$ in Equations (8-37) can be obtained from corresponding Airy theory kinematical quantities as follows: Multiply $u_{max_{Airy}}^2/\dot{u}_{max_{Airy}}$ by 0.86 to get $U_{max}^2/\dot{U}_{max}$ and multiply $\dot{u}_{max_{Airy}}$ by 1.36 to obtain $\dot{U}_{max}$.

If a particular sediment size is shown to be unstable under particular water motion, there is naturally the question of the depth of erosion. It would appear that there is considerable research that still needs to be done in this area as there is little related information available. In the interim, however, some pertinent data are presented by Carstens et al. (1969).

Backfilling with Rock

The burial and protection of the Sand Island No. 2 outfall was discussed in Section 8-1. The situation in that case, namely using sizable *armor rock*, is a standard solution for outfalls built in nonsheltered marine waters (see Figure 8-20).

The breaking of big waves on the face of a breakwater is a dramatic sight, and it is not surprising from the power exhibited that large rock or concrete

Figure 8-20 Armor stone over outfall

units making up such breakwaters can be thrown up and over the back side. It is much more difficult to visualize that a large wave passing through perhaps 30 or 40 ft (9 or 12 m) of water can readily push large blocks and rock around on the sea bed.

As an illustration of this situation, consider the situation near two submarine pipelines used for crude oil offloading to a refinery at Barbers Point, Hawaii. During original construction, 8-ton concrete clump anchors were used and were left behind when the construction work was completed. Large swell moves through the area from time to time. It has been found by divers who inspect the lines each month that the blocks cover a considerable amount of territory, frequently fetching up against the pipelines and necessitating their being dragged away. This movement is in water depths of about 40–50 ft (12–15 m). There is clear proof, then, that exposed blocks of even substantial weight can be moved by ocean wave action.

Although Equations (8-37) from the HRS data were obtained for simulated particle sizes only up to 1.88 in. (48 mm), these equations might also be considered for use in determining the stability of rock. It was found in the tests that, for $d > 2$ mm, there was no Reynolds number effect apparent for the data. However, this could be a situation extending up to a limiting Reynolds number not attained in the particular experiments. An equivalent situation occurs, for example, for steady flow-induced forces on a smooth, free sphere where the drag coefficient is independent of Reynolds number for values of that parameter between 2×10^3 and 2×10^5 for flows of low turbulence intensity.

Another possible problem of extending the HRS data to larger particle sizes concerns the thickness of the boundary layer and whether or not particles are embedded in it.

The Coastal Engineering Research Center Approach and Its Verification

"Shore Protection, Planning and Design" (1966) contains a figure for the steady flow speed V_i necessary to cause incipient motion of a particle of size (diameter) d. Different lines are given for different bottom slopes, and both "embedded" and "nonembedded" conditions are included. The former applies to a matrix of particles, the latter to single, isolated particles. For embedded particles on a level bed, the relationship between particle size and incipient-motion (steady) flow speed has been given in Equation (8-17). The nonembedded-particle incipient-motion flow speed for a level bed is consistently 72% of that given by Equation (8-17).

During the spring of 1976 an associate and I ran a preliminary experiment in the sea on wave-induced instability of exposed concrete cubes. The cubes were made up in nominal sizes of 1, 2, 3, and $4\frac{1}{2}$ in. (25, 51, 76, and 114 mm) with various inserts and admixtures providing a range of bulk specific gravities. The cubes were left on the bottom and were thus saturated with sea water for any tests.

During data-gathering sessions the cubes were placed on a concrete slab that was designed and fabricated especially for the experiments. One-third of the slab involved an open area for the cubes with the remaining two-thirds available for an observer diver. Various steel inserts were built into the latter portion so that the diver could hold on and not be swept away by water motion. A short vertical pipe adjacent to the area for the cubes was used to mount a ducted velocity sensor.

The site was in 37 ft (11.3 m) of water. Cables ran from the velocity sensor and from a diver-actuated switch to power supply and recording instruments in a boat overhead. When cubes moved, the diver would close the switch, actuating a marker pen on the flow speed record. He would then record relevant information regarding cube identities and motion on a pad of underwater-writing paper mounted on the base under him. The difficulty of doing this during heavy wave surge can scarcely be imagined.

Our test waves were long, having periods between 12 and 17 s. Wave heights were such that peak bottom flow speeds over 5 fps (1.5 m/s) were frequently experienced, and all test cubes moved at one time or another. We found that the predictions of the curve in "Shoreline Protection, Planning and Design" for nonembedded rock and a level bottom came very close to our results. This *should* be the case since inertial force terms were very small for our test conditions, but the correspondence lent confidence to the published information for wave situations in which inertial effects are unimportant—i.e., for situations where the amplitude of wave-induced water motion greatly exceeds the size of the rock. Extrapolating, then, Equation (8-17) would appear reasonable for judging the stability of embedded rock under waves where, again, the constraint on inertia force contributions is maintained.

Another Approach

The sizing of protective rock for the Orange County No. 2 outfall in California (#35 in Table 10-1), the two Hawaiian outfalls, Sand Island No. 2 (#44) and Barbers Point (#49), and the Burwood Beach line at Newcastle, Australia, all made use of the same general approach. The analysis in each case proceeded from the so-called "Shields curve" for sediment instability in steady, uniform flow.

Consider a rock or sediment particle of diameter d surrounded by other similar rocks or particles lying on the bottom of a wide channel carrying a steady, uniform flow (velocity V) of liquid. Let the specific weight of the rock or sediment particle be γ_s (specific gravity s_s) and that of the liquid γ. Furthermore, represent the (mass) density and kinematic viscosity of the liquid by ρ and ν, respectively. Let the shear stress exerted by the flow on the boundary material be τ, with that stress causing incipient motion of the rock or sediment particle concerned τ_d.

The Shields curve (e.g., Vanoni, 1975) is a plot of two dimensionless parameters composed from the list of variables above. These parameters are a dimensionless shear stress

$$\mathbf{T_*} = \frac{\tau_d}{(\gamma_s - \gamma)d} \tag{8-39}$$

and a boundary Reynolds number

$$\mathbf{R_*} = \frac{(\tau_d/\rho)^{1/2}d}{\nu} \tag{8-40}$$

For $\mathbf{R_*} \gtrsim 500$, the fully turbulent flow regime, $\mathbf{T_*} = 0.06$. The Shields curve indicates a minimum value of $\mathbf{T_*}$ between 0.03 and 0.04, for $\mathbf{R_*} \approx 10$.

The analysis for the size of armor rock necessary for stability for the four outfalls listed above all followed the same general steps. A constant value of $\mathbf{T_*}$ was first assumed: this was 0.03 for the two Hawaiian outfalls, 0.04 for the California line, and 0.06 for the Australian one. The lower values are, of course, the more conservative. The latter appears reasonable in light of the likelihood of fully turbulent flow conditions; however, effects of flow unsteadiness might mean that the lower numbers were better.

Boundary shear stress can be computed from the equation

$$\tau = \frac{f}{4}\left(\rho\frac{V^2}{2}\right) \tag{8-41}$$

where f is the Darcy–Weisbach friction factor and V is the steady, uniform flow speed. The remainder of the analysis for the referenced outfalls deals with translation of f and V into sediment size-related and wave-related equations. The friction factor disappears through use of boundary layer equations (e.g., Schlichting, 1960), specifically,

$$\frac{1}{\sqrt{f}} = 2\log_{10}(y/d) + 2.11 \tag{8-42}$$

where y is the depth of flow with d the particle size. The flow speed is converted in the case of the Australian outfall (Noda and Collins, 1974) to wave conditions by assuming the validity of the solitary wave theory, i.e.,

$$V = u_s = \frac{H}{y}\sqrt{gy}\left(1+\frac{H}{y}\right)^{1/2} \tag{8-43}$$

where H is the wave height and g is the acceleration of gravity. The solitary wave is a very long, period-independent concept (e.g., Ippen and Kulin, 1955). To allow for situations wherein the solitary wave peak velocity predictions are too high, let

$$V = \beta_s u_s \tag{8-44}$$

β_s can be computed using $V = 1.4\, u_{\text{max}_{\text{Airy}}}$. Figure 8-21 shows the results of this analysis for various assumed values of T_*/β_s^2 and for $s_s = 2.65$. The $T_*/\beta_s^2 = 0.06$ ($T_* = 0.06$, $\beta_s = 1$) line not only conforms to the results for the Australian line but also very closely to the results of the analyses for the two Hawaiian lines, even though T_* was taken as 0.03, since in those cases V was computed by an approach that differed from Equation (8-43).

Using equations given by Ippen and Mitchell (1957) for the solitary wave it can be shown that the parameter

$$\frac{u_{\text{max}}^2}{\dot{u}_{\text{max}}d} = 1.5\frac{y}{d}\sqrt{H/y}\,\frac{(1+0.8H/y)}{(1+1.5H/y)} \tag{8-45}$$

Insertion of reasonable values of y/d and H/y into the right-hand side of Equation (8-45) yields sufficiently large values of $u_{\text{max}}^2/\dot{u}_{\text{max}}d$ to indicate clearly, as one would expect, that inertial effects are negligible for the solitary wave.

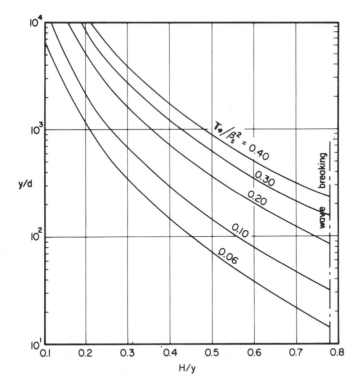

Figure 8-21   Design curves for stone instability

Other Considerations

Up to this point the only requirement considered for the protective rock for an outfall is that the individual rocks should not be unstable under design wave action. There are two other important considerations involved:

1. The entire matrix of pipe and rocks must be resistant to sliding and shear failure.
2. The rock arrangement must prevent loss of finer pipe bedding material underneath it.

There is little information for guidance in the first case. Horizontal and vertical inertia coefficients, 1.7 and 1.4, respectively, were presented by R. M. Towill Corp. (1972) for a two-dimensional trapezoidal object set on the sea floor. The test object simulated a cross section of a pipe protected by rock where the top and bottom widths were 22.0 and 48.5 ft (6.7 and 14.8 m) and the height was 9.1 ft (2.8 m). Limited additional information appears in Beckmann and Thibodeaux (1962).

Rules that can be applied to a rock layer ("filter") in order to prevent escape of underlying material ("base") have been given by Posey (1971) as follows:

$$d_{15}(\text{filter}) < 5d_{85}(\text{base})$$
$$4d_{15}(\text{base}) < d_{15}(\text{filter}) < 20d_{15}(\text{base})$$
$$d_{50}(\text{filter}) < 25d_{50}(\text{base})$$

In this list d corresponds to the nominal sediment particle size with the subscript indicating percent finer than by weight. Thus d_{15} is the size of a standard mesh through which 15% by weight of the material will pass.

8-11 OTHER DESIGN ISSUES

Buried Pipes

Wave Force Problems during Construction: During the laying of an outfall that is to be buried, it is wise to have the rocking operation keep up with the laying. The sudden appearance of a storm or heavy swell could pose problems for an outfall lying exposed in a trench. The engineer wishing to check on the stability of an outfall lying in a trench will find some pertinent information provided by Brater and Wallace (1972).

Wave Forces on Buried Pipe: Pressures associated with waves can penetrate into the sea bed and loads on a buried pipe can result. This matter has received the attention of Lai et al. (1974) who applied both finite difference and finite element models. The significance of such loading is that the pipe could gradually work itself up toward the sediment surface and into a potentially dangerous position with respect to the waves.

Earthquake Design

In the stability analysis of bare or ballasted outfalls vis-à-vis earthquakes, initial information required involves design values of horizontal and vertical accelerations. The latter is considered by increasing (upward acceleration) or decreasing (downward acceleration) the normal force between a part of the outfall system that could slide and the base under it. Various failure surfaces are assumed and a factor of safety against failure computed for each. Earthquake-induced changes in pore pressures within the soil may need to be considered.

Failure occurs in shear if the force imposed on the pipe and rock by the water, as the sea floor bearing the outfall is accelerated, is larger than the maximum frictional resistance force that can be mobilized along the potential failure surface. In general this resistance force would reflect both interlocking in the soil or rock matrix (friction angle) and cohesion.

The hydrodynamic force F_I exerted on the outfall and rock is of inertial form,

$$F_I = C_I \rho \forall a_H \qquad (8\text{-}46)$$

where C_I is the inertial coefficient of the configuration above the failure surface; ρ is water (mass) density; $\forall$ is the volume of the assembly considered; and a_H is the horizontal acceleration.

For bare outfalls of diameter D, supported various distances h above the bottom, C_I can be obtained from Figure 8-18. For a trapezoidal cross section, such as for a fully ballasted outfall, tests in conjunction with the Sand Island No. 2 outfall set $C_I = 1.7$ as outlined earlier.

8-12 APPURTENANCES

Bends

Three separate ways of achieving a bend in the Orange County No. 2 outfall (#35 in Table 10-1) were considered: the *pulled joint*, beveled joints, or mitered pipe sections. The extended bell type of RCP can allow an angular misalignment of up to one-half a degree per joint without the tightness of a joint being impaired, and this approach was used for small changes in alignment. For larger bends this approach (combining many sections) was considered impractical because additional movement at any joint beyond the one-half degree would have caused leakage.

The *beveled joint*, with bell and spigot not parallel but the pipe section straight, was preferred for bends of large total deflection. Joints can be beveled up to 2° per joint, and long, sweeping curves can be created where the weight of the pipe itself plus the ballast are sufficient to withstand lateral forces due to the change in alignment. The additional cost of beveled ends can be reduced somewhat in the overall sense by using the same sections (and forms) over and over. *Mitered joints* (essentially flat angles) are expensive and difficult to construct as well as requiring massive thrust-resisting structures (Harper, 1971).

Manholes

Provision should be made in an ocean outfall for the possibility of internal inspection and/or maintenance (see Chapter 11). To enable divers to carry out either type of mission some means of access is provided to the inside of the line. The structure that houses the opening is the *manhole*, and the cover over the opening is called the *manhole cover* (see Figure 8-22). A discussion of manhole covers can be found in Harper (1971).

Figure 8-22 Manhole cover on installed outfall (before installation of riser and placement of armor rock)

The spacing between manholes along an outfall has been set, for a number of outfalls (e.g., Orange County No. 2 and Sand Island No. 2) at 600 ft (180 m), since divers prefer to have escape within 300 ft (90 m). A 400-ft (120-m) spacing may be used in the diffuser. The opening is normally 36 in. (914 mm), the minimum size that divers recommend. The manhole cover has been made of various materials such as cast iron (Orange County No. 2), which requires some corrosion allowance, and Ampcoloy (Sand Island No. 2). Two manhole covers for the latter outfall are shown (upside down) in Figure 8-23.

Some of the Sand Island manhole covers were a composite of monel plate and a concrete filler plug. Monel had been initially specified for all covers at Sand Island, but the nonavailability of this material in sufficient quantities necessitated a redesign. Incidentally, seven small inspection openings, 10 inches (254 mm) in diameter, were included in the Sand Island diffuser to facilitate inspection of the pipe interior without removal of the large manhole covers.

During inspection or maintenance, effluent may be diverted through manholes, the egress structure (which will be considered below), or perhaps another (older) outfall.

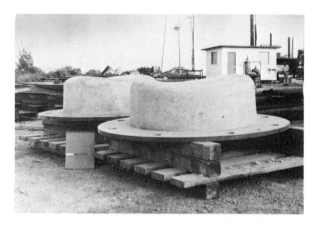

Figure 8-23 Two manhole covers for Sand Island No. 2 outfall

Egress Structures

If an ocean outfall is cleaned by use of a mechanical device passing down the line (Chapter 11), there must either be a downstream location where this device can be removed, or it must be pulled back to the starting point. Removing the cleaning device offshore is done at an *egress structure*, usually immediately upstream of the diffuser, where a full-diameter opening in the line is provided but covered with a removable cover. The egress structure in the 120-in.-diam (3048-mm) Orange County No. 2 line was 14 ft (4.3 m) long, considered adequate for removing any reasonable size of cleaning device.

Flapgate Structures

At the offshore end of an outfall, an end structure with a full-diameter gate is installed so that hydraulic flushing of the line can be carried out. Harper (1971) has discussed various design aspects of such features and the *flapgate structure* used for the Sand Island No. 2 outfall is shown in Figure 8-24. A flow velocity of 3 fps through the pipe is normally considered a minimum flushing rate. See Chapter 7.

Figure 8-24 Flapgate structure for Sand Island No. 2 outfall

8-13 REFERENCES

A. H. GLENN AND ASSOCIATES (n.d.): "Meteorological-Oceanographic Conditions Affecting Design and Operation of Offshore Tanker Terminal Facilities: New Plymouth Power Project," prepared at New Orleans, La., for Taranaki Harbours Board, New Zealand.

ADE, G. J. (1970): "Nelson Prestressed Flexible Submarine Outfall," *New Zealand Engineering* (August 15), pp. 205–208.

ALTERMAN, I. (1962): Discussion of "Wave Force Coefficients for Offshore Pipelines," by H. Beckmann and M. H. Thibodeaux, ASCE, *Journal of the Waterways and Harbors Division*, Vol. 88, No. WW4, pp. 149–150.

AMERICAN WATER WORKS ASSOCIATION (1965): "AWWA Standard for Reinforced Concrete Water Pipe—Noncylinder Type, Not Prestressed," AWWA C302-64, New York.

BARRIENTOS, C. S., and C. P. JELESNIANSKI (1976): "SPLASH—A Model for Forecasting Storm Surges," *Proceedings*, Fifteenth Coastal Engineering Conference, Honolulu, Hawaii (July), Vol. 1, pp. 941–958.

BEA, R. G. (1971): "How Sea Floor Slides Affect Offshore Structures," *The Oil and Gas Journal*, Vol. 69, No. 48 (November 29), pp. 88–92.

BEA, R. G., et al. (1975): "Soil Movements and Forces Developed by Wave-Induced Slides in the Mississippi Delta," *Journal of Petroleum Technology* (April), pp. 500–514.

BEATTIE, J. F., L. P. BROWN, and B. WEBB (1971): "Lift and Drag Forces on a Submerged Circular Cylinder," *Preprints*, Third Annual Offshore Technology Conference, Houston, Texas (April), Vol. 1, pp. 319–328.

BECKMANN, H., and M. H. THIBODEAUX (1962): "Wave Force Coefficients for Offshore Pipelines," ASCE, *Journal of the Waterways and Harbors Division*, Vol. 88, No. WW2 (May), pp. 125–138.

BENJAMIN, J. R., and C. A. CORNELL (1970): *Probability, Statistics, and Decision for Civil Engineers*, McGraw-Hill, New York.

BLUMBERG, R. (1964): "Hurricane Winds, Waves and Currents Test Marine Pipe Line Design, Part 3," *Pipe Line Industry*, Vol. 21, No. 2 (August), pp. 34–39.

BLUMBERG, R. (1966a): "Design for Environmental Extremes," *Pipe Line Industry*, Vol. 25, No. 4 (October), pp. 31–34.

BLUMBERG, R. (1966b): "Introduction—Offshore Risks," Hurricane Symposium, American Society for Oceanography, *Publication No. 1* (October), pp. 294–303.

BLUMBERG, R., and A. M. RIGG (1961): "Hydrodynamic Drag at Supercritical Reynolds Numbers," *ASME*, Petroleum Session, Los Angeles, Calif. (June 14).

BODINE, B. R. (1971): "Storm Surge on the Open Coast: Fundamentals and Simplified Prediction," U.S. Army, Corps of Engineers, Coastal Engineering Research Center, Washington, D.C., *Technical Memorandum No. 35* (May).

BORGMAN, L. E. (1973): "Probabilities for Highest Wave in Hurricane," ASCE, *Journal of the Waterways, Harbors and Coastal Engineering Division*, Vol. 99, No. WW2 (May), pp. 185–207.

BRATER, E. F., and R. WALLACE (1972): "Wave Forces on Submerged Pipe Lines," *Proceedings*, Thirteenth International Coastal Engineering Conference, Vancouver, Canada (July), Vol. 3, pp. 1703–1722.

BRETSCHNEIDER, C. L. (1967a): "Storm Surges," in *Advances in Hydroscience*, edited by V. T. Chow, Vol. 4, pp. 341–418, Academic Press, New York.

BRETSCHNEIDER, C. L. (1967b): "Estimating Wind Driven Currents over Continental Shelf," *Ocean Industry*, Vol. 2, No. 6 (June), pp. 45–48.

BRETSCHNEIDER, C. L. (1967c): "How to Calculate Storm Surges over the Continental Shelf," *Ocean Industry*, Vol. 2, No. 7 (July), pp. 31–34; No. 8 (August), pp. 50–57; No. 11 (December), pp. 42–47; Vol. 3, No. 1 (January 1968), p. 46.

BRETSCHNEIDER, C. L. (1967d): "Wave Forecasting," *Ocean Industry*, Vol. 2, No. 10 (October), pp. 53–60; Vol. 2, No. 11 (November), pp. 38–46.

BRETSCHNEIDER, C. L. (1967e): "On Wind Tides and Longshore Currents over the Continental Shelf due to Winds Blowing at an Angle to the Coast," University of Hawaii, Dept. of Ocean Engineering, unpublished paper.

BRETSCHNEIDER, C. L. (1968): "Decay of Wind Generated Waves to Ocean Swell by Significant Wave Method," *Ocean Industry*, Vol. 3, No. 3 (March), pp. 36–51; No. 4 (April), pp. 45–50; No. 5 (May), pp. 54–62; No. 6 (June), pp. 99–105.

BRETSCHNEIDER, C. L. (1970): "Forecasting Relations for Wave Generation," University of Hawaii, J.K.K. Look Laboratory of Oceanographic Engineering, Honolulu, Hawaii, *Look Lab Hawaii*, Vol. 1, No. 3 (July), pp. 31–34.

BRETSCHNEIDER, C. L. (1972a): "A Non-Dimensional Stationary Hurricane Wave Model," *Preprints*, Fourth Annual Offshore Technology Conference, Houston, Texas (May), Vol. 1, pp. 51–68.

BRETSCHNEIDER, C. L. (1972b): "Revisions to Hurricane Design Wave Practices," *Proceedings*, Thirteenth Coastal Engineering Conference, Vancouver, Canada (July), Vol. 1, pp. 167–195.

BRETSCHNEIDER, C. L. (1973): "Design Hurricane Waves for the Island of Oahu, Hawaii, with Special Application to Sand Island Ocean Outfall System," *Look Lab/ Hawaii*, Vol. 3, No. 2 (July), pp. 35–53.

BRETSCHNEIDER, C. L., and R. O. REID (1954): "Modification of Wave Height due to Bottom Friction, Percolation, and Refraction," U.S. Army, Corps of Engineers, Beach Erosion Board, *Technical Memorandum No. 45* (October).

BROWN, R. J. (1967): "Hydrodynamic Forces on a Submarine Pipeline," ASCE, *Journal of the Pipeline Division*, Vol. 93, No. PL1, pp. 9–19, with Discussion, Vol. 93, No. PL3, pp. 75–81.

BROWN, R. J. (1971): "How Deep Should an Offshore Line Be Buried for Protection?" *Oil and Gas Journal*, Vol. 69, No. 41 (October 11), pp. 90–92, 97, 98.

BROWN, R. J. (1972): "Pipelines Can be Designed to Resist Impact from Dragging Anchors and Fishing Boards," *Preprints*, Fourth Annual Offshore Technology Conference, Houston, Texas (May), Vol. 1, pp. 579–586.

BROWN, R. J. (1973): "Pipeline Design to Reduce Anchor and Fishing Board Damage," ASCE, *Transportation Engineering Journal*, Vol. 99, pp. 199–210.

CAMPBELL, S. (1976): "Savings Expected from Polyethylene Pipe Use," *World Dredging and Marine Construction*, Vol. 12, No. 6 (May), pp. 39–40.

"Canadian Tide and Current Tables: 1976" (1975): Environment Canada, Fisheries and Marine Service, Canadian Hydrographic Service, Ottawa, Ontario, Canada.

CARDONE, V. J., W. J. PIERSON, and E. G. WARD (1976): "Hindcasting the Directional Spectra of Hurricane-Generated Waves," *Journal of Petroleum Technology*, Vol. 28, No. 4 (April), pp. 385–394.

CARSTENS, M. R., F. M. NEILSON, and H. D. ALTINBILEK (1969): "Bed Forms Generated in the Laboratory under an Oscillatory Flow: Analytical and Experimental Study," U.S. Army, Corps of Engineers, Coastal Engineering Research Center, *Technical Memorandum No. 28* (June).

CARSTENS, T., S. P. KJELDSEN, and O. GJORSVIK (1977): "The Conflict between Pipelines and Bottom Trawls—Some Results from Laboratory and Field Tests," River and Harbour Laboratory at the Norwegian Institute of Technology, Trondheim, Norway, *Bulletin No. 16E*, pp. 187–214.

CHRISTIAN, J. T., et al. (1974): "Large Diameter Underwater Pipeline for Nuclear Power Plant Designed against Soil Liquefaction," *Preprints*, Sixth Annual Offshore Technology Conference, Houston, Texas (May), Vol. 2, pp. 597–606.

COLÓN, J. A. (1966): "Some Aspects of Hurricane Carla (1961)," American Society for Oceanography, *Publication No. 1*, Hurricane Symposium, Houston, Texas (October), pp. 1–33.

DALRYMPLE, R. A. (1974): "A Finite Amplitude Wave on a Linear Shear Current," *Journal of Geophysical Research*, Vol. 79, No. 30 (October 20), pp. 4498–4504.

DALTON, C., and R. A. HELFINSTINE (1971): "Potential Flow Past a Group of Circular Cylinders," ASME, *Transactions, Journal of Basic Engineering*, Vol. 93D (December), pp. 636–642.

DEAN, R. G. (1965): "Stream Function Representation of Nonlinear Ocean Waves," *Journal of Geophysical Research*, Vol. 70, pp. 4561–4572.

DEAN, R. G. (1974): "Evaluation and Development of Water Wave Theories for Engineering Application," U.S. Army, Corps of Engineers, Coastal Engineering Research Center, Fort Belvoir, Va., *Special Report No. 1*, 2 vols. (November).

DEDOW, H. R. S. (1966): "A Pulsating Water Tunnel for Research in Reversing Flow," *La Houille Blanche*, No. 7, pp. 837–841.

DEMARS, K. R., V. A. NACCI, and W. D. WANG (1977): "Pipeline Failure: A Need for Improved Analyses and Site Surveys," *Proceedings*, Ninth Annual Offshore Technology Conference, Houston, Texas (May), Vol. 4, pp. 63–70.

EMERY, K. O., and R. H. TSCHUDY (1941): "Transportation of Rock by Kelp," *Bulletin of the Geological Society of America*, Vol. 52, pp. 855–862.

FREEMAN, J. C., L. BAER, and G. H. JUNG (1957): "The Bathystrophic Storm Tide," *Journal of Marine Research*, Vol. 16, No. 1, pp. 12–22.

GARRISON, C. J., and V. S. RAO (1971): "Interaction of Waves with Submerged Objects," ASCE, *Journal of the Waterways, Harbors, and Coastal Engineering Division*, Vol. 97, No. WW2, pp. 259–277.

GERRARD, J. H. (1961): "An Experimental Investigation of the Oscillating Lift and Drag of a Circular Cylinder Shedding Turbulent Vortices," *Journal of Fluid Mechanics*, Vol. 11, pp. 244–255.

GJORSVIK, O., S. P. KJELDSEN, and S. LUND (1975): "Influences of Bottom Trawl Gear on Submarine Pipelines," *Proceedings*, Seventh Annual Offshore Technology Conference, Houston, Texas (May), pp. 337–346.

GODA, Y. (1964): "Wave Forces on a Vertical Circular Cylinder: Experiments and a Proposed Method of Wave Force Computation," Port and Harbour Technical Research Institute, Yokusuka, Japan, *Report No. 8* (August).

GOLDSMITH, V., et al. (1974): "Wave Climate Model of the Mid-Atlantic Shelf and Shoreline (Virginian Sea): Model Development, Shelf Geomorphology, and Preliminary Results," U.S. National Aeronautics and Space Administration, *Publication SP-358*.

GRACE, R. A. (1973): "Available Data for the Design of Unburied, Submarine Pipelines to Withstand Wave Action," Institution of Engineers, Australia, *Preprints*, First Australian Conference on Coastal Engineering, Sydney (May), pp. 59–66.

GRACE, R. A. (1976): "Near Bottom Water Motion under Ocean Waves," *Proceedings*, Fifteenth International Conference on Coastal Engineering, Honolulu, Hawaii (July), Vol. 3, pp. 2371–2386.

GRACE, R. A., and J. CASTIEL (1975): "Non-Breaking and Breaking Wave Forces on Pipelines," *Proceedings*, Civil Engineering in the Oceans III, University of Delaware, Newark, Del. (June), pp. 523–534.

GRESLOU, L., and Y. MAHE (1954): "Etude du Coefficient de Reflexion d'une Houle sur un Obstacle Constitué par un Plan Incliné," *Proceedings*, Fifth Conference of Coastal Engineering, Grenoble, France (September), pp. 68–84.

GRICE, C. F. (1968): "Finding Underwater Objects; Part 3—Case Histories, Underwater Search Techniques," *Ocean Industry*, Vol. 3, No. 2 (February), pp. 34–39.

GRISWOLD, G. M. (1963): "Numerical Calculation of Wave Refraction," *Journal of Geophysical Research*, Vol. 68, pp. 1715–1723.

GUERRY, T. L. (1976): "Highlights of 1975 North Sea Pipe Line Operations," *Ocean Industry*, Vol. 11, No. 3 (March), pp. 23–28.

HARPER, F. A. (1971): "1971 Installation of Orange County Sanitation District's 5-Mile Ocean Outfall," Cement and Concrete Products Industry of Hawaii, Subaqueous Pipeline Installation Seminar (October 26), Honolulu, Hawaii.

HARTLEY, R. A. (1971): "Coatings and Corrosion," *Preprints*, Third Annual Offshore Technology Conference, Houston, Texas (April), Vol. 2, pp. 461–492.

HENKEL, D. J. (1970): "The Role of Waves in Causing Submarine Landslides," *Geotechnique*, Vol. 20, No. 1 (March), pp. 75–80.

HERBICH, J. B. (1977): "Wave-Induced Scour around Offshore Pipelines," *Proceedings*, Ninth Annual Offshore Technology Conference, Houston, Texas (May), Vol. 4, pp. 79–90.

HO, F. P. (1976): "Hurricane Tide Frequencies on the Atlantic Coast," *Proceedings*, Fifteenth Coastal Engineering Conference, Honolulu, Hawaii (July), Vol. 1, pp. 886–905.

HO, F. P., and R. J. TRACEY (1975): "Storm Tide Frequency Analysis for the Gulf Coast of Florida from Cape San Blas to St. Petersburg Beach," U.S. Dept. of Commerce, National Oceanic and Atmospheric Administration, National Weather Service, Office of Hydrology, Silver Spring, Md., *Technical Memorandum NWS HYDRO-20* (April).

HOERNER, S. (1965): *Fluid Dynamic Drag*, Hoerner Fluid Dynamics, Brick Town, N.J.

HOUGH, B. K. (1957): *Basic Soils Engineering*, Ronald Press, New York.

HUMPHREYS, J. S. (1960): "On a Circular Cylinder in a Steady Wind at Transition Reynolds Numbers," *Journal of Fluid Mechanics*, Vol. 9, pp. 603–612.

HYPERION ENGINEERS (1957): *Ocean Outfall Design*, joint venture of Holmes and Narver, Inc., Daniel, Mann, Johnson and Mendenhall, and Koebig and Koebig, Los Angeles, Calif. (October 15).

"Improper Design Charged on $18-Million Outfall" (1972): *Engineering News-Record*, Vol. 189, No. 24 (December 14), p. 15.

IPPEN, A. T., Ed. (1966): *Estuary and Coastline Hydrodynamics*, McGraw-Hill, New York.

IPPEN, A. T., and G. KULIN (1955): "Shoaling and Breaking Characteristics of the Solitary Wave," Massachusetts Institute of Technology, Cambridge, Mass., Hydrodynamics Laboratory, *Technical Report No. 15* (April).

IPPEN, A. T., and M. M. MITCHELL (1957): "The Damping of the Solitary Wave from Boundary Shear Measurements," Massachusetts Institute of Technology, Cambridge, Mass., Hydrodynamics Laboratory, *Technical Report No. 23* (June).

JANSON, L-E. (1975): "Submerged Polyethylene Pipes in Lakes, Rivers, and the Sea," *Journal of the Water Pollution Control Federation*, Vol. 47, No. 4 (April), pp. 869–873.

JELESNIANSKI, C. P., and A. D. TAYLOR (1974): "SPLASH Program User's Guide," U.S. National Oceanic and Atmospheric Administration, National Weather Service, Silver Spring, Md., Techniques Development Laboratory (November) (*NTIS Publication COM-75-10181/6WO*).

JONES, G. W., JR. (1968): "Unsteady Lift Forces Generated by Vortex Shedding about a Large, Stationary, and Oscillating Cylinder at High Reynolds Number," ASME Symposium on Unsteady Flow, Philadelphia, Pa. (May), *Paper 68-FE-36*.

JONES, W. T. (1970): "Forces on a Transverse Circular Cylinder in the Turbulent Boundary Layer of a Steady Flow," Rice University, Houston, Texas, Ph.D. Thesis (May).

JONES, W. T. (1971): "Forces on Submarine Pipelines from Steady Currents," *ASME, Paper 71-UnT-3*, Petroleum Mechanical Engineering with Underwater Technology Conference, Houston, Texas (September 19–23).

JONES, W. T. (1976): "On-bottom Pipeline Stability in Steady Water Currents," *Proceedings*, Eighth Annual Offshore Technology Conference, Houston, Texas (May), Vol. 2, pp. 763–778.

KEELING, H. J. (1959): "Corrosion Protection Features of the Hyperion Ocean Outfall," *Corrosion* (NACE), Vol. 15, No. 7 (July), pp. 41–46.

KOSTER, J. (1974): "Digging in of Anchors into the Bottom of the North Sea," Delft Hydraulics Laboratory, The Netherlands, *Publication No. 129* (June).

KRIEG, J. L. (1966): "Hurricane Risks as They Relate to Offshore Pipelines," Hurricane Symposium, American Society for Oceanography, *Publication No. 1* (October), pp. 304–313.

KUDRASS, H-R (1974): "Experimental Study of Nearshore Transportation of Pebbles with Attached Algae," *Marine Geology*, Vol. 16, pp. M9–M12.

LAI, N. W., R. F. DOMINGUEZ, and W. A. DUNLAP (1974): "Numerical Solutions for Determining Wave Induced Pressure Distributions around Buried Pipelines," Texas A&M University, College Station, Texas, *Sea Grant Publication 75-205* (December).

LEMÉHAUTÉ, B., D. DIVOKY, and A. LIN (1968): "Shallow Water Waves: A Comparison of Theories and Experiments," *Proceedings*, Eleventh Conference of Coastal Engineering, London, England (September), pp. 86–107.

LEMÉHAUTÉ, B., and R. C. Y. KOH (1967): "On the Breaking of Waves Arriving at an Angle to the Shore," *Journal of Hydraulic Research*, Vol. 5, No. 1, pp. 67–88.

LITTLEJOHNS, P. S. G. (1972): "Submarine Pipeline Research: Review of Literature of Lift and Drag of Cylinders in Unidirection Flow," Hydraulics Research Station, Wallingford, England, *Report No. INT 109* (October).

LYONS, C. G. (1973): "Soil Resistance to Lateral Sliding of Marine Pipelines," *Preprints*, Fifth Annual Offshore Technology Conference, Houston, Texas (April), Vol. 2, pp. 479–484.

McCORMICK, M. E. (1973): *Ocean Engineering Wave Mechanics*, Wiley-Interscience, New York.

MES, M. J. (1976): "Vortex Shedding Can Cause Pipe Lines to Break," *Pipeline and Gas Journal*, Vol. 203, No. 10 (August), pp. 29, 30, 32, 34, 40, 42.

MILZ, E. A., and D. E. BROUSSARD (1972): "Technical Capabilities in Offshore Pipeline Operations to Maximize Safety," *Preprints*, Fourth Annual Offshore Technology Conference, Houston, Texas (May), Vol. 2, pp. 809–826.

NIPPON STEEL CORPORATION (1973): "Submarine Pipeline Engineering," Tokyo, Japan, Catalog No. EXE 109 (July).

NODA, E. K., and J. I. COLLINS (1974): "Extreme and Normal Wind and Wave Conditions at Burwood Beach, Newcastle, Australia," Tetra Tech Inc., Pasadena, Calif., Final Report for Contract No. TC 425 (July).

O'DONNELL, J. P. (1971): "Mammoth 'Tool' Repairs Damaged Offshore Pipeline," *Oil and Gas Journal*, Vol. 69, No. 1 (January 4), pp. 63–70.

O'DONNELL, J. P. (1977): "Coal-Tar Enamel Still Most Widely Used Pipeline Coating in Industry," *The Oil and Gas Journal*, Vol. 75, No. 4 (January 24), pp. 58–62.

PAGENKOPF, J. R., and B. R. PEARCE (1975): "Evaluation of Techniques for Numerical Calculation of Storm Surges," Massachusetts Institute of Technology, Cambridge, Mass., Ralph M. Parsons Laboratory for Water Resources and Hydrodynamics, *Report No. 199* (February).

PEARCE, B. R., and J. R. PAGENKOPF (1975): "Numerical Calculation of Storm Surges: An Evaluation of Techniques," *Proceedings*, Seventh Annual Offshore Technology Conference, Houston, Texas (May), Vol. 2, pp. 887–901.

PEARSON, E. A. (1956): "An Investigation of the Efficacy of Submarine Outfall Disposal of Sewage and Sludge," California State Water Pollution Control Board, *Publication No. 14*, Sacramento, Calif.

PECK, R. B., W. E. HANSON, and T. H. THORNBURN (1953): *Foundation Engineering*, Wiley, New York.

"Polyethylene Pipe Used for 36-inch Sewage Outfall" (1972): *Civil Engineering*, Vol. 42, No. 10 (October), pp. 58–60.

POSEY, C. J. (1971): "Protection of Offshore Structures against Underscour," ASCE, *Journal of the Hydraulics Division*, Vol. 97, No. HY7 (July), pp. 1011–1016.

QUAYLE, R. G., and D. C. FULBRIGHT (1975): "Extreme Wind and Wave Return Periods for the U.S. Coast," *Mariners Weather Log*, Vol. 19, No. 2 (March), pp. 67–70.

R. M. TOWILL CORPORATION (1972): "Sand Island Ocean Outfall System," Job No. 29–71: Final Design Report, prepared for the City and County of Honolulu, Dept. of Public Works, Division of Sewers, at Honolulu, Hawaii (September 27).

RANCE, P. J. (1969): "The Influence of Reynolds Number on Wave Forces," *Proceedings*, Symposium on Wave Action, Delft, The Netherlands (July), Vol. 4, Paper 13.

RANCE, P. J., and N. F. WARREN (1968): "The Threshold of Movement of Coarse Material in Oscillatory Flow," *Proceedings*, Eleventh Conference on Coastal Engineering, London, England (September), Vol. 1, pp. 487–491.

ROBERTSON, J. M. (1965): *Hydrodynamics in Theory and Application*, Prentice-Hall, Englewood Cliffs, N.J.

RUDNICK, P., and R. W. HASSE (1971): Extreme Pacific Waves, December 1969," *Journal of Geophysical Research*, Vol. 76, No. 3 (January 20), pp. 742–744.

SAROYAN, J. R. (1969): "Protective Coatings," in *Handbook of Ocean and Underwater Engineering*, edited by J. J. Myers, C. H. Holm, and R. F. McAllister, pp. 7-37 to 7-75, McGraw-Hill, New York.

SARPKAYA, T. (1976): "In-Line and Transverse Forces on Smooth and Sand-roughened Cylinders in Oscillatory Flow at High Reynolds Numbers," Naval Postgraduate School, Monterey, Calif., *Report No. NPS-69SL76062* (June 4).

SCHLICHTING, H. (1960): *Boundary-Layer Theory*, 4th ed., McGraw-Hill, New York.

SEED, H. B., and K. L. LEE (1966): "Liquefaction of Saturated Sands During Cyclic Loading," ASCE, *Journal of the Soil Mechanics and Foundations Division*, Vol. 92, No. SM6 (November), pp. 105–134.

"Shore Protection, Planning and Design" (1966): U.S. Army, Corps of Engineers, Coastal Engineering Research Center, Washington, D.C., *Technical Report No. 4*, 3rd ed. (June).

SILVESTER, R. (1974): *Coastal Engineering*, 2 vols., Elsevier, Amsterdam.

SMALL, S. W., and L. B. SERPAS (1972): "Submerged Weight Control for Submarine Pipeline Construction," *Preprints*, Fourth Annual Offshore Technology Conference, Houston, Texas (May), Vol. 1, pp. 595–606.

SMALL, S. W., R. D. TAMBURELLO, and P. J. PIASECKYJ (1972): "Submarine Pipeline Support by Marine Sediments," *Journal of Petroleum Technology*, Vol. 24, No. 3 (March), pp. 317–322.

SNODGRASS, F. E., et al. (1966): "Propagation of Ocean Swell across the Pacific," *Philosophical Transactions of the Royal Society of London*, Vol. A259, pp. 431–497.

STORRIE, W. (1947): "Toronto Waste Works Extensions—A General Description of the System with Details of Some Outstanding Features," *Journal of the New England Water Works Association*, Vol. 61, No. 1, pp. 1–33.

"Submarine Pipeline Wins 1967 Award" (1967): *Australian Civil Engineering* (December 5), pp. 11–15.

The Great Alaska Earthquake of 1964: *Engineering* (1973): National Academy of Sciences, Washington, D.C.

THOM, H. C. S. (1971): "Asymptotic Extreme-Value Distributions of Wave Heights in the Open Ocean," *Journal of Marine Research*, Vol. 29, No. 1 (January), pp. 19–27.

THOMPSON, R. M. (1976): "European Pipeliners Try New Coatings," *The Oil and Gas Journal* (July 5), pp. 92–93.

"Threshold Movement of Shingle Subjected to Wave Action" (1968): Hydraulics Research Station, Wallingford, England, *Annual Report*, pp. 13–14.

"Threshold Movement of Shingle Subject to Wave Action" (1969): Hydraulics Research Station, Wallingford, England, *Notes*, No. 15 (December), pp. 5–6.

"Underwater Oil Pipeline Floats to Surface" (1975): *Engineering News-Record*, Vol. 195, No. 18 (October 30), p. 13.

VALENTINE, H. R. (1959): *Applied Hydrodynamics*, Butterworths, London.

VANONI, V. A., Ed. (1975): *Sedimentation Engineering*, American Society of Civil Engineers, New York.

WALKER, J. R. (1974): "Wave Transformations over a Sloping Bottom and over a Three-Dimensional Shoal," University of Hawaii, J. K. K. Look Laboratory of Oceanographic Engineering, Honolulu, Hawaii, *Miscellaneous Report No. 11* (May).

"Welding 'Hut' Used to Fix Damaged Line in 60-Foot Gulf Waters" (1968): *Oil and Gas Journal*, Vol. 66, No. 46 (November 11), p. 61.

WIEGEL, R. L. (1964): *Oceanographical Engineering*, Prentice-Hall, Englewood Cliffs, N.J.

WILLIAMS, H. C. (1966): "The Gisborne Submarine Sewer Outfall," *New Zealand Engineering* (March 15), pp. 110–120, with discussion, *New Zealand Engineering* (May 15, 1967), pp. 201–203.

WILSON, B. W. (1955): "Graphical Approach to the Forecasting of Waves in Moving Fetches," U.S. Army, Corps of Engineers, Beach Erosion Board, *Technical Memorandum No. 73*.

WILSON, B. W. (1957): "Hurricane Wave Statistics for the Gulf of Mexico," U.S. Army, Corps of Engineers, Beach Erosion Board, Washington, D.C., *Technical Memorandum No. 98* (June).

WILSON, J. C. (1970): "Engineering Properties of Surficial Marine Sediments," *Preprints*, Second Annual Offshore Technology Conference, Houston, Texas (April), Vol. 1, pp. 25–36.

WORTHINGTON, H. W., and J. B. HERBICH (1971): "Computer Prediction of Wave Heights in Coastal Areas," *Preprints*, Third Annual Offshore Technology Conference, Houston, Texas (April), Vol. 2, pp. 147–162.

WRIGHT, S. G., and R. S. DUNHAM (1972): "Bottom Stability under Wave Induced Loading," *Preprints*, Fourth Annual Offshore Technology Conference, Houston, Texas (May), Vol. 1, pp. 853–862.

8-14 BIBLIOGRAPHY

General

"A Predesign Report on Marine Waste Disposal," Vol. 4 (1975): Report prepared by Brown and Caldwell, Consulting Engineers, Walnut Creek, Calif., for the City and County of San Francisco (October).

ALVY, R. R. (1960): "Engineering Ocean Outfalls," *Consulting Engineer*, Vol. 14, No. 3 (March), pp. 96–104.

BECKMAN, W. J. (1970): "Engineering Considerations in the Design of an Ocean Outfall," *Journal of the Water Pollution Control Federation*, Vol. 42, No. 10 (October), pp. 1805–1831.

BERRY, W. H. (1970): "Activities Associated with the Laying of Submarine Pipelines," *Pipes and Pipelines International* (October), pp. 13–16, 18–20.

BRUUN, P., and P. JOHANNESSON (1971): "The Interaction between Ice and Coastal Structures," *Proceedings*, First International Conference on Port and Ocean Engineering under Arctic Conditions, Trondheim, Norway (August), Vol. 1, pp. 683–702.

CAST IRON PIPE RESEARCH ASSOCIATION (1967): *Handbook of Cast Iron Pipe*, 3rd ed., Chicago, Ill.

COMMONWEALTH DEPT. OF WORKS (1973): "Marine Outfalls: Public Health and Environmental Influences on Design," Sixteenth Conference, Water Supply and Sewerage Authorities, Australia, pp. 255–270, with discussion.

JOSA Y CASTELLS, F. (1975): "Criteria for Marine Waste Disposal in Spain," *Marine Pollution and Marine Waste Disposal*, edited by E. A. Pearson and E. De F. Frangipane, Pergamon Press, New York. [Proceedings of Second International Congress, San Remo, Italy (December 1973), pp. 33–51.]

LAWRENCE, C. H. (1962): "Sanitary Considerations of Five-Mile Ocean Outfall," ASCE, *Transactions*, Vol. 127, Pt. 3, pp. 294–325.

MATLOCK, H., W. R. DAWKINS, and J. J. PANAK (1971): "Analytical Model for Ice Structure Interaction," ASCE, *Proceedings of the Engineering Mechanics Division*, Vol. 97, No. EM4, pp. 1083–1092.

MICHEL, B. (1970): "Ice Pressures on Engineering Structures," U.S. Army, Corps of Engineers, Cold Regions Research and Engineering Laboratory, Hanover, N.H., Report (June) (NTIS *Publication AD 709 625*).

Pipeline Design for Water and Wastewater (1975): American Society of Civil Engineers, Pipeline Division, New York.

REID, R. O. (1952): "Some Oceanographic and Engineering Considerations in Marine Pipe Line Construction," *Proceedings*, Second Conference on Coastal Engineering, Council on Wave Research, Engineering Foundation, Berkeley, Calif., p. 749.

REYNOLDS, J. M. (1968): "Submarine Pipelines," *Pipes and Pipelines International* (May), pp. 22–36.

Shore Protection Manual (1973): U.S. Army, Corps of Engineers, Coastal Engineering Research Center, Fort Belvoir, Va., 3 vols.

General Submarine Pipeline Design

HEANEY, F. L. (1960): "Design, Construction, and Operation of Sewer Outfalls in Estuarine and Tidal Waters," *Journal of the Water Pollution Control Federation*, Vol. 32, No. 6 (June), pp. 610–621.

JOHN CAROLLO ENGINEERS (1970): "Final Design Report on Ocean Outfall No. 2," prepared for County Sanitation Districts of Orange County, Fountain Valley, California.

KARAL, K. (1977): "Lateral Stability of Submarine Pipelines," *Proceedings*, Ninth Annual Offshore Technology Conference, Houston, Texas (May), Vol. 4, pp. 71–78.

KOENIG, H. L. (1977): "The Submersible as a Pipeline Tool," paper presented at the ASCE Annual Convention (October 17–21), San Francisco.

LAI, N. W., S. J. CAMPBELL, R. F. DOMINGUEZ, and W. A. DUNLAP (1973): "A Bibliography of Offshore Pipeline Literature," Texas A&M University, *Report No. TAMU-SG-74-206* (June).

MOHR, H. O. (1977): "Subsea Pipeline Connectors: A Look at What's Available," *Pipe Line Industry*, Vol. 46, No. 1 (January), pp. 34–39.

PARKHURST, J. D., L. A. HAUG, and M. L. WHITT (1967): "Ocean Outfall Design for Economy of Construction," *Journal of the Water Pollution Control Federation*, Vol. 39, No. 6 (June), pp. 987–993.

SMALL, S. W. (1970): "The Submarine Pipeline as a Structure," *Proceedings*, Second Annual Offshore Technology Conference, Houston, Texas (April), Vol. 1, pp. 735–746.

SNOOK, W. G. G. (1968): "Marine Disposal of Trade Wastes," *Chemistry and Industry*, No. 46 (November 16), pp. 1593–1598.

Materials, Coatings, and Corrosion

BERGER, D. M. (1976): "How to Select Coatings for Buried Steel Pipe," *Pipeline and Gas Journal*, Vol. 203, No. 2 (February), pp. 14–18, 20.

BOSICH, J. F. (1970): *Corrosion Prevention for Practicing Engineers*, Barnes and Noble, New York.

BROWN, B. F. (1969): "Corrosion" and "Corrosion Behavior," in *Handbook of Ocean and Underwater Engineering*, edited by J. J. Myers, C. H. Holm, and R. F. McAllister, pp. 7-2 to 7-11 and 7-26 to 7-30, McGraw-Hill, New York.

DAVIS, J. G. (1972): "Cathodic Protection of Offshore Facilities," *World Dredging and Marine Construction*, Vol. 8, No. 5 (April), pp. 14–17, 37.

DAVIS, J. G., G. L. DOREMUS, and F. W. GRAHAM (1971): "The Influence of Environmental Conditions on the Design of Cathodic Protection Systems for Marine Structures," *Preprints*, Third Annual Offshore Technology Conference, Houston, Texas (April), Vol. 2, pp. 425–432.

DAVIS, J. G., G. L. DOREMUS, and F. W. GRAHAM (1972): "The Influence of Environment on Corrosion and Cathodic Protection," *Journal of Petroleum Technology*, Vol. 24, No. 3 (March), pp. 323–328.

DI GREGORIO, J. S., and J. P. FRASER (1974): "Corrosion Tests in the Gulf Floor," in *Corrosion in Natural Environments*, American Society for Testing and Materials, Philadelphia, Pa., *Special Technical Publication 558*, pp. 185–208.

LaQue, F. L. (1975): *Marine Corrosion: Causes and Prevention*, Wiley, New York.

LaQue, F. L. (1968): "Materials Selection for Ocean Engineering," in *Ocean Engineering*, edited by J. F. Brahtz, pp. 588–632, Wiley, New York.

Lennox, T. J., Jr., R. E. Groover, and M. H. Peterson (1970): "Electrochemical Characteristics of Six Aluminum Galvanic-Anode Alloys in the Sea," National Association of Corrosion Engineers, 26th National Conference, Philadelphia, Pa. (March), *Preprint No. 104*.

Littauer, E. L. (1964): "The Cathodic Protection of Buoys and Offshore Structures," *Transactions of the 1964 Buoy Technology Symposium*, Marine Technology Society, Washington, D.C. (March), pp. 325–338.

"Marine Construction Report" (1976): *Offshore*, Vol. 36, No. 12 (November), pp. 105, 106, 111, 113.

Miller, D. R. (1956): "Engineering Goes to Sea," *Public Works*, Vol. 87, No. 10 (October), pp. 136–137.

National Association of Corrosion Engineers (1976): "Control of Corrosion on Steel, Fixed Offshore Platforms Associated with Petroleum Production," NACE, *Standard RP-01-76*, Houston, Texas (April).

Peabody, A. W. (1967): *Control of Pipeline Corrosion*, National Association of Corrosion Engineers, Houston, Texas.

Rizzo, F. E., R. Wilson, and D. P. Bauer (1977): "Test Methods for Cathodically Protected Marine Pipelines," *Proceedings*, Ninth Annual Offshore Technology Conference, Houston, Texas (May), Vol. 4, pp. 437–442.

Tuthill, A. H., and C. M. Schillmoller (1967): "Selection of Marine Materials," *Journal of Ocean Technology*, Vol. 2, No. 1, pp. 6–36.

Storms and Storm Surges

Abbott, M. B., A. Damsgaard, and G. S. Rodenhuis (1973): "System 21, 'Jupiter' (A Design System for Two-Dimensional Nearly-Horizontal Flows)," *Journal of Hydraulic Research*, Vol. 11, pp. 1–28.

Bodine, B. R. (1969): "Hurricane Surge Frequency Estimated for the Gulf Coast of Texas," U.S. Army, Corps of Engineers, Coastal Engineering Research Center, *Technical Memorandum No. 26* (February).

Bruun, P., et al. (1962): "Storm Tides in Florida as Related to Coastal Topography," University of Florida, Gainesville, Coastal Engineering Laboratory, Bulletin Series No. 109 (January).

Ho, F. P., et al. (1976): "Storm Tide Frequency Analysis for the Open Coast of Virginia, Maryland, and Delaware," U.S. Dept. of Commerce, National Oceanic and Atmospheric Administration, Washington, D.C., NOAA-TM-NWS-HYDRO-32, NOAA-76111033 (August).

MARINOS, G., and J. W. WOODWARD (1968): "Estimation of Hurricane Surge Hydrographs," ASCE, *Journal of the Waterways and Harbors Division*, Vol. 94, No. WW2 (May), pp. 189–216.

MYERS, V. A. (1970): "Joint Probability Method of Tide Frequency Analysis Applied to Atlantic City and Long Beach Island, N.J.," U.S. Dept. of Commerce, ESSA *Technical Memorandum WBTM HYDRO 11* (April).

PETRAUSKAS, C., and L. E. BORGMAN (1971): "Frequencies of Crest Heights for Random Combinations of Astronomical Tides and Tsunamis Recorded at Crescent City, California," University of California, Berkeley, Hydraulic Engineering Laboratory, *Technical Report HEL 16-8* (March).

REID, R. O., and B. R. BODINE (1968): "Numerical Model for Storm Surges in Galveston Bay," ASCE, *Journal of the Waterways and Harbors Division*, Vol. 94, No. WW1 (February), pp. 33–57.

General Waves

ABERNETHY, C. L., and G. GILBERT (1975): "Refraction of Wave Spectra," Hydraulics Research Station, Wallingford, England, *INT 117* (May).

BEA, R. G. (1975): "Gulf of Mexico Hurricane Wave Heights," *Journal of Petroleum Technology* (September), pp. 1160–1172.

BATTJES, J. A. (1974): "Computation of Set-up, Longshore Currents, Run-up and Overtopping due to Wind-Generated Waves," Delft University of Technology, Dept. of Civil Engineering, *Report No. 74-2*.

BESSE, C. P., and N. F. LE BLANC (1966): "An Application of Oceanographic Data in Offshore Structural Design," Society of Petroleum Engineers of AIME, Symposium on Offshore Technology and Operations, New Orleans, La. (May), *Paper No. SPE-1419*.

BRETSCHNEIDER, C. L. (1968): "Significant Waves and Wave Spectrum," *Ocean Industry*, Vol. 3, No. 2 (February), pp. 40–46.

BRETSCHNEIDER, C. L. (1973): "Prediction of Waves and Currents," University of Hawaii, J.K.K. Look Laboratory of Oceanographic Engineering, Honolulu, Hawaii, *Look Lab Hawaii*, Vol. 3, No. 1 (January), pp. 1–17.

GAUGHAN, M. K., and P. D. KOMAR (1975): "The Theory of Wave Propagation in Water of Gradually Varying Depth and the Prediction of Breaker Type and Height," *Journal of Geophysical Research*, Vol. 80, No. 21 (July 20), pp. 2991–2996.

IWAGAKI, Y. (1968): "Hyperbolic Waves and Their Shoaling," *Proceedings*, Eleventh Conference on Coastal Engineering, London, England (September), Vol. 1, pp. 124–144.

JAHNS, H. O., and J. D. WHEELER (1973): "Long-Term Wave Probabilities Based on Hindcasting of Severe Storms," *Journal of Petroleum Technology*, Vol. 25 (April), pp. 473–486.

KOH, R. C. Y., and B. LEMÉHAUTÉ (1966): "Wave Shoaling," *Journal of Geophysical Research*, Vol. 71, No. 8 (April 15), pp. 2005–2012.

LEMÉHAUTÉ, B. (1976): *An Introduction to Hydrodynamics and Water Waves*. Springer-Verlag, New York.

LONGUET-HIGGINS, M. S. (1970): "Longshore Currents Generated by Obliquely Incident Sea Waves," Parts 1 and 2, *Journal of Geophysical Research*, Vol. 75, No. 33, pp. 6778–6801.

MCCLENAN, C. M. (1975): "Simplified Method for Estimating Refraction and Shoaling Effects on Ocean Waves," U.S. Army, Corps of Engineers, Coastal Engineering Research Center, Fort Belvoir, Va., *Technical Memorandum No. 59* (November).

RESIO, D. T., and C. L. VINCENT (1976): "Design Wave Information for the Great Lakes, Report 1, Lake Erie," U.S. Army Engineer Waterways Experiment Station, Vicksburg, Miss., *Technical Report H-76-1* (January).

SAVILLE, T., JR. (1951): "A Method for Drawing Orthogonals Seaward from Shore," U.S. Army, Corps of Engineers, Beach Erosion Board, *Bulletin*, Vol. 5, No. 4 (October 1), pp. 1–6.

SKOUGAARD, O., et al. (1975): "Computation of Wave Heights due to Refraction and Friction," ASCE, *Journal of the Waterways, Harbors, and Coastal Engineering Division*, Vol. 101, No. WW1, pp. 15–32.

WARD, E. G., D. J. EVANS, and J. A. POMPA (1977): "Extreme Wave Heights along the Atlantic Coast of the United States," *Proceedings*, Ninth Annual Offshore Technology Conference, Houston, Texas (May), Vol. 2, pp. 315–324.

WILSON, W. S. (1966): "A Method for Calculating and Plotting Surface Wave Rays," U.S. Army, Corps of Engineers, Coastal Engineering Research Center, *Technical Memorandum No. 17* (February).

ZOBEL, R. F., and R. DIXON (1970): "Note on the Forecasts of Wave Height Made by the Meteorological Office," *Meteorological Magazine*, Vol. 99, pp. 177–183.

Wave Energy Dissipation

HUNT, J. N., and BRAMPTON, A. H. (1972): "Effect of Friction on Wave Shoaling," *Journal of Geophysical Research*, Vol. 77, No. 33 (November 30), pp. 6558–6564.

LIU, P. L-F. (1973): "Damping of Water Waves over Porous Bed," ASCE, *Journal of the Hydraulics Division*, Vol. 99, No. HY12 (December), pp. 2263–2271.

PUTNAM, J. A. (1949): "Loss of Wave Energy due to Percolation in a Permeable Sea Bottom," *Transactions*, American Geophysical Union, Vol. 30, No. 3, pp. 349–356.

PUTNAM, J. A., and J. W. JOHNSON (1949): "The Dissipation of Wave Energy by Bottom Friction," *Transactions*, American Geophysical Union, Vol. 30, No. 1, pp. 67–74.

REID, R. O., and K. KAJIURA (1957): "On the Damping of Gravity Waves over a

Permeable Sea Bed," *Transactions*, American Geophysical Union, Vol. 38, pp. 662–666.

Wave Forces and Related Studies

DALRYMPLE, R. A. (1975): "Waves and Wave Forces in the Presence of Currents," *Proceedings*, Civil Engineering in the Oceans III, University of Delaware, Newark, Del., June, pp. 999–1018.

DENSON, K. H., and M. S. PRIEST (1974): "Effect of Angle of Incidence on Wave Forces on Submerged Pipelines," ASCE, National Meeting on Water Resources Engineering, Los Angeles, Calif. (January), *Meeting Preprint 2120*.

FUNG, Y. C. (1960): "Fluctuating Lift and Drag Acting on a Cylinder at Supercritical Reynolds Numbers," *Journal of the Aerospace Sciences*, Vol. 27, No. 11, pp. 801–804.

GRACE, R. A. (1971): "The Effects of Clearance and Orientation on Wave-Induced Forces on Pipelines: Results of Laboratory Experiments," University of Hawaii, J.K.K. Look Laboratory of Oceanographic Engineering, *Report No. 15* (April).

GRACE, R. A., and S. A. NICINSKI (1976): "Wave Force Coefficients from Pipeline Research in the Ocean," *Proceedings*, Eighth Annual Offshore Technology Conference, Houston, Texas (May), Vol. 3, pp. 681–694.

HOERNER, S. F., and H. V. BORST (1975): *Fluid-Dynamic Lift*, Hoerner Fluid Dynamics, Brick Town, N.J.

McNOWN, J. S., and G. H. KEULEGAN (1959): "Vortex Formation and Resistance in Periodic Motion," ASCE, *Journal of the Engineering Mechanics Division*, Vol. 85, No. EM1, pp. 1–6.

WILSON, B. W. (1963): "Anticipated Hurricane Effects on a Submarine Pipeline," *Proceedings*, Congress of the International Association for Hydraulic Research, London, England, pp. 327–350.

WILSON, J. F., and H. M. CALDWELL (1970): "Force and Stability Measurements on Models of Submerged Pipelines," *Preprints*, Second Annual Offshore Technology Conference, Houston, Texas (April), Vol. 1, pp. 747–758.

Local Scour

CHAO, J. L., and P. V. HENNESSY (1972): "Local Scour under Ocean Outfall Pipelines," *Journal of the Water Pollution Control Federation*, Vol. 44, No. 7 (July), pp. 1443–1447.

CHEE, S. P. (1975): "Resistance of Rocks to Movement," International Association for Hydraulic Research, *Proceedings*, 16th Congress, São Paulo, Brazil (July-August), Vol. 2, pp. 448–452.

HYDRAULICS RESEARCH STATION (1973): "A Study of Scour around Submarine Pipelines: Field Tests on the Behavior of Pipes when laid on the Sea Bed and Subjected to Tidal Currents," *INT 113*, Wallingford, England (March).

KJELDSEN, S. P., et al. (1973): "Local Scour near Offshore Pipelines," *Proceedings*, Second International Conference on Port and Ocean Engineering under Arctic Conditions, University of Iceland (August), pp. 308–331.

KJELDSEN, S. P., et al. (1974): "Experiments with Local Scour around Submarine Pipelines in a Uniform Current," Technical University of Norway, River and Harbor Laboratory, Trondheim, Report (March).

KOMAR, P. D., and M. C. MILLER (1973): "The Threshold of Sediment Movement under Oscillatory Water Waves," *Journal of Sedimentary Petrology*, Vol. 43, No. 4 (December), pp. 1101–1110.

NAHEER, E. (1977): "Stability of Bottom Armoring under the Attack of Solitary Waves," California Institute of Technology, W. M. Keck Laboratory of Hydraulics and Water Resources, Pasadena, Calif., *Report No. KH-R-34* (January).

Foundation Stability

BIJKER, E. W. (1974): "Coastal Engineering and Offshore Loading Facilities," *Proceedings*, Fourteenth Coastal Engineering Conference, Copenhagen, Denmark (June), pp. 45–65.

BONAR, A. J. (1971): "Research on Pipe Flotation," report prepared for ASCE Research Council at the University of Houston, Houston, Texas (June).

BONAR, A. J., and O. I. GHAZZALY (1973): "Research on Pipeline Flotation," ASCE, *Transportation Engineering Journal*, Vol. 99, pp. 211–233.

DE ALBA, P., C. K. CHAN, and H. B. SEED (1976): "Determination of Soil Liquefaction Characteristics by Large-Scale Laboratory Tests," prepared by Shannon and Wilson, Inc. and Agbabian Associates, Seattle, Wash., for U.S. Nuclear Regulatory Commission, *NUREG-0027, NRC-6* (September).

DE ALBA, P., H. B. SEED, and C. K. CHAN (1976): "Sand Liquefaction in Large-Scale Simple Shear Tests, " ASCE, *Journal of the Geotechnical Engineering Division*, Vol. 102, No. GT9 (September), pp. 909–927.

GHAZZALY, O. I., L. M. KRAFT, JR., and S. J. LIM (1975): "Stability of Offshore Pipe in Cohesive Sediment," *Proceedings*, Civil Engineering in the Oceans III, University of Delaware, Newark, Del. (June), pp. 490–503.

LEDFORD, R. C. (1953): "Design of Submarine Pipe Lines for Stability," *Petroleum Engineer*, Vol. 25, No. 5 (May), pp. D-70, D-72, D-74, D-76.

LEE, K. L., and H. B. SEED (1967): "Cyclic Stress Conditions Causing Liquefaction of Sand," ASCE, *Journal of the Soil Mechanics and Foundations Division*, Vol. 93, No. SM1 (January), pp. 47–70.

MANLEY, R. N., and J. B. HERBICH (1976): "Foundation Stability of Buried Offshore Pipelines—A Survey of Published Literature," Texas A&M University, College Station, Texas, *Report No. COE 174* (February).

PEACOCK, W. H., and H. B. SEED (1968): "Sand Liquefaction under Cyclic Loading Simple Shear Conditions," ASCE, *Journal of the Soil Mechanics and Foundations Division*, Vol. 94, pp. 689–708.

SEED, H. B., and I. M. IDRISS (1971): "Simplified Procedure for Evaluating Soil Liquefaction Potential," ASCE, *Journal of the Soil Mechanics and Foundations Division*, Vol. 97, No. SM9 (September), pp. 1249–1273.

SEED, H. B., and W. H. PEACOCK (1971): "Test Procedure for Measuring Soil Liquefaction," ASCE, *Journal of the Soil Mechanics and Foundations Division*, Vol. 97, No. SM8 (August), pp. 1099–1119.

SELLMEIJER, J. B. (1976): "Wave Induced Pore Pressures in Sandbeds," Delft Hydraulics Laboratory, Delft, Netherlands (October), [unpublished manuscript].

WILSON, B. W. (1961): "Foundation Stability for a Submarine Liquid Sulphur Pipeline," ASCE, *Journal of the Soil Mechanics and Foundations Division*, Vol. 87, No. SM4 (August), pp. 1–37.

YEN, B. C., R. L. ALLEN, and H. H. SHATTO (1975): "Geotechnical Input for Deep-Water Pipelines," *Proceedings*, Civil Engineering in the Oceans III, University of Delaware, Newark, Del. (June), pp. 504–522.

Earthquakes and Tsunamis

CLOUGH, R. W., and J. PENZIEN (1975): *Dynamics of Structures*, McGraw-Hill, New York.

COX, D. C., and J. F. MINK (1963): "The Tsunami of 23 May 1960 in the Hawaiian Islands," *Bulletin of the Seismological Society of America*, Vol. 53, No. 6 (December), pp. 1191–1209.

COX, D. C., G. PARARAS-CARAYANNIS, and J. P. CALEBAUGH (1976): "Catalog of Tsunamis in Alaska," revised 1976, World Data Center A for Solid Earth Geophysics, *Report SE-1* (March).

CROSS, R. H. (1967): "Tsunami Surge Forces," ASCE, *Journal of the Waterways and Harbors Division*, Vol. 93, No. WW4 (November), pp. 201–231.

"Earthquake Engineering Research Center Library Printed Catalog" (1975): University of California, Berkeley, Earthquake Engineering Research Center, *Report No. EERC 75-12,* (May).

GARRISON, C. J., and R. B. BERKLITE (1972): "Hydrodynamic Loads Induced by Earthquakes," *Preprints*, Fourth Annual Offshore Technology Conference, Houston, Texas (May), Vol. 1, pp. 429–442.

NEWMARK, N. M., and E. ROSENBLUETH (1971): *Fundamentals of Earthquake Engineering*, Prentice-Hall, Englewood Cliffs, N.J.

RASCON, O. A., and A. G. VILLARREAL (1975): "On a Stochastic Model to Estimate Tsunami Risk," *Journal of Hydraulic Research*, Vol. 13, No. 4, pp. 383–403.

SPAETH, M. G., and S. C. BERKMAN (1965): "The Tsunami of March 28, 1964, as Recorded at Tide Stations," U.S. Coast and Geodetic Survey, Dept. of Commerce, Rockville, Md. (April).

VAN DORN, W. G. (1965): "Tsunamis," *Advances in Hydroscience*, Vol. 2, pp. 1–48, Academic Press, New York.

WIEGEL, R. L., Ed. (1970): *Earthquake Engineering*, Prentice-Hall, Englewood Cliffs, N.J.

Fluid Mechanics

CHOW, V. T. (1959): *Open-Channel Hydraulics*, McGraw-Hill, New York.

DAILY, J. W., and D. R. F. HARLEMAN (1966): *Fluid Dynamics*, Addison-Wesley, Reading, Mass.

DAUGHERTY, R. L., and J. B. FRANZINI (1965): *Fluid Mechanics with Engineering Applications*, 6th ed., McGraw-Hill, New York.

HENDERSON, F. M. (1966): *Open Channel Flow*, Macmillan, New York.

ROUSE, H. (1946): *Elementary Mechanics of Fluids*, Wiley, New York.

ROUSE, H., Ed. (1950): *Engineering Hydraulics*, Wiley, New York.

SHAPIRO, A. H. (1961): *Shape and Flow*, Anchor Books, Garden City, N.Y.

STREETER, V. L. (1971): *Fluid Mechanics*, 5th ed., McGraw-Hill, New York.

9

The Backdrop for Construction

9-1 INTRODUCTION

Project Design

The two previous chapters have considered the technical aspects of marine outfall design. Although all of the design considerations and calculations are important, and appear in a document called a *Design Report*, the purpose of the project design is not the Design Report per se but the preparation of Plans and Specifications.

The drawings, or *plans*, for a project depict the physical aspects of the structure, showing the arrangement, dimensions, construction details, materials, profiles, and other information necessary for both *cost estimating* the project and in *building* it. Each outfall project has its associated roll of pertinent drawings.

Specifications are written instructions about the project; generally they describe how the project is to be constructed and what results are to be achieved. The specifications are customarily divided into two parts, one having to do with technical information, the other with nontechnical matters.

The Designer and the Construction Contractor

The prospective owner of an ocean outfall, such as a municipality or an industrial enterprise, usually employs an outside consulting engineering firm to carry out the outfall design referred to above. An *engineer-designer* is initially selected from among several such companies that have been invited to submit information related to their qualifications and experience in designing similar facilities. The owner reviews the information submitted and usually conducts interviews before choosing one firm that he considers best qualified. Negotiations are then conducted with the firm chosen to arrive at a mutually acceptable contract, including a fee, for the design. Of course, if an agreement cannot be reached at this point, the owner is free to enter negotiations with another firm. The fee is based on the anticipated costs plus a profit for the designer, and will amount to approximately 4–8% of the construction cost, depending on the size and complexity of the project. Since the purpose of this contract is to obtain professional design services, it is customary that the consultant (*designer*) is selected primarily on the basis of his professional qualifications, and not on the cost of the services alone.

Price competition, on the other hand, is the major consideration in the selection of the construction contractor. In a process known as *competitive bidding*, construction firms are invited to submit sealed bids for the construction phase of the project, based on the plans and specifications prepared by the designer. Bids are opened at an appointed time and publicly announced. Assuming that the owner has the requisite financial resources, and the low bid appears reasonable and is from a competent firm, a contract is then awarded to the low bidder.

Where the owner executes a single contract for the entire construction phase, as is generally the case, the contractor is known as a *general contractor*. The general contractor is legally responsible to the owner for the entire construction, although the contractor may in turn *subcontract* with other companies to carry out selected parts of the work.

Contracts awarded on the basis of competitive bidding are for a fixed price and tend to place the principals involved in an adversary relationship. On the one hand, the owner will interpret the plans and specifications in such a manner as to obtain the highest performance possible, while on the other hand, the contractor limits his effort to meeting minimum requirements in order to maximize his profit. This can lead to disagreements that are resolved through negotiation between the parties or by arbitration or litigation.

The contractor also assumes most of the risks that may be encountered during construction. These risks are not negligible in light of the circumstances he faces. The contractor is confronted with the task of building a novel and complex facility, within a given period of time and for a fixed sum of money, and under severe environmental conditions over which he has little control. It is not unusual at all for him to face numerous technical difficulties that necessitate costly changes in his plans. There are other potential extra costs that may arise as a result of weather, labor strikes, or price escalation that must be absorbed by the contractor.

The role of the designer during construction varies depending on the owner's in-house capability for administering the construction contract. He may be called upon to act as the owner's technical representative, or agent, in carrying out many of the owner's obligations under the construction contract. In this role, he oversees the progress of the work, interprets the requirements of the contract, and inspects and approves the work performed.

The Construction Manager

In another variation, an owner may employ a *construction manager*, in addition to the engineer-designer, to provide professional services during the planning, design, and construction phases of the project. A construction manager will typically have extensive experience and knowledge concerning construction methods and management and can make a valuable contribution to the owner in the control of cost, time, and quality of the facility constructed. The construction manager is employed under a professional services contract, in approximately the same way as the designer, and may be an engineering firm or a construction contractor.

During planning, the construction manager can provide input on programming and budgeting, cost analysis, and value engineering as well as scheduling. In the design phase, he can provide the designer with guidance on construction

Figure 9-1 Inspector diver leaving diving platform on trestle

materials, methods, and estimates of cost.* He can also formulate the construction plan, coordinate construction documents, manage bidding and negotiations, and handle contract awards for the owner. During construction, the construction manager can assist the owner with coordination, inspection (Figure 9-1), and other functions related to the administration and supervision of the work.

Outfall design is based on site data that are often incomplete or different from conditions encountered once construction is underway. Changes in design may be required under these circumstances, or for other unforeseen reasons that affect the progress of the work. While a design objective should be the minimization of perturbations arising from such effects, procedures for changes must be clearly established so that the team of owner, designer, contractor, and construction manager, as applicable, may work closely and expeditiously to see the project through to conclusion.

Orientation

This chapter is designed first to provide the bridge between the design and construction phases of an ocean outfall project. It is intended in addition that the chapter provide a brief outline of the "behind the scenes" activities involved in an outfall construction project. Chapter 10 deals almost exclusively with the actual field operations involved in outfall construction, whereas the remainder of this chapter deals with management and administrative considerations.

9-2 DETAILS CONCERNING PLANS AND SPECIFICATIONS

Introduction

The nontechnical (contractual and administrative) portion of the specifications, referred to in Section 9-1, generally appears under the heading of "General Provisions" or "General Conditions" with the technical matter grouped under a heading such as "Detailed Specifications." Municipal agencies can eliminate some duplication in effort in the preparation of specifications by using a standard format for the General Provisions. The City and County of Honolulu, Department of Finance, as an example, has its "General Provisions of Construction Contracts of the City and County of Honolulu." Some saving in design effort can be effected through the use of standardized drawings for

---

*It is virtually imperative that the designer get feedback from experienced outfall construction personnel before issuing plans and specifications. Construction feasibility and costs are major issues. The latter depends heavily on grade and alignment tolerances—which should be reasonable.

standard design details used throughout the system. Again as an example, the City and County of Honolulu, Department of Public Works, has a publication entitled "Standard Details for Roads, Storm Drains, and Sewers" that supplements the actual outfall drawings.

General Provisions in Specifications for Ocean Outfalls

The publication by the American Society of Civil Engineers (1966) lists 42 possible chapters within General Provisions for *engineering construction* projects. Engineering construction can be divided into highway, utility, and *heavy construction*. The latter includes the construction of marine outfalls. The following is a considerably shortened list of topics within General Provisions that might be considered suitable for outfall construction projects:

1. Notice inviting bids (invitation to bid)
2. Abbreviations and definitions
3. Proposal requirements and conditions (competency of bidders; disqualification of bidders; examination of plans, specifications, and site of work; proposal guarantee)
4. Award and execution of contract (comparison of bids; rejection of bids; award of contract; execution of contract; insurance; construction schedule; formal contract, notice to proceed)
5. Scope of the work (general; interpretation of plans and specifications; construction of roads and storage yards; removal of obstructions; restoration; changes; final cleaning up)
6. Performance of work (commencement of work; temporary suspension of work; contract time; extension of time for delay; additional and emergency work; disputes; performance schedule; value engineering incentive; suspension of contract; failure to complete work on time; final settlement of contract)
7. Control of material (samples and tests; defective materials; storage of materials; standards; material and equipment; drawings)
8. Control of the work (authority of the engineer-designer; supervision by contractor; inspection; provisions for inspection; removal of defective and unauthorized work; maintaining alignment and grade; methods and appliances)
9. Legal regulations and responsibilities (observance of laws and ordinances; rights-of-way; permits and licenses; U.S. Coast Guard regulations; sanitary provisions; safety; responsibility for loss or damage; government access to work; labor standards; certified payrolls; first aid; pollution control; security)

10. Protection of utilities and facilities (precautions; coordination with utility agencies)

11. Method of payment (progress payments; payment for delivered material; suspension of payments; final inspection and final payment; extra work)

Detailed Specifications in Specifications for Marine Outfalls

The listing below is a very general breakdown of the components in a perhaps typical set of detailed specifications for an outfall project. Considerable detail is provided on many of these separate subtopics in Chapter 10.

1. General (general description of work and location; datum; soils investigations; topography; electrical power; water; docking facilities; rail and highway facilities; material to be supplied by the owner; safety; mitigative measures; bypassing of flows during construction; abbreviations)

2. Materials (aggregate for pipe concrete, mortar or grout; water for pipe concrete, mortar or grout; cement; tremie concrete; reinforcing steel; gravel for pipe bedding; ballast and armor stone; reinforced concrete pipe and appurtenances; pipe fabrication; miscellaneous)*

3. Construction onshore (clearing and grubbing; excavation; control of water; reinforced concrete work; pipe foundation stabilization; pipe installation; backfilling and grading; piping; valves and fittings; painting and galvanizing; welding; electrical work; fence construction; demobilization and cleanup)*

4. Construction offshore (construction of trestle; sheet pile installation; barge anchoring; excavation for pipe; disposal of dredged material; blasting prior to pipe laying; pipe laying; testing pipe joints in place; installing diffuser; placing ballast and armor stone; placing underwater concrete; cathodic protection; demobilization and cleanup)

9-3 BUDGET ESTIMATE PREPARATION

Introduction

The plans and specifications discussed in the previous section provide the basis for estimating the construction costs of the project. A budget estimate is important both in assessing (and applying for) financial needs but also to judge

---

*Use of Standards of the American Society for Testing and Materials (Philadelphia, Pa.), as appropriate, is advised.

the reasonableness of ultimate bids. Accurate cost estimating for construction projects in general requires a very special skill. Cost estimating for marine work is an even more specialized area. Cost estimators for building construction, for example, are not normally adequately qualified to provide accurate estimates for ocean outfall jobs.

In this section the cost estimating being considered is that carried out by the owner (or his agent), prior to bidding on the project, to enable the owner to judge his financial requirements and to make arrangements accordingly. It is very important that this *budget estimate* be accurate. The cost estimating done by individual construction companies for the preparation of their bids is not specifically considered. This is because large construction companies with experience in ocean outfall work will have information available to them, through analysis of their past related jobs by their Cost Accounting Departments, that are not available to owners or designers in general.

When plans and specifications are vague and not specific in every detail, in part perhaps because the designer wished to give as much latitude as possible in construction techniques by the contractor, a wide range of contractor bids can be expected. Some of the resulting low bids may come from contractors with the expectation of increasing their compensation through "extras" made a part of the contract by "change orders."

There are three dimensions to construction cost estimating: equipment, materials, and labor. There are in turn two components for equipment and labor that go together to make the estimated cost: the cost per unit time (e.g., days) and the amount of time required to complete a particular task. Cost estimates for the material component of overall costs also involve two considerations: unit cost and quantity. The element of time can be associated with materials through the production rates of equipment and labor.

The experience of the cost estimator is called for in determining the mix of equipment and personnel to combine on a particular activity within the construction project. It may be necessary to consult union rules as well since, as an example, a power shovel not only requires an operator but also an oiler. The blend of equipment and personnel required to carry out a particular function is known as a *spread*. The overall daily cost of this array is then the sum of the daily costs of all components.

Wage rates can be obtained from appropriate tabulated union data for the work area. One option for computing charges for equipment is to assume that all such equipment required is rented.

A very useful booklet for heavy construction rental rates is the "Blue Book Rental Rates for Construction Equipment" (n.d.) divided into 22 separate chapters. A list of these chapter headings is of interest because it sets out the various activity areas in heavy construction projects:

1. Introduction
2. Index

3. Air Tools
4. Crushing and Conveying
5. Asphalt and Bituminous
6. Compaction
7. Concrete
8. Drilling
9. Tractors and Earthmoving
10. Excavating
11. Motors and Generators
12. Hoists and Derricks
13. Lifting
14. Marine
15. Pile Driving
16. Pumping
17. Road Maintenance
18. Shop Tools
19. Trailers
20. Trucks
21. Tunnel
22. Miscellaneous

The book gives rates that are averages for the United States and then provides corrections for different geographical regions of the country. The separate chapters are in continuous revision and are sent to subscribers of the publication as completed.

Chapter 14, Marine, in the "Blue Book..." is a very short one and does not contain a great deal of information. This is not a shortcoming of the book but an indication that costs of marine work vary widely with many factors as follows: geographical location, water routes, sea conditions, staging areas on land, local labor market (especially for divers), environmental controls, water depths involved, sea floor conditions, and of course the type of outfall and its burial requirements. These factors will be discussed in context in Chapter 10 of this volume, but they are variables in a marine outfall system that make it impossible to give general and meaningful cost figures in a cost-estimating handbook.

Outfall Construction Costs

Construction costs are continually in a state of flux, and even for a cost-estimating book that is frequently updated there are time lags to the preparation of the cost estimate and of course to the date at which construction

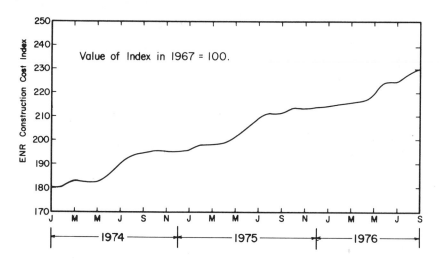

Figure 9-2   Increase in Engineering News-Record construction cost index during 1974, 1975 and 1976

is anticipated, the time for which construction costs have to be estimated. A heavily used indicator of overall construction costs is the Engineering News-Record (ENR) construction cost index. One form of this index was set at 100 in 1913, but a later version was set at 100 in 1967. The variation in this (1967) index for the three years 1974 to 1976 is shown for illustrative purposes in Figure 9-2.*

Two cost estimates for the Sand Island No. 2 outfall (#44 in Table 10-1) are presented in Table 9-1. The figures presented are summaries of detailed analyses of costs. Revision of the initial estimate was necessary because of a several-month delay in obtaining adequate project funding during a period of rapidly escalating construction costs. The 4.5-month, 23% jump in estimated total project cost shown in Table 9-1 is indicative of the rapid rate of cost escalation at that time, although in this case two different cost estimators were involved in preparing the two sets of numbers.

Mobilization/Demobilization is necessary because of the high initial cost of moving the plant to the project, as well as setting up to commence work, and because of the subsequent removal of equipment from the site and towing it away.

---

*During 1976, projected costs of U.S. wastewater treatment plants, force mains, and outfalls were made using 1973 costs and an annual 12.5% cost escalation. This is greater than the industry-wide construction rate of increase shown in Fig. 9-2.

Table 9-1: Estimated Cost for Construction of Sand Island No. 2 Outfall

Item		*Estimated Cost**	
		Sept. 27, 1972	*Feb. 13, 1973*
Mobilization/Demobilization		$ 860,000	$ 1,172,000
Excavation		1,347,819	2,420,521
Backfill		2,160,106	2,327,320
Pipe and Pipe-laying		3,313,435	3,404,028
Structures		8,850	40,820
	Subtotal	7,690,210	9,364,689
Contingency (5%)		384,510	468,234
	Subtotal	8,074,720	9,832,923
Estimated Profit (15%)		1,153,531	1,474,938
	Subtotal	9,228,251	11,307,861
State Tax (4%)		369,130	508,853
Total		$9,597,381	$11,816,714

*The low bid, received Oct. 11, 1973, was $13,574,571.

For the Feb. 13, 1973, cost estimate for the Sand Island No. 2 outfall in Table 9-1, a rough breakdown of the first four cost groups is as follows:

Mobilization and Demobilization: trestle, 49%; towing in and out of laying barge, 17%; preparing and dismantling rock barges, 12%; dock facilities, 4%; clearing site, 4%; bonds, insurance, permits, etc., 14%.*

Excavation: driving and pulling sheet piling on land, 7%; excavation on land, 3%; driving and leaving sheet piling along trestle, 42%; excavation from trestle, 18%; excavation offshore, 28%; miscellaneous, 2%.

Backfill: gravel bedding, 17%; stone, 43%; underwater concrete, 32%; other, 8%.

Pipe and pipe laying: on land, 7%; from trestle, 13%; from barge, 80%.

In the case of the Barbers Point outfall (#49 in Table 10-1) the summary of the cost estimate was presented in a somewhat different format in that costs were given for segments of the line rather than by the functional category of work. In that case, it was then possible to identify the total cost of excavation, backfill, etc., associated with each physical segment.

*Overhead as such, not an entry in Table 9-1, had this explicit part included under Mobilization and Demobilization.

9-4 BID PREPARATION, BIDDING, AND CONSTRUCTION CONTRACT AWARD

Introduction

Once the Final Design Report, the cost estimates, and the plans and specifications have been prepared by the engineer-designer, and after financial arrangements have been made by the owner, the next step is for a construction contractor to be chosen to carry out the work.

The notice that bids will be accepted for the work is normally placed in local newspapers and in engineering and trade publications. It is also given through direct contact with qualified construction companies. The plans and specifications are made available to contractors who express their interest so that they can prepare their bids. The Final Design Report is either distributed to each interested contractor or, more likely, made available in the offices of the designer or owner for perusal by the contractor's representative. For the Barbers Point outfall (#49 in Table 10-1) there were three reports in addition to the Final Design Report made available to prospective contractors for consultation: one on oceanographic data and its analysis, one on soil and foundation conditions, and an underwater photo reconnaissance report. Cores were also made available. Any questions by the contractors can be answered on a one-to-one basis by the designer, but a preferred method for clarifying the technical and administrative requirements of the proposed contract is a prebid meeting open to all contractors.

As an example, the Barbers Point outfall in Hawaii had an advertised bid-opening day of March 25, 1976. On Feb. 12, 1976, a prebid meeting was held. Besides having a number of the owner's and designer's representatives in attendance, all documents concerned with the outfall were available in the meeting room for study. There were also enough copies of the drawings so that each person in attendance could study his own during the course of the meeting as questions were being asked and fielded. Several days after the meeting a copy of the proceedings and a list of those who had attended were sent to those on the list.

A complete set of *Contract Documents* for a construction project normally consists of the following:

1. Notice to Contractors (Advertisement)
2. Instructions to Bidders
3. Bid Form
4. Contract (Agreement)
5. General and Special Conditions of Contract
6. Plans
7. General Specifications

8. Special Specifications
9. Addenda

When projects are competitively bid, it is customary that the required bidding and contract forms accompany the specifications. In large projects, such as those involved with major ocean outfalls, such contract forms may be bound separately.

Contract Forms

For a nominal sum ($50 in the case of the Barbers Point job) a prospective contractor can obtain the plans and specifications for an outfall project as well as the contract forms. The latter package typically contains various documents, the primary one of which is the proposal or bid form ultimately to be used by the contractor.

Also included is information on a "bid bond" (also called "bidder bond" or "proposal bond") and forms pertaining to it. Such bonds, payable to the owner, are used to ensure that the successful bidder will enter into a contract with the owner for the amount of his bid and will provide contract bonds and services as required. Alternates to the bid bond in whole or in part are certified or cashier's checks made payable to the owner. For the Barbers Point project (#49 in Table 10-1) the minimum fraction of the bid price required for the bid bond was 5% as specified by EPA. The bid bonds are returned at the bid opening or very shortly thereafter for the unsuccessful bidders, and after contract signing for the successful bidder.

Other items included in the contract forms package can include the following: blank copy of an affidavit swearing that bid was prepared in the absence of any collusion between or among prospective bidders; blank copy of a statement of nondiscrimination in employment; information on (minimum) wage rates; blank lists for proposed pipe manufacturers, other suppliers, and subcontractors; a Contractor's Qualification Statement, a form (detailed later) to provide details of contractor experience, financial data, and other pertinent information; a plan to be used in the proposed work; specified time of project completion [790 calendar days for Orange County No. 2 (#35) and 540 for Barbers Point (#49)] and daily penalty for extra time; and a blank copy of the proposed contract.

The Agreement for a lump sum construction contract can be subdivided into the following sections:

1. Identification of the Owner, Contractor, and Engineer
2. Scope of the Work
3. Time of Completion
4. The Contract Sum
5. Progress Payments
6. Extra Work

At the time of award of a contract, the contractor is usually required to furnish contract bonds. These bonds are issued by a *surety*; this refers to the party that assumes legal liability for the debt, default, or failure in duty of another. A *surety bond* is then the written document that describes the conditions and obligations of such an agreement. Two important classes of contract bond involved in marine outfall construction projects are the *performance* bond and the (labor and material) *payment* bond. Blank copies of these forms may be included with the contract forms.

The construction contractor's responsibilities involve the satisfactory completion of the work on the one hand and the (prompt) payment of all costs associated with the work on the other. Performance and payment bonds, respectively, are designed to assure these ends. Both have face values, often the full amount of the contract amount, that serve as upper limits to the expense that surety might incur in finishing the contract or in meeting the contractor's debts.

When a contractor approaches a surety company for coverage for an outfall project, that company will thoroughly investigate the contractor's financial and technical capability for carrying out the project. Surety companies consider various factors before underwriting a contract bond, and contractors applying for such bonds usually must present the following classes of information on the company, its personnel and its resources:

1. general historical background of the firm
2. recent (related) projects completed
3. financial status, including bank credit and secondary assets
4. equipment and machinery, with an indication of that immediately available
5. detailed information on company organization and the skills, responsibilities, and experience of key personnel

A further important consideration, in this case relating to the present project, is the amount of money "left on the table" by the contractor's bid. This refers to the difference between the low bid (assumed accepted) and the second-low bid. Clough (1975) suggests that a spread of more than 5–6% is cause for concern. The difference for the Barbers Point outfall (#49 in Table 10-1) was 6.5%, for Sand Island No. 2 (#44) 17%, but the figure for Mokapu (#47) was 47%.

The final contract forms item to be considered is the Contractor's Qualification Statement. The information called for includes details on the following: history of corporation; principals, their names, addresses, and construction experience; percent of construction work normally performed by the company itself (i.e., not subcontracted); any defaults by company or any other where principal officer involved; projects underway—name, owner, designer, contract amount, percent complete, scheduled completion date; major projects of company in past 5 years—name, owner, designer, contract amount, completion date, percent of work with own forces; states and categories in which the organization is legally qualified to do business; trade references; bank references; bonding company name and address; latest financial statement.

Bidding and Construction Contract Award

The *lump-sum contract* is the usual one for marine outfalls. In this case, the contractor agrees to complete the specified work for a fixed sum of money. It is the responsibility of the contractor to finish the project for the stipulated amount whatever difficulties he experiences. He has to absorb whatever losses he incurs unless the information he was given in conjunction with the preparation of his bid was clearly in error. The lump-sum contract is preferred by owners at the least because the total cost of the project is known in advance.*

The (lump-sum) proposal form for the Barbers Point outfall (#49 in Table 10-1) contained the following passage:

> "The undersigned hereby proposes and agrees, if this proposal is accepted, to furnish and pay for all labor, materials, tools and equipment and incidental work necessary to construct or install, in place complete, the work called for under and in accordance with the true intent of the Contract Documents...."

It is customary that contractors planning to bid a particular project so advise the owner by means of a written "Notice of Intention to Bid." A bid of a contractor is an offer to perform the specified work for a specified fee and is not a contract to actually perform the work. After the public bid opening, each bid is examined to determine if all the bidding conditions as specified in the General Conditions of the Specifications have been met. Any discrepancies that occur in the bid can result in disqualification of the bidder. Usually, the low bidder remaining after this filtering is selected to perform the work consistent with the plans and specifications.[†] The formal contract, or agreement, is then signed between the successful construction company and the owner. It should be noted

---

*The *unit price contract* is a common one in heavy construction. Because of uncertainties regarding quantities, only the price per unit weight or volume is set initially. During work, quantities are measured and the contractor is paid accordingly. This approach is unworkable in marine outfall projects because of the difficulty of determining actual quantities—dredged material, for example.

[†]In the United States, it is standard practice to select the low bidder to carry out a construction project. There are other approaches in some other countries (Clough, 1975). One of these involves the elimination of the lowest and highest bids and then the selection of the contractor with bid closest to the average of the remaining bids.

that if the owner is dissatisfied with the bids submitted, the bidding process can be repeated, with or without revised design plans.

I attended the bid opening for the Barbers Point outfall (#49 in Table 10-1) on March 25, 1976. The affair was over in three minutes, with the clerk in charge simply slitting open the sealed envelopes and reading the three bids as follows:

1. Dillingham Corporation, Hawaiian Dredging
    and Construction Co. (Division) $12,610,577
2. Healy Tibbitts Construction Co. $11,840,500
3. Morrison-Knudsen Co., Inc. $14,693,000

The types of developments leading to the disqualification of bidders are as follows:

1. Noncompliance with requirements for submitting "competency/qualifications" forms
2. Evidence of collusion between or among bidders
3. Defaults as well as lack of responsibility and cooperation in past work
4. Lack of proper or available equipment and/or sufficient experience
5. No contractor's license or a license that does not cover the type of work contemplated
6. More than one bid from the same organization
7. Late delivery of bid
8. Irregularities in the proposal (e.g., miscalculations, noninitialed erasures)

There are many contractors who do not have adequate marine construction experience to be able to properly bid an ocean outfall job. Their bids will probably be considerably too low as a result, but that will likely mean that they are the low bidders. If the owner blindly selects the contractor submitting the lowest bid, he may be in trouble. An example is illustrative.

In the design of an ocean outfall for Napier, New Zealand, the consultants gave an estimate of $500,000 for its construction. When bids were received for the work, three (all from experienced contractors) were in the range of $450,000 to $550,000, whereas one from an inexperienced contractor was for $176,000. The Napier City Council decided to give the job to the low bidder despite the consultant's adverse recommendation. The result was regrettable. The work was imperfectly executed and, with about one-quarter of it completed, the contractor abandoned the job with the owner's full agreement. The owner then undertook to finish the work, and the final cost was about $2.0 million.

9-5 PLANNING, SCHEDULING, CONTROLLING, AND MONITORING CONSTRUCTION

The General Concept of CPM

Introduction: Large projects, such as the construction of a large ocean outfall system, particularly if a sewage treatment plant facility is involved, are so involved in terms of separate jobs to be done and complicated in terms of the many people and organizations included, that some comprehensive way of managing the entire project is almost mandatory. In general, construction management, whether by the contractor or the construction manager, can be carried out using some approach such as CPM (Critical Path Method or sometimes referred to as Complete Project Management). The following paragraphs are devoted to an introduction to the general concept of CPM. In a later subsection, an example of its application to a particular ocean outfall project is presented.

Before introducing the general ideas behind CPM, it is appropriate to stress one outstanding feature of such a planning approach: the project is "built on paper." CPM forces people to plan; the whole of the project is completely thought through from start to finish. Another outgrowth of considerable importance involves the clear identification of individual components that are vitally important in terms of the construction sequence and timing.

There are three basic phases to the CPM procedure: planning, scheduling, and the control-monitor phase. The planning phase consists of breaking the project down into separate work elements called *activities* and establishing the logic or sequence of these activities. These are put together in an *arrow diagram* or *network* that shows the logic or flow of the project from start to finish, depicting how the separate activities that comprise the project influence and are influenced by each other. For example, the activity of "purchase concrete pipe" must precede another activity "lay concrete pipe in trench," and this activity in turn must have been accomplished for the activity "backfill trench" to be carried out. The activity arrows start and end at unnamed *events*, represented in the arrow diagram by numbers within circles.

Scheduling Phase: The very early stages of CPM planning are perhaps critical. During this period, the basic manner in which the construction is to be accomplished is established. Input from a variety of people is desirable in order to ensure that the arrow diagram is complete and correct.

Following the division of the project into activities and the drawing of an accurate arrow diagram depicting the logical flow from the beginning of the undertaking to the end, the scheduling phase begins. The time duration required for each activity is estimated, and the durations of all activities are printed

437

beside the appropriate arrows in the arrow diagram. The analyst then methodi-
cally computes the time it will take to proceed from start to finish. Obviously
different times would be computed along different branching parts of the
network. But there is one particular sequence of activities that gives the longest
time. This sequence constitutes the *critical path* for the project and obviously
governs the time it will take to complete it. Only by decreasing the duration of a
critical activity (one on the path) can the project time be reduced.

CPM continues with the estimation of the equipment and manpower
resources required for each activity as well as a cost estimation for each activity
(see Figure 9-3). Comparison of the estimated time of project completion with
the project completion date specified by the owner may reveal an inconsistency
which means that the contractor must in some way speed up his initial concept
of project construction timing. The contractor could also come to the same
conclusion for other reasons such as insufficient time between his own estimated
time of project completion and the specified time to allow for contingencies. In
marine outfall work, where weather and sea conditions can exert a tremendous
influence, this is an important consideration. During the construction of the
Sand Island No. 2 outfall (#44 in Table 10-1), for example, from January
18–February 3, 1975, there were no pipes laid from the trestle because existing
sea conditions caused heavy surge in the area.

The duration and cost aspects of an activity are combined in a sense since
both *normal* and *crash* durations can be assigned to each activity. The former is
considered to involve the least direct cost; the latter involves the least amount of
time computed to be possible for that activity.*

With reductions in total work time the intent, the scheduling phase of CPM
then continues, as necessary, to wittle down the time of completion of the
project by shortening the cheaper critical activities.† For a particular logic
diagram the project time of completion can be no shorter than when all the
critical activities are listed in terms of their crash durations.

When a final arrow diagram is completed, then various times associated
with any single activity can be computed. So as not to hold up a following
(dependent) activity, a preceding activity must be finished before a certain date
(*late finish*). That means it must have started before another date, the *late start*.
If a short-time activity parallels another (longer) one up to a certain point, it has
a certain length of time (*float*) to play with. If that activity starts right away
when it can, it starts at the *early start* date, and then it finishes at the *early finish*.
For critical activities, the "early" and "late" items must be the same.

---

*The consideration of indirect costs (e.g., salaries, rents, utilities, main office costs) may mean a
minimum *total* cost at a duration intermediate between these two.

†Overall time reductions may also be possible, and at no cost, by changing the network logic.

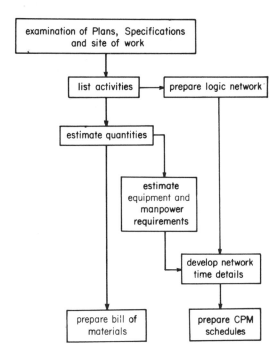

Figure 9-3 Evolution of construction plan

The contractor can develop the timing of his noncritical activities by using *resource leveling*. This involves the adjustment of start times across the spectrum from "early" to "late," so that the time-use of resources (e.g., men and equipment) is reasonably consistent throughout the project.

The final network diagram provides a clear and certain means of communication and coordination among all parties involved in the project.

Control-Monitor Phase: Things seldom go completely as planned in the construction industry where there are various factors that may completely alter a well-planned construction program. These include strikes, unusually bad weather, sudden material shortages, and unforeseen job conditions. The control-monitor phase of CPM is included to deal with just such a possibility. Job information, which is fed into this phase, includes the following: additions to project (new activities); deletions from the project; changes to durations, descriptions, trade indicators, cost estimates, and resources estimates; actual start dates; actual finish dates. The outputs of this phase are these: time status reports; revised schedules; revised resource analysis; revised cash flow predictions; cost status reports.

CPM in Practice

The owner may have had a prebid CPM (or other) construction plan developed that he expects a contractor to adopt and follow, but it is more usual for the contractor to develop his own plan after he has been awarded the contract. It is doubtful that he would have gone through the time and expense of developing a *detailed* CPM plan (which he truly expects to follow) before bid-opening, since contractors usually expect to secure contracts for only 5–30% of the work they bid on. It would not be unusual in such cases for the owner to review the contractor's CPM plan, for differences to be ironed out, and then for that plan to be adopted as a road map by the two parties. Both can then prepare cash flow projections.

There are various interesting features to an agreed-upon plan. When a particular event has been reached on the diagram, the contractor has obviously completed all the activities that lead up to that point and should then be paid for all of them. For another thing, the owner has a clear view of the critical items in the network and can disallow claims that pertain to noncritical items. Finally, the contractor, on the other hand, has a very clear way of charting extra activities carried out for which he would expect to be paid.

Each activity in the arrow diagram can be placed in a list and identified by its preceding and following event numbers, described briefly according to its nature, located in time according to its intended start and completion dates and activity duration, and represented in terms of cost.

Example of CPM Use in Outfall Construction

It is instructive to consider the details concerning the performance schedule for the Sand Island No. 2 outfall (#44 in Table 10-1). It was specified that the contractor had to submit his performance schedule for approval by the engineer within 30 days of the Notice to Proceed. As an interim measure, however, the contractor had to submit, within 10 days of the Notice to Proceed, a preliminary performance schedule covering the first 90 days of the work. This preliminary schedule, which could be either in bar (Gantt) chart form or CPM form,* was specifically to include information on the mobilization of plant and equipment and the sequence of all operations, including procurement of equipment and materials to complete the work. The general approach to be followed by the contractor through the balance of the project was also to be laid out. The costs of activities expected to be completed or partially completed before submission and approval of the final performance schedule were also to be part of the preliminary schedule.

---

*The bar chart is inferior to the CPM diagram at the least because it does not display interdependencies.

It was stated that the (final) performance schedule had to be a time-scaled network analysis system in CPM form showing the sequence and interdependence of all activities required for complete performance of the construction contract. Except for nonconstruction activities (e.g., procurement of materials, delivery of equipment, concrete curing) the detail of information was to be such that the duration of no activity exceeded 30 days.

It was stipulated that a mathematical analysis in computer-produced form had to accompany the network. This had to present the following data for each activity: beginning and ending event numbers, description, duration, cost,* earliest start and finish dates (calendar), latest start and finish dates, scheduled start and finish dates, total float, and general bid category of which activity was a part.

Two sets of data were to accompany the performance schedule and the computer-produced schedule. On the one hand the following information was required: proposed number of working days per week, planned number of shifts per day, the number of hours per shift, and the usage on the project of major construction equipment. In addition, cash flow curves had to be provided, based on both early and late finish activity dates.

The contractor was called upon to present monthly reports to the engineer. Items called for included the following:

1. The percentage completion of each activity as of the report date
2. Revised early and late start cash flow envelope projections
3. The percentage of total work completed
4. Progress along the critical path in terms of days ahead of or behind schedule
5. Discussion of problem areas, current and anticipated delaying factors and their impact, and an explanation of corrective actions to be taken or proposed

With respect to the last item, the following passage appeared in the specifications for the Barbers Point outfall (#49 in Table 10-1), for which CPM was used.

> "The Contractor acknowledges and agrees that actual delays in activities which, according to the computer-produced calendar-dated schedule, do not affect any contract completion date shown by the critical path, do not have any effect on the contract completion date or dates and therefore will not be the basis for a change therein."

Progress payments to the Barbers Point and Sand Island No. 2 contractors

---

*The sum of all applicable activity costs had to equal the total contract bid price. Overhead and profit were to be prorated throughout all applicable activities.

were based on the total value of activities completed or partially completed and shown on the updated diagram and computer-produced schedule. Underwater inspections to verify satisfactory completion of work can be done by the construction manager himself, as for Sand Island No. 2, or by a diving subcontractor, as at Mokapu, who reports to the construction manager.

9-6 VARIOUS TOPICS

Insurance and Safety

It is standard practice that construction contracts require the contractor to provide certain insurance coverages. Possible inclusions are the following: project property insurance, contractor's property insurance, insurance against flood-caused damages, liability insurance, employee insurance, motor vehicle insurance, business insurance, accident insurance, and life insurance. Some forms of insurance coverage are also advisable for the owner and designer.

Safety is an important aspect of construction work. The work associated with ocean outfalls is particularly dangerous, both above and below water. During the construction of the Point Loma outfall at San Diego (#18 in Table 10-1), two men lost their lives, one of whom was a diver. In the United States, above-water safety measures must conform to the Construction Safety Act of 1969 and to the Occupational Safety and Health Act (OSHA) of 1970.* OSHA inspectors make unannounced visits to construction sites and civil penalties are assessed for major violations, with citations issued for minor ones. By all accounts, the OSHA inspections, when they take place, are tough and thorough. The barge used in laying the Sand Island No. 2 outfall (#44) was visited once by an OSHA representative and also once by the U.S. Coast Guard. The shore area and the trestle for a project involving the laying of twin cooling water outfalls at Kahe, Oahu, Hawaii, in 1976, was once visited by federal and State of Hawaii inspectors together.

Change Orders

It may become apparent during heavy construction projects that portions of the plans and specifications are unworkable. For this reason, the owner reserves the right to make alterations to the plans and specifications as he sees fit. Such alterations can involve deviations, additions, or omissions. Such changes are officially set forth in a *change order*. This describes in detail the modifications to be made and the reasons for such changes; it also spells out the change in contract amount and any authorized extension of contract time.

---

*Dive regulations will be considered in Chapter 10. OSHA does consider one item that is directly diver-related—underwater blasting.

On marine outfall projects, change orders normally involve what is classified as *additional work* (extras). As an example, during the first half of the construction of the Mokapu outfall (#47 in Table 10-1) there were four change orders involving an additional $1.2 million. In each of these cases, the change order amount resulted from negotiations among the owner, the construction manager, and the contractor, following the preparation of separate cost estimates by the construction manager and the contractor for the net project cost increase associated with the change.

The type of situation that can lead to the issuance of a change order can be illustrated by a situation that has actually occurred in marine outfall construction. Insufficient information on subbottom conditions was obtained for a project partly because of the cost and partly due to the delays that would be involved in preparing and having reviewed an EIS for setting off exploratory explosive charges at various stations along the intended line. The owner took the calculated risk that the subbottom conditions were such that the driving of sheet piling along the outfall in nearshore waters, called for by the designer, would be possible. When it turned out that the subbottom material was a very dense rock and sheet piling could not be driven, a change had to be made.

Where change orders must receive the approval of various governmental agencies, such as in the United States where municipal, county, state, and federal monies may be involved, as well as regulations at all levels, long delays are customary. It is, of course, out of the question to hold up a project for several months, so that work normally goes on according to the altered plans, simply from the written order of the owner.

Value Engineering

Many construction contracts include what are known as *value engineering* or *value analysis* incentive clauses. Such contract inclusions have been used particularly by public agencies, and since October 1, 1976, the U.S. Environmental Protection Agency has made value engineering studies mandatory on its construction grants projects of wastewater treatment facilities over $10 million. Although value engineering can be aimed at reducing the cost of ownership, it is normally involved in decreasing the cost of performance of the construction project. This must be achieved without impairing any of the essential functions of the facility being built or jeopardizing characteristics such as service life, reliability, economy of operations, ease of maintenance, or desired appearance.

A value engineering change proposal (VECP) originates with the contractor, to change drawings, specifications or other requirements of the contract. The reduction in contract performance costs must be predicted. If the owner accepts the VECP, a change order is executed. The contract price is reduced by the total estimated decrease in the cost of performance less 50% of the difference between the total amount of such total estimated decrease and any owner costs involved in applying VECP to the contract.

Value engineering can also be practiced in planning and design of construction projects. A team of specialists and/or the construction manager can go over the system proposed by the designers with the objective of eliminating excess costs while not compromising the quality or function of the various system components.

9-7 REFERENCES

"Blue Book Rental Rates for Construction Equipment" (n.d.): compiled by National Research and Appraisal Company, published by Equipment Guide-Book Co., Palo Alto, Calif.

CLOUGH, R. H. (1975): *Construction Contracting*, 3rd ed., Wiley, New York.

"General Provisions of Construction Contracts of the City and County of Honolulu" (1967): City and County of Honolulu, Hawaii, Dept. of Finance (July).

"Standard Details for Roads, Storm Drains and Sewers" (1973): City and County of Honolulu, Hawaii, Dept. of Public Works (February).

9-8 BIBLIOGRAPHY

ADRIAN, J. J. (1973): *Quantitative Methods in Construction Management*, American Elsevier, New York.

ADRIAN, J. J. (1976): *Business Practices for Construction Management*, American Elsevier, New York.

ALDRIDGE, R. G., J. G. BOMBA, and W. G. BROWNFIELD (1971): "Offshore Construction Specifications—An Art in Itself," *Pipeline and Gas Journal*, Vol. 198, No. 2, (February), pp. 41–43.

(The) American Institute of Architects (1969): "Recommended Guide for Bidding Procedures and Contract Awards," Washington, D.C. (November).

(The) American Institute of Architects (1970): "General Conditions of the Contract for Construction," AIA Document A201, Washington, D.C. (April).

American Society of Civil Engineers (1966): "Form of Contract for Engineering Construction Projects," ASCE Form JCC-1, New York.

American Society of Civil Engineers (1970): *Design and Construction of Sanitary and Storm Sewers*, New York.

(The) Associated General Contractors of America, Inc. (1962): "Processing Change Orders and Disputes on Federal Construction Contracts," Washington, D.C.

(The) Associated General Contractors of America, Inc. (1965): *CPM in Construction*, Washington, D.C.

(The) Associated General Contractors of America, Inc. (1966): "Contractor's Equipment Ownership Expense," Washington, D.C.

(The) Associated General Contractors of America, Inc. (1968): *Cost Control and CPM in Construction*, Washington, D.C.

(The) Associated General Contractors of America, Inc. (n.d.): "Insurance and Bond Check List," Washington, D.C. (November).

AYERS, C. (1975): *Specifications: for Architecture, Engineering, and Construction*, McGraw-Hill, New York.

BONNY, J. B., and J. P. FREIN, Eds. (1973): *Handbook of Construction Management and Organization*, Van Nostrand Reinhold, New York.

BUSH, V. G. (1975): *Safety in the Construction Industry: O.S.H.A.*, Reston Publishing Co., Reston, Va.

CLOUGH, R. H. (1972): *Construction Project Management*, Wiley (Interscience), New York.

"Contract Documents for the Construction of an Ocean Outfall and Appurtenances" (1972): prepared by CH_2M-Hill, Corvallis, Oregon, for the Port of Coos Bay, Oregon (March).

County Sanitation District No. 2 of Los Angeles County, California (1964): "Specifications and Drawings for the Construction of Joint Outfall 'A' Unit 1-D, 4th Ocean Outfall," Los Angeles, California (March 25).

DAY, D. A. (1973): *Construction Equipment Guide*, Wiley (Interscience), New York.

DERK, W. T. (1974): *Insurance for Contractors*, 4th ed., Fred S. James & Co., Inc., Chicago, Illinois.

DOUGLAS, J. (1975): *Construction Equipment Policy*, McGraw-Hill, New York.

DUNHAM, C. W., and R. D. YOUNG (1971): *Contracts, Specifications, and Law for Engineers*, 2nd ed., McGraw-Hill, New York.

FOSTER, N. (1972): *Construction Estimates from Take-off to Bid*, McGraw-Hill, New York.

Green Guide for Construction Equipment (1977): Equipment Guide-Book Co., Palo Alto, Calif.

GROW, T. A. (1975): *Construction: A Guide for the Profession*, Prentice-Hall, Englewood Cliffs, N.J.

HAVERS, J. A., and F. W. STUBBS, JR., Eds. (1971): *Handbook of Heavy Construction*, 2nd ed. McGraw-Hill, New York.

JOHN A. CAROLLO, Consulting Engineers (1968): "Specifications and Plans for Ocean Outfall No. 2, Marine Section," prepared for County Sanitation Districts of Orange County, Fountain Valley, California.

LEWIS, J. R. (1975): *Construction Specifications*, Prentice-Hall, Englewood Cliffs, N.J.

O'BRIEN, J. J. (1971): *CPM in Construction Management; Project Management with CPM*, 2nd ed., McGraw-Hill, New York.

O'BRIEN, J. J. (1974): *Construction Inspection Handbook*, Van Nostrand–Reinhold Co., New York.

OPPENHEIMER, S. P. (1971): *Directing Construction for Profit: Business Aspect of Contracting*, McGraw-Hill, New York.

PAGE, J. S. (1977): *Cost Estimating Manual for Pipelines and Marine Structures*, Gulf Publishing Co., Houston, Texas.

PARKER, H. W., and C. H. OGLESBY (1972): *Methods Improvement for Construction Managers*, McGraw-Hill, New York.

PEURIFOY, R. L. (1958): *Estimating Construction Costs*, 2nd ed., McGraw-Hill, New York.

PEURIFOY, R. L. (1970): *Construction Planning, Equipment and Methods*, 2nd ed., McGraw-Hill, New York.

Portland Cement Association (1975): *Administrative Practices in Concrete Construction*, Wiley, New York.

RADCLIFFE, B. M., D. E. KAWAL, and R. J. STEPHENSON (1967): *Critical Path Method*, Cahners Publishing Company, Inc., Chicago, Ill.

"Safety and Health Regulations for Construction" (1972): Occupational Safety and Health Administration, U.S. Dept. of Labor, *Federal Register*, Vol. 37, No. 243 (Dec. 16), pp. 27503–27598.

"Standard Specifications for Public Works Construction" (1975): City and County of Honolulu, Counties of Kauai, Maui, and Hawaii, Dept. of Public Works (May).

"The 1976 Dodge Guide for Estimating Public Works Construction Costs" (1976): Annual Edition No. 8, McGraw-Hill Information System Co., New York.

"Value Engineering Workbook for Construction Grant Projects" (1976): U.S. Environmental Protection Agency, Washington, D.C., *MCD-29, EPA-430/9-76-008* (July).

Wheel and Crawler Loaders: Specifications (n.d.): Equipment Guide-Book Co., Palo Alto, Calif.

WIEST, J. D., and F. K. LEVY (1969): *A Management Guide to PERT/CPM*, Prentice-Hall, Inc., Englewood Cliffs, N.J.

10

Construction of Outfall

10-1 INTRODUCTION

Hundreds and hundreds of outfalls have been constructed in marine waters. Some of these have been included in Table 10-1, which names and briefly outlines the main characteristics of 50 outfalls from various parts of the world. This chapter is designed to present the procedures involved in outfall construction with direct reference to specific outfalls where the particular procedures were employed.

The excavation of a trench for an outfall will be considered in Section 10-2, with Sections 10-3–10-5 devoted to the placing of an outfall in such a trench or directly on the sea floor. The bedding of the outfall with gravel and the protecting of the pipe with armor rock occupy Section 10-6.

Underwater operations in general, and diving in particular, play vital roles in marine outfall construction and are detailed in Section 10-7. Section 10-8 outlines an outfall construction method where excavation accompanies or follows pipe laying. The final section of text in this chapter, 10-9, considers miscellaneous items related to outfall construction but not directly related to the topics of earlier sections.

10-2 TRENCHING

Introduction

There are two general approaches to the burying of a submarine pipe. In the first case, the pipe is laid in a preformed trench that is then either mechanically backfilled or allowed to fill in naturally as waves and currents

Table 10-1: Fifty Selected World Outfalls

Number	Indicator Year for Construction	Name and Location	General Outfall Characteristics				
			Pipe Size and Material	Approximate Length, ft (m)	Approximate Maximum Water Depth, ft (m)	Receiving Water	References*
1	1939	Mornington, Port Philip Bay, Vic., Australia	12-in. pipe (300-mm)	2600 (800)	33 (10)	ocean embayment	80
2	1942	Deal, New Jersey	17.5-in. wrought iron pipe (440-mm)	1100 (350)	— —	—	70
3	1942	Glenelg, Adelaide, S.A., Australia	24-in. steel pipe (610-mm)	1100 (350)	13 (4)	large ocean embayment	73
4	1946	Ashbridge's Bay, Toronto, Ontario, Canada	108-in. RCP (2740-mm)	3300 (1000)	47 (14)	large lake	74
5	1947	Whites Point, near Los Angeles, California	72-in. RCP (1830-mm)	5000 (1500)	110 (34)	open coast-ocean	15 22
6	1948	Hyperion, Los Angeles, California	144-in. RCP (3660-mm)	5300 (1600)	60 (18)	open coast-ocean	54
7	1949	Sand Island No. 1, Honolulu, Hawaii	78-in. RCP (1980-mm)	3700 (1100)	40 (12)	open coast-ocean	35
8	1949	Watsonville, California	16-in. wrought iron pipe with concrete jacket (400-mm)	1400 (400)	— —	open coast-ocean	32
9	1953	Orange County No. 1 south of Los Angeles, California	78-in. RCP (1980-mm)	7000 (2100)	55 (17)	open coast-ocean	28 75
10	1955	Palm Beach, Florida	30-in. pipe (760-mm)	5800 (1800)	90 (27)	open coast-ocean	52
11	1957	Hyperion (sludge line), Los Angeles, California	22-in. steel pipe (560-mm)	36,000 (10,900)	300 (91)	open coast-ocean	48 50 56

Table 10-1 (cont.)

12	1959	Hyperion, Los Angeles, California	144-in. RCP (3660-mm)	27,500 (8400)	200 (61)	open coast-ocean	5 33 36 50 65 23
13	1959	plant at Pittsburgh, Suisun Bay, California	18-in. steel pipe (460-mm)	360 (110)	36 (11)	estuary	
14	1959	Port Lincoln, S.A., Australia	12-in. steel pipe (300-mm)	560 (170)	10 (3)	ocean embayment	73
15	1960	Swanbourne, W.A., Australia	30-in. steel pipe with concrete jacket (760-mm)	3500 (1100)	36 (11)	open coast-ocean	73
16	1960	Watsonville, California	39-in. steel pipe (990-mm)	3900 (1200)	60 (18)	open coast-ocean	42
17	1961	Sandy Bay, Hobart, Tas., Australia	20-in. steel pipe with concrete jacket (510-mm)	2440 (700)	100 (30)	estuary	73
18	1962	Point Loma, San Diego, California	108-in. RCP (2740-mm)	14,000 (4400)	215 (65)	open coast-ocean	60
19	1963	International Paper Co., Gardiner, Oregon	30-in. (760-mm)	3000 (900)	30 (9)	open coast-ocean	6
20	1963	North Miami Beach, Florida	—	10,000 (3000)	60 (18)	open coast-ocean	25
21	1964	Whites Point, near Los Angeles, California	120-in. RCP (3050-mm)	11,900 (3600)	190 (58)	open coast-ocean	61
22	1964	Woodman Point, Fremantle, W. A., Australia	30-in. steel pipe with concrete jacket (760-mm)	6000 (1800)	55 (17)	partial ocean embayment	73
23	1964	Portsmouth, New Hampshire	—	—	—	—	59

Table 10-1: Fifty Selected World Outfalls (cont.)

			General Outfall Characteristics				
Number	Indicator Year for Construction	Name and Location	Pipe Size and Material	Approximate Length, ft (m)	Approximate Maximum Water Depth, ft (m)	Receiving Water	References
24	1965	West Point, Seattle, Washington	96-in. RCP (2440-mm)	3900 (1200)	240 (73)	ocean embayment	11 16
25	1965	Gisborne, New Zealand	30-in. concrete pipe (760-mm)	6000 (1800)	16 (5)	ocean embayment	83
26	1968	Hollywood, Florida	60-in. concrete pipe (extension) (1500-mm)	12,000 (3600)	90 (27)	open coast-ocean	—
27	1968	Hastings, United Kingdom	28-in. steel pipe (710-mm)	10,500 (3200)	60 (18)	open coast-ocean	12 27
28	1969	San Francisco Bay, California	54-in. RCP (1370-mm)	2700 (800)	—	large estuary	3
29	1969	Ingoldmells, Linc., United Kingdom	24-in. steel pipe with concrete jacket (610-mm)	8100 (2500)	—	—	18
30	1969	Nelson, New Zealand	36-in. concrete pipe (910-mm)	1500 (450)	45 (14)	ocean embayment	2
31	1969	Straight Point, S. Devon, United Kingdom	24-in. PVC pipe (610-mm)	—	—	—	—
32	1970	Macauley Point, Victoria, B.C., Canada	36-in. concrete encased steel pipe (910-mm)	6000 (1800)	200 (60)	ocean strait	—

Table 10-1 (cont.)

33	1970	Rochester, New York	120-in. steel pipe (3050-mm)	18,000 (5400)	100 (30)	large lake	71
34	1970	Prince Rupert, B.C., Canada	32-in. high-density polyethylene pipe (810-mm)	480 (150)	85 (26)	ocean embayment	1
35	1970	Orange County No. 2, south of Los Angeles, California	120-in. RCP (3050-mm)	27,400 (8400)	195 (59)	open coast-ocean	26 53
36	1970	Lions Gate STP, Vancouver, B.C., Canada	54-in. FRP (1370-mm)	700 (210)	66 (20)	ocean strait	—
37	1970	Baglan, South Wales, United Kingdom	48-in. steel pipe with concrete jacket (1220-mm)	13,300 (4100)	—	—	77
38	1971	Port Fairy, Vic., Australia	15-in. pipe (380-mm)	5000 (1500)	66 (20) 35 (11)	open coast-ocean	80
39	1972	Monmouth County, New Jersey	48-in. concrete encased steel pipe (1220-mm)	—		—	—
40	1973	Five Finger Island Outfall, Nanaimo, B.C., Canada	36-in. steel pipe (910-mm)	6600 (2000)	250 (75)	ocean strait	—
41	1973	Nassau County, South Oyster Bay, Long Island, New York	84-in. RCP (2130-mm)	36,000 (11,000)	100 (30)	—	46
42	1973	Cannes, France	53-in. concrete encased steel pipe (1350-mm)	4000 (1200)	330 (100)	open coast-sea	38
43	1974	Powell River, B.C., Canada	24-in. high-density polyethylene pipe (610-mm)	2000 (600)	200 (61)	ocean strait	—

Table 10-1: Fifty Selected World Outfalls (cont.)

			General Outfall Characteristics				
Number	Indicator Year for Construction	Name and Location	Pipe Size and Material	Approximate Length, ft (m)	Approximate Maximum Water Depth, ft (m)	Receiving Water	References
44	1974	Sand Island No. 2, Honolulu, Hawaii	84-in. RCP (2130-mm)	12,500 (3800)	235 (72)	open coast-ocean	31
45	1975	Harmac Pulp Ltd., Nanaimo, B.C., Canada	48-in. steel pipe (1220-mm)	3800 (1200)	350 (107)	ocean strait	58 43
46	1975	Ipanema Beach, Rio de Janeiro, Brazil	102-in. prestressed concrete pipe (2440-mm)	14,200 (4300)	92 (28)	open coast-ocean	55
47	1976	Mokapu, Kailua, Hawaii	48-in. RCP (1220-mm)	5000 (1500)	105 (35)	open coast-ocean	—
48	1976	Santa Barbara, California	48-in. RCP (1220-mm)	8700 (2700)	—	open coast-ocean	40 67
49	1976	Barbers Point, Hawaii	78-in. RCP (1980-mm)	10,500 (3200)	200 (61)	open coast-ocean	—
50	1977	Waitara, New Zealand	30-in. post-tensioned RCP (760-mm)	3600 (1100)	37 (11)	open coast-ocean	—

*Many references in this chapter are anonymous. For ease in referring to this material, particularly in this table, all references in Chapter 10 have been numbered.

move material over the sea floor. Trenching as a part of this approach will be considered in this section. On the other hand, the pipe may be laid directly on the sea floor, and trenching, lowering, and backfilling are essentially done in one overall operation. This procedure will be considered in Section 10-8.

The excavation of a trench on the sea floor for a pipeline is often done with a dredge. There are two major classifications of offshore dredges: mechanically operating and hydraulically operating. The mechanical type can be further divided into repetitive and continuous types. The hydraulic class of dredge is sometimes called a *suction dredge*.

Mechanical Dredges

There are several types of continuous mechanical dredge, such as the bucketline dredge, but this overall type of dredge sees little use in trenching for submarine pipes and will not be considered further. There are, however, three types of repetitive mechanical dredges that can be used in trenching for ocean outfalls. These are the dipper, dragline, and clamshell dredges, and these will be briefly described in turn below.

The *dipper* is the floating equivalent of the land-based excavating shovel. In fact, it could be viewed as such a shovel mounted on a barge. Because of the limited length of the shovel arm, dredging is limited to a depth of about 100 ft (30 m). The rigid connection between the shovel and the barge, which is good for forcing the shovel into bottom material, means considerable problems in excavating in even small seas, so this dredge is usually used only in protected waters. The best use for the dipper is for excavating hard compact materials, rock, and other solid formations after blasting.

The *dragline* dredging system makes use of an open bucket (Figure 10-1) that is dragged by cables over the ocean bottom such that material is scraped

Figure 10-1 Dragline bucket (Courtesy, The Yaun-Williams Bucket Co.)

into the bucket. The dragline can be operated from a ship or it can be controlled from the shore using a winch and a return line running through a pulley connected to an offshore anchoring system. This method is simple and involves a low capital cost, but even with a large bucket the throughput of material is low. There is also a real problem in controlling the depth of cut using this technique, although the alignment of a prospective line can be very well followed. A decided advantage to the dragline technique in some cases is that it can be used in deep water, into the hundreds of feet if required. A dragline is shown coming ashore in Figure 10-8. Much of the dredging for the Georgia-Pacific outfall at Newport, Oregon, was done using a dragline following breakup of the bottom material by a heavy steel chisel.

The *clamshell*, or grapple or grab, dredge has jaws that open near the sea floor and take a large bite out of the bottom material when the jaws are reclosed (see Figure 10-2). It works best as an excavator in soft materials, although not so soft that material washes out of the clamshell. The grab dredge is suspended from, and raised and lowered by, wire rope. The four-rope heavy digging type of clamshell dredge has been termed one of the "world's foremost digging tools" (Welling and Cruickshank, 1966). Thirty-ton grabs are possible with such a large dredge, and the material, when raised to the surface, is usually *sidecast* (dropped back into the water some distance to the side as in Figure 11-1) or dropped into an attending barge, as would be the case for the dipper and possibly the dragline dredge. The clamshell dredge can work in depths of approximately 60–200 ft (20–60 m). One of its major advantages is that it can be opened underwater in the event that it becomes caught on an underwater obstruction. Clamshells on barges were used for excavation offshore for the Orange County No. 2, Sand Island No. 2, and Mokapu outfalls (Table 10-1). A clamshell was used to

Figure 10-2 Clamshell bucket (Courtesy, The Yaun-Williams Bucket Co.)

excavate the trench between rows of sheet piling under the trestle for the Sand Island No. 2 outfall. On the Mokapu job, a trestle-mounted clamshell (Figure 10-3) removed debris from blasting.

In general, mechanical dredges are all characterized by their inability to transport the dredged materials for long distances, their lack of self-propulsion, and their relatively low production. They can, however, work well in restricted waters and in relatively compact formations (Murphy and Herbich, 1969).

Figure 10-3 Clamshell bucket used from trestle during Mokapu outfall construction

Hydraulic Dredges

Hopper and Dustpan Dredges: Hydraulic dredges work by using pumps and a pipe extending to the sea floor to suck bottom material and water up toward the dredge. Hydraulic dredges are self-contained in that they dig and dispose of the excavated material, unlike the mechanical dredges. Disposal is effected either by pumping a slurry through a pipeline to a *spoil* (storage) area or else pumping the sediment into internal hoppers, or bins, which are later emptied over a designated area. Generally speaking, the hydraulic dredge is more efficient, versatile, and economical than the mechanical dredges in situations where large quantities of material must be excavated.

There are three types of hydraulic dredges. The first of these is the self-propelled *hopper dredge*, or simply the *trailer suction dredge*. This dredge looks like an ocean ship except that it has a drag-arm on either side of the vessel.

These arms, which have suction dragheads at their bottom ends, are raised and lowered with hoisting tackle and winches. The pump discharge passes into hoppers where the sediment settles, and the excess water passes overboard through overflow troughs. Centrifuges can be used to strip this overflow of fine suspended material. Once the hoppers are full of material, the ship proceeds to the spoil area and the hoppers are bottom dumped. The dredge then returns to the dredging location.

The trailer suction dredge is particularly good for working in traffic. It moves at approximately 2–3 mph (3–5 km/h) and has twin propellers and twin rudders for maneuverability. Such dredges have lengths of approximately 180–550 ft (55–170 m) with hopper capacities from 500–8000 cubic yards (400–6000 m³). These dredges can work to a depth of approximately 70 ft (21 m).

The *dustpan dredge* is the second type of hydraulic dredge. In this case, the suction head is about as wide as the vessel, and there are high-velocity jets to stir up bottom material before it is sucked into the head. The dustpan dredge is good for moving high volumes of soft material to a spoil area by means of a pipeline. It is sometimes paired with a trailer suction dredge, pumping material from the mobile dredge's spoil area to another one on shore.

Cutterhead Dredge: The third type of hydraulic dredge is the hydraulic pipeline *cutterhead dredge*, or more simply the *cutter suction dredge*, which is widely accepted as being the most efficient and versatile dredge available. A major feature of this type of dredge is that there is a rotating cutter apparatus (Figure 10-4) surrounding the intake end of the suction pipe. These dredges can excavate in all types of alluvial materials, clay, and hardpan. The larger and

Figure 10-4 Cutter for cutterhead dredge

more powerful dredges can even excavate softer basalt, limestone, and coral without blasting.

The cutter is powered, through a shaft, by a motor located above water. The motor-shaft-suction pipe setup is mounted on a type of ladder at the bow of the dredge, and this ladder can be tilted up or down by cables from an "A" frame (see Figure 10-5). In some cases, the suction pipe may double as the motor shaft. The dredge is also fitted out with a large boom, e.g., 180 ft (55 m) long, that can be used to drop anchors well out to the sides of the dredge.

A large cutter suction dredge could have a discharge pipe size of 42 in. (1067 mm). With 5000 to 10,000 hp on the pump and 2500 hp on the cutter, a big dredge can pump 2000 to 4500 yards^3/h (1500 to 3500 m^3/h) of soft material, or 200 to 2000 yards^3/h (150 to 1500 m^3/h) of soft to medium hard rock through 15,000 ft (4.6 km) of pipeline. Such a line is usually laid on pontoons and in a wide arc so that it is not pulled tight as the dredge moves forward.

This moving is normally done using *spuds* and using a swing wire system tied in to the dredge's anchoring system. The spuds are essentially long retractable piles, supported by a gantry crane at the stern of the dredge (see Figure 10-5). One spud, say the starboard one, is dropped to the sea floor and the dredge swung about this point as a pivot. This is done by winching against the anchor. The port stern corner of the dredge has moved ahead somewhat, and the port spud is then driven and the starboard one retracted, the swing of the dredge then being back the other way, this time about the port spud. But if weather

Figure 10-5   Cutterhead dredge (Courtesy, Ellicott Machine Corporation)

conditions become too adverse, the spud system cannot be employed and the dredge is pulled about by winching using a 3-point mooring system. Tugs are used, in conjunction with anchor barges, to relocate the dredge's anchors.

Cutter suction dredges of small size have been used in 3 ft (1 m) of water. Big dredges with special ladders are capable of working in up to 200 feet (60 m) or so of water. Box cuts are normally made by such dredges for pipeline trenches since slope failure along the sides can customarily be counted upon to open out the trench. The cutter is sometimes removed for cleaning out a cut.

McConnell (1967) describes the trenching for an offshore pipeline in New Zealand using a cutter suction dredge. This was an 18-in. suction/14-in. (460 mm/360 mm) delivery dredge with a rated output of 360 yards^3/h (280 m^3/h). It was employed to excavate 300,000 yards^3 (230,000 m^3) of sand for a trench 80 ft (24 m) wide×35 ft (11 m) deep at the deepest burial point at the shore end. With a 72-ft (22 m) ladder, the dredge could only excavate to 52 ft (16 m) below the water line. This particular dredge used the suction pipe as the rotating drive for an integral cutter. The sand was discharged away from the trench using a delivery pipe up to 600 ft (180 m) long.

The open ocean exposure of the New Zealand excavation site ruled out the use of spuds for positioning, since wave action would make such an operation very difficult. Rather, winch and wire controls were employed. The seas exerted a real influence on the efficiency of the dredging operations. Whereas 3-ft (1-m) swells could be endured in shallow water, even 1-ft (0.3-m) waves in deeper water caused the dredge to "pole vault" on the ladder. Cutterhead dredges developed since the New Zealand work can tolerate 7-ft (2.1-m) seas.

Blasting

Introduction: Explosives are the conventional technique when the natural bottom material is simply too hard to be excavated or when the trench called for is so deep that rock is reached under granular overburden. I am aware of one 78-in. (1980-mm) submarine pipe where a trench depth of 40 ft (12 m) through sandstone had to be established by blasting. One solution to breaking or crushing hard formations is the classical method of drilling bore holes in the hard material, such as rock or coral, placing an explosive charge in each hole before debris fills them in, and blasting (References 60, 61). But this is a slow and costly process, and faster, more economical techniques have been developed, viz. *shaped charges* (e.g., Penzias and Goodman, 1973). One of these, which has been frequently used in oil pipeline work, is Quick Ditch (Jet Research Center, Inc., Arlington, Texas) which uses shaped charges to concentrate the explosive energy where wanted. No holes are drilled. Trenching to 12 ft (4 m) below the bottom can be carried out (Brown and Grundy, 1973).

The chemicals for this technique are shipped in nonexplosive form and then mixed on location. Special charge containers are filled with the explosive. See the foreground in Figure 10-34. These can be floated out on rafts to the desired location and there either positioned from a small boat in shallow water or through use of divers in deeper depths. When the containers have been positioned, a firing cord is attached to arm the charges. Detonation is electronic. A dragline, for example, can be used to clean out the trench after blasting has been completed.

The trenching for the Mokapu outfall (#47 in Table 10-1) was done using shaped charges set in a concrete form (Figure 10-6) that was lifted into position by a crane on the trestle. This form was stable under the heavy wave action in the area. An adequate trench was frequently excavated at Mokapu, in softer formations, with just the shaped charges. When the trench had to pass through a formation of very hard black rock, the shaped charges cleared off about the top 2 ft (0.6 m) of rock but fractured some of the deeper rock sufficiently that it could be loosened with a cable-tool type of chipper (foreground of Figure 10-7) and then the debris removed by a clamshell (background).

Various outfalls listed in Table 10-1 were laid in trenches blasted out of the sea floor. These include San Diego (#18) and Whites Point (#21) as well as

Figure 10-6   Shaped charges set in concrete form

Figure 10-7   Chipping tool in use from Mokapu outfall trestle

Sand Island No. 2 (#44) and Barbers Point (#49) to be described in the following two subsections.

Sand Island No. 2 Outfall: In the case of the Sand Island No. 2 outfall (#44), the bottom consisted for the most part of 2 ft (0.6 m) of hard coral crust overlying approximately 5 ft (1.5 m) of sand which itself overlay more coral crust. Sometimes there was sand over the top coral horizon. Both horizons of coral crust had to be blasted for the line. Early experience using shaped charges proved to leave much to be desired. Considerable diver work was required to move these shaped charges into proper position. Then, when the charges were set off, a deep, narrow vertical hole developed rather than the desired extensive cavity. The soft overburden appeared to compound the problem. It was also found to be more expensive to use this technique than to use the approach ultimately adopted.

This second system involved tying together, at 5-ft (1.5-m) intervals, using explosive primer cord, bags containing 25 lb (110 N) of dynamite to a limit of 500 lb (2200 N). A weight at one end of the string was dropped on centerline first and then the bags slid down a sloping board set on the stern of a barge, following the weight toward the bottom as the barge followed the centerline. Once the string was laid out on the sea floor, primer cord was attached to the end bag and run to the surface, the barge moved away from the site, and the charge was set off. This technique ultimately managed to excavate an adequate trench for the outfall. However, sometimes up to five explosive passes had to be made over the same location to get the 15-ft-deep trench (4.5 m) down "to grade," i.e., to the elevation desired. A clamshell removed debris and carried out excavation of soft material. It also had to be used later to clean debris out of the trench after currents eroded the soft material underneath the coral crust causing this crust to tumble into the trench. In some locations the trench walls were virtually vertical, in many other places at approximately 2 on 1 (steep) slopes.

Barbers Point Outfall: The blasting for the trench for the Barbers Point outfall (#49) was carried out under the supervision of the company Jet Research Center, Inc. Most of the work was carried out in water depths of 20–85 ft (6–26 m). Some areas of hard coral required multiple blasting to achieve proper grade.

The aluminum shaped charges used on the work were "made up" by mixing a nitromethane liquid explosive with liquid initiator in the container. The charge was completed by capping the top opening with a rubber stopper through which detonating cord was passed. The charges were set in position by divers on the seabed using a 60×10-foot (18×3-m) template. This was set on line using a red ruby laser from a shore station. The individual charges were set four across in the transverse direction with 6-ft (1.8-m) longitudinal spacing. There were 40 charges per "shot."

The barge would back off for the shot, then move 60 ft (18 m) inshore of the previous position along the outfall line to repeat the process. The amount of movement was determined from a Hewlett-Packard (Hollywood, California) distance-measuring system.

The number of shots per day depended on the water depth: 8 shots/day in 20 ft (6 m) of water; 3 shots/day in 85 ft (26 m) of water. The latter involved two divers to their optimum time of 90 min for a depth of 90 ft (27 m).

Other Blasting Approaches: An approach, but a difficult one, that can be used for blasting for an outfall in shallow, nearshore waters is that used for the Otter Rock, Oregon, outfall in early 1973. Single sticks of dynamite were drilled with two holes and primer cord was passed through these holes. A bundle of three sticks was made up by taping two sticks without primer cord to the one with cord. The primer cord for one bundle was left on a reel having sufficient capacity to reach the blasting area whereas about 10-foot (3 m) primer cord lengths were used with the other bundles.

Each diver of the pair working took out 5 to 6 bundles, about 25 lbs. (110 N), in a burlap sack while a worker on shore held the unspooling primer cord reel. When all the bundles had been placed and all tied in to the main primer cord line the divers returned to shore and the charge was set off.

The work at Otter Rock was difficult and hazardous because of the persistent heavy surf.* The work came close to costing the life of one of the divers who became hopelessly tangled in primer cord and was rescued after a time by his buddy diver.

A novel means of excavating a trench through a region of heavy surf has been described in Reference 76. A 30-foot-high (9 m) platform on skids was pulled seaward, along the line for a 40-in. (1020-mm) submarine pipe, by a winch mounted on a barge offshore. The platform was returned to shore by means of a land winch. On each trip offshore, two frames containing explosive charges were set. Each frame had 27 shaped charges and five *bangalores* (tubes full of explosives). To eliminate the shock absorber effect of a sand layer over the conglomerate being blasted, charges were jetted down to the rock before being set off. Maximum penetration was 7 ft (2.1 m). The desired trench dimensions were 3 ft×6 ft (approximately 1×2 m).

Three example references that describe blasting operations used in conjunction with the laying of submarine pipelines are Bowman and Bemis (1961), Chabert (1969), and Love (1971). In situations where a rock bottom is involved and heavy surf, a relatively inexpensive technique of long-line-blasting used in New Zealand is a possibility. Here plastic pipes filled with liquid explosive

---

*Divers working in shallow water with heavy surf should tape on masks and flippers. It is also a good idea to have a line attached to mask and snorkel, with rubber "keepers" holding fins onto the foot. These measures were followed at Otter Rock. One diver during this project had a knife and sheath ripped from his leg by wave action.

(which can be mixed on site) are pulled out along the proposed trench line; the explosive is then detonated. Repeated pulling and detonation are used until the desired trench depth is reached.

Additional Considerations

It has already been outlined that the U.S. Coast Guard should be notified, for U.S. waters, of construction involving outfalls. The following passage appeared in the "Local Notice to Mariners" issued by the Fourteenth Coast Guard District, Honolulu, Hawaii, on June 22, 1976.*

HAWAIIAN ISLANDS—OAHU—BARBERS POINT

Work has commenced on a sewage outfall pipeline into the ocean commencing at a point on Oneula Beach in approximate position 21-18-33N 158-01-53W. Blasting operations will be conducted during July and August. Three long blasts will be sounded prior to detonation. A walking platform will be utilized out to a depth of approximately 20 feet: Four barges and work boats will be in the area. Project duration is anticipated to be 18 months.

Ref: Chart No. 19357

Figure 10-8 Dragline bucket used from shore

The customary floating dredging methods may not be effective for large operations in at least two situations. First, they cannot all be used when the water depth becomes too shallow; second, they cannot be employed when the sea floor is too hard. In the first case, as in preparing trenching for the marine outfall at Nelson, New Zealand, a dragline from the shore can be used to advantage (Ade, 1970). Basically, with this technique, an open bucket is drawn backward to sea through use of a winch on shore and an anchor pulley offshore. The bucket can be pulled up on shore and dumped as in Figure 10-8. Another

---

*The same or a similar paragraph was repeated in each weekly issue of the "Local Notice to Mariners" throughout the duration of the project.

method of trenching in shallow water is to use a trestle, a type of temporary construction pier that will be described in the following general section. In this case, a land-type shovel or clamshell with suitable undercarriage can be moved along the trestle rails to work at whatever location is necessary. A type of cutterhead dredge was used from the trestle for the Orange County No. 2 outfall (#35 in Table 10-1) to excavate the trench.

If there is loose material along the trench wall and/or the original ocean bottom, and if there is almost any level of wave action, a trench may rapidly fill with this material. This may be useful after a pipe is laid and rocked, but it is troublesome if it occurs before laying.* For this reason, trenching should not normally get too far ahead of the pipelaying and complete, not just partial, rocking afterward. Sufficient distance should be allowed, of course, if blasting is being done, so that pipe damage does not result. A similar consideration applies to a trestle. Figure 10-9 shows an "H" pile for a trestle that was torn loose by too close a blast.

Figure 10-9   Trestle "H" pile damaged by trench blasting

It may be of interest that excavator vehicles working entirely underwater are available, e.g., Reference 21. Outfalls have even been installed in trenches created by the directed propeller slip stream of a specially-fitted salvage tug (Bowlus et al., 1964; Klopfenstein, 1974). In the second of these cases, the ship could work at high tide where land-based equipment could reach at low tide.

If an outfall has to be constructed across a broad and shallow foreshore consisting of hard material such as rock or coral, it might well be cheaper to investigate the use of anchors for the line rather then to trench by blasting, excavate the broken material, lay the pipe, and then backfill. Basically such an anchor consists of a bracket that fits over the pipe and some means of attachment to the sea floor.

---

*It is the purpose of a diving inspector (Figure 9-1) at this stage not only to check on the dimensions of the trench as excavated but also to check for refilling of the trench.

If the material is really hard, then anchors can be set up by drilling holes into the bottom, dropping expandable anchor bolts into the holes, and then (possibly) filling with cement grout. Galvanized bolts and straps complete the arrangement. An outfall at Depoe Bay, Oregon, was secured in this way—but for that installation a certain amount of blasting was carried out to partly recess the 10-inch (254-mm) line in the substrate. However, this weakened the rock in one area that was to serve to anchor the pipe. In this case divers did not take pains with the blasting as they were unaware of the plan for drilling anchor bolt holes. There are special underwater drilling rigs available that can expedite such drilling operations for anchor bolt holes (Chabert, 1969), but pneumatic drills held by divers have been more usual. Drilling holes in the surf zone is extremely difficult and dangerous, but has been managed by experienced divers—as at Depoe Bay. If the bottom is somewhat softer, then there are commercially available, special drill-in anchors with multiple helical blades that can be combined with an anchor bracket over the pipe to keep it in place (Holland, 1962; Short, 1967a, b).

10-3   INSTALLING THE OUTFALL: BOTH INSHORE AND OFFSHORE WATERS

Introduction

There are several general methods available for laying ocean outfalls, and these will be examined in this section. The method to be used depends upon many factors which include the local topography and bathymetry, the size and type of pipe, and environmental conditions such as the tide and wave climate.

It is possible to divide pipelaying techniques into three classes depending upon the water depths involved. In the first class, there are those methods of laying that are adapted to shallow water, less than approximately 20 ft (6 m), or shoreward of the region in which a barge or ship can work. Although this region is sometimes called the *surf zone*, it will be referred to herein as the *inshore zone*. Seaward of the 20-ft (6-m) depth contour, it is possible to use barges and ships and the laying methods applicable here will be listed under *offshore zone*. Finally there are those special techniques that can be used for both zones, and such methods will be described as part of a third class, which, because of its generality, will be discussed first.

Bottom Pull Method

This method is normally used for relatively short outfalls, although there are exceptions (Reference 56). It requires that there be a substantial area back of the coast where long lengths of the pipe to be laid can be stocked on a bed parallel to the intended direction of the line (see Figure 10-10). Basically, the

Figure 10-10 Pipes arranged in parallel strings prior to pulling (Courtesy, Nippon Steel Corporation)

pipeline is pulled out segment by segment, where the individual lengths may be on the order of 1500 ft (450 m), and can go to 5000 ft (1500 m). When a section has been pulled down a special launchway into the water, it is stopped (Figure 10-11) and the next section is rolled behind it and connections are made. This combined length is then pulled out and the jointing process is repeated. The jointing of the strings may be relatively straightforward, but in the case of concrete-clad steel pipes it is an involved operation. The joint is welded, then checked by radiography (Figure 10-12), then an external protective coat is added, and, following that, wire mesh and a concrete weight coating follow.

Once a pipe length has been pulled off the bed, the next length is rolled onto the launchway itself or onto dollies that support the pipe and provide easier movement for the pipe. The dollies drop into a trench after use and are retrieved. There are various methods used for pulling this pipe out along the bottom. One such method makes use of a heavy duty winch on the shoreline that pulls out the pipe through a pulley mounted on an offshore anchor block and over a shoreline tower. Operation of the pulley can be monitored by TV (Bowman and Bemis, 1961).* Usually, however, an offshore barge on the pipe centerline provides the pulls, either by anchoring itself and pulling the pipe to it by means of a winch or by pulling itself ahead on its anchors, thereby dragging

---

*An offshore, *underwater* winch is a conceivable alternative, but one that worked very badly in the only outfall job where I know of its being used.

Figure 10-11 Pipe entering water (Courtesy, Nippon Steel Corporation)

Figure 10-12 Checking a pipe weld (Courtesy, Nippon Steel Corporation)

the pipe along. The length of the pull is limited by the winch power, the allowable pipe tension, and the weight of the line. As an indication of pull capacity, the bottom pulling of one 42-in. (1070-mm) line 16,000 ft (5 km) long required a winch having a capacity of 300,000 lb (1.3 million N). In this case, sectional pulls were 5000 ft (1500 m).

The pipe route, whether trenched or not, should at least be marked with buoys and thoroughly inspected by divers. Any dragging obstructions should be removed as should sediments moved into the trench by wave and/or current action. Practically speaking, however, the latter may be difficult, and some outfalls have been laid too high. Nevertheless, a special dragging head at the offshore end of the line should be used to ease the dragging process and prevent snagging. Such a structure will be detailed later.

A pipe's stability once in place depends very heavily upon its bulk specific gravity. Whereas a heavier pipe tends to be more stable, it is also far harder to pull along the sea floor. For this reason pipes are not only pulled when empty but are often fitted with buoyancy devices that relieve some of the normal force existing between the pipe and the bottom. One source of buoyancy (and protection for the dragged pipe) is timber. This has been used to advantage, for example, in British Columbia (MacKay et al., 1974). But it is much more usual to use some form of buoyancy tanks or pontoons. Such tanks can be enclosed or of controllable buoyancy type using ports and valves. The controllable buoyancy technique was employed for the Five Finger (#40 in Table 10-1) and Harmac (#45) lines, but there are difficulties in the operation. Release of buoyancy devices on a pipe bottom-pulled to depths of 200–300 ft (60–90 m) can be a tricky business and should (like everything else) be thoroughly thought out by the contractor before proceeding. A system for cutting buoy lines which is used mounted under a submersible is shown in Figure 10-13.

Figure 10-13 Cutter system for removing pipe buoyancy tanks (Courtesy, HB Contracting Ltd.)

It is generally felt to be advisable to have about 5–10 lb/ft (75–150 N/m) of negative buoyancy for a pipeline when it is being bottom pulled. But when pontoons are attached to a line and when it is lightened, waves and currents can more readily push the pipe off course. In fact, if environmental conditions become too severe it is necessary to be able to quickly flood the line. For this reason the pulling head (Figure 10-14) becomes a rather complicated structure because it may house pumps for forcing out leakage or ballast water. Or it may contain a plug which, when difficulties are encountered, moves shoreward as water is allowed to enter the head through a valve, then is forced back to the head again through use of compressed air when pulling is to be resumed. To decrease leakage into the pipe a pressure of, say, 20 psi (1.4 bar) may be maintained within the pipe while it is being pulled by means of an air pressure line. Water may be admitted to the line when strings are being joined and then expelled again when pulling is restarted. Stability of the line, when it is being pulled, may be increased by using outrigger pontoons to prevent the pipe from rotating more than 20°. It may also be enhanced, and should be, by making use of a hold-back winch or tension shoe on the beach to keep tension in the line so that it moves off course only with difficulty. The tension shoe can also be used to give a push to the pipe if pulling difficulties are experienced, but this is not a preferred technique, since it could lead to buckling of the pipe.

A cable is usually used to pull the pipe, although the use of a small pipe has at least two advantages. First, the buoyancy of the pulling string can then be adjusted by filling or emptying the pulling pipe; second, the pipe has less tendency to rotate than the cable. A cable can be very difficult to get moving once it becomes embedded in a sandy bottom. Maximum pulling speed is of the

Figure 10-14 Pulling head (Courtesy, Greater Vancouver Sewerage and Drainage District)

order of 20 ft/min (6m/min). It is a good idea to have divers inspect a pipe as it is pulled.

Details for the bottom-pull method for submarine pipes have been provided for individual projects by the following: Bowman and Bemis (1961), Lyon (1961), Bowlus et al. (1964), Williams (1966), Reference 4, Chabert (1969), Ade (1970), Goss (1970), Love (1971), Dean et al. (1972), Sanders (1972), and MacKay et al. (1974). The last reference is particularly valuable; it reports on experience with a number of submarine lines in the Vancouver, Canada, area. The largest single pipe pulled was 78-in. (1980-mm) steel. However, in one case a bundle consisting of two 66-in. (1680-mm) and one 48-in. (1220-mm) steel pipe was pulled. It is normally steel pipe that is installed using the bottom pull technique. However, both post-tensioned RCP and cast iron outfalls have been installed in this way. The outfall at Otter Rock, Oregon, provides an example of the latter. In that case the bell-and-spigot joints were held together by collars and chain.

Examples of outfalls in Table 10-1 that were pulled are the Hyperion sludge line (#11), Watsonville (#16), Gardiner (#19), Hastings (#27), Harmac (#45) and Waitara (#50). The latter ran into various problems, one of which involved breaking of the tow line. Also, the contractor was unable to flood the pipe and the inshore section floated out of the trench—which was nominally 6 ft (2 m) deep.

The bottom pulling technique can work for other than straight outfalls. If an outfall with an L- or T-shaped diffuser is involved, hinges in the junction area can be installed so that the diffuser leg(s) can be folded back along the outfall during laying and then reoriented upon completion of the pulling operation.

Flotation Method

As in the preceding method, the pipe is put together in long strings and laid on a bed parallel to the direction of the line. Buoyancy pontoons are attached to these strings usually by means of straps, although floats can be attached with rope just before a part of the pipe moves into the water. Each string is pulled into the water and towed into position as a floating unit. Men and equipment on a tie-in barge, holding the offshore end of the previous string, make the connection, and then the pontoons are released except for those near the offshore end of the just-connected string. Another is pulled into the water, floated to the site and connected, and so on.

This method can be a hazardous operation, being troubled by even moderate seas. Galloping of the pipe can occur even for small waves. Currents also tend to push the pipe off line. For these reasons this method is generally restricted to use in protected waters, and it is usually employed only for long lines. Since the pipe effectively hangs between pontoons, large pipe stresses can be built up if this spacing is too great, especially during lowering. The Monmouth County outfall (#39 in Table 10-1) was laid in this way.

A variation on the above procedure has the joining of sections done on shore before the completed, floating portion of the line is dragged further seaward. This particular approach was employed for the following outfalls listed in Table 10-1: Macauley Point (#32), Prince Rupert (#34), and Powell River (#43). Considerable difficulty was experienced controlling the floating Macauley Point line when a storm passed through the area. Control was maintained using two tugs along the sides of the pipe with a bigger ship at the offshore end.

The source of buoyancy used in this technique varies from installation to installation. In the case of Macauley Point, the natural buoyancy of the air-filled pipe was sufficient. For the Prince Rupert line, polystyrene donuts provided buoyancy, and for the Powell River line logs were used.*

The winch capacity to pull a floating string into the water is normally far less than that required for the same pipe if bottom-pulled. Waves and currents could conspire to greatly increase forces, however. In the Macauley Point project the pipe was actually pushed out, by a bulldozer.

10-4 INSTALLING THE OUTFALL: OFFSHORE WATERS

Lay Barge Method

This method of laying submarine pipe is used extensively in the offshore oil and gas industry (e.g., Lee and Bankston, 1967; Nippon Steel Corporation, 1973; Deason, 1974), but it can also be employed in laying sewer outfalls. Cast iron pipe and concrete-coated steel pipe have been laid in this way.

Pipe lengths are put together in assembly line fashion on the lay barge which is supplied with pipe by an attending pipe barge. As the pipe is put together, the lay barge moves itself forward on its anchors, which may themselves be progressively moved forward by an attending tug, and the pipe moves into the water. The pipe moves down a ramp above water and usually down a long ramp, or *stinger*, which is submerged and extends to near the bottom so that large stresses are not built up in the unsupported pipe between the end of the stinger and the seabed. Even then, very large stresses may be experienced in the pipe; 85% of the minimum specified yield stress in a steel pipe has been quoted. The radius of curvature of the pipe is very important.

Overall, this laying method appears to be the fastest and most economical method of laying certain pipe materials offshore as long as weather and sea conditions are not too severe and as long as depths do not exceed approximately

---

*When a pipe is floated into position it should be sunk from one end. I have been told the strange tale of a PVC outfall that was filled with water from both ends. Sinking of the pipe was finally accomplished by shooting holes into the hogged central section to release the trapped air.

200 ft (60 m). This technique can even be used in waters as shallow as a fathom in depth. Pipe length is not a limiting factor and large pipe diameters can be handled.

Floating Crane

This method is in a sense similar to the preceding technique in that a large barge and accompanying pipe barge are used. But in this case, a crane on the laying barge lowers a string of pipe, which is held horizontally, down to the bottom where the new section is joined to the existing line. This is not a rapid system, and its use involves high costs. The spectrum of uses for this method runs from the very large pipe system described by McConnell (1967) to much smaller lines.

There have been numerous ocean outfall undertakings in protected waters where an adequate job of construction has been done through use of a crane barge (Figure 10-15) with a *strongback* or *stiffleg* (heavy beam) supporting the length of a pipe to be linked to the already-laid line. The new pipe section is pulled into the old bell through use of cables. Final seating of the new spigot in the old bell may be done by creating a vacuum in the space between them when

Figure 10-15 Crane barge, tug and trestle

they have been only partially pushed together (Etheridge, 1974). Or it can be done using bolts to pull the new section into the old, as in the laying of the Hollywood, Florida, extension (#26 in Table 10-1). In rough waters, however, it is virtually impossible to work with the pipe section itself on the end of cables. There is simply insufficient control over the position of the section to be laid, with the result that banging of pipe cannot be avoided; also, the lack of control makes it extremely difficult to properly insert a new section into the already-laid line. A system has been developed that allows the construction firm to exert the requisite control over the pipe section being laid. This device is called "The Horse."

The general idea for the Horse apparently developed during the laying of the Hyperion outfall sewer, and the concept has been used in Oregon (Georgia-Pacific Paper Co., Newport) and on various jobs in California (e.g., Orange County No. 2) since that time, as well as on the laying of the Sand Island No. 2 outfall near Honolulu in 1974–1975. Although the operational concept was similar, actually Horses of different design were used on the Orange County and Sand Island outfalls. Descriptive material on the two Horses is available in References 30 and 31. It is understood that the second Horse cost approximately US $130,000 to build (in 1973). Details concerning operations of the Horse on the Sand Island job in the next few paragraphs will serve to illustrate the type of operations generally involved with big (reinforced concrete) outfalls.

Davy Crockett and the Horse

The "barge" used on Sand Island No. 2 with the Horse was actually a 440-ft (135-m) World War II Liberty ship that had the superstructure cut off and the engines removed. This vessel, called the Davy Crockett, featured a thick steel decking, pipe rollways, a 130-ton-lifting-capacity crane amidships for heavy work, and a smaller crane aft for general work. It drew only 12 ft (3.6 m) of water. Anchor winches were mounted on deck for control of barge movements.

A technique that proved, after several trials, to be workable for the Sand Island No. 2 outfall work had the barge moored through six cables to six separate cylindrical spring buoys. Heavy chain was led from these buoys to the anchors. Much of this chain lay along the bottom so that both a horizontal pull was maintained on the anchor and the spring buoy wasn't pulled under water. A cable extended from the anchor vertically upward to a spherical marker buoy. This cable could be pulled through the buoy so that the anchor could be lifted by an anchor scow. This scow was equipped with winch and "A" frame. A tug towed the scow and its load to the new anchoring location.

The Horse used for the Sand Island No. 2 outfall is pictured in Figure 10-16. There were two major parts to the Horse: the main frame and the traveller that moved under the frame. The main frame was 40 feet by 40 feet (12 × 12 m) in plan and was supported by four legs each having a foot at its lower end (see Figure 10-17). The traveller was basically a bridge crane, moved

Figure 10-16 "Horse" entering water

Figure 10-17 Foot on "Horse"

by hydraulic rams along rails suspended underneath the main frame. The maximum excursion from a central initial position was 10 ft (3 m) in any one of the four possible directions. The traveller provided for horizontal movement whereas vertical movement was accomplished by having the four legs (differentially) extendable. The maximum possible adjustment was 8 ft (2.5 m). The control over the rams moving the traveller and the four legs was provided by a long bundle of hydraulic lines from the deck of the Davy Crockett where a control console was located.

When a pipe section (or "joint") was to be laid, the rubber O rings were put in place in the grooves provided in the pipe (Figure 8-9) and these were lubricated with Crisco.* The pipe was then rolled down the rollway on the ship under the Horse. The method by which the traveller gripped the pipe in the Sand Island No. 2 project was through use of two pairs of hydraulically controlled arms having curved steel pads measuring 4×4 ft (1220×1220 mm) and 1-in. (25-mm) thick (see Figure 10-18). No slings were used as in earlier work. One arm on one side of the traveller could be folded back to permit a new 84-in.-diam (2134-mm) RCP section to be rolled underneath the Horse.

---

*This is the commercial name for a particular semisolid vegetable shortening.

Figure 10-18 Gripping pad on "Horse"

Once everything was ready, the Horse was lifted through a four-cable bridle by the big crane (see Figure 10-19). Depending upon whether work was being done in or out of a deep trench, up to 15 ft (4.5 m) deep on Sand Island No. 2 (Figure 8-4), the legs of the Horse were either initially fully extended or not extended, respectively. In some cases where the trench was overly wide, a beam had to be welded across transverse pairs of feet for support purposes. This is shown in Figure 10-19. During the Sand Island No. 2 work, there were some 15 separate changes to the Horse that had to be effected to allow for changes in the trench geometry and materials. There were also delays to repair damage to the Horse.

A marine superintendent standing on a plank extending off the side of the Davy Crockett directed the lowering of the Horse. He aligned the hoist cable with a laser alignment system (Appendix A) on shore. A contractor diver then entered the water and provided instructions through his tender to the crane operator who lowered the Horse and its pipe to its "landing" on the bottom. The Horse had to be landed (straddling the trench) so that, with the possible excursions mentioned previously, the new section could be linked up with the old.

The contractor diver, shown working in Figure 10-20, then relayed the instructions to the operator of the hydraulic control console who moved legs up and down, the traveller back and forth, until the joint was finally made. The diver had to be careful to keep his hoses free of the Horse hydraulics. Maneuvering instructions from a diver in such situations often came from his sense of feel rather than his sense of sight. This was because of the usually marginal visibility as a result of all the activity in the area. A major problem in this regard concerned the stirring up of fine materials in the trench. Great care regarding

Figure 10-19 "Horse" and crane

incoming pipe alignment had to be maintained as breaking and spalling of the pipe material could occur, and *did* occur from time to time during the Sand Island No. 2 work.

After an inspector diver inside the pipe had checked the gap clearance as satisfactory,* usually with a stepped wood or aluminum block inserted at various points around the pipe periphery, and had checked that clearance to the trench wall would permit retraction of the Horse clamps, bedding stone was dumped over the new section from a bucket operated by the small crane. The diver directed this stone under the pipe using a 2-inch (50-mm) water jet nozzle operating off the Horse. This tee-shaped fitting actually had two jets (Figure 10-21) to eliminate reaction forces.

When a suitable bedding base had been built up to the required height (usually indicated by a white, *skunk*, line painted on the pipe, e.g., 20 in. or 500 mm below the spring line), the double gasket joint was pressure tested[†] to 120% of the design pipe pressure, as measured above the ambient pressure. If this test

---

*Allowable pipe joint gaps on the Sand Island No. 2 outfall were expressed as a maximum sum of the gaps on any two opposite sides of the joint. These maxima were set as follows: 2 inches (51 mm) for the 84-inch (2134-mm) pipe, $1\frac{1}{2}$ inches (38 mm) for the 66-inch (1676-mm) pipe, and 1 inch (25 mm) for the 48-inch (1219-mm) pipe. The latter two pipe sizes were used in the offshore end of the diffuser.

[†]Details of this test are included in Sec. 10-9.

Figure 10-20 Contractor diver "making joint" (Courtesy, John Goode)

Figure 10-21 Tee-shaped nozzle for washing bedding under pipe

was satisfactory, then the Horse was removed and returned to the deck of the Davy Crockett. The gap was then again checked. It was usual on the Sand Island No. 2 job for the top of the joint to open out from $\frac{1}{8}$–$\frac{1}{4}$-in. (3–6 mm) when the pipe section was released.

When ocean outfall work moves into depths greater than about 100 ft (30 m), diver work times are so limited because of decompression schedules that it is desirable to have another method of making and inspecting joints than having divers in the water. This was accomplished, in the case of the Horse, by mounting a 30-in.-diam (760-mm) horseshoe-shaped observation chamber (bell) right on the traveller (see Figure 10-16). This chamber, which was maintained at atmospheric pressure, was equipped with numerous viewing ports (on top and on the sides) and was long enough to house both the contractor's man (diver) on one side and the inspector diver on the other. The chamber was really too confined to have one man moving back and forth from one side to the other. Again, the contractor's man sent instructions to the operator at the hydraulic control console. It was felt better to do this than to have the contractor's diver operate the controls himself: the bell was dark for one thing, to facilitate viewing outside; it was confined; and there were so many things to observe and sufficient movement required inside the chamber that it would be asking too much to also require the observer to look down and operate a system of buttons. Air was supplied to the two observers in the bell through a line bundled with the hydraulic lines. The amount of penetration of the pipe spigot into the bell was charted by noting the number of $\frac{1}{2}$-in. (12-mm) white stripes painted on the spigot that were visible outside the limit of the bell.

The maximum rate of pipe-laying in the trench for the Sand Island No. 2 outfall was five 84-in.-diam (2134-mm), 24-ft-long (7.2-m) sections per day. This was in a depth of about 60 ft (18 m). In this work, laying/bedding of the pipe was carried out during the day shift with rocking at night. Outside of the trench, however, progress was much faster with the peak laying speed being three sections per hour at an approximate depth of 200 ft (60 m).

The Davy Crockett, owned by Peter Kiewit, Inc., was leased to the Morrison-Knudsen Company Inc., contractors for the Sand Island No. 2 outfall, for $45,000 per month.

Towers

The offshore sections of two large, early outfalls, those at Hyperion (#12 in Table 10-1) and San Diego (#18), were installed using what amounts to an oil-field jackup drilling platform. In fact, after the San Diego job, the platform was shifted to oil field work so that it was unavailable for the later Whites Point project (#21). As a result, incidentally, the Horse was developed for use at Whites Point.

The platform at Hyperion and San Diego was enormous, with a deck that measured $190 \times 122.5 \times 17$ ft (deep) ($58 \times 37 \times 5$ m) and long legs permitting work in depths up to 200 ft (60 m). Details are provided in Reference 65. The platform could move from one location to another by jacking its buoyant deck down until the whole structure floated, then pulling itself via anchor line to the desired new position.

Other towers have been used for outfalls but in considerably shallower water, e.g., for Honolulu's Sand Island No. 1 (Reference 35).

10-5 INSTALLING THE OUTFALL: INSHORE WATERS

Trestle

Building the Trestle: The inshore sections of many outfalls have been laid from a trestle, which is effectively a temporary pier extending out into the ocean along the outfall line. Examples of outfalls in Table 10-1 where trestles were employed are Orange County No. 1 (#9), Palm Beach (#10), San Diego (#18), Whites Point (#21), Orange County No. 2 (#35), Sand Island No. 2 (#44), Mokapu (#47), Santa Barbara (#48), and Barbers Point (#49). The trestle is essentially used in waters that are too shallow for pipelaying ships or barges or that are shallow enough that amplified waves and swell would mean considerable motion of such a floating platform. Although the trestle itself by no means eliminates wave surge problems in inshore waters, it ensures that work can be done from a fixed platform.

Basically, the principle behind the construction of a trestle is that pipe or "H" piles are driven into the sea floor at a prescribed longitudinal and lateral spacing. They are usually not grouted, so that the piles can be retrieved later. However, piles set close inshore may be grouted, as at San Diego (#18 above). In that project pile penetration varied from 4.5-8 feet (1.4-2.4 m).

A *pile bent* is made in the lateral direction by setting beams across pairs of piles. Bolting rather than welding is typical, again to ease recovery upon work completion. Longitudinal beams are then laid across the pile bents,* and on top of these rails for the crane/pile-drive/pipelay carriage that progressively advances seaward as it completes the pile driving and followup work. A companion carriage may also be involved for work and/or transport. Walkways and water and compressed air supply pipes follow the crane out. When the work is finally over, the crane returns to shore, taking up the piles as it progresses.

A comparatively easy way of quickly positioning four piles in a trestle is to use a template that is cantilevered out seaward from the already constructed portion of the trestle. Piles are lifted by the crane and dropped through appropriate openings in the template as shown in Figure 10-22. They are then driven later to the desired penetration. It was necessary for some piles on the Sand Island No. 2 outfall to be 70 ft (21 m) long where runs of 10–15 ft (3–5 m) in cavities were not unusual.

Excavating under the Trestle: While the trestle is being extended by one crane, another crane can be busy further inshore driving sheet piling, preferably with a vibratory hammer (Figure 10-23), and/or excavating with a clamshell, readying the sea floor for the laying of the pipe. Excavation for the shallow-water pipe trench from the trestle can also be done using a hydraulic dredge

---

*Central support for these beams between pile bents may be provided by single intermediate piles.

Figure 10-22 "H" pile being lowered through slot in cantilevered template

Figure 10-23 Vibratory hammer driving sheet piling

mounted on a carriage that moves along the trestle. Control of trench depth is normally done using a lead-line. The minimum depth in the Sand Island No. 2 case was 14 ft 10 in. (4.5 m). The minimum width was specified to be 13 ft 4 in. (4.1 m) within 3 ft (0.9 m) of the bottom of the trench (see Figure 8-3). In many cases the trench was overdug, resulting in a need for additional amounts of rock. Incidentally, it is easier to excavate a trench within sheet piling, such as one encounters under trestles, by excavating just inside the sheet pile line first, then along the trench centerline, rather than doing the centerline first.

Pipe Laying from a Trestle: A *gantry crane* (Figures 10-24 and 10-25) can be used to lay the pipe. This is basically a hoisting arrangement shaped like an inverted "U" that rolls along the track and straddles the pipe it is carrying and then laying. It is also possible for one or two standard cranes to carry out the pipe-laying operation. In either case, the pipe section is supported by two or three *slings*, cables passed under the pipe.

The laying of the Sand Island No. 2 outfall in waters shallower than about 20 ft (6 m) was done from a trestle and the pipe was placed within a continuous sheet pile wall. The existence of this wall allowed for the use of a novel type of laterally stabilized system to prevent the pipe from transverse movement due to the persistent wave surge during laying. This system was developed after it was found that air-powered tugger winches attached to the four corners of the laying structure plus inhaul and outhaul tuggers gave insufficient control. Among other things, there was a tendency for the new pipe section to bang the last-laid section, resulting in damage. It was also a nightmare of criss-crossing cables for the contractor and inspector divers. A simple system consisting only of inhaul and outhaul tuggers gives woefully inadequate control in wave surge, and since

Figure 10-24 Gantry crane and 12-foot-diameter reinforced concrete pipe section held by slings

Figure 10-25 12-foot-diameter reinforced concrete pipe section being lowered by gantry crane

sheet pile walls tend to focus wave energy, this is usually a problem under trestles.

The structure developed for Sand Island No. 2, named the "Pony," consisted of a rectangular steel frame lifted, through eyes in the four corners, by the crane on the outer of two travelling carriages equipped with cranes. Two slings passed around any pipe section and held the section tight under the frame. Connected to the sides of the frame were two arms, each of which was controlled by two hydraulic rams (see Figure 10-26). When the Pony was underwater and close to the final laying position, these arms were extended to stabilize it between the two rows of sheet piling. Lateral movement of the pipe could be obtained by retracting one arm and extending the other. Inshore and offshore movement of the pipe section was either effected by moving the whole carriage, for relatively large motion, or by making use of two tuggers, pulling through blocks attached to the trestle, for smaller movements. Instructions were relayed to the surface by the contractor diver on the bottom, and the foreman on the trestle directed the proceedings with hand signals.

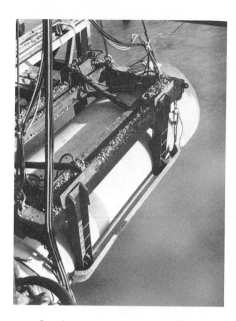

Figure 10-26   The "Pony"

On the Sand Island job, pipe was laid shoreward from the end of the trestle, and the inshore end of a freshly laid pipe section was set on a support called a sleeper, such as a cut-off concrete pile cap,* as a temporary measure until inspection of the joint and the joint pressure test, detailed in Section 10-9, were completed. After bedding of the pipe was completed, the whole weight of the pipe section was set down on the temporary support and the bedding, the slings were undone, and then carefully withdrawn so as not to roll the pipe, as the Pony returned to the surface for another section. The pipe section normally dropped about 0.1 ft (0.03 m), but sometimes dropped up to 0.2 ft (0.06 m), due to settlement of the bedding material, after it was released by the Pony. A similar amount of such settlement was found in the Orange County No. 2 outfall construction job.

The elevation of the freshly laid pipe section was checked by a man on the trestle using a calibrated line attached to a "C" clamp connected to the top of the pipe section.

Miscellaneous Considerations: Between days of pipelaying, if there is much wave action, great changes can take place that must be remedied before pipelaying can continue. Bedding stone can be eroded and already laid pipe sections displaced; sand and boulders can be driven inside pipe already installed. Both of these things happened under the trestle at Mokapu (#47), in

*During the construction of the Hollywood, Florida, line, sleepers were made of a wood and concrete composition that had a slight negative buoyancy so that they could be easily handled by divers. Burlap bags full of cement were used during construction of the Georgia-Pacific Paper Co. outfall at Newport, Oregon, for this purpose.

late 1975, when the wave environment was severe (see Figure 10-27). Later a temporary steel bulkhead was placed over the end of the outfall, at the end of a working day, to keep debris out of the pipe.

On Sand Island No. 2 a concrete block wall was built inside the first pipe laid near the end of the trestle to stop surge from travelling along the pipe. Sheet piles extending above the requisite height in the Sand Island No. 2 outfall work were cut off underwater at a nominal distance of 4 ft (1.2 m) below the original ocean bottom. Details on underwater cutting tools are available in various references, e.g., Reference 78 and Cayford (1966). A major reason that the piles were left in place was that pulling of the piles in the immediate vicinity of the laid and rocked pipe could have partially opened a joint or chipped a pipe.

Figure 10-27   Waves breaking on Mokapu outfall trestle

Trestles against the Sea: The height of the deck of a trestle above the water surface must be adequate to ensure that it clears high waves. In the construction of the Hawaii Kai outfall on Oahu in the mid 1960s, for example, heavy wave action damaged the deck of a trestle being used in the work. This deck was 14 ft (4.3 m) above the water surface. Big swells during the construction of the Sand Island No. 2 outfall in early 1975 broke over the trestle, 15 ft (4.6 m) above the still water level, dislodged four intermediate 12 × 12-in. (305 × 305-mm) piles, and damaged a structure set on the trestle. Moderate swell from a storm in late 1975 sent water over the Mokapu trestle whose deck level was 14.2 feet (4.4 m) above the mean water surface (see Figure 10-27). Work had to be done on the trestle's cross bracing before the trestle was serviceable for trolley (carriage) traffic again.

During the last week of September, 1976, Hurricane Kate developed in the Pacific and moved in a general northwest direction past the Hawaiian Islands.

On September 28 it passed 300 miles (about 500 km) east of the island of Hawaii. Central pressure of the hurricane was about 984 mb, the maximum wind speed 70 knots, and the estimated radius to maximum winds 25 nautical miles.

On the island of Oahu, 200 miles (320 km) northwest of the island of Hawaii, the construction crew at the Mokapu outfall site (#47 in Table 10-1) was involved in placing rock over the outfall and in removing trestle bents as the rock cover was inspected and passed. Two bents had been removed when the first large swell from the hurricane arrived during the afternoon of September 29. The crew watched helplessly as the surf grew and started to tear the trestle apart. There was reportedly green water well above the top of the trestle 14.2 ft (4.3 m) above mean sea level.* Walkway boards were thrown high in the air and the whole end of the trestle was seen to move appreciably as the waves passed. The air was filled with spray. Some bents had already gone down by the time darkness fell, and the morning's light showed that nine bents had been knocked down overall, reducing the length of the trestle by 180 ft (55 m).

A diving inspector entered the water on October 1 to survey the damage, and I did the same the following day. There were piles, beams, and rails twisted (Figure 10-28) and scattered among piles of redistributed armor rocks. In some

Figure 10-28   Bent "H" pile from Mokapu trestle

cases, the $12 \times 12 \times 1/2$-in. ($305 \times 305 \times 13$-mm) "H" piles had been bent so severely that they were squashed flat. I found one long length of rail about 100 ft (30 m) from its former location.† Enormous ("10-ton") armor stone had been appreciably displaced, and one large ("1-ton") boulder had been moved to a position over the end of the pipe (see Figure 10-29). The bulk of the pipe itself, heavily buried, was not harmed. The short, exposed part of the pipe, nestled deep in a trench, was also unharmed.

Several trestles along the American west coast have reportedly suffered severe damage from heavy seas. The trestle for the inshore portion of the

---

*When the swell hit it was low tide.

†Several days later a longitudinal "I" beam for the trestle was found washed up on a gravel bar a minimum of 500 ft (150 m) from its original position. The beam was 20 ft (6 m) long, of nominal size 24×12 in. (610×305 mm), weighing 100 lb/ft (1460 N/m).

Figure 10-29   Rocks near temporary terminus of Mokapu outfall

Georgia-Pacific Paper Co. outfall at Newport, Oregon, experienced difficulties in mid-June, 1965. Heavy seas damaged the end of the trestle and dumped a $100,000 pile driver into the water. A $150,000 crane sent out to retrieve the pile driver several days later caused part of the trestle to fail, and the crane joined the pile driver in the sea. The surf was sufficiently severe that it was over a week before divers would enter the water to start salvaging the equipment.

Spider

In 1970, the firm of Healy Tibbitts Construction Company fabricated a special construction rig called Spider I for use in the building of a fishing pier. This was a movable platform to work in water up to 28 ft (8.5 m) deep (smooth water). Shortly later a second, bigger movable platform, Spider II, was built for up to a 40-ft (12-m) depth. In 1972, these two rigs together constructed the inshore portion of the Hawaiian Independent Refinery, Inc. (HIRI), three-pipe system extending out to an offshore oil tanker mooring in 120 ft (36 m) of water off Barbers Point, 10 miles (16 km) from Honolulu.

Both rigs consisted basically of two steel platforms, one above the other (see Figure 10-30). Spider I weighed 200 tons, Spider II 400 tons. Spider I, for example, was designed for a deck load of 100 tons. It was supported by telescoping 3-ft-diam (914-mm) pipe piles having 5-ft-diam (1524-mm) feet. The upper platform had plan dimensions of 50×52 ft (15.3×15.9 m) whereas the lower one was 50×72 ft (15.3×22.0 m). Both rigs moved as follows:

1. In the working position all eight legs of the rig were on the bottom and the platforms were close together vertically (see Figure 10-31). The first stage in movement was for the top platform to be raised while its feet were still on the sea floor. While this was happening the lower deck was also raised, but its legs were brought up off the bottom.

Figure 10-30   Spider II platform at Santa Barbara

Figure 10-31   Spider II and provisioning trestle

2. The lower deck was then moved in the direction of intended motion a distance up to 15 ft (4.6 m) and dropped so that its legs came to rest on the bottom.

3. The legs of the upper platform were then retracted, this platform moved over the lower one and then, when in position, its legs extended to meet the sea floor. The vertical position of its surface was then adjusted as desired.

4. The procedure was repeated until the Spider reached the work site.

The maximum rate of movement was 450 ft/h (140 m/h), and the foregoing steps were controlled by one man in a control room. Each rig was fitted with a crane to be used for various operations such as driving sheet piling, dredging, or laying pipe.

Spider II was used during the construction of the Santa Barbara outfall (#48 in Table 10-1) in 1975–76. Its crane was used to offload sheet piles from trucks, stack these, and later drive them flanking the outfall line. Spider II was also involved in excavating within the sheets, in laying pipe, and backfilling. The use of the Spider platform cut out the need for an *extensive* trestle. The short trestle involved, only 700 ft (210 m) long, was built *on* the sheet piling and was used to carry a provisioning tram (pipe and rock) out to the Spider (see Figures 10-31 and 10-32). Beyond the short trestle, provisioning of the Spider II was done by barge.

Spider I was put to work for the Barbers Point outfall (#49) in 1976–77.

Figure 10-32 Detail of Santa Barbara provisioning trestle

Outfalls across Tidal Flats

Building an outfall across an extensive tidal flat can be difficult problem. It is instructive to briefly consider two submarine pipes that were extended across tidal flats with disastrous consequences, because there are some lessons. One line was pulled to lie *on* the bottom during low tide. As the tide rose, the pipe acted as a dam, until it finally failed due to the unbalanced head of water. In another case, several sections of pipe were put together in a trench during low tide, the end of the pipe capped, and the trench refilled with native material, 1.5 ft (0.5 m) over the pipe crown, before the tide rose again. On two occasions the pipe had risen to the surface of the trench by the next low tide, presumably because of buoyancy of the pipe in the liquefied sand surrounding it. On each occasion a half-dozen pipes had broken.

10-6   BEDDING AND PROTECTING THE OUTFALL

Placing Bedding Stone

"Pea gravel" is composed of small stones about the size of peas. It should never be considered as bedding material for an outfall unless the water is forever calm. It is much too small and can easily be eroded by even token wave action. The bedding stone used on the Sand Island No. 2 outfall is typical of that used as foundation material for ocean outfalls, viz., $1\frac{1}{2}$-in. (38-mm) nominal. Such material has such a low fall velocity in water, approximately 3 fps (1 m/s), that it can be dropped right on the diver's bubbles. A fairly standard technique then is for the contractor's diver to sluice this rain of gravel under and partway around the pipe using the water jet from a special system, mentioned earlier and shown in Figure 10-21, that consists of a tee (to eliminate a flow-induced reaction force on the diver) and a hose leading to a suitable supply pump on the surface.

Different techniques can be used to drop the bedding stone in relatively

Figure 10-33   Hopper and trough for bedding stone

Figure 10-34 Dump bucket for bedding stone

shallow water. On the barge, a clamshell can be employed; this can also be done from a trestle, but it is perhaps more usual in the latter case to make use of some form of dumping hopper and an inclined trough (see Fig. 10-33). In either case, great care must be taken that a stray piece of large rock is not released over the diver with the bedding stone. Care must also be exercised that the diver's hose isn't buried by the rock. A bucket used for dumping bedding stone is another possibility (see Fig. 10-34).

When the water is deeper than 40–50 ft (12–15 m) and/or there are currents to contend with, it is more usual that the bedding stone be *directed* to the work site, as scatter would be too great if the simple dropping approach were employed. This can be done by feeding the stone into the top of a (telescoping) tremie (for underwater concrete) pipe that extends down to just above the bedding area. This technique was employed for Orange County No. 2 (#35), Sand Island No. 2 (#44), and Barbers Point (#49), for example.

Armor or Ballast Rock

Sizes Involved: Armor rock is big enough that it cannot be simply dropped on either the diver or the outfall. The following information for the Sand Island No. 2 outfall will illustrate the point. There the various rock sizes were classified as follows:

Table 10-2: Rock Sizes for Sand Island No. 2 Outfall

Rock Designation	bedding stone	C	B	A	A1	A2
Nominal Size, inches (mm)	$1\frac{1}{2}$	6	12	18	24	30
	(38)	(150)	(300)	(460)	(610)	(760)

The grading of rock sizes A1 and A2 is shown below. Size A2 is shown stockpiled in Figure 10-35. The nominal sizes 24- and 30-in. (610- and 760-mm) correspond to (spherical) weights of 680 and 1327 lb (3025 and 5903 N), respectively, using a rock specific gravity of 2.6 as assumed in the Sand Island No. 2 outfall specifications. The correspondence between rock size and weight is shown in Figure 10-36.

Figure 10-35 A2 armor rock for Sand Island No. 2 outfall

Table 10-3: Grading of A1 Rock

Weight of rock, W, lb (N)	1316±10% (5854±10%)	704 (3132)	300 (1334)	160 (712)	80±10% (356±10%)
Mix %<$\dot{W}$	100	30–50	15–40	5–30	0–20

Table 10-4: Grading of A2 Rock

Weight of rock, W, lb (N)	2571±10% (11,436±10%)	1316 (5854)	704 (3132)	300 (1334)	160±10% (712±10%)
Mix %<W	100	30–50	15–40	5–30	0–20

Placing the Rock: It is possible to direct small armor rock sizes, such as 6 in. (150 mm), to the work site using a tremie pipe, but even this is a marginally effective method since the larger members of the rock can jam in the pipe. In the Orange County No. 2 outfall construction, rock was dumped on a special deflecting plate/splitter arrangement that directed the rock into approximate position. Considerable detail on this operation is provided by Harper (1971). A

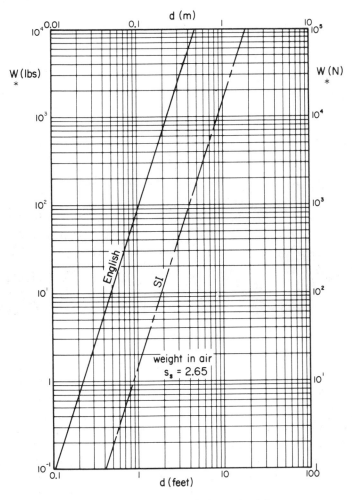

Figure 10-36 Relationship between rock size and weight

technique used in the Sand Island No. 2 work was to use large "skips," steel truck tanks cut on the diagonal, to carry the rock to the bottom where they were carefully tipped to end dump their loads. Strings of old rubber tires attached to the skips cushioned the (prescribed maximum) fall of the rock over the pipe (see Fig. 10-37).

Whatever the means used, it is impossible, practically speaking, to rigidly adhere to rock specifications for an outfall, particularly in deeper water. Therefore, minimum dimensions are customarily used for the trench and backfill materials, including armor stone. The cross sections for inshore and offshore waters for the Sand Island No. 2 outfall are shown in Figures 8-3 and 8-4. The upper boundary of the class B stone shown in Figure 8-4 is my idea of an

Figure 10-37 Rock "skips" and armor rock on rock barge

unachievable requirement.* I believe it would be more realistic to indicate on a plan a band indicating acceptable upper boundaries of the class B stone. The design of armor rock protection blankets is not that much of a science anyway.

The rock placed during outfall construction inevitably mounds up in certain zones, in certain cases above the sea floor when the pipe is so-called "over-rocked." It is essential that the diving inspector, see Figure 10-38, ensure that there is the required minimum rock cover over the pipe. One practical method of checking this cover, albeit somewhat approximately, is to use a precision depth gage set on the pipe before rocking and atop the rock cover after rocking. But a real difficulty for the diver inspector in such a case is knowing what station he is working at once the rock has been placed. Before the rock is placed he can determine his position from markings on the pipe sections; afterwards he has no such guides.

Even if the diver inspector can accurately determine the depth of rock cover, it is still difficult to tell if bridging within the matrix has occurred, leading to large voids. One of the functions of using a well-graded rock mix, of course, is to avoid this possibility. Gaging of rock cover by using measurements from the

---

*Outfall construction divers with whom I have talked have repeatedly expressed their exasperation with the designs and specifications of the engineers. The divers feel that there is minimal appreciation for the real world of ocean outfall construction on the part of the standard designer. Consider, for instance, the specification of *five* different grades of rock in the cross-section for one buried outfall.

Figure 10-38   Diving inspector making notes on armor rock cover

ocean bottom is a very inexact procedure since the undulatory nature of the sea floor makes it a poor reference level.

Checking the adequacy of a rock cover for an outfall can best be done when the water is relatively clear, before operations start for the day. Such inspection can be done with swimming divers for water depths less than approximately 100 feet (30 m), beyond which decompression problems develop. To avoid such difficulties for deeper locations, inspections can be made from some form of atmospheric pressure diving bell suspended from the ship (see Figure 10-39).* The inspector in the bell is moved back and forth or along the line by relaying instructions to the crane operator and/or the anchor winches operator on the barge.

There is a real problem of ballasting a standard diffuser section with rock, especially in deeper water where one puts swimming divers into the water only with reluctance, because the rock tends to mound over the ports. The pipe section around the port must not be rocked because this could inhibit the desired initial jet dilution mechanism, if not block the opening completely. The ports themselves should have temporary covers in many cases to keep stone from moving inside the pipe. Such covers should be numbered, incidentally, especially if some of the ports are to be left covered for a time, and divers removing covers should bring them to the surface for checking unless an inspector makes his own underwater inspection.

Rocking of a pipe should keep up with the laying so that forces from large

---

*There are various such atmospheric pressure bells on the market. For example, the Reading and Bates (Tulsa, Okla.) commercial design features a sphere 60 in. (1520 mm) in inside diameter with five 12-inch-diameter (305-mm) viewing ports and a 24-in.-diam (610-mm) access port closed by a lock. This bell has a 450-ft (137-m) depth capability. Communication lines and breathing gas hoses connect the bell with the barge deck as does a support cable from a suitable crane. Great care must be taken when putting such a bell into the water or removing it since banging against the barge or ocean bottom because of wave-caused ship motion is a real possibility. For protection, the Reading and Bates spherical bell is enclosed by a grillwork of pipes.

Figure 10-39 Diving bell

waves do not open out the unprotected line. Divers have reported that failure to do this during the construction of an outfall at Huntington Beach, California, caused the line to open up at a joint about twenty sections shoreward of the laying position.

Proper pipe grade to avoid high points is important. During construction of the Sand Island No. 2 line a high point led to the removal of eleven 24-ft-long (7.3-m), 84-in.-diam (2134-mm) pipe sections as well as the overlying armor rock since there was about an 18-in. (450 mm) air pocket in the line. Two pipe bells were broken in the process. A delay of 2 weeks was involved for an activity on the critical path. If rock must be removed from an outfall, air lifts are capable of removing material up to about 9-in. size (230-mm).

When bedding and rocking are being done offshore, it is usual for a gravel/rock barge to be towed to the lay barge. There, either the gravel/rock barge is tied alongside (Figure 10-37) and its load taken off by a clamshell as needed, or the barge is unloaded, the material being stored in hoppers on the lay barge. A standard method for the gravel/rock barge to be loaded ashore is for conveyors to be used at a suitable pier bordering a storage yard. Large front-end loaders can also be used to place material on the gravel/rock barge.

10-7 UNDERWATER OPERATIONS

Orientation

The work of both contractor and inspector divers during the construction of marine outfalls has been briefly sketched from time to time in the previous sections of this chapter. It is intended that this section round out the overall construction functions of these two classes of divers plus add supplementary, pertinent information about underwater operations.

Safety

Appendix B provides background on diving and includes a discussion of the dangers of diving *per se*. Divers on underwater construction projects also run many additional risks that include the following: cave-ins, strong currents, dangerous marine animals, opaque water, heavy equipment, and sloppy tenders.* Even above water divers may be in danger if the diver's station on deck is located so that the rock bucket is swung directly over them.

"Buddy" diving, wherein two divers remain together throughout a dive, is widely preached as a prudent safety measure. But this approach is virtually never practiced on marine outfall construction where *one* man enters the water to work or to inspect. Part of the reason for this concerns the self-reliant nature of most commercial divers. Part reflects the cost-consciousness of the diver's employer. Divers in the water are expensive as will be illustrated later. Even divers on "standby," ready to enter the water rapidly in the event of an emergency, are expensive and have seen infrequent use.

The primary U.S. federal law dealing with safety, and *generally* applicable to marine outfall construction, is the Occupational Safety and Health Act of 1970, Public Law (PL) 91-596, mentioned in Chapter 9. The Occupational Safety and Health Administration (OSHA) was set up within the Department of Labor to administer the Act.

The United Brotherhood of Carpenters and Joiners of America, a union representing divers, petitioned OSHA on August 8, 1975, to issue an Emergency Temporary Standard (ETS) for diving operations in United States coastal waters. This was in response to heavy diver injury and mortality in North Sea oil operations, and it considered the potential for similar problems in U.S. coastal waters.

Hearings were held by OSHA to obtain oral testimony on the petition and possible standards. Written testimony was also obtained. A federal agency task force with consultants then put together Reference 19, which was to become effective July 15, 1976.

The response to a serious, or potentially serious, situation is often a measure of overkill. Although there was general agreement that the ETS would save diver lives, firms that employed divers complained loudly about the needless additional complexities and costs that the legislation would mean to them. Some of these views were put forward in Reference 72. A suit by the Association of

---

*Two examples where such dangers led to injuries that kept individual construction divers out of the water for several weeks are the following: During the construction on the Sand Island No. 2 (#44) a man lost the end of one finger when checking the pipe gap and the section being inserted was rammed "home." The leg of a "Horse" being used on Mokapu (#47) caused part of the trench to cave in, and a dislodged chunk of bottom material broke the foot of a diver.

Diving Contractors resulted in a "stay" of the ETS by the U.S. Court of Appeals for the Fifth District, New Orleans, La. Following this setback of August 11, 1976, OSHA withdrew the ETS. Hearings in New Orleans in December 1976 and January 1977 resulted in considerable input concerning diving regulations, and OSHA issued its final commercial diving standard in the *Federal Register* on July 22, 1977. This was to become effective on October 20, 1977.

Diver Decompression

Two deck decompression chambers (DDCs) were maintained on the deck of the Davy Crockett during construction of the Sand Island No. 2 outfall. One chamber was for use by contractor divers and the other by the inspectors. Control was normally maintained by an occupant of the chamber, but control could be taken over outside the DDC via a control panel. Both DDCs were of double-lock design. This means that there were two separate chambers.

In the Reading and Bates (Tulsa, Okla.) commercial double-lock decompression chamber, for example, the inside diameter is 55 in. (1400 mm). The inner chamber is 84 in. (2134 mm) long and the outer one 50 in. (1270 mm) in length. Entry to the outer chamber from the outside is through a 24-in.-diam (610-mm) port that mates, if necessary, with a port on a diving bell.

The DDCs were used on Sand Island since a decompression system that avoided excessive underwater decompression stops was employed. The diver's last water decompression stop was at 30 ft (9 m). He was then brought promptly to the deck of the barge on a small elevator and within 3 min was partly undressed and entered the outer lock of the DDC with his tender.* The inner lock had been set at a pressure equivalent to 80 ft (24 m) of water before the diver surfaced, and when the outer door was closed the pressure of the two equal-size locks was equalized to 40 ft (12 m), which is the standard depth equivalent for this technique. The breathing gas was oxygen. When the diver had been stripped of his gear, he entered the inner lock and closed the door between the locks. There he remained for the required decompression time; see Table 10-5. The tender, on the other hand, lowered the pressure of the outer lock to atmospheric and departed. He then waited, as a standby tender, in case he was needed by the decompressing diver.

The surface decompression technique permits the diver a measure of comfort during decompression. If he is cold or exhausted and/or the sea rough, staying in the water to decompress could cause considerable discomfort. In addition, the time of decompression is shortened because the diver goes onto O_2 in the DDC (see Table 10-5). Use of the DDC for decompressing divers also clears the deck for immediate use of the next team of divers. However, questions

---

*The time from a diver's leaving the 30-ft (9-m) water station to his reaching a depth equivalent of 40 ft (12 m) in the DDC should not exceed 5 min.

Table 10-5: Surface Decompression Table, using Oxygen, for U.S. Navy—Defined Optimum Conditions*

Work Depth (ft) (m)		Optimum Exposure Time (min)	Decompression Time Starting at 40 ft (12 m) in DDC on O_2 (min)	Total Decompression Time from Bottom until Exit from DDC (min)	Total Ascent Time Breathing Air and Using Water Stops (min)
70	21	120	23	32	52
80	24	115	31	40	71
90	27	90	30	40	68
100	30	80	32	42	73
110	33	70	33	44	74
120	37	60	32	45	71
130	40	60	37	53	86
140	43	55	38	57	87
150	46	50	38	66	89
160	49	45	38	73	85
170	52	40	36	83	82

*U.S. Navy Dept. (1970).

have been raised concerning the possibility of physiological damage to divers decompressing using this approach.

The maximum effective (because of repetitive dives) amount of time that can be spent at various depths over a 12-hr time period, using this surface oxygen-decompression technique, is called the *optimum exposure time* by the U.S. Navy. This time is supposed to "represent for the average diver the best balance of safety, length of work period, and amount of useful work" (U.S. Navy Department, 1970). When a diver has spent the optimum working time on the bottom he is said to be "burnt out." The U.S. Navy Department (1970) stressed that diver exposure beyond the "optimum period" should be permitted only "under special conditions."

Diver Pay

Diver and tender conditions and pay rates are very clearly spelled out, such as in Reference 17 for Southern California (1968–1973) and Reference 44 for Hawaii (1975–1978). A diver is required to make only one (U.S. Navy) optimum dive within any 24-hr period if the depth is 100 ft (30 m) or more. The limit is one such dive in any 12-hr period for depths of less than 100 ft (30 m). Adequately equipped double-lock decompression chambers are necessary on the diving barge.

Although diver pay rates themselves change from contract to contract, it is instructive to list rates taken from References 17 and 44 to illustrate the diver payment format. In Southern California between 1968 and 1973, a diver received $100/day or any part thereof. If the diver furnished diving gear, he got $130 more; if he supplied gear, tender, and insurance the figure was $230. A standby diver got $54 per day or part thereof plus $173 if he furnished gear, tender, and insurance.

A diver receives extra pay (double-time) for working night shifts. Overtime is anything over eight normal-time hours per day or 40 h/week. In addition he gets extra pay for diving in pipes, tunnels, or other enclosed areas. But primarily he gets extra ("depth") pay when diving deeper than 50 ft (15 m). The shallower water rates in the 1975–1978 Hawaii schedule were as follows: 50–100 ft (15–30 m) ($1.35/ft over 50 feet); 100–150 feet (30–45 m) ($65 + $1.85/ft over 100 ft); 150–200 ft (45–60 m) ($160 + $2.75/ft over 150 ft).

10-8 POST-LAYING EXCAVATION

It was outlined in Section 10-2 that there are two general methods available for completing a submarine pipeline installation. In the first case, there is preliminary excavating, and then the pipe is laid in the excavated trench and the trench backfilled in separate operations. This is the method that has been described in some detail in previous sections, particularly for the Sand Island No. 2 outfall. However, there is another installation method wherein the trenching and backfilling follow the laying of the pipe on the sea floor. The Nippon Steel Corporation (1973) has termed this the *post-laying excavation method.*

The basic concept involved in this method of installing submarine pipelines is that the line is first laid on the bottom along the desired route using one of the techniques described in earlier sections. A special travelling trencher is then mounted on the pipe, and this device is made to move along the pipe and the line sags down into the ditch as the trencher proceeds.* Backfilling may take one of the forms discussed earlier, but more usually the trench is allowed to backfill by natural means. Travelling trenchers for very small diameter lines may actually lower and cover the pipe in one operation. This is the "jetted-in" arrangement depicted in Figure 8-5(f).

The post-laying approach would appear to be limited to small to moderate-sized outfalls made of steel pipe. The Seamole travelling trencher,† for example, can handle pipe sizes from 2 in. (50 mm) to 42 in. (1070 mm) in water depths up to 600 ft (180 m). It can apparently handle either hard or soft bottom conditions. This device rides on rollers contoured to fit the pipe, and cutting under the pipe is accomplished by two counter-rotating cutter heads driven by hydraulic motors. Control is from an attending barge by way of cables and hoses. The angle of the sides of the trench can be varied through the angle of the cutters, and spoil is discharged to the sides of the trench by suction dredge pumps mounted on the carriage. In one job of trenching under a pipe 23 in. (584 mm) in outside diameter, the Seamole progressed 2 ft/min (0.6 m/min) through clay

*The problem of sea floor obstructions can be particularly acute in such work.

†Oceanonics, Inc. (Houston, Texas).

with a shear strength of 2000 lb/ft^2 (approximately 10^5 N/m^2) cutting a trench 5 ft (1.5 m) across the top, 2 ft (0.6 m) across the bottom, and 5 ft (1.5 m) deep (Ives, 1969).

Another type of travelling trencher has been described by Ward (1966) and in this case high-pressure water jets loosen the material, which is then removed from the trench by a suction dredge pump. This unit can cut a 4-ft-wide×6-ft-deep (1.3×1.8 m) trench from between 2 ft/min (0.6 m/min), for harder materials, to 10 times this figure in soft muds and loose sands. It will not adequately deal with formations of large gravel. The pressure on the jets is phenomenal—1000 psi (68 bar)—and the pumps are rated at 5400 hp.

Six thousand feet (1800 m) of the Hyperion sludge line (#11 in Table 10-1) were trenched for using this general approach, and the trench was approximately 17 ft (5 m) deep in the surf zone. In this case, pumps and compressors were mounted on an attending barge and a water/air stream was used to carry away the bottom material.

Finally, a travelling trencher that loosens bottom material through fluidization is described by Reynolds and van der Steen (1974). A complete dredge cutter/ship system is described in Reference 63.

10-9 ADDITIONAL CONSIDERATIONS

Orientation

Various topics of importance in marine outfall construction do not fit properly into the preceding sections of this chapter. These topics have been grouped together in this section.

Potential Problems with Reinforced Concrete Pipe

Harper (1971) has summarized the types of problems encountered with large-diameter RCP in the casting process, in being handled in the pipe yard, in being loaded aboard the laying barge, and in being laid. This was in conjunction with the Orange County No. 2 outfall (#35 in Table 10-1). In the casting process, common imperfections were the following: rock pockets on various parts of the pipe due to leaking of forms, failure of a form vibrator, and other factors; air pockets caused by trapped air on the inside face of the bell; shrinkage cracks around the circumference on the outside where the slope of the extended bell met the barrel of the pipe. Pipe damage in the pipe yard resulted from the following: improper removal of the gasket groove snap rings; gouging by the pusher tractor system; breaking out of pieces of concrete in the bell section when core forms were not lifted out carefully. Because of such problems, consideration should be given to having a full-time inspector employed to observe each phase in the manufacture of the pipe.

Inspections should also be made prior to placement from the laying ship to ascertain if the pipe has sustained damage during transportation, loading, and unloading. If such damage is found, it should be repaired. Inspection by a diver-inspector is also advisable, as outlined earlier, once the pipe is in position on the sea floor. During the work on the Orange County No. 2 line, pipe was damaged during unloading as follows: differential motions of the laying ship and supply barge resulted in bumping of the pipe sections on the deck; the rock hopper swung across the deck collided with a pipe section. During the placing of the pipe on the same job, many pipes were damaged when wave action caused the pipe section being laid to bump into the already installed mating section and when the pushing board on the Horse caused damage to the bell either by rubbing or from an excess of pressure applied to an irregular surface.

The joint area of one of the RCP sections used on the Mokapu job is shown in Figure 8-9. A problem on that job, and with others such as Orange County No. 2 mentioned above, is that almost any banging of the spigot will mean that the 1-in. (25-mm) concrete tip protecting the "O"-ring will be broken, exposing the "O"-ring and promoting a leaky joint. It has been suggested that the protection distance for the "O"-ring be at least doubled.

In one instance each during Sand Island No. 2 and Barbers Point work, the heavy clamshell, involved in rocking the outfall at night, struck the end of the pipe, breaking off part of the bell including reinforcing steel. The damage was sufficiently severe that the section had to be removed and replaced in both cases.

Damage to RCP, which amounts to chipping away of part of the concrete, can be repaired using commercial products such as Sika-Plug C (used on the Sand Island No. 2 outfall) or Water Plug (used on San Diego). If the damage is minor, the repairs can be effected on the bottom; both Sika-Plug C and Water Plug will set up underwater. These products can be carried to the bottom, in powder form, in a closed plastic bag. At the repair site the bag is punctured or opened, the sea water mixed with the powder, and then the mixture placed on the damaged location and finished. This must be done promptly since these products will harden in several minutes. If the damage occurs above water, or if damage sustained underwater is not minor and the pipe must be brought to the surface, then Sika-Plug C and Water Plug may be used there, assuming approval by the project engineer.

Harper (1971) reported that pipe gaskets came off from time to time as a pipe section was lowered through the water during construction of Orange County No. 2. It was a costly delay, then, to bring the section back aboard ship and replace the gasket. Better results were obtained with the use of highly elastic gaskets to which unmelted shortening had been applied.

During the laying of the beveled pipe sections in the curve upstream of the diffuser for the Barbers Point outfall (#49) the joints opened out on the upslope side and the pipe angled down the slope in an offshore direction. Eight 24-ft long (7.3-m) sections had to be pulled out to correct a 54-ft (16.5-m) deviation from the intended line. In this connection it is appropriate to note that maintaining line along a curve at sea is difficult in itself. Even straight lines may be bothersome to follow if their extensions do not reach the shore where a laser alignment system can be located. An L-shaped diffuser is a case in point.

Testing Joints in Double-Gasket Reinforced Concrete Pipe

The type of double-gasket pipe joint shown in Figure 8-9 permits the inspector to test the "tightness" of the joint after the spigot of a new pipe section has been inserted into the bell of the last-laid section. A split or extruded "O"-ring can mean a leaky joint. The test involved is made possible by the monel tube, shown in Figure 8-9, that extends from a point between the two "O"-rings to a point out of the joint area. The latter opening is normally closed by a monel plug when not being used in the testing. During testing, it receives a pipe nipple on the end of a hose extending up to a pressurizing station on the trestle or barge.

Before a pipe section is lowered into the water, the nipple on the end of the test hose is inserted into the entry to the monel tube and the tube is flushed with water to assure no blockage. If this weren't done, a pipe joint could be judged satisfactory simply because water couldn't move through the tube.

For testing, after the pipe section has been inserted, an air compressor line is used to apply the requisite test pressure to the surface of the water column extending down to the pipe joint. This water surface is typically in a transparent chamber so that any fall in level can be charted, and the chamber is calibrated so that the actual volume of water passing down the hose to the pipe joint can be determined.

During the construction of the Sand Island No. 2 outfall, joints were tested with pressures about 20% greater than the calculated maximum inside-outside pressure differentials for the outfall in service.* The test lasted for 5 min, and if the leakage down the hose (and out either way through the "O"-rings) was less than 0.1 gpm (0.38 ℓ/min), the joint passed. Joints failing this type of test during the Mokapu work (#47) were completely encased in concrete and then passed.

The removal of the hose nipple from the monel tube after testing was carried out by a diver until the depths became too great. After that point a special fitting was used so that the hose could be released when the Horse was removed and brought back to the surface. The deepest pipe sections, making up the diffuser for the Sand Island No. 2 outfall, had only one gasket and no joint pressure test was intended or applied.

Some submarine pipe joints are not pressure-tested, e.g., those in which flexible plastic gaskets such as Ram-Nek are used. In this case, the jointing areas are primed, the prime coat is allowed to dry, and strips of the material are laid

---

*The total applied pressure for all sections under the trestle was 52 psig (3.5 bars gage).

along the joint areas. Ram-Nek meets Ram-Nek when the pipe sections come together underwater, and the joint is sealed.

Miscellaneous

Standard Inspection: Spot checks, using standardized testing techniques, on all construction materials are mandatory. Examples are gaskets, concrete aggregate, gravel bedding material, ballasting stone, steel components such as reinforcing bar and plate, sheet piling, and cement. The results of laboratory tests on certain items, such as cement, welding rods, concrete cylinders, gravel, reinforcing steel, stone and aggregate, and gaskets, should be submitted to the owner or his representative, by the contractor or the designated testing agency. Occasional sections of pipe should be bulkheaded and hydrostatically pressure tested at the casting yard. Manhole sections in particular should be tested. Pipe bells should be checked since a rough inside surface can lead to leaking and/or tearing of gaskets.

Potential Problems with Weight Coatings on Steel Pipe: Concrete weight coatings can spall off steel pipes during construction or shortly thereafter. One instance was during the laying of the Hawaiian Independent Refinery, Inc. lines in 1972–73. Usually the concrete coating develops minute cracks at bends during laying, leading to spalling unless the coating is well-designed and executed. One thing to check to prevent separation of the concrete, if applicable, is that the coal tar wrap between concrete and pipe not be too thick and that it is firmly bonded to the pipe.

Marking Manholes: One problem with outfalls buried under armor rock is how to mark manholes. One approach has been to leave a cone-shaped cavity over the manhole free of armor rock and to place much smaller material, such as bedding stone, over the manhole. The smaller stone helps the diver to locate the manhole and then can be readily removed by an air-lift if access to the inside of the pipe is desired.

Use of Weather and Sea Forecasts by Contractors: At some outfall job sites, weather conditions and/or wave action can be a big problem. It is suggested that contractors avail themselves of the services of the National Weather Service in the U.S., or equivalent organizations in other countries, so that work can be properly scheduled around imminent wind, weather, and wave conditions. In the event of predicted high wind and/or wave conditions, floating equipment can be moved to a safe anchorage and movable items on trestles can be either transported to shore or securely lashed down. Figure 10-40 shows a crane barge, used offshore for an outfall project, after strong winds and heavy wave action drove it onto a nearby wave-cut terrace. Not only can wind and wave predictions minimize the possibility of such an occurrence, but workers and inspectors can be told ahead of time that their services will not be required on a given day or series of days.

Figure 10-40   Crane barge driven ashore by high winds and waves

10-10 REFERENCES

1. "A Polyethylene Pipe Replaces Steel for Sanitary Sewage Outfall" (1971): *Civic Administration* (January).

2. ADE, G. J. (1970): "Nelson Prestressed Flexible Submarine Outfall," *New Zealand Engineering* (August 15), pp. 205–208.

3. BEZZANT, R. (1970): "How to Make an Outfall in San Francisco Bay Mud," *The American City*, Vol. 85, No. 5 (May), pp. 83–85.

4. "Big Pipeline Laid Underwater" (1967): *Australian Civil Engineering and Construction* (March 6), pp. 51, 53, 55, 57.

5. "Biggest Ocean Pipelining Job Under Way" (1959): *Engineering News-Record*, Vol. 163, No. 1 (July 2), pp. 38–40, 42.

6. BOWLUS, F., H. F. LUDWIG, and L. MELBERG (1964): "Placing Oregon Outfall Sewer," *Western Construction* (March), pp. 61–64.

7. BOWMAN, R. L., and J. H. BEMIS (1961): "Pipeliners Whip Heavy Surf and Coral to Lay Offshore Hawaiian Lines," *Pipe Line Industry*, Vol. 14, No. 1 (January), pp. 33–37.

8. BROWN, C., and C. GRUNDY (1973): "Shaped Charges Blasting Technique Simplifies, Speeds Dredging Operations," *World Dredging and Marine Construction*, Vol. 9, No. 9 (July), pp. 31–33.

9. CAYFORD, J. E. (1966): *Underwater Work*, Cornell Maritime Press, Cambridge, Maryland.

10. CHABERT, G. (1969): "Constructing a New Sea-Loading System off Egypt," *Ocean Industry*, Vol. 4, No. 4 (April), pp. 28–35.

11. CHIN, A. G., and M. C. DIRKS (1967): "Placing Pipelines and Outfalls under Water for Seattle Metro Trunk Sewers," *Civil Engineering*, Vol. 37, No. 12 (December), pp. 54–56.

12. CRISP, E. W., H. M. STEWART, and S. J. N. FLETCHER (1970): "Design and Construction of a Submarine Sea Outfall at Hastings," *Proceedings*, The Institution of Civil Engineers, Vol. 47 (October), pp. 121–143.

13. DEASON, D. (1974): "Lake Leman Gas Line Laid in 1000-foot Water Depth," *Pipe Line Industry*, Vol. 40, No. 3 (March), pp. 43, 46.

14. DEAN, J., et al. (1972): "Pipeline Construction at Kharg Island Ends up Quickly," *Offshore*, Vol. 32, No. 12 (November), pp. 61–62, 64, 69.

15. "Deepest Ocean Outfall for Sewage Diffusion" (1953): *Engineering News-Record*, Vol. 151, No. 26 (December 24), pp. 38–39.

16. "Deep-Sea Divers Position Metro Outfall" (1965): *Pacific Builder and Engineer*, Vol. 71, No. 6 (June), pp. 71–73.

17. "Divers Agreement, 1968–1973" (n.d.): Piledrivers, Divers and Tenders Local No. 2375, United Brotherhood of Carpenters and Joiners of America, Wilmington, California.

18. "East Coast Resort Solves Seasonal Sewage Disposal Problem" (1970): *Civil Engineering and Public Works Review*, Vol. 65, pp. 501–503.

19. "Emergency Temporary Standards for Diving Operations" (1976): Dept. of Labor, Occupational Safety and Health Administration, *Federal Register*, Vol. 41, No. 116 (June 15), pp. 24272–24292.

20. ETHERIDGE, D. C. (1974): "Barge Cluster Shields Underwater Pipelaying," *Construction Methods and Equipment*, Vol. 56, No. 8 (August), pp. 61–64.

21. "Excavators Dig in 40 ft. of Water on Both Sides of the Atlantic" (1967): *Construction Methods and Equipment*, Vol. 49, No. 12 (December), pp. 68–70.

22. "Experience Guides Ocean Outfall Design" (1948): *Engineering News-Record*, Vol. 140, No. 10 (March 4), pp. 88–91.

23. GILL, J. M., J. H. HUGUET, and E. A. PEARSON (1960): "Submarine Dispersal System for Treated Chemical Wastes," *Journal of the Water Pollution Control Federation*, Vol. 32, No. 8 (August), pp. 858–867.

24. GOSS, W. M. (1970): "Big 14,400-ft Subsea Line Pulled," *Oil and Gas Journal*, Vol. 68, No. 47 (November 23), pp. 87–91.

25. GREENLEAF, J. W., JR., and B. A. McADAMS (1964): "Designing an Ocean Outfall

for North Miami Beach," *Journal of the Water Pollution Control Federation*, Vol. 36, No. 9 (September), pp. 1107–1115.

26. HARPER, F. A. (1971): "1971 Installation of Orange County Sanitation District's 5-Mile Ocean Outfall," Cement and Concrete Products Industry of Hawaii, Subaqueous Pipeline Installation Seminar (October 26), Honolulu, Hawaii.

27. "Hastings Sea Outfall" (1968): *Consulting Engineer*, Vol. 32, No. 9 (September), p. 63.

28. HELEN, R. R. (1955): "Outfall Runs 7,000 Feet into Pacific," *Civil Engineering*, Vol. 25, No. 3 (March), pp. 33–36.

29. HOLLAND, S. M. (1962): "Screw Anchors Hold Wandering Submarine Line," *Pipe Line Industry*, Vol. 17, No. 6 (December), p. 54.

30. "'Horse' Helps in 5-Mile Ocean Outfall Construction" (1971): *California Builder and Engineer*, Vol. 77, No. 6 (March 26), pp. 18–19.

31. "'Horse' Saddles Up for Hawaii Deepwater Pipe Job" (1974): *The eM-Kayan* (publication of Morrison-Knudsen Co., Inc.) (July).

32. "How Watsonville, Calif., Protects Sewer Outfall Line" (1950): *The American City*, Vol. 65, No. 1 (January), p. 15.

33. Hyperion Engineers (1957): *Ocean Outfall Design*, Los Angeles, California (October).

34. IVES, G. O. (1969): "Line Travelling, Self-Propelled Mole Digs Underwater Trench," *Pipeline Engineer* (October), reprint.

35. "Jumbo Places Ocean Outfall for Honolulu Sewage" (1951): *Engineering News-Record*, Vol. 146, No. 20 (May 17), p. 32.

36. KESLER, J. (1957): "Major New Outfall Sewer to Serve Los Angeles," *Public Works*, Vol. 88, No. 2 (February), pp. 93–96.

37. KLOPFENSTEIN, D. (1974): "Salvage Vessel Overcomes Pipeline Sand Problems," *Ocean Industry*, Vol. 9, No. 4 (April), pp. 229, 231, 233, 235, 237.

38. "Large Diameter Line Laid Offshore Cannes" (1972): *Ocean Industry*, Vol. 7, No. 12 (December), pp. 22–23.

39. LEE, G. C., and C. L. BANKSTON, JR. (1967): "Pipelining Offshore," *Offshore*, Vol. 27, No. 6 (June), pp. 36–39, 45.

40. LEES, J. (1976): "Walking Platform, Pipelay 'Horse' Used in Outfall Construction," *Ocean Industry*, Vol. 11, No. 3 (March), pp. 29–31.

41. LOVE, F. H. (1971): "Singapore Long Pull," *Pipeline and Gas Journal*, Vol. 198, No. 2 (February), pp. 44–46.

42. LYON, F. D. (1961): "Placing a Submerged Ocean Outfall," *Water and Sewage Works*, Vol. 108, No. 6 (June), pp. 246–247.

43. "MacBlo Switches from Log Haul to Steel Haul" (1975): *Trail Times* (June 30), Trail, B.C., Canada.

44. "Master Agreement for Hawaii 1975–1978" (n.d.): between General Contractors Labor Association and Operating Engineers Local Union No. 3 of the International Union of Operating Engineers, AFL-CIO.

45. McConnell, A. M. (1967): "Offshore Circulating Water System, Marsden Power Station," *New Zealand Engineering* (October 15), pp. 397–406, with discussion, *New Zealand Engineering* (May 15, 1968), pp. 207–208.

46. McKain, D. W. (1973): "An 84-inch Outfall Laid off Long Island," *Ocean Industry*, Vol. 8, No. 7 (July), pp. 40–41.

47. McKay, D. L., D. A. Gillis, and R. Durward (1974): "Large Diameter Submarine Steel Pipeline Crossings," ASCE, National Meeting on Water Resources Engineering, Los Angeles, California (Jan. 21–25), *Preprint 2149*; abridged version in *Civil Engineering*, Vol. 44, No. 10 (October), pp. 74–77.

48. Miller, D. R. (1956): "Engineering Goes to Sea," *Public Works*, Vol. 87, No. 10 (October), pp. 136–137.

49. Murphy, H. D., and J. B. Herbich (1969): "Suction Dredging Literature Survey," Texas A&M University, Center for Dredging Studies, *Report No. 104-CDS* (June).

50. Narver, D. L., and E. H. Graham, Jr. (1958): "Two Long Ocean Outfalls Constructed," *Civil Engineering*, Vol. 28, No. 1 (January), pp. 38–43.

51. Nippon Steel Corporation (1973): "Submarine Pipeline Engineering," *Catalog No. EXE 109,* New York (July).

52. "Ocean Outfall Slits Coral Reefs" (1956): *Construction Methods and Equipment*, Vol. 38, No. 11 (November), pp. 114, 117, 120, 125, 126.

53. Otto, R. (1971): "Orange County Outfall," *Concrete Pressure Pipe Digest*, Vol. 1, No. 3 (May).

54. "Outfall Has Precast Units 100 ft Long" (1948): *Engineering News-Record*, Vol. 141, No. 5 (August 5), pp. 64–67.

55. "Outfall, Plagued by Endless Surf, Finally Makes It to Sea" (1975): *Engineering News-Record* (January 30), pp. 20–21.

56. "Pacific Ocean Outfall Pulled 7 Miles in Only 7 Days" (1957): *Construction Methods and Equipment*, Vol. 39, No. 8 (August), pp. 140, 141, 143, 144, 148–150, 154, 155, 157–158, 163, 164, 167, 170.

57. Penzias, W., and M. W. Goodman (1973): *Man beneath the Sea*, Wiley (Interscience), New York.

58. "Pipe Handling 'Seahorse' and Divers Install Ocean Outfall System" (1975): *The eM-Kayan* (publication of Morrison-Knudsen Co., Inc.) (May).

59. "Pipe Rides into Place on an Underwater Track" (1965): *Engineering News-Record*, Vol. 174, No. 8 (February 25), pp. 26–28, 30.

60. "Pipe Valve and Vortex Drop Bubble-Free Effluent to Outfall" (1963): *Engineering News-Record*, Vol. 171, No. 23 (December 5), pp. 42–43, 46.

61. "Placing 10-ft. Diameter Outfall Pipe" (1965): *Western Construction* (March), pp. 72–75.

62. "Polyethylene Pipe Used for 36-inch Sewage Outfall" (1972): *Civil Engineering*, Vol. 42, No. 10 (October), pp. 58–60.

63. "Remote Controls Direct Pipe Line Trenching Dredge" (1975): *Ocean Industry*, Vol. 10, No. 1 (January), pp. 72, 77.

64. REYNOLDS, J. M., and A. R. F. VAN DER STEEN (1974): "Device Buries Pipe in 164 ft of Water," *The Oil and Gas Journal*, Vol. 72, No. 23 (November 11), pp. 182–186.

65. "Rig Wades in Water 200 Feet Deep to Lay Offshore Pipe" (1959): *Construction Methods and Equipment*, Vol. 41, No. 5 (May), pp. 96–98, 101, 102–104.

66. SANDERS, W. M. (1972): "Building the Zueitina Terminal," *Marine Technology Society Journal*, Vol. 6, No. 6 (November–December), pp. 17–24.

67. "Santa Barbara Outfall Underway" (1976): *California Builder and Engineer*, Vol. 82, No. 3 (February 13), pp. 18–20.

68. SHORT, T. A. (1967a): "Offshore Pipeline Anchoring System," *Proceedings of the Offshore Exploration Conference* (OECON), pp. 602–624.

69. SHORT, T. A. (1967b): "Pipeline Anchoring System," *Offshore*, Vol. 27, No. 6 (June 2) pp. 57–61.

70. "Sliding an Outfall Sewer into the Ocean" (1942): *Engineering News-Record*, Vol. 128, No. 1 (January 8), pp. 4–5.

71. "Story-High Steel Outfall Runs 3 Miles Under Lake" (1970): *Engineering News-Record*, Vol. 185, No. 12 (September 17), pp. 82–83.

72. "Struggle for Regulation of Commercial Diving Continues" (1976): *Offshore*, Vol. 36, No. 11 (October), pp. 108, 110.

73. "The Design and Construction of Submerged Ocean Outfalls, Including the Quality of Effluent for Discharge to the Ocean off Bathing Beaches" (1965): contributed by the Metropolitan Water Supply, Sewerage and Drainage Board, Perth, to the Twelfth Conference, Water Supply and Sewerage Authorities, Hobart, Tas., Australia, pp. 134–151, with discussion.

74. "Toronto Sewage-Treatment Plant" (1946): *Water and Sewage* (September), pp. 21–24.

75. "Tower Places Ocean Outfall" (1953): *Engineering News-Record*, Vol. 151, No. 8 (August 20), pp. 36–38.

76. "Trenching through Heavy Surf" (1973): *Offshore Services*, Vol. 6, No. 7 (September), pp. 54, 62.

77. "2.5 Mile Sea Outfall in South Wales" (1970): *Civil Engineering and Public Works Review*, Vol. 65, p. 1204.

78. "Underwater Welding, Cutting and Hand Tools" (1969): *Symposium Proceedings* (October, 1967), Marine Technology Society, Washington, D.C.

79. U.S. Navy Department (1970): *U.S. Navy Diving Manual*, Washington, D.C. (March).

80. "Victoria's Ocean Outfall Sewers" (1973): *Aqua* (Winter), pp. 15–18.

81. WARD, D. R. (1966): "Marine Pipe-Laying Techniques," *Pipeline Engineer* (December), pp. 3–7.

82. WELLING, C. G., and M. J. CRUICKSHANK (1966): "Review of Available Hardware Needed for Undersea Mining," Marine Technology Society, *Transactions*, Exploiting the Ocean, pp. 79–115.

83. WILLIAMS, H. C. (1966): "The Gisborne Submarine Sewer Outfall," *New Zealand Engineering* (March 15), pp. 110–120, with discussion, *New Zealand Engineering* (May 15, 1967), pp. 201–203.

10-11   BIBLIOGRAPHY

General Submarine Pipeline Design and Construction

ALDRIDGE, C. (1956): "What's Involved in Planning and Constructing an Offshore Pipeline," *Oil and Gas Journal*, Vol. 54, No. 59 (June 18), pp. 174–179.

BLUMBERG, R., et al. (1971): "Analysis of Ocean Engineering Problems in Offshore Pipelining," *Preprints*, Third Annual Offshore Technology Conference, Houston, Texas (April), Vol. 1, pp. 297–308.

BOWIE, G. L., and R. L. WIEGEL (1977): "Marine Pipelines: An Annotated Bibliography," U.S. Army, Corps of Engineers, Coastal Engineering Research Center, Fort Belvoir, Va., *Miscellaneous Report 77-2* (March).

BROWN, R. J. (1971): "Rational Design of Submarine Pipelines," *World Dredging and Marine Construction*, Vol. 7, No. 3 (February), p. 17.

BROWNFIELD, W. G. (1973): "Using Geology to Lay Subsea Lines," *Offshore*, Vol. 33, No. 10 (September), pp. 52–57, 60.

CRANE, R. E. (1970): "Cast Iron Subaqueous Crossings," *Cast Iron Pipe News*, (January/February), pp. 1–4.

D'ANGREMOND, K., and J. A. HUIJSSON (1976): "Shore Approaches Need Special Care," *Oil and Gas Journal*, Vol. 74, No. 36 (September 6), pp. 142–144, 147, 148.

DEASON, D. (1970): "The Status of Weight Coating," *Pipe Line Industry*, Vol. 32, No. 3 (March), pp. 51–53.

EATON, J. R. (1977): "Pipeline Construction in Cook Inlet by the Pulling Method," *Journal of Petroleum Technology* (March), pp. 242–248.

HAAGSMA, S. C. (1973): "Offshore Pipeline Burial," ASCE, *Transportation Engineering Journal*, Vol. 99, No. TE4 (November), pp. 980–984.

HEANEY, F. L. (1960): "Design, Construction, and Operation of Sewer Outfalls in Estuarine and Tidal Waters," *Journal of the Water Pollution Control Federation*, Vol. 32, pp. 610–621.

HOBBS, H. (1966): "Criteria for the Design and Construction of Submarine Pipelines," *Pipes and Pipelines International* (July), pp. 25–27.

JOHNSON, S. J., J. R. COMPTON, and S. C. LING (1972): "Control for Underwater Construction," *Underwater Soil Sampling, Testing, and Construction Control*, American Society for Testing and Materials, Philadelphia, Pennsylvania, *Special Technical Publication 501*, pp. 122–180.

KOOPMAN, R. T. (1976): "OSHA and Its Effects on Water Works Construction," *Journal of the New England Water Works Association*, Vol. 90, No. 2 (June), pp. 93–99.

KRIEG, J. L. (1965): "Criteria for Planning an Offshore Pipeline," ASCE, *Journal of the Pipeline Division*, Vol. 91, No. PL1, pp. 15–37.

KRIEG, J. L. (1966): "Hurricane Risks as They Relate to Offshore Pipelines," Hurricane Symposium, American Society for Oceanography, *Publication No. 1* (October), pp. 304–313.

LAMB, M. J. (1966): "Underwater Pipelines," *Exploiting the Ocean*, Marine Technology Society (June), p. 293.

"Laying Outfalls with Offshore Platforms" (1973): *Western Construction*, Vol. 48, No. 5 (May), p. 36.

MILLER, D. R. (1956): "Engineering Goes to Sea," *Public Works*, Vol. 87, No. 10 (October), pp. 136–137.

MILLER, D. R. (1966): "Marine Studies for the Design and Construction of Offshore Pipelines," *Proceedings*, Specialty Conference on Coastal Engineering at Santa Barbara (October 1965), pp. 991–1006, American Society of Civil Engineers, New York.

Offshore Platforms and Pipelining (1976): Petroleum Publishing Co., Book Division, Tulsa, Oklahoma.

PARKHURST, J. D., L. A. HAUG, and M. L. WHITT (1967): "Ocean Outfall Design for Economy of Construction," *Journal of the Water Pollution Control Federation*, Vol. 39, No. 6 (June), pp. 987–993.

"Polyethylene Pipe Relines 32-inch Outfall on California Coastline" (1974): *Civil Engineering*, Vol. 44, No. 1 (January), pp. 18, 20.

REID, R. O. (1952): "Some Oceanographic and Engineering Considerations in Marine

Pipe Line Construction," *Proceedings*, Second Conference on Coastal Engineering, Council on Wave Research, Engineering Foundation, Berkeley, California, p. 749.

STRUNG, C. (1970): "Submarine Placing of Concrete by the Tremie-Method," *Preprints*, Second Annual Offshore Technology Conference, Houston, Texas (April), Vol. 2, pp. 813–818.

WRIGHT, R. R. (1977): "Proper Inspection Methods Minimize Pipeline Failures," *Oil and Gas Journal*, Vol. 75, No. 21 (May 23), pp. 51–56.

Dredging

"Air-powered Jet Line Trenching for North Sea Pipe Lines" (1975): *Ocean Industry*, Vol. 10, No. 4 (April), pp. 355, 358.

"Deep Sea Trencher Buries Oil and Gas Lines by Remote Control Equipment" (1976): *Offshore*, Vol. 36, No. 12 (November), p. 231.

HERBICH, J. B. (1975): *Coastal and Deep Ocean Dredging*, Gulf Publishing Co., Houston, Texas.

HUSTON, J. (1967): "Dredging Fundamentals," ASCE, *Journal of the Waterways and Harbors Division*, Vol. 93, No. WW3 (August), pp. 45–69.

HUSTON, J. (1970): *Hydraulic Dredging: Theoretical and Applied*, Cornell Maritime Press, Inc., Cambridge, Maryland.

MOHR, A. W. (1974): "Development and Future of Dredging," ASCE, *Journal of the Waterways, Harbors and Coastal Engineering Division*, Vol. 100, No. WW2 (May), pp. 69–83.

"New Trenching System Buries Japanese Pipeline" (1977): *The Oil and Gas Journal*, Vol. 75, No. 1 (January 3), pp. 65, 66, 68.

"Ocean Bottom Dredge Speeds Offshore Pipe Line Trenching" (1972): *Pipe Line Industry*, Vol. 36, No. 5, (May), p. 62.

ROORDA, A., and J. J. VERTREGT (1963): *Floating Dredges*, The Technical Publishing Co., H. Stam N. V., Haarlem, The Netherlands.

SANTI, G. (1971): "Trenching and Dredging in Deep Water," *Ocean Industry*, Vol. 6, No. 6 (June), pp. 30–31.

SCHEFFAUER, FREDERICK C. (1954): *The Hopper Dredge: Its History, Development and Operation*, U.S. Government Printing Office, Washington, D.C.

SCHWARTZ, H. I. (1971): "Hydraulic Trenching of Submarine Pipeline," ASCE, *Transportation Engineering Journal*, Vol. 97, No. TE4 (November), pp. 723–728.

The British Hydromechanics Research Association (1976): *Proceedings of the First International Symposium on Dredging Technology*, Canterbury, England (September 1975), Cranfield, Bedford, England.

TURNER, T., and V. FAIRWEATHER (1974): "Dredging and the Environment: The Plus Side," *Civil Engineering*, Vol. 44, No. 10 (October), pp. 62–65.

Blasting

"British Firm Takes a Rig Down to Seafloor Drilling Shot Holes in Shallow Water" (1975): *Offshore*, Vol. 35, No. 12 (November), pp. 54–56.

BROWN, C. (1969): "Quick Ditch Method Blasts Pipeline Trenches," *World Dredging and Marine Construction*, Vol. 5, No. 11 (October).

Canadian Industries Ltd., Explosives Division (1968): *Blasters' Handbook*, 6th ed., Montreal, Canada.

COOK, M. A. (1974): *The Science of Industrial Explosives*, IRECO Chemicals, Salt Lake City, Utah.

DICK, R. A. (1968): "Factors in Selecting and Applying Commercial Explosives and Blasting Agents," U.S. Dept. of the Interior, Bureau of Mines, *Information Circular 8405*.

E.I. du Pont de Nemours and Co., Inc. (1966): *Blasters' Handbook*, 15th ed., Wilmington, Delaware.

GREGORY, C. E. (1973): *Explosives for North American Engineers*, Trans Tech Publications, Cleveland, Ohio.

GUSTAFSSON, R. (1973): *Swedish Blasting Technique*, SPI, Gothenburg, Sweden.

HALLANGER, L. W. (1976): "Interim Field Guide to Nearshore Underwater Explosive Excavation," U.S. Navy, Civil Engineering Laboratory, Port Hueneme, Calif., *Technical Report R843* (June).

JOHANSSON, C. H., and P. A. PERSSON (1970): *Detonics of High Explosives*, Academic Press, New York.

JOHNSON, S. M., Ed. (1971): "Explosive Excavation Technology," U.S. Army Engineer Nuclear Cratering Group, Livermore, California, *Technical Report No. 21* (June).

LANGEFORS, U., and B. KIHLSTRÖM (1967): *The Modern Technique of Rock Blasting*, 2nd ed., Wiley, New York.

LENZ, R. R. (1965): *Explosives and Bomb Disposal Guide*, Charles C. Thomas, Springfield, Illinois.

NICHOLLS, H. R., C. F. JOHNSON, and W. I. DUVALL (1971): "Blasting Vibrations and Their Effects on Structures," U.S. Dept. of the Interior, Bureau of Mines, Washington, D.C., *Bulletin 656*.

PAGE, G. L. (1974): "Modification of a Pneumatic Track Drill for Underwater Use by Divers," U.S. Navy, Civil Engineering Laboratory, Port Hueneme, Calif., *Technical Note N-1339* (April).

U.S. Army Corps of Engineers (1969): "Systematic Drilling and Blasting for Surface Excavations," *Engineer Manual EM 1110-2-3800*, Draft (June 30).

Constructing Specific Outfalls

BARBIER, J. -M. (1975): "Réalisation de l'Emissaire de La Salie au Sud d'Arcachon," *La Houille Blanche*, Vol. 30, pp. 567–573.

BUTLER, J., and D. JAMISON (1968): "Laying Concrete Pipes through Atlantic Surf," *Concrete Pressure Pipe Journal*, Vol. 10, No. 4 (December), pp. 6–11.

DROSSEL, M. R. (1973): "Prestressing Jacks Set Underwater Pipeline Precisely on Line," *Construction Methods and Equipment*, Vol. 55, No. 6 (June), pp. 62–65.

DUGDALE, J. (1975): "Edinburgh's New Disposal Scheme," *Water and Waste Treatment*, Vol. 18, No. 12 (December), pp. 16, 18.

FITZPATRICK, E. B., JR., and W. J. BECKMAN (1974): "Constructing an Ocean Outfall," *Civil Engineering*, Vol. 44, No. 9 (September), pp. 83–85.

HARDY, B. E. (1973): "Design, Construction and Launching of a Submarine Outfall Sewer at Whitstable," *Journal of the Institution of Municipal Engineers*, Vol. 100 (July), pp. 201–207.

"Hydraulic Platform Walks in Water to Place Outfall Pipeline" (1976): *Construction Methods and Equipment*, Vol. 58, No. 12 (December), pp. 46–48.

MacPHERSON, A. R. (1950): "Deep Outfalls for Sewage and Sludge Disposal," *Public Works*, Vol. 81, No. 3 (March), pp. 30–31.

"Ocean Outfall Project Features Control by Helicopters, Radio, and Television" (1957): *Western Construction*, Vol. 32, No. 11 (October), p. 52.

"Outfall Built End-First Under Stormy Lake" (1970): *Engineering News-Record*, Vol. 185, No. 26 (December 24), pp. 14–15.

"Outfall Goes Out to Sea" (1970): *California Builder and Engineer*, Vol. 76, No. 7 (April 10), pp. 28–29.

"Outfall Project Depends on the Spider and the Horse" (1973): *Western Construction*, Vol. 48, No. 5 (May), pp. 31–34.

"Pipelaying 'Horse' Preassembles and Helps Install Ocean Outfall" (1976): *Water and Sewage Works*, Vol. 123, No. 2 (February), pp. 50–51.

"Plastic Pipeline Requires New Methods" (1968): *Engineering News-Record*, Vol. 181, No. 19 (November 7), p. 81.

"Polyethylene Outfall Sewer Line Towed 60 Miles on Lake Ontario" (1977): *Water and Pollution Control*, Vol. 115, No. 5 (May), pp. 11, 13.

"Revised Outfall Construction Method Conquers Surf and Sand" (1975): *Construction Methods and Equipment*, Vol. 57, No. 7 (July), pp. 68–69.

"Special Rigs Build Ocean Outfall" (1976): *Western Construction*, Vol. 51, No. 4 (April), pp. 27–28.

"Synchronized Team Handles Heavyweight Sewer Pipe Combinations" (1972): *Construction Methods and Equipment*, Vol. 54, No. 12 (December), pp. 40–42.

WELLS, D. R. (1976): "Polyethylene Pipe Solutions to Ocean Sewer Outfalls," *Civil Engineering*, Vol. 46, No. 9 (September), pp. 62–64.

Constructing Specific Submarine Pipelines Other Than Outfalls

"Biggest-Yet Polyethylene Pipe for North Bay Intake" (1974): *Water and Pollution Control*, Vol. 112, No. 3 (February).

EMERSON, G. G. (1975): "A Fail-Safe Subaqueous Concrete Sewer," *Civil Engineering*, Vol. 45, No. 9 (September), pp. 77–78.

"Fast Pipe Bedding Methods Surge Marine Job Ahead" (1974): *California Builder and Engineer*, Vol. 80, No. 11 (June), pp. 16–18.

GARDNER, J. R. (1945): "Submarine Pipe Lines in Deep Water at Portland, Me.," *Journal of the New England Water Works Association*, Vol. 59, pp. 148–162.

"Gas Pipe Sinks to New Depths" (1975): *Construction Methods and Equipment*, Vol. 57, No. 1 (January), pp. 42–45.

"Gas Pipeline Pulled through Dutch Dunes" (1975): *International Construction* (November), pp. 46, 48, 49, 51.

GOUDY, A. P. (1965): "Shek Pik Submarine Pipeline, Hong Kong Water Supply," *Proceedings*, The Institution of Civil Engineers, Vol. 30, pp. 531–555.

LAMB, B. (1971): "Long-Legged Platform Walks into Ocean Job," *Construction Methods and Equipment*, Vol. 53, No. 4 (April), pp. 81–84.

MOORE, W. D., III (1977): "Semisubmersible Lays Large-Diameter Brent Pipeline across Surf Zone," *The Oil and Gas Journal*, Vol. 75, No. 23 (June 6), pp. 124, 129.

"N. S. Gas will be Stored at Honsea" (1974): *Offshore*, Vol. 34, No. 10 (September), pp. 113–114.

"Offshore Intake Line Utilizes Waterside and Onshore Skills" (1975): *Construction Methods and Equipment*, Vol. 57, No. 5 (May), pp. 49–50.

"Oil Pipelines Laid in Trench across Canal" (1974): *World Dredging and Marine Construction*, Vol. 10, No. 3 (February), p. 17.

"Pipeline Pulls Completed in the Orkneys" (1976): *The Oil and Gas Journal*, Vol. 74, No. 18 (May 3), pp. 188, 193, 194.

"Pulling a Huge Pipe Line across the Sea Floor" (1971): *Ocean Industry*, Vol. 6, No. 6 (June), pp. 32–33.

"Submarine Pipeline Wins 1967 Award" (1967): *Australian Civil Engineering* (December 5), pp. 11–15.

"Swift Response Limits Problem of Reinstalling Buoyant Section" (1976): *Pipe Line Industry*, Vol. 45, No. 3 (March), pp. 52–53.

"Two 40-in. Pipe Lines Will be Pulled across Bosporus Straits" (1977): *Ocean Industry*, Vol. 12, No. 1 (January), pp. 64, 66.

WARD, D. R. (1967): "Laying Large Diameter Offshore Pipelines," *Offshore*, Vol. 27, No. 6 (June 2), pp. 52–56.

Diving

"Commercial Diving Operations: Proposed Rulemaking and Hearing" (1976): *Federal Register*, Vol. 41, No. 215 (November 5), pp. 48950–48969.

HAMILTON, R. W., JR., and H. R. SCHREINER (1968): "Putting and Keeping Man in the Sea," *Chemical Engineering*, Vol. 75, No. 13 (June), pp. 263–270.

"Man Goes to Work Beneath the Sea" (1969): *Construction Methods and Equipment*, Vol. 51, No. 7 (July), pp. 188–204 and 212–216.

"New Concept: Diver-Operated Underwater Cranes" (1976): *Ocean Industry*, Vol. 11, No. 11 (November), pp. 89, 91.

SEIB, J., JR. (1976): "How Divers Assist in Laying Pipe in the Deep Waters of the North Sea," *Offshore*, Vol. 36, No. 12 (November), pp. 121–122.

"Special Underwater Operations" (1974): *Construction Methods and Equipment*, Vol. 56, No. 8 (August), pp. 41–42.

Tarrytown Labs, Ltd. and Undersea Medical Society, Inc. (1976): "Recommended Medical and Operating Standards for Divers," prepared for the National Institute for Occupational Safety and Health, Rockville, Md. (August 16).

TITCOMBE, R. M. (1973): *Handbook for Professional Divers*, J. B. Lippincott, Philadelphia.

WARNER, S. A. (1977): "Diving Fatalities Lead to Corrective Action," *Ocean Industry*, Vol. 12, No. 5, (May), pp. 124, 126.

ZINKOWSKI, N. B. (1971): *Commercial Oil-Field Diving*, Cornell Maritime Press, Cambridge, Maryland.

Related to Design of Trestles

AAGAARD, P. M., and C. P. BESSE (1973): "A Review of the Offshore Environment—25 Years of Progress," *Journal of Petroleum Technology*, Vol. 25, No. 12 (December), pp. 1355–1360.

DEAN, R. G., and P. M. AAGAARD (1970): "Wave Forces: Data Analysis and Engineering Calculation Method," *Journal of Petroleum Technology*, Vol. 22, No. 3, pp. 368–375.

HUNT, HAL W. (1974): *Design and Installation of Pile Foundations*, Associated Pile and Fitting Corp., Clifton, N.J.

McCLELLAND, B., J. A FOCHT, JR., and W. J. EMRICH (1967): "Problems in Design and Installation of Heavily Loaded Pipe Piles," *Proceedings of the Conference on Civil Engineering in the Oceans*, ASCE (September), pp. 601–634.

SAINSBURY, R. N., and D. KING (1971): "The Flow Induced Oscillation of Marine Structures," *Proceedings*, The Institution of Civil Engineers, Vol. 49 (July), pp. 269–302.

WIEGEL, R. L., K. E. BEEBE and J. MOON (1957): "Ocean Wave Forces on Circular Cylindrical Piles," ASCE, *Journal of the Hydraulics Division*, Vol. 83, No. HY2, Paper 1199.

General Construction

DICKIE, D. E. (1975a): *Rigging Manual*, 1st ed., Construction Safety Association of Ontario, Toronto, Canada (October).

DICKIE, D. E. (1975b): *Crane Handbook*, 1st ed., Construction Safety Association of Ontario, Toronto, Canada (October).

11

Construction Effects and

Post-Construction Activities

11-1 CONSTRUCTION EFFECTS ON THE MARINE BIOTA

In Chapter 4 considerable attention was paid to the protection of the marine environment from high concentrations of deleterious components in wastewaters. Relative to the attention focussed on such matters by water-pollution-control agencies, I believe that insufficient attention has been paid to the possibly harmful effects of construction practices on the marine biota during the 1–2 years, or occasionally more, required to build a major outfall.

Blasting during outfall construction is known to result in the direct mortality of fish and other organisms.* Blasting can also have secondary effects. Illustrative of these is an observation by marine biologists studying effects of the construction of a thermal effluent outfall at Kahe, Hawaii, in 1976. Blasting loosened bases of branching-type corals such as *Pocillopora meandrina* up to 150 ft (45 m) from the explosion site. These damaged corals would then be readily knocked over by the occasional heavy wave surge appearing in the Kahe area.

Despite the possible importance of blasting effects, this section primarily concerns both the acute and chronic effects on marine organisms of sediment-producing operations associated with marine outfall construction. Sediments enter the water column when a trench is excavated by clamshell and sidecast beside the trench (see Figure 11-1). In addition, bedding and rocking an outfall can cause a heavy sediment burden on the water. I can attest to the latter personally, having dived under the Davy Crockett during bedding and rocking

*Underwater blasting may also bring sharks into the area.

Figure 11-1   Clamshell sidecasting dredge spoil

operations associated with the Sand Island No. 2 outfall off Honolulu in early 1975. There was greatly impaired water visibility throughout the water column, especially toward the bottom. This extended over an area of several acres.

Sometimes, the material dredged from the trench for an outfall is not clean natural material but contains constituents of wastewater origin: heavy metals, pesticides, foreign inert material, and biodegradable organic matter. The reason for this is that new outfalls are frequently located adjacent to older (raw effluent) ones. Even if they are not physically close, they may be positioned so that prevailing currents sweep diluted effluent from the old line over the newer one, and the sediments receive the fallout. This latter situation was precisely the case, for example, with the Sand Island No. 2 outfall.

The process of dredging results in direct removal of benthic plants and animals and very likely leads to their subsequent death. Coupled with this is the loss of habitat for whatever benthic organisms remain in adjacent areas. In some cases, dredging may expose benthic deposits with high oxygen demand. In others, the nature of the particle size of the dumped spoil material may be altered. Fine material entering the water column may settle out well downstream of the work site whereas the coarse material remains close to the work area. The coarser material may then be unsuitable for its former inhabitants.

Material washes out of a bucket of clamshell whenever a bite is taken out of the bottom. Sediment also enters the water column when the dredge spoil is dumped. This can result in the release to the water column of part of whatever organic components, nutrients, heavy metals, or hydrocarbons were contained in the sediments. The organic components can cause an oxygen demand on the water; the nutrients released can have a biostimulatory result; the heavy metals and hydrocarbons can have a toxic effect on the local plankton and nekton. Some inorganic materials such as sulfides, if released, can have an oxygen demand, and hydrogen sulfide is toxic to many marine organisms. Another undesirable feature of silt and sediments suspended in the water column concerns a possible reduction in that water's ability to assimilate oxygen-demanding wastes.

Excess amounts of sediment in the water column can result in fish suffocation as gills and gill chambers become coated or clogged with material. Secondary infections may be a possibility. However, it is to be expected that most fish species would avoid localized areas of heavy sediment load.

A major problem concerning sediments entering the water column during dredging or subsequent dredge spoil disposal is the decrease in the depth of the euphotic zone caused by the blockage of the sun's light. Both the phytoplankton and benthic plants are affected directly by this development. There can be direct effects on animal organisms as well through the loss in food matter or of habitat. Corals may be affected through decreased production by their zooxanthellae and perhaps even by the elimination of these algal symbionts from the coral tissue as a result of the stress caused by the suspended material.

When the silt and sediments reach the bottom, they can blanket fish nests, eggs, and food organisms; habitat spaces can be destroyed. Although many benthic organisms such as polychaetes may be able to tolerate being buried by sediments without adverse effects, many attached organisms such as oysters cannot tolerate such deposition. Direct death due to smothering is one possibility, but more subtle alterations are likely for small sediment loads. Feeding activities can be influenced; impairment of proper respiratory and filter feeding functions can occur. The immediate result may be a reduction in growth with the end result an early death.

Many organisms, requiring a hard substrate for attachment, will not settle on a soft, shifting substratum. Infaunal organisms may find the soft consistency of the dredge spoil unsuitable as a habitat even if the particle size distribution remains virtually intact. Suspended material can also have an abrasive effect on members of the benthic community when it is swept by currents back and forth past sessile organisms.

The types of problems sketched here are not entirely preventable during the construction of marine outfalls, and some organisms will undoubtedly be killed. Some care during construction, however, can be exercised to reduce the degree and extent of such effects. One such possibility involves dumping dredge spoil into bins, for ultimate disposal on land, or in diked enclosures near shore, rather than side-casting it.

11-2 OUTFALL INSPECTION AND REPAIR

Introduction

Besides preparing the plans and specifications and the final design report, the outfall designer sometimes prepares an operation and maintenance manual, basically a set of instructions for the client to ensure the continued satisfactory operation of the facility for its design life. The manual would include recommendations for periodic inspection of the outfall to check for corrosion, joint leakage, buildup of foreign materials along flow surfaces, and other factors. The manual should also include details on recommended maintenance procedures plus emergency procedures in case of accident or natural disaster.

External Inspection and Repair

Many outfalls have been built, put into operation, and never again inspected. One reason may be that operators have been unaware of the potential marine problems. Another possibility is that operators have been somewhat afraid of what problems might be found.

I believe that all outfalls should be inspected at least twice a year to ensure that no damage or breakage has occurred or that no potentially damaging situation has developed. Additional inspections may be warranted after a time of heavy wave action or after an earthquake. A pair of largely exposed submarine pipes extending out from shore to a tanker mooring at Barbers Point, Hawaii, is inspected by divers monthly. Various problems with the lines were discovered during the inspections and remedied before they became serious. In one case, after a time of heavy wave action, it was discovered that part of one pipe had been thrown up and over the other. See page 358.

Inspection can also indicate situations that impair diffuser performance. A diver who had inspected the diffuser of Seattle's West Point outfall (#24 in Table 10-1) related that sea anenomes had blocked off some ports. Allen et al. (1976) reported dense sea anenome concentrations around ports in the Hyperion effluent line but no blockage. A diver inspection of the Menasha Corporation outfall at Coos Bay, Oregon, in June 1977, showed that only ten of 32 riser ports were above the sand and functioning. Three of the ten ports were barely showing. The designers had anticipated some problems of this nature, and all risers were fitted with rubber-faced check valves to keep sand out of the line.

Outfall inspections should not be limited to the zone in which the pipe was originally laid exposed on the sea bed. The whole alignment should be checked since erosion can bring an outfall to the sediment surface. Inspections in the surf zone should not be dispensed with simply because of the difficulties in working through that region. The surf zone is the very area within which trouble might

519

develop. Even on coasts beset with difficult wave conditions there are calm times when an end-to-end outfall inspection is possible.

Divers have inspected many outfalls. Virtually all outfalls in Hawaiian waters have been inspected at one time or another, but not on a regular basis. The Southern California outfalls have similarly been inspected from time to time. It doesn't take much, sometimes, to materially aid divers during outfall construction and later inspection. As an example, specifications for the Lion's Gate FRP diffuser (#36 in Table 10-1) called for it to be colored yellow, to help divers in locating it. Pingers were specified for the Humboldt Bay Wastewater Authority, California, outfall to aid in relocating the line.*

The report by James M. Montgomery, Consulting Engineers (1973), documents diver inspections of a 5800-ft-long (1770-m) steel outfall into Goleta Bay near Santa Barbara, California. Both motion picture films and photographs were taken to document observations of satisfactory conditions and operations.

When outfalls enter deeper waters, their inspection by divers is not practical, and various outfalls have been inspected by submersible (La Cerda, 1974). These include the Hyperion 5-mile (8.1-km) effluent and 7-mile (11.3 km) sludge lines (#11 and #12 in Table 10-1) beyond a depth of about 100 ft (30 m). They also include the outfalls at Oxnard, California, at San Diego, California (#18) and at Powell River, B.C., Canada (#43). A camera on a towed sled is another possibility for examining the integrity and functioning of outfalls in deeper waters. Such a system has been used to record the operation of the Orange County No. 2 outfall diffuser (#35) off California.

A potential problem for outfalls, mentioned in Chapters 6 and 8, concerns differential erosion leading to unsupported pipe spans. Underscour along the Menasha Corporation line mentioned above has resulted in unsupported lengths to 50 ft (15 m). Unsupported lengths mean potential failure because of static overstressing, joint pullout, and/or flow-induced oscillations. In the North Sea, where there is a great deal of movement of bottom materials, many lines are checked twice yearly by side-scan sonar (Ells, 1975). Suspensions are indicated on the side-scan record by clear separations between the dark line of the pipe and its light shadow. Unduly long suspensions in the North Sea are remedied by supporting the pipe at intermediate points by means of sandbags placed by divers. Concrete-filled tires have been used in some cases under unsupported lengths of outfall pipe.

The threat to submarine pipelines of ship anchors has been outlined in Chapter 8. During the time of construction, official steps should be taken to have the outfall alignment shown clearly on the next issue of the chart (Appendix A) of the general area. The drawing of such a line by itself is probably insufficient as the following indicates.

---

*A magnetometer towed by a ship may assist in finding buried steel lines.

The "Notice to Mariners" issued by the Fourteenth Coast Guard District, Honolulu, Hawaii, on August 3, 1976, contained the following passage:

HAWAIIAN ISLANDS—OAHU—SAND ISLAND

The Sand Island Ocean Outfall extending into Mamala Bay is shown as a dashed purple line. The legend "Do not anchor within 600 yards on either side of the sewer" should be placed next to the dashed purple line.

Ref: Chart No. 19359 (C&GS 4110), 19364 (C&GS 4132), 19367 (C&GS 4109)

If there *is* a damaged outfall, repair work may very well be necessary. With this in mind, careful thought should be given about who does the detailed inspection of the damage. In one instance of a broken outfall, two different firms were hired by the owner, one to inspect the damage and the other to design the remedial measures. The small photographic company hired for the inspection constructed a scale model of the outfall and took some brilliant color photographs of it. That was the extent of their report. The designer then had to send his own crew out to obtain the appropriate pre-redesign information, something that should have been done in the first place.

There are commercially available devices or techniques that can be used to rejoin a steel pipe at an underwater break. Such devices, good up to a pipe size of 48-in. (1219-mm) do not require that the pipe ends match or even that the two lengths are parallel. Pertinent references are "Welding 'Hut'..." (1968), Powell and Van Heuit (1968), White (1969), O'Donnell (1971), and "Simple Technique..." (1975).

If the outfall is made of reinforced concrete, then repair of a broken line will require that replacement sections be installed. An example of such a situation occurred in the largely exposed Hilo, Hawaii 48-in. (1219-mm) reinforced concrete outfall in early 1972 when 10 pipe sections were displaced in a water depth of 25 ft (7.5 m). The cost of repair in late 1972 was approximately $80,000. A barge was moored over the site, and the damaged sections were removed. Bedding material was then put in place. New pipe sections were installed using a small version of a Horse as a diver-assist. Partial backfilling was carried out in areas where the pipe penetrated rock outcroppings. Tremie concrete was poured along the repaired pipe segment, inside forms, up to the springline.

11-3 OUTFALL OPERATION AND MAINTENANCE

It is usually figured that there are no operation and maintenance costs for an outfall *per se* when constructed. The calculations that went into Table 4-5, for example, assumed no recurring annual costs for the outfalls themselves, although pumping costs were figured.

Internal problems with outfalls can certainly develop. An example concerns the outfall at San Diego, California, where various items lost down the pipe, principally vinyl and stainless steel liners that had become detached, caused one major blockage which fortunately ultimately cleared itself, but not before the high pressures caused had blown open some components at the treatment plant. It is this type of problem that makes a treatment plant operator nervous with respect to the legislation growing out of PL 92-500 (Chapter 4). With the provision for bypassing both treatment plant and outfall with raw sewage physically prevented, a blockage such as that referred to above could have disastrous consequences, causing far more problems than a temporary shoreline discharge of raw effluent.

A not uncommon development with marine outfalls is that the *effective* friction factor (Chapter 7) increases as time passes. This increase in resistance to flow passage occurs because of corrosion, and resultant roughening, of the pipe walls and/or because of deposits on the flow surfaces.*

Although it would not be expected that effluent lines would suffer wall deposits to the same degree as a sludge line, it is instructive first to consider experience with the Hyperion sludge line (#11 in Table 10-1). It was suspected that grease had been deposited along the inner walls of this line, in part through cooling of the sludge as it passed through the outfall. The effective Darcy-Weisbach friction factor f (given in Chapter 7) for the line had had the following history following completion of the outfall (Hume et al., 1961): 0.0137–0.0157 after 9 months; 0.0188 after 12 months; 0.0260 after 16; and 0.0334 after 23 months, just before the line was cleaned and the friction factor dropped to 0.0152.

An object that is pushed or dragged through a pipe to block off one part from another or to clean it is known as a "pig." An articulated pig was made up to clean the pipe walls of the Hyperion sludge line, and it was pushed through the line by water pressure after a "prover" was sent through first. The object of using the prover, consisting mainly of two plywood discs, was to see if an object 20 in. in diam (508 mm) would travel the whole length of the line. The pig itself was articulated so as to pass around various bends at the upstream end of the line. The pig carried brush and scraper elements, details of which are provided by Frank (1961). Both the prover and pig were followed acoustically as they each made the 3.5-h journey.

An early example of outfall cleaning concerns the experience of the South Essex Sewerage District with its 54-in. (1371-mm) line, 8300 ft (2500 m) long, offshore from Salem, Mass. A 2.5-in. (64-mm) coat of hard calcium carbonate scale developed on the inner walls of the line apparently due to heavy concentrations of caustic lime in some industrial wastewaters. A pig was obtained that consisted of six steel "heads" with spring steel scrapers. This pig was put in

---

*Pipe area decreases caused by deposits manifest themselves at the upstream end of the line as friction factor increases. They will be considered as such in what follows.

the line and the pumps then started up. Approximately 2.5 h later, the pig emerged from the end of the pipe and was retrieved by a diver. The progress of the pig was monitored through use of a sound emitter in the cleaning device and two directional listening devices in the water. After the cleaning there was a very appreciable decrease in effective friction coefficient (Nyman, 1939).

An extensive discussion of conditions leading to the cleaning of the Orange County No. 1 outfall (#9 in Table 10-1), of the background to cleaner selection, and to the use of the cleaner finally selected is contained in Galloway (1964). The pig used was an airplane tire sawed in half and reinforced with steel straps. Flow acting on the cupped part of the tire forced it through the line, and a cable not only provided a brake but also allowed the tire to be retrieved after it stalled in the line just downstream of the first diffuser port.

Since the cleaning of the lines referred to above, various commercial enterprises have developed standard pigs for use in scraping pipelines. For example, Knapp, Inc. (Houston, Texas) makes pipe pigs, for various functions including cleaning, in diameters of 8–48 inches (203–1219 mm). The pigs are designed to pass through curves and narrow sections such as at valves. Helle Engineering, Inc. (San Diego, Calif.) makes "pig pingers" and systems to locate such pingers in a pipe. Not only is it possible to use pigs to clean lines, but also to assist in relining them internally, in-place, with an epoxy-type material, such as that manufactured by Standard Pipeline Coating Co., Inc. (Dallas, Texas).

An old outfall with rough walls and poor flow characteristics, but with continuing good wall strength, can be improved by inserting another pipe. An example of this is given in "Polyethylene Pipe..." (1974) where a 32-in.-diam (812-mm) high-density polyethylene pipe was used to reline a leaking 36-in.-diam (914-mm) PVC-lined concrete outfall.

11-4 MONITORING

Introduction

Effluent limitations and receiving water standards are imposed to preserve man's beneficial uses of the receiving waters and to assure the protection and propagation of a balanced and indigenous community of organisms. Monitoring is in part designed to check if such effluent and receiving water standards are in fact maintained. However, the more fundamental purpose of monitoring is to check if in fact man's beneficial uses are being preserved and the marine biota are suffering no ill effects, standards or no.

Effluent standards are largely imposed as a matter of convenience. The numbers used are back-figured from the pollutant levels supposed to be noninhibitory to the marine biota in the receiving waters coupled with supposed minimum levels of effluent dilutions. It would be grossly inadequate simply to monitor effluent characteristics, and not those of the receiving waters, since the same assumptions would be involved in connecting these characteristics to corresponding ones in the receiving waters as were involved in composing the somewhat arbitrary effluent standards in the first place.

But it is vastly more difficult and costly to monitor receiving water characteristics than it is to monitor the characteristics of outflows from wastewater treatment plants. Thus, although the "proof of the pudding" lies offshore, *continuous* monitoring can realistically be carried out only onshore. Thus comprehensive monitoring programs involve two components as do standards: one regarding effluents, the other concerning receiving waters. Examples of each type of undertaking will be presented in the two following subsections.

A thorough receiving water monitoring program will involve the determination of pertinent water column parameters, the study of bottom sediments, the taking and examination of local organisms, and the consideration of public health questions and aesthetics.

If monitoring shows, for example, that many fish close to the outfall (Figure 11-2) are building up dangerous concentrations of heavy metals and/or pesticides in their tissues, or developing abnormalities or contracting diseases, something should be done. Turning off the flow completely may be possible when some industrial concerns are involved, but it is out of the question for municipal outfall sewers. In the latter case, changes onshore in terms of source control or functions at the wastewater treatment plant may be indicated, but of course there are sizable time lags associated with such alterations. The impracticality of effecting changes to a wastewater treatment-disposal system means that the initial design must be adequate with a fair factor of safety.

At the other end of the scale, monitoring could conceivably indicate that

Figure 11-2   Fish congregating near raw sewage outflow (Courtesy, Richard W. Grigg)

effluent limitations need not be as strict as practiced to adequately protect the marine environment and/or to comfortably meet specified maximum pollutant concentrations in the receiving waters. This was mentioned in Chapter 4 concerning water quality regulations in British Columbia waters.

When regulatory agencies dictate monitoring schedules and procedures, the questions of the scale of the monitoring and how much a municipal body or industrial organization should pay for monitoring are irrelevant. However, when monitoring is not regulated, the above question must be answered. Ludwig and Onodera (1962) have considered the matter, and they derived an equation to relate proposed monitoring cost to wastewater discharge. This cost increased as the 8/10 power of the discharge, indicating that relative monitoring costs for small dischargers may be large compared to those for large dischargers.

Effluent Monitoring

The publication "Pollution Control Objectives for Municipal Type Waste Discharges in British Columbia" (1975), discussed in Chapter 4, divided wastewater discharges to receiving waters into four groups dependent on flow rate: <0.12, 0.12–1.2, 1.2–6.0, and ≥6.0 Mgd.* The frequency of effluent monitoring was dependent on the group, and was more frequent the higher the flow. For example, 96-h TL_{50}'s (Chapter 4) were required annually for the lowest flow group, monthly for the highest when nonshellfish waters were involved. If shellfish receiving waters were involved, the frequency of each could be increased.

The parameters to be monitored could extend to the full lists in Tables 4-1 and 4-2 depending upon the situation and the dictates of the responsible governmental agency. However, BOD_5, SS, fecal coliforms, and total phosphorus were included in all such monitoring. Both BOD_5 and SS were required to be monitored (from grab samples) daily for flows ≥6.0 Mgd and quarterly for those <0.12 Mgd. In the latter case, more frequent sampling could be called for if analysis of SS showed unsatisfactory results with respect to the effluent standards.

Monitoring Receiving Water Quality

The County Sanitation Districts of Orange County in California (CSDOC) operates its major outfall (#35 in Table 10-1) that in 1976 discharged about 180 Mgd (7.9 m^3/s) through a mile-long diffuser into depths to about 200 ft (60 m). Quarterly monitoring reports for the receiving waters were required by the responsible local state agency, the California Regional Water Quality Control Board, which also dictated the parameters to be obtained and the locations of

---

*Volumes in the B.C. standards were given in Imperial gallons but have been converted here to U.S. gallons.

stations. This monitoring was divided into five groups as follows:

1. water quality monitoring
2. floatables and discoloration
3. bottom sampling
4. rig fishing
5. benthic trawling

The water-quality parameters selected were DO, pH, fecal coliforms, SS, temperature, and light transmittance. Values for these were required at 6-m intervals down through the water column at six permanent stations flanking the outfall diffuser and at one control station.* Samples were taken using a string of Van Dorn water samplers (Chapter 6), and DO, pH, fecal coliform, and SS determinations were done in the laboratory, using methods given in the latest edition of *Standard Methods...* (1976). DO was also obtained on the boat. The temperature was obtained using a bathythermograph, the light transmittance from a photometer.

At the time of the report CSDOC (1975), methods for determining discoloration were still being investigated, and within "floatables and discoloration" only the former was analyzed. Related sampling could only be done in calm weather. Floating particulates were collected from the water surface using a special skimming sampler towed to one side of a survey boat. Oil and grease were collected from the surface with fiberglass cloth capillary screen. Total grease and oil were then expressed as the sum of the hexane extractables (after laboratory tests) from the trawled particulates and the screened oil and grease. Both control stations and stations over the diffuser were used, and results were expressed in milligrams of hexane extractable floatables per square meter of water surface.

The bottom sampling by CSDOC involved eighteen stations including one control station. At least three sediment grab samples, of a minimum size of 1 ℓ, were taken at each station; these samples were screened with a 1 mm^2 mesh, for examination of organisms. This involved identification, counting, and the calculation of a diversity index for coelenterates, polychaetes, macrocrustaceans, molluscs, ectoprocts, echinoderms, and algae.[†]

The sediments themselves were analyzed for particle size distribution, hydrogen sulfide, and chlorinated hydrocarbons. Both the sediments and two representative species of associated organism (tubiculous polychaetes and bivalve molluscs) were analyzed for silver, cadmium, chromium, copper, nickel, lead, and zinc using established analytical methods.

---

*Sampling at 6-m intervals was also done at and downcurrent of the discharge, distances being 0, 30, 100, 300, 1000, and 3000 m. In CSDOC (1975), the downcurrent-station-sampling ("directional water column stations") was carried out the day after the permanent station sampling.

[†]Details on procedures, the diversity index, and types of benthic organisms are given in Chapter 6.

The "rig fishing" involved hook and line fishing over the diffuser during daylight hours to provide a catch similar to that of commercial or sportsfishing results during the day. One hundred fish were caught, and each individual was measured, weighed, identified, and examined for deformities. A diversity index and biomass figures were calculated from the results. A species list was prepared as well. Tissues from representatives of two important species from this list [in the CSDOC (1975) report the white croaker and vermilion rockfish] were analyzed for heavy metals and pesticides.

Trawling was done at seven stations, each for a duration of 10 min, using an otter trawl with a 25-ft (7.5-m) headrope, 1.5-in. (38-mm) mesh body, and 0.5-in. (13-mm) cod end liner. Boat tow speed was 1.5 to 2.0 knots. All vertebrates and invertebrates were examined for deformities such as tumors and fin rot.* A species list was prepared and total biomass and species diversity were computed. The tissues of several species of fish were analyzed for selected heavy metals and pesticides. The internal organs of some individuals were also examined.

The report by CSDOC (1975) not only contained tabulated data relating to all of the foregoing, but an extensive cross-listing of common scientific names of vertebrates and invertebrates so that the data tables themselves did not have to be cluttered by such information.

Although not used in the CSDOC (1975) work, it is possible in monitoring investigations to take fish in fish traps or by spearing. The latter is effective when the tissues of specific fish are wanted for heavy metal and pesticide analysis.

The fish trap can be used in a fish capture-mark-release-recapture sequence. With this approach, it is possible to pinpoint those fish (species *and* individuals) that are residents of the area so that future studies can be focused on these organisms.

Benthic Animals as Pollution Indicators

Because benthic organisms are basically prisoners of one specific geographical location, they can be regarded as reflecting the condition of the water column above them as well as fallout from it in the past and present. Benthic organisms are then basically integrators of water quality conditions and can serve as very satisfactory indicators of such conditions. This information should be reflected by their diversity, species composition, and/or standing crop.

On the other hand, chemical surveys only indicate water-quality conditions at the time of sampling. Furthermore, the chemical characterization of conditions in a water column at a particular time is necessarily always incomplete since, even in the most thorough of studies, some parameters remain unknown.

---

*Of the 4188 individuals reported taken in CSDOC (1975), eight had fin rot and four tumors.

Different species of benthic organisms have different tolerance ranges to pollutants, and various investigators have attempted to give certain organisms an index relative to their pollution tolerance. However, there is often considerable disagreement about the true tolerance partly because of regional differences in species and environments. Enumeration of the species and numbers of organisms in each species, on the other hand, provides a massive list that is difficult to interpret. Many investigators, in the interest of being succinct, have represented the benthic community by a number or numbers to represent the diversity of the community, interspecies associations, or a link between the makeup of the community and some water-quality parameter or parameters. These latter techniques have not always been completely successful.

There may be important suggestions or evidence for biologists in a listing of species abundance data that are masked by the integrative nature of an index. Thus it may be of use to give both the index and the data on which it is based, although, as intimated above, this may result in very long lists of tabulated data.

The polychaete *Capitella capitata* has been proposed as an indicator species for pollution (see Chapter 4). It is able to tolerate low-salinity, low-oxygen, and high-organic-content conditions typical of wastewater effluent locations and has often been found in great numbers near outfalls. However, *C. capitata* is also found in areas remote from outfalls—where, for example, there are rich organic muds or underwater freshwater seeps. SCCWRP (1973) found that the numbers of *C. capitata* near the Orange County No. 1 outfall (#9 in Table 10-1) decreased drastically following termination of discharge through that line, suggesting that this species is particularly tolerant of high stress but does not compete as successfully in other types of communities. SCCWRP preferred consideration of associated polychaete species groups as pollution indicators.

Southern California Coastal Water Research Project Monitoring Experience

SCCWRP (1973) has reported on extensive monitoring of the constituents in sediments due to nearby outfalls. Both trace metals (cadmium, chromium, cobalt, copper, iron, lead, manganese, nickel, silver, and zinc) and chlorinated hydrocarbons (total DDT and total PCB) were studied. Samples were taken using either stainless steel Phleger corers with plastic liners, box corers, or diver hand-held corers. Analysis of the samples enabled the researchers to draw concentration contours for specific contaminants. These showed clearly that higher concentration in surface sediments were associated with the area of wastewater discharge. The outfalls at Oxnard, Whites Point (#21), Hyperion (#12), Orange County (#35), and Point Loma (#18) were analyzed in this way.

SCCWRP also did analyses to determine levels of the contaminants listed above in fish taken from areas near the outfalls. There was no clear association of fish muscle tissue concentrations of trace metals and chlorinated hydro-carbons with proximity to a wastewater discharge.

SCCWRP (1973) discussed diseases in nearshore demersal fishes. They referenced various studies that have shown that wild fish populations show diseases and abnormalities. Furthermore, they found no support for statements that *cancerous* lesions occur on some fish in wastewater-tainted waters (Young, 1964).

Between 1969 and 1972, SCCWRP examined fish taken in otter trawls near the major Southern California outfalls and elsewhere for abnormalities. They focused on three disease syndromes: tumors in dover sole (*Microstomus pacificus*), fin erosion in dover sole and white croaker (*Genyonemus lineatus*), and structural anomalies in three actively-swimming species.

It was found that the average proportion of dover sole with tumors was 1.3% (range 0.4–2.4%) and that the incidence of such disease bore no relation-ship to proximity of wastewater discharge. Fin erosion, however, definitely appeared to be related to fishes' nearness to such discharges. Abnormalities seemed more prevalent in Southern California fishes than in those from Baja California.

Consideration was also given to disease and abnormalities in benthic algae (kelp), crabs, and sea urchins. In 1971–1972, SCCWRP carried out an experi-mental program to study the influence of Southern California wastewater discharges on marine phytoplankton. Included were measurements near outfalls of chlorophyll, nutrients, and primary production, and laboratory nutrient-en-richment experiments. In the field work, it was not possible to separate effluent enrichment effects from those due to upwelling.

The recent extensive SCCWRP experience with monitoring conditions on, in, and above soft bottoms is documented in numerous brief papers in the organization's annual reports (1974, 1975, 1976).

The marine biota associated with the rocky sea floor flanking the Whites Point outfall were studied by scientific divers in 1977. Six stations along each of the 20- and 50-ft (6- and 15-m) depth contours were involved as were two control stations at La Jolla 100 miles (160 km) to the south. It was found that the effluent exerted a significant influence on the types of species present and their abundance, as well as the number of species, for a distance of 6 km (to the north). However, improvements since the observations of Grigg and Kiwala (1970) were apparent. In general, results obtained by SCCWRP appear to indicate improving conditions within the areas of influence of the various effluents under study.

11-5 REFERENCES

ALLEN, M. J., H. PECORELLI, and J. WORD (1976): "Marine Organisms around Outfall Pipes in Santa Monica Bay," *Journal of the Water Pollution Control Federation*, Vol. 48, pp. 1881–1893.

COUNTY SANITATION DISTRICTS OF ORANGE COUNTY, CALIFORNIA (1975): "Offshore Monitoring Report: October-December 1975," Fountain Valley, Calif.

ELLS, J. W. (1975): "Scours and Spanning Threaten Sea Lines," *Oil and Gas Journal*, Vol. 73, No. 27 (July 7), pp. 67–71.

FRANK, J. A. (1961): "Experiences in Cleaning the Hyperion Sludge Outfall Line," *Journal of the Water Pollution Control Federation*, Vol. 33, No. 11 (November), pp. 1199–1201.

GALLOWAY, R. N. (1964): "Marine Outfall Cleaning," *Journal of the Water Pollution Control Federation*, Vol. 36, No. 1 (January), pp. 80–83.

GRIGG, R. W., and R. S. KIWALA (1970): "Some Ecological Effects of Discharged Wastes on Marine Life," *California Fish and Game*, Vol. 56, pp. 145–155.

HUME, N. B., et al. (1961): "Operation of a 7-mile Digested Sludge Outfall," ASCE, *Transactions*, Vol. 126, Part 3, pp. 306–311.

JAMES M. MONTGOMERY, Consulting Engineers, Inc. (1973): "Ocean Waters Waste Discharge Technical Report," prepared for Goleta Sanitary District, California, at Pasadena, Calif. (January 15).

LACERDA, J. (1974): "New Day Dawns for Submersibles," *Ocean Industry*, Vol. 9, No. 5 (May), pp. 25–28.

LUDWIG, H. F., and B. ONODERA (1962): "Scientific Parameters of Marine Waste Discharge," *Advances in Water Pollution Research*, Proceedings of the International Conference, London (September), Vol. 3, edited by E. A. Pearson, pp. 37–49, with discussion and reply, pp. 50–56.

NYMAN, C. L. (1939): "Water Main Cleaning Methods Used in Outfall Sewer," *American City*, Vol. 54, No. 6 (June), pp. 71–73, 109.

O'DONNELL, J. P. (1971): "Mammoth 'Tool' Repairs Damaged Offshore Line," *Oil and Gas Journal*, Vol. 69, No. 1 (January 4), pp. 63–70.

"Pollution Control Objectives for Municipal Type Waste Discharges in British Columbia" (1975): Province of British Columbia, Dept. of Lands, Forests and Water Resources, Water Resources Service, Victoria, B.C., Canada (September).

"Polyethylene Pipe Relines 32-in. Outfall on California Coastline" (1974): *Civil Engineering*, Vol. 44, No. 1 (January), pp. 18, 20.

POWELL, R. M., and R. E. VAN HEUIT (1968): "Ocean Outfall Maintenance and Repair," *Journal of the Water Pollution Control Federation*, Vol. 40, No. 11 (November), pp. 1900–1904.

"Simple Technique Expedites Pipe Line Repairs" (1975): *Ocean Industry*, Vol. 10, No. 1 (January), pp. 35–36.

Southern California Coastal Water Research Project (1973): "The Ecology of the Southern California Bight: Implications for Water Quality Management," El Segundo, Calif., *TR 104* (March).

Southern California Coastal Water Research Project (1974): Annual Report, El Segundo, Calif.

Southern California Coastal Water Research Project (1975): Annual Report, El Segundo, Calif.

Southern California Coastal Water Research Project (1976): Annual Report, El Segundo, Calif.

Standard Methods for the Examination of Water and Wastewater, 14th ed. (1976): American Public Health Association, American Water Works Association, and Water Pollution Control Federation, Washington, D.C.

"Welding 'Hut' Used to Fix Damaged Line in 60-Foot Gulf Waters" (1968): *Oil and Gas Journal*, Vol. 66, No. 46 (November 11), p. 61.

WHITE, W. E., JR. (1969): "Underwater Pipeline Repairs—Conventional and New," ASME, *Paper 69-UnT-11*.

YOUNG, P. H. (1964): "Some Effects of Sewer Effluent on Marine Life," *California Fish and Game*, Vol. 50, pp. 33–41.

11-6 BIBLIOGRAPHY

Construction Effects

"Dredge Disposal Study: San Francisco Bay and Estuary" (1977): U.S. Army Corps of Engineers, San Francisco District, *Main Report* (February).

"Dredging: Environmental Effects and Technology" (1976): *Proceedings of WODCON VII*, World Dredging Conference, San Francisco and San Pedro, Calif. (July).

"Dredging in Estuaries—Technical Manual: A Guide for Review of Environmental Impact Statements" (1977): Oregon State University, Corvallis, Oregon.

HUBBS, C., and A. RECHNITZER (1952): "Report on Experiment Designed to Determine Effects of Underwater Explosions on Fish Life," *California Fish and Game*, Vol. 38, pp. 333–366.

JEANE, G. S., II, and R. E. PINE (1975): "Environmental Effects of Dredging and Spoil Disposal," *Journal of the Water Control Federation*, Vol. 47, No. 3 (March), pp. 553–561.

LEVIN, J. (1970): "A Literature Review of the Effects of Sand Removal on a Coral Reef Community," University of Hawaii, Honolulu, Hawaii, Sea Grant Programs, *Publication TR-71-01* (December).

McCAULEY, J. E., D. R. HANCOCK, and R. A. PARR (1976): "Maintenance Dredging and Four Polychaete Worms," ASCE, *Proceedings*, Specialty Conference on Dredging and its Environmental Effects, Mobile, Ala. (January 28), pp. 673–683.

MORTON, J. W. (1976): "Ecological Impacts of Dredging and Dredge Spoil Disposal: A Literature Review," Cornell University, Master's Thesis (January).

O'CONNOR, J. M., D. A. NEUMANN, and J. A. SHERK, JR. (1976): "Lethal Effects of Suspended Sediments on Estuarine Fish," U.S. Army, Corps of Engineers, Coastal Engineering Research Center, Fort Belvoir, Va., *Technical Paper No. 76-20* (December).

OLIVER, J. S., and P. N. SLATTERY (1976): "Effects of Dredging and Disposal on Some Benthos at Monterey Bay, California," U.S. Army, Corps of Engineers, Coastal Engineering Research Center, Fort Belvoir, Virginia, *Technical Paper No. 76-15* (October).

O'NEAL, G., and J. SCEVA (1971): "The Effects of Dredging on Water Quality in the Northwest," U.S. Environmental Protection Agency, Seattle, Washington, (July).

PEDDICORD, R., and V. McFARLAND (1976): "Effects of Suspended Dredge Material on the Commercial Crab, *Cancer Magistel*," ASCE, *Proceedings*, Specialty Conference on Dredging and its Environmental Effects, Mobile, Ala. (January 28), pp. 633–644.

ROUNSEFELL, G. A. (1972): "Ecological Effects of Offshore Construction," Marine Science Institute, Bayou la Batre, La., Report (*NTIS Publication AD-739-704*).

SHERK, J. A., JR. (1971): "The Effects of Suspended and Deposited Sediments on Estuarine Organisms," University of Maryland, Chesapeake Biological Laboratory, *Contribution No. 443* (February).

SLOTTA, L. S., et al. (1973): "Effects of Hopper Dredging and In Channel Spoiling in Coos Bay, Oregon," Oregon State University, Corvallis, Oregon, Schools of Engineering and Oceanography (July).

SLOTTA, L. S., et al. (1974): "An Examination of Some Physical and Biological Impacts of Dredging in Estuaries," Oregon State University, Corvallis, Ore., Interim Progress Report for National Science Foundation, R.A.N.N. Grant GI 34346 (December).

SMITH, D. D. (1976): "New Federal Regulations for Dredged and Fill Material," *Environmental Science and Technology*, Vol. 10, No. 4 (April), pp. 328–333.

STANLEY, D. J., and D. J. P. SWIFT (1976): *Marine Sediment Transport and Environmental Management*, Wiley, New York.

SULLIVAN, S. P., and F. GERRITSEN (1972): "Dredging Operation Monitoring and Environmental Study, Kawaihae Harbor, Hawaii," University of Hawaii, J. K. K. Look Laboratory of Oceanographic Engineering, *Technical Report No. 25* (September).

THOMPSON, J. R. (1973): "Ecological Effects of Offshore Dredging and Beach Nourishment: A Review," U.S. Army, Corps of Engineers, Coastal Engineering Research Center, *Miscellaneous Paper No. 1-73* (January).

"Untangling Dredging Regulations" (1976): U.S. Dept. of Commerce, Maritime Administration, Western Region, San Francisco, Calif. (June).

WINDOM, H. L. (1976): "Environmental Aspects of Dredging in the Coastal Zone," C. R. C., *Critical Reviews in Environmental Control*, Vol. 6, Issue 2 (March), pp. 91–109.

Operation and Maintenance

ELLISON, W. F. and R. D. ROBUCK (1976): "Split Sleeves Reinforce Subsea Lines," *The Oil and Gas Journal*, Vol. 74, No. 46 (November 15), pp. 60–62, 67, 68.

"New Cleaning Pig Has Floating Action" (1970): *Pipe Line Industry*, Vol. 32, No. 3 (March), p. 71.

Monitoring

BALCH, N., et al. (1976): "Monitoring a Deep Marine Wastewater Outfall," *Journal of the Water Pollution Control Federation*, Vol. 48, No. 3 (March), pp. 429–457.

BANDY, O. L., J. C. INGLE, JR., and J. M. RESIG (1965): "Modification of Foraminiferal Distribution by the Orange County Outfall, California," Marine Technology Society, *Ocean Science and Ocean Engineering*, Vol. 1, pp. 54–76.

BURGESS, F. J., and W. P. JAMES (1971): "Airphoto Analysis of Ocean Outfall Dispersion," U.S. EPA, Water Pollution Control Research Series, 16070ENS06/71 (June).

California State Water Resources Control Board (1974): "The Demonstration and Standardization of a Method for Monitoring the Ecological Effects of Marine Waste Discharges," Sacramento, Calif., *Publication No. 54* (April).

CARLISLE, J. G., JR. (1969): "Results of a Six-Year Trawl Study in an Area of Heavy Waste Discharge: Santa Monica Bay, California," *California Fish and Game*, Vol. 55, pp. 24–46.

CARLISLE, J. G., JR. (1972): "A Trawl Study in an Area of Heavy Waste Discharge: Santa Monica Bay, California," in *Marine Pollution and Sea Life*, edited by M. Ruivo, Food and Agriculture Organization of the United Nations, Conference, Rome, December 1970, Fishing News (Books) Ltd., London, England, pp. 417–421.

COPELAND, B. J., and T. J. BECHTEL (1971): "Species Diversity and Water Quality in Galveston Bay, Texas," *Water, Air, and Soil Pollution*, Vol. 1, pp. 89–105.

COVILL, R. W. (1975): "Bacteriological, Biological and Chemical Parameters Employed in the Forth Estuary," in *Pollution Criteria for Estuaries*, edited by P. R. Helliwell and J. Bossanyi, Pentech Press, London, Chapter 9, with discussion.

CRANE, J. D., and R. H. JONES (1976): "Results of Ocean Diffusion and Biological Studies of the Hollywood, Florida, Ocean Outfall," U.S. EPA-600/3-76-003 (January) (*NTIS Publication PB-247-684*).

D'AMATO, R. (1973): "The Movement of Effluent from the City of Miami Sewage Ocean Outfall," University of Miami Sea Grant Program, Coral Gables, Fla., *Sea Grant Technical Bulletin No. 27*.

"Ecological Responses to Ocean Waste Discharge: Results from San Diego's Monitoring Program" (1970): Water Resources Engineers, Inc., Walnut Creek, Calif., report to the California State Water Resources Control Board and the San Diego Regional Water Quality Control Board (July).

EGANHOUSE, R. P., JR. (1975): "The Measurement of Total and Organic Mercury in Marine Sediments, Organisms, and Waters," Southern California Coastal Water Research Project, El Segundo, Calif., *TM 221* (September).

FENG, S. Y., and G. M. RUDDY (1975): "Concentrations of Zinc, Copper, Cadmium, Manganese, and Mercury in Oysters (*Crassostrea Virginica*) along the Connecticut Coast," The Third International Ocean Development Conference, Tokyo, Japan (August), *Preprints*, Vol. 4, pp. 109–129.

FILICE, F. P. (1959): "The Effect of Wastes on the Distribution of Bottom Invertebrates in the San Francisco Bay Estuary,:' *The Wasman Journal of Biology*, Vol. 17, pp. 1–17.

FUKAI, R., and W. W. MEINKE (1959): "Trace Analysis of Marine Organisms: A Comparison of Activation Analysis and Conventional Methods," *Limnology and Oceanography*, Vol. 4, pp. 398–408.

GRIGG, R. W. (1972): "Some Ecological Effects of Discharged Sugar Mill Wastes on Marine Life along the Hamakua Coast, Hawaii," University of Hawaii, Honolulu, Hawaii, Water Resources Research Center, Water Resources Seminar Series No. 2 (December), pp. 27–45.

GRIGG, R. W. (1975): "Environmental Impact of Raw Sewage on an Hawaiian Coral Reef," paper presented at the 13th Pacific Science Congress, Vancouver, B.C. (August 18–30), *Abstracts*, Vol. 1, pp. 125–126.

JAMES, W. P., and F. J. BURGESS (1971): "Pulp Mill Outfall Analysis by Remote Sensing Techniques," *Journal of the Technical Association of the Pulp and Paper Industry* (Tappi), Vol. 54, No. 3 (March), pp. 416–418.

JAMES, W. P., F. J. BURGESS, and D. J. BAUMGARTNER (1971): "An Aerial Photographic Study of Waste Field from Three Ocean Outfalls," *Preprints*, Offshore Technology Conference, Houston, Texas (April), Vol. 1, pp. 483–498.

JAMES, W. P., D. J. BAUMGARTNER, and F. J. BURGESS (1973): "Airphoto Analysis of Ocean Outfall Dispersion," in *Advances in Water Pollution Research*, edited by S. H. Jenkins, pp. 881–891, Pergamon Press, New York.

JOKIEL, P. L., and S. L. COLES (1974): "Effects of Heated Effluent on Hermatypic Corals at Kahe Point, Oahu," *Pacific Science*, Vol. 28, pp. 1–18.

KENIS, P. R., M. H. SALAZAR, and J. A. TRITSCHLER (1972): "Environmental Study of the Sewage Outfall Area at San Clemente Island," U.S. Naval Undersea Center, San Diego, Calif., TP 292 (June) (*NTIS Publication AD-745-298*).

KUENZLER, E. J., et al. (1971): "Structure and Functioning of Estuarine Ecosystems Exposed to Treated Sewage Wastes," University of North Carolina, Chapel Hill, N.C., Report (February) (*NTIS Publication COM-71-00688*).

LYNN, W. R., and W. T. YANG (1960): "The Ecological Effects of Sewage in Biscayne Bay: Oxygen Demand and Organic Carbon Determinations," *Bulletin of Marine Science of the Gulf and Caribbean*, Vol. 10, No. 4, pp. 491–509.

MATSUDO, H., and D. W. CHAMBERLAIN (1976): "Notes on Mycoplasma-like Organisms in Skin Tumors of Dover Sole, *Microstomus Pacificus*, Taken Near Sewage Outfalls," *Bulletin of the Southern California Academy of Sciences*, Vol. 75, No. 3 (December 23), pp. 270–273.

McCAIN, J. C., and J. M. PECK, JR. (1973): "The Effects of a Hawaiian Power Plant on the Distribution and Abundance of Reef Fishes," University of Hawaii Sea Grant Program, *Publication 73-03* (June).

McDERMOTT, D. J., et al. (1976): "Metal Contamination of Flatfish around a Large Submarine Outfall," *Journal of the Water Pollution Control Federation*, Vol. 48, pp. 1913–1918.

McKEE, H. C., and D. S. TARAZI (1974): "Development of Sample Preparation Methods for Analysis of Marine Organisms," U.S. Environmental Protection Agency, Publication EPA-660/3-74-026 (January).

McNULTY, J. K. (1961): "Ecological Effects of Sewage Pollution in Biscayne Bay, Florida: Sediments and the Distribution of Benthic and Fouling Macroorganisms," *Bulletin of Marine Science of the Gulf and Caribbean*, Vol. 11, pp. 394–447.

MEARNS, A. J., and C. S. GREENE, Eds. (1974): "A Comparative Trawl Survey of Three Areas of Heavy Waste Discharge," Southern California Coastal Water Research Project, El Segundo, Calif., *TM 215* (September).

MOTT, P. G. (1975): "Airborne Sensors for Monitoring Pollution," in *Pollution Criteria for Estuaries*, edited by P. R. Helliwell and J. Bossanyi, Pentech Press, London, Chapter 10, with discussion.

MURRAY, S. N., et al. (1974): "Biological Features of Intertidal Communities near the U.S. Navy Sewage Outfall Wilson Cove, San Clemente Island, California," U.S. Naval Undersea Center, San Diego, Calif., Research Report (July) (*NTIS Publication AD-783-029*).

ORLOB, G. T., and D. A. O'LEARY (1976): "Impact on Marine Benthos of Waste Water Discharge," ASCE, National Water Resources and Ocean Engineering Convention, San Diego, Calif. (April), *Preprint 2658*.

PLUHOWSKI, E. J. (1976): "Remote Sensing of Turbidity Plumes in Lake Ontario," ASCE, *Transportation Engineering Journal*, Vol. 102, pp. 475–488.

"Remote Sensing Study, Eastern Florida Coast, Dade County, Florida" (1972): National Field Investigations Center, Denver, Colorado (April) (*NTIS Publication PB-229-807*).

"Report of the Seminar on Methods of Detection, Measurement and Monitoring of Pollutants in the Marine Environment" (1971): Food and Agriculture Organization of the United Nations, Rome, Fisheries Reports, No. 99, Suppl. 1 (August).

ROWNEY, J. V. (1973): "Gradient Analysis of Phytoplankton Productivity and Chemical Parameters in Polluted and other Nearshore Habitats," Naval Postgraduate School, Monterey, Calif., Master's Thesis (March) (*NTIS Publication AD-761 466*).

SELLECK, R. E. (1975): "The Significance of Surface Pollution in Coastal Waters," in *Discharge of Sewage from Sea Outfalls*, edited by A. L. H. Gameson, pp. 143–153, Pergamon Press, Oxford, England.

SIDWELL, V. D., et al. (1974): "Composition of the Edible Portion of Raw (Fresh and Frozen) Crustaceans, Finfish, and Mollusks. I. Protein, Fat, Moisture, Ash, Carbohydrate, Energy Value, and Cholesterol," *Marine Fisheries Review*, Vol. 36, No. 3 (March), pp. 21–35.

SMITH, K. L., JR., G. T. ROWE, and J. A. NICHOLS (1973): "Benthic Community Respiration near the Woods Hole Sewage Outfall," *Estuarine and Coastal Marine Science*, Vol. 1, pp. 65–70.

SMITH, R. W., and C. S. GREENE (1976): "Biological Communities Near Submarine Outfall," *Journal of the Water Pollution Control Federation*, Vol. 48, pp. 1894–1912.

STEWART, R. E. (1973): "Unusual Plume Behavior from an Ocean Outfall off the East Coast of Florida," *Journal of Physical Oceanography*, Vol. 3, No. 2 (April), pp. 241–243.

STEWART, R. E., et al. (1974): "Diffusion of Sewage Effluent from Ocean Outfall," ASCE, *Journal of the Sanitary Engineering Division*, Vol. 97, No. SA 4 (August).

TURNER, C. H., A. R. STRACHAN, and C. T. MITCHELL (1967): "Survey of a Marine Environment Subsequent to Installation of a Submarine Outfall," *Marine Resources Operation Reference No. 67-24*, Resources Agency of California Dept. of Fish and Game (October 30).

TURNER, C. H., E. E. EBERT, and R. R. GIVEN (1966): "The Marine Environment in the Vicinity of the Orange County Sanitation District's Ocean Outfall," *California Fish and Game*, Vol. 52, pp. 28–48.

YOUNG, D. R., D. J. McDERMOTT, and T. C. HEESEN (1976): "DDT in Sediments and Organisms around Southern California Outfalls," *Journal of the Water Pollution Control Federation*, Vol. 48, pp. 1919–1928.

Appendices

Accurate Positioning
of Marine Vessels

The obvious choice for a pilot was Enos. Like most Newfoundland seamen he possessed, we presumed, special senses which are lost to modern man. He had sailed these waters all his life, often without a compass and usually without charts. When you asked him how he managed to find his way to some distant place he would look baffled and reply: "Well, me son, I knows where it's at."

Farley Mowat, *The Boat Who Wouldn't Float*,
McClelland and Stewart Ltd.,
Toronto, Canada (1969), with permission.

A-1 INTRODUCTION

An ocean outfall project is a complex undertaking that involves a myriad of marine tasks, from the planning stages through the design, construction, operation, and maintenance. Most related tasks are done from a small boat, tug, ship, or barge, and in virtually all cases the actual location of the vessel out on the water when the work is being done is important. For example, a water depth by itself is an essentially useless piece of information unless it is coupled with a particular location; if current data are to be obtained from an appropriate meter, the position of the meter is important not only for showing on some form of plan where the data are applicable, but also to aid a boat operator in relocating the meter if a marker buoy disappears. During the construction of the outfall, for instance, it is critical that dredging be done along the outfall line, and some form of precise positioning must be used. And of course this same trench must be relocated when the pipe is laid. It is, in general, important to be able to do two things in ocean outfall work:

1. locate data-taking stations and work sites on an overall paper plan of the operational area, and

2. locate specified stations and work sites out on the water.

In this Appendix, I will endeavor to briefly lay out the techniques of positioning and plotting for coastal waters that find application at various stages through an ocean outfall project. This material is particularly applicable to the contents of Chapter 5, "Obtaining Pertinent Physical Oceanographic Data," and Chapter 10, "Construction of Outfall." The material to be presented is of a substantially superficial nature, and the interested reader is referred to the following excellent reference books for more in-depth treatments: Bowditch (1962), Brinker (1969), Chapman (1967), and Dunlap and Schufeldt (1969).

A-2 THE EARTH

Introduction

The earth is a body that moves through space as it spins about an axis which we call the north–south axis. The earth moves around the sun, describing a closed path every 365.25 days. The orbit of the earth around the sun is elliptical, but the ellipse does not have a large eccentricity and the average distance between the two is about 93×10^6 miles (150×10^6 km). The time that the earth takes to complete one rotation on its own axis relative to the sun is 1 day or 24 h. A natural satellite of the earth is the moon, a planet that completes an apparent circuit of the earth approximately every 24 h 50 min. The moon is, on the average, approximately 240,000 miles (390,000 km) distant from the earth.

The points on the earth marking the axis of rotation are known as the north and south poles. The earth is not a perfect sphere, although very close to one, and the diameter of the earth that contains both poles is 7,900 (*statute* or ordinary) miles (12,717 km). The diameter of the earth lying in a plane perpendicular to the axis and halfway between the poles is 7,926 miles (12,759 km). The line of circumference imagined here is known as the equator. Both the equator and the circumference through the poles are known as *great circles* (making the assumption that the earth *is* spherical); any circle on the face of sphere that has the center of the sphere in its plane is known as a great circle. Any other circle is known as a *small circle*.

Coordinates on the Earth

Imagine any great circle drawn on the surface of the earth through the poles. Such a line is known as a *meridian*. The meridian that passes through Greenwich, England, is known as the prime meridian and is taken as the origin of a coordinate system on the earth's surface that is based on meridians. Three hundred sixty degrees of arc are subtended at the center of the earth by the equator. These 360° are divided into two equal portions, one of which is called *west longitude* and the other *east longitude*. Thus, a point on the equator at west longitude 30° is directly opposite a point on the equator that has position 150° east longitude.

The position of a point above or below the equator is given in terms of *latitude* which is also divided into two equal parts, north and south latitude depending upon the pole nearer to the point in question. The origin of the latitude system is at the equator, and thus a latitude of 90° represents one of the poles. Note that the meridians of longitude are all great circles but that all circles of constant latitude (or *parallels* of latitude) are small circles except for the equator. One minute of arc on the earth's surface along a meridian is equal to the unit of distance measurement known as the nautical mile which is 1.15 (statute) miles, 6080 ft or 1854 m.*

Although position specification through use of latitude and longitude is the universally accepted method for marine vessels, boats in coastal waters (the area in which boats associated with ocean outfall projects would be working) may specify their position by using a local coordinate system. This can be (in U.S. waters) the State Plane Coordinate System which assumes that an overall limited area is made up by different Zones over which the earth can be assumed to be flat with only minor error. California, for example, has its California Lambert Zones and Hawaii its Hawaiian Plane Coordinate Grid System.

Direction

An observer on the earth's surface finds north and south defined along his meridian, and east and west by his parallel. East is assigned the directional representation 90°, south, 180°; west, 270°; and north, 360°. Since such an observer cannot *see* the poles, he must use another method of finding, say, north. For many years this has been done by noting the position of the pole star (north star) or through use of a *magnetic compass*. The former approach may cause errors of up to approximately 1° in the assignment of true north, is only appropriate in the Northern Hemisphere, and of course can only be used at night. The principle employed in the second case is that the earth possesses a magnetic field, and a magnetic needle will tend to align itself with the earth's lines of force. Unfortunately, the magnetic poles of the earth do not correspond to the geographic poles, the distance between the two north poles being about 1000 miles (1610 km), that between the two south poles approximately 1560 miles (2510 km). Thus, in either hemisphere it is possible to visualize both geographic and magnetic meridians. The horizontal angle between two such meridians for any location on the earth's surface is known as either (magnetic)

---

*Actually, one minute of arc at the equator is 1.153 statute miles; it is 1.149 statute miles along any meridian.

variation or *declination*, and the value for any such location is readily available on charts, which will be discussed presently. Since many boats contain ferrous materials, the compass indication on such a boat must be corrected for *deviation*, or the difference between actual and indicated magnetic directions, to obtain, for example, magnetic north. A modern device that enables the boat operator to determine true north directly is the *gyro compass* which aligns itself with the earth's axis of rotation.

A-3 CHARTS

A *map* is a representation of some part of the earth's surface showing political boundaries, physical features, cities and towns, and other geographic information. A *chart* is also a representation of a portion of the earth's surface, but has been specifically designed specially for convenient use in navigation. It is intended to be worked upon…and must readily admit…position determination in terms of latitude and longitude. A *nautical chart* is primarily concerned with navigable water areas. It includes information such as coastlines and harbors, channels and obstructions, currents, depths of water, and aids to navigation.*

The earth's surface is, of course, curved, and various *projections* have been devised to represent some portion of the earth on a flat piece of paper. Where relatively small areas are involved, such as several miles by several miles as in the cases of coastal charts and harbor charts, little distortion is involved. The scale of such charts, useful in the marine work associated with outfalls, is 1 : 100,000 or larger. A scale of 1 : 100,000 means that one foot on the charts corresponds to 100,000 feet in the field. A larger scale, for example, is 1 : 5,000. Charts for United States coastal waters are prepared by the National Ocean Survey (U.S. Department of Commerce, National Oceanic and Atmospheric Administration) and are sold to the public, in most major coastal cities, by marine equipment dealers.

Part of any nautical chart is a *compass rose* which has two concentric circles, the outer representing true directions for the area shown on the chart, the other showing magnetic directions. On large-scale maps too, there is a distance scale (e.g., in nautical miles) marked, and this is possible because of the very minor real changes in scale, due to projection distortion, across the area depicted. Each chart displays major parallels of latitude as well as meridians (of longitude).

For the purpose of ocean outfall undertakings certain features shown on the chart are very important. The distribution of depths is perhaps the focus, and normally smooth curves are drawn on a chart to show lines of constant depth. These are generally marked for the following depths (in fathoms) on coastal

---

*From *Dutton's Navigation and Piloting*, 12th edition, by Dunlap and Shufeldt. Copyright © 1969, U.S. Naval Institute.

charts: 1, 2, 3, 5, 10, 20, 50, 100, 200, and 300. Whether depths are expressed in fathoms, feet, or meters should always be carefully checked. Another important type of information shown on coastal charts concerns the nature of the bottom; at various points over the chart there are notations concerning the type of bottom material. Other interesting bits of information in the water area concern the location of navigation buoys, submarine pipelines and cables, wrecks, areas of heavy wave action, anchorages, dumping areas, and prohibited areas such as military target zones. But an important part of any nautical chart is the land area shown, not only for the shape of the coastline and location of and entrances to harbors, for example, but mainly for the location of landmarks that are plainly visible from the offshore waters within the region shown on the chart. This is because these landmarks provide a means of positioning out on the water that will be examined presently.

Marine survey information gathered at some indeterminate spot in a general coastal area is next to useless. The location of the survey point must be established and plotted on a suitable chart so that observations can be properly ordered. The methods by which this can be accomplished will be examined in the following two sections. First, it is required that the position at sea be established; second, this point must be placed on the appropriate chart.

A-4 POSITIONING AND ALIGNING

Introduction

There are two general classes of situations associated with marine outfall sewers where an accurate method of positioning is required. On the one hand there are those investigations in which the boat is continuously on the move; examples concern the monitoring of fluorescent dye concentrations in making studies of diffusion and dispersion, undertaking bathymetric surveys, and doing geophysical profiling. These types of studies are described in Chapters 5 and 6. On the other hand one has a sizable list of activities that are done at one location that must either be accurately established at one point in time or else maintained for a period of time. In the first of these cases, one must know the position of his current meter, wave measuring device or buoy. In the second case one must remain in the proper position when drilling (Chapter 6) or when dredging out the outfall trench, placing rock, or laying pipe (Chapter 10).

Celestial positioning is a technique whereby the approximate position of a vessel on the water is derived first by measuring the vertical angle between certain heavenly bodies, such as stars and planets, and the horizon. Then appropriate tables are consulted to determine the position on the earth's surface that one would have to occupy to obtain such an angle at the particular time at which the observations were made. The measurement of such angles is carried out through use of a small hand-held optical instrument called a *sextant* which is

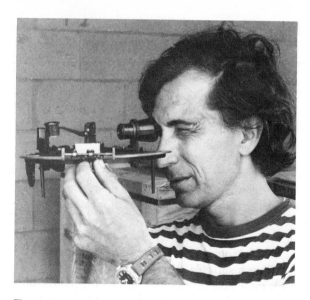

Figure A-1 Use of sextant for measuring horizontal angles

shown being used in Figure A-1. There are various disadvantages to this positioning method, among which one could single out the necessity of having a clear sky and an accurate chronometer. The modern-day method of positioning marine vessels is through use of electronic methods. However, the sextant can be used in another manner as will be explained later.

Electronic Positioning Methods

There are numerous types of electronic positioning systems in existence. Each has its particular advantages in terms of range or accuracy, or a blend of the two, and most are best suited to use on the high seas where it is unnecessary to have a position established within a few tens of feet.

The Cubic Corporation (San Diego, California) Autotape DM-40A (formerly DM-40) Electronic Positioning System has seen considerable use in marine survey work. Examples of its use on ocean outfall jobs include the Orange County 5-mile (8-km) outfall (1971) where it was used in placing buoys for soil sampling and the Mokapu outfall in Kailua, Hawaii, where it was used in the bathymetric surveys in 1971 and for resurveys in 1975. Advantages to this system are its automatic features and its accuracy and portability. A similar radio line-of-sight commercial system is the Motorola Mini-Ranger III used to position the crane barge during construction of the Barbers Point (Hawaii) outfall.

The general idea behind the Autotape system is that two shore stations are established, preferably at elevated points for best coverage of offshore vessel tracks. The station elevations are accurately measured, the distance and the line of bearing between them are obtained, and special compact DM-40A responder units are placed at the two locations. A two-range interrogator* is mounted on the survey vessel, and the Autotape system uses microwaves to measure the slope distances between the boat and the two responders at the same time. These two distances are normally updated every second and displayed on the face of the interrogator to the nearest 0.1 m. However, the advertised probable range accuracy is .0.5 m + one-one hundred thousandth of the range. This error would amount, for example, to only 0.6 m for a range of 10 km. Sea state, boat stability, and antenna height will, however, exert some influence on errors.

The Autotape system has a maximum range of 150 km, far greater than would be necessary in ocean outfall-related work, and can work under any weather conditions. Since the Autotape system is very precise and versatile, it is very expensive to buy (about $100,000) although it can be leased through a surveying equipment rental firm such as Lewis and Lewis (Ventura, California) for a fee of (1976) $300 per day for the first ten days, $150 per day between ten and ninety days, and $100 per day thereafter. Rental on the Motorola Mini-Ranger III from the same source is about two-thirds of the above rates.

Before the Autotape interrogator unit is loaded aboard a boat for a marine survey, a chart can be prepared with concentric slope distance rings around each responder station. The surveyor on the boat can then stop the interrogator display from time to time out on the water, record the values shown, and then consult the prepared chart in order to ascertain the vessel's position and its relationship to desired positions or tracks.

There are, however, optional instruments—printers, recorders, and plotters that can be used with the Autotape system that largely eliminate the middle-man. Recording can be on paper tape or on magnetic tape. A very desirable feature is that concurrent information on the two ranges, time, and another channel of data, typically that from a depth sounder, can be logged.

Position along a Line by Using Lasers

Where position along a line has to be rigidly maintained, such as for a pipelaying ship along the outfall trench, a very accurate means of maintaining line is desired. A system that works in many types of weather conditions (but not heavy fog) is the laser alignment system.† In this case the position of a

---

*The Autotape DM-43 system deals with three ranges.

†Normal white light derived from a heated filament lamp contains various proportions of all the colors of the visible spectrum. The refractive index of the atmosphere is different for each color with the result that the constituent colors in a white light beam become separated and the beam widens. Laser light, however, has only one wavelength and a laser beam does not suffer the dispersion effect experienced by white light. Laser beams are very intense, narrow, and well *collimated* (have parallel beam boundaries) and may be used over long optical paths. However, laser light, like white light, is scattered by particles of dust, smoke, or water droplets. Thus it may see only limited use as an alignment medium in areas of frequent coastal fog such as mid-California in the summer.

continuous laser beam falling on a target on the marine vessel can be continuously electronically monitored and the departure of the center of the target from the desired line shown on two meters, one for horizontal departures (alignment) and the other for vertical (grade). The accuracy of this system is reputed to be extraordinary: for example 0.002 inches at ranges of 150 to 300 ft (0.05 mm at 45 to 90 m). A very accurate electronic method of measuring distance along an established line is the Electrotape system, also from Cubic Corporation, available from Lewis and Lewis or similar leasing companies. Hewlett-Packard (Hollywood, California) also manufactures electronic distance-measuring instruments.

Other Approaches

Substantial marine survey work has been done using adaptations of precise on-land surveying techniques. A favorite approach is to use two *theodolites* (a precision-type transit) on shore, at fixed stations, to position a boat. In such a case radio contact is maintained between the vessel and the two theodolite operators. In cases where the position of a boat is required, the angles between the base line and the two lines of sight to the boat are read and the position of the vessel established by solving the resulting triangle (three angles, one side). When it is desired that the boat take up a specific station, the desired angles are set on the theodolites and the boat positioned through radio contact. Normally, when the boat moves into the proper position, an anchor-buoy arrangement is dropped on the spot and the vessel then maneuvers with respect to that position. In many cases, since work is generally done along a straight outfall line, one theodolite can be positioned at the shoreward end of the centerline; there are then no angle adjustments or measurements to be made by this instrument.

Close to shore and during clear weather conditions, a technique that has been used to advantage is the *range-azimuth approach*. In this case an operator on shore maintains an electronic distance-measuring device and a transit (or equivalent) at a shore station of known coordinates lying on a specified baseline. At selected times the operator obtains the distance to the vessel and the angle of its bearing with the baseline. Although the position of the boat could then be rapidly determined by suitable hand-calculator operations, a superior method is for the operator to feed the information into an adjacent automatic plotter and then radio course corrections to the boat as the results appear.

A final means of establishing position at sea is through use of the sextant, as mentioned earlier, but to measure horizontal angles rather than the vertical ones used in celestial positioning. Three distinct objects on shore are required for this technique, and if the position of the boat is to be later plotted on a chart, these must be objects shown on the chart. Where suitable objects do not exist naturally, towers or flags may have to be placed and their positions fixed beforehand on the chart by surveying on shore.

When positioning using a sextant, two horizontal angles are measured, one between the left-hand object and the center one, the other between the center object and the right-hand one. Angles are read by a sextant to an accuracy higher than the angles can be plotted on a chart, so sextants have seen considerable use in marine survey work. Disadvantages to this method of positioning are that it takes a few moments to execute, and in fact may be very difficult to carry out on rough water, and that it cannot be used a long distance from the coast where the sighting objects are located. Precision of the method may therefore be highly variable even without considering the skill of the operator, which can be a significant factor.

The establishment of a position on a chart where angles have been measured by a sextant is most easily carried out through use of a three-arm protractor. Two angles between three radial arms are set to correspond to the angles measured in the field. The protractor is moved around until lines along these arms pass through the chart positions of the three sighting points. A pencil is then inserted into a hole at the common base of the three arms and a mark made on the chart and appropriately labeled.

Use of Floating Aids-to-Navigation to Fix a Navigational Position

When sextants are employed in an effort to establish the positions of a boat, or when a *range* is found to provide a line of position for such a vessel, a marker buoy should never be employed as one of the objects used.* The following is taken from the U.S. Coast Guard, Fourteenth District, Honolulu, Hawaii, "Local Notice to Mariners," May 4, 1976.

> The aids to navigation depicted on charts comprise a system consisting of fixed and floating aids with varying degrees of reliability. Therefore, prudent mariners will not rely solely on any single aid to navigation, particularly a floating aid.
>
> The buoy symbol is used to indicate the approximate position of the buoy body and the sinker which secures the buoy to the seabed. The approximate position is used because of practical limitations in positioning and maintaining buoys and their sinkers in precise geographical locations. The limitations include, but are not limited to, inherent imprecisions in position fixing methods, prevailing atmospheric and sea conditions, the slope of and the material making up the seabed, the fact that buoys are moored to sinkers by varying lengths of chain, and the fact that buoy body and/or sinker positions are not under continuous surveillance but are normally checked only during periodic maintenance visits which often occur more than a year apart. The position of the buoy body can be expected to shift inside and outside the charting symbol due to the forces of nature. The

---

*The word "range" is used here in the sense of a lineup of two objects, not of distance. This double use of the same word in seagoing parlance is somewhat confusing.

mariner is also cautioned that buoys are liable to be carried away, shifted, capsized, sunk, etc. Lighted buoys may be extinguished or sound signals may not function as the result of ice, running ice or other natural causes, collisions, or other accidents.

For the foregoing reasons, a prudent mariner must not rely completely upon the position or operation of floating aids to navigation, but will also utilize bearings from fixed objects and aids to navigation on shore. Further, a vessel attempting to pass close aboard always risks collision with a yawing buoy or with the obstruction the buoy marks.

A-5 RADAR

The word *radar* is derived from *ra*dio *de*tecting *a*nd *r*anging. The basic principle behind the system involves the measurement of the time taken between the transmission of an electromagnetic signal and its return as an echo after being reflected back to point of origin by a *target*. An antenna is used both to transmit the signal and to receive the echo. The signal itself consists of a burst of high-frequency waves, and these are rapidly repeated. Different commercial radar systems have different signal characteristics, but frequencies are typically from 3000 to 10,000 MHz, bursts from 0.05 to 1 μs in duration, with from 900 to 3600 pulses emitted each second.

The antenna does not remain stationary but rotates 360° in order to survey the surroundings of the boat. A typical rotational speed of such an antenna is 30 rpm. Radar systems are designed so that, in step with the antenna's rotation, a radial line sweeps around a circular screen in the pilot house of the boat. This screen, of from 7- to 16-in. (178- to 406-mm) diameter depending upon the type of radar set, is called a *plan position indicator* (PPI). In the standard case, the center of the screen corresponds to the boat's position.

The screen is the face of a cathode-ray tube, and as the radial line on the screen describes its continuing path around the screen a glow is left at scaled distances from the boat along particular bearings where echoes were received. The screen phosphor is selected to have some persistence so that echoes continue to glow for a time even after the radial line has passed them. Thus a chart-like representation of the surrounding area is obtained on the screen.

The top of the screen can be made to conform to magnetic or true north or else to the heading of the boat. Angular position (*bearing*) of a target from the top of the screen is made possible by a graduated degree ring around the edge of the screen. The scaled distance from the center of the screen (boat) to the edge of the screen (*range*) can be set at different values using a switch. In turn, intermediate circular rings on the screen take on different values. The Raytheon Marine Co. (Manchester, New Hampshire) 3100 Mariners Pathfinder Radar, for example, has the following range scales (range-ring spacings) in nautical miles: 1/2 (1/4); 2 (1/2); 4 (1); 8 (2); 16 (4); 32 (8).

Radar can be used to establish the position of the boat (*fix*) with respect to charted features in two ways. First, the bearing and range of such a feature on the screen can be used in reverse to establish the boat's position on a chart where the specific target is shown. Second, the bearing of two or more charted features on the screen can be used in reverse on an appropriate chart to locate the boat. Checks on distances can be made by relating the charted spacing between two features and their spacing on the screen.

Radar can be useful in establishing the position of a floating object with respect to its location and heading. For example, radar can be invaluable in locating buoys or current drogues, if they are mounted with adequate radar reflectors that make them much more visible on the screen. A real problem with this method concerns *sea return*. In the presence of even moderate wave action, the PPI is filled with bright flecks as the electromagnetic waves are reflected from properly oriented facets in the waves. Experienced observers are often able, however, to separate out the desired target due to its steady echo rather than the transient ones provided by the waves.

Radar systems, because of their bulk, cannot be used in small boats. When mounted in suitable vessels, they can be temperamental installations subject to electrical and mechanical malfunctions. If properly working, however, such systems provide fixes quickly, and can be used at night and during periods of otherwise low visibility when visual systems are not possible. Fixes are also possible at some distance from land where visual techniques break down.

A-6 REFERENCES

BOWDITCH, N., (1962): *American Practical Navigator*, U.S. Naval Oceanographic Office, Washington, D.C.

BRINKER, R. C. (1969): *Elementary Surveying*, 5th ed., International Textbook Co., Scranton.

CHAPMAN, C. F. (1967): *Piloting, Seamanship and Small Boat Handling*, Motor Boating, New York.

DUNLAP, G. D., and H. H. SHUFELDT (1969): *Dutton's Navigation and Piloting*, U.S. Naval Institute, Annapolis.

A-7 BIBLIOGRAPHY

ANDERSON, N. M., et al. (1973): "An Evaluation of the Mini-Ranger Positioning System," Environment Canada, Fisheries and Marine Service, Marine Sciences Directorate, Pacific Region, Victoria, B.C., Canada, *Pacific Marine Science Report 73-11* (unpublished manuscript).

DALLAIRE, G. (1974): "Electronic Distance Measuring Revolution Well Underway," *Civil Engineering*, Vol. 44, No. 10 (October), pp. 66–71.

HARRISON, P. W., F. R. TOLMON, and B. M. NEW (1972): "The Laser for Long Distance Alignment—a Practical Assessment," *Proceedings*, Institution of Civil Engineers, Vol. 52, pp. 1–24.

HOBBS, R. R. (1974): *Marine Navigation*, 2 vols, Naval Institute Press, Annapolis.

INGHAM, ALAN E., Ed. (1975): *Sea Surveying*, Wiley, New York.

SCHNETLER, F. (1972): "Position Fixing for Submarine Pipelines," *World Dredging and Marine Construction*, Vol. 8, No. 2 (February), pp. 20–21.

WOODS, M. V., et al. (1973): "An Evaluation of the Trisponder 202A with Model 210 Transponders," Environment Canada, Fisheries and Marine Service, Marine Sciences Directorate, Pacific Region, Victoria, B.C., Canada, *Pacific Marine Science Report 73-8* (unpublished manuscript).

B

‖‖

Diving

‖‖

B-1   INTRODUCTION

The importance of divers in a complete marine disposal of wastewater scheme cannot be overemphasized. An initial sea floor reconnaissance in the general area being considered for an ocean outfall is a must, as stressed in Chapter 1. During the formal data-gathering phase (Chapters 5 and 6) scientific divers play a front-line role in securing biological and geological oceanographic data and participating as well in securing some types of physical oceanographic information. The marine biologist again enters the diving picture when monitoring the performance of ocean outfalls (Chapter 11).

The foregoing types of studies are carried out by divers using SCUBA* equipment. Although divers outfitted in this way are also used in construction activities as inspectors (Chapter 10), the divers doing the construction work, such as inserting and bedding pipe on the sea floor, use an entirely different system that links them to the support vessel on the surface by hoses. Because diving and divers play such vital roles in the overall subject matter of this volume, an appendix providing background on diving and divers is regarded as an integral part of it.

---

*Self-contained underwater breathing apparatus.

Pressures and Gases

The average pressure at sea level on the earth's surface due to the weight of the atmosphere above it is one atmosphere absolute, abbreviated atma, equivalent to 14.69 psia, 10.13 N/cm^2 or 1013 millibar (mb).

An arbitrary volume of atmospheric air contains primarily nitrogen and oxygen. The concentrations of these two main gases are by convention given in volumetric terms—by imagining that all but one of these two gases is removed from a chamber and the chamber volume is reduced to maintain that gas at atmospheric pressure. Volumetric concentrations of nitrogen and oxygen in an average parcel of air are respectively 78.08% and 20.95%. By multiplying either of these quantities in fractional terms times atmospheric pressure one obtains the *partial pressure* of either gas.

When a gas and liquid are in contact, gas molecules tend to enter the liquid, and in turn to be expelled by the liquid. An equilibrium condition is reached when the rate at which gas molecules enter the liquid equals that of their expulsion from the liquid. The equilibrium concentration of gas in the liquid increases as the partial pressure of that gas increases—due either to an overall pressure change or to a greater proportion of that gas in the gas mixture.

Suitable conditions for the transfer of gases to and from the blood are provided in the lungs. The venous blood, high in carbon dioxide, releases the excess of this gas to the gas mixture in the lungs—to be subsequently exhaled; the blood in turn receives a dose of oxygen from the inhaled air to compensate for the deficiency due to its usage as the blood passed through the body. Nitrogen has no physiological function. Thus for given pressure conditions there is no net transfer of this gas between the blood and the air in the lungs. Equilibrium conditions have been established. Nitrogen is sometimes referred to as a "carrier gas."

Oxygen

Oxygen is used by the body in respiration, and moderate reductions in partial pressure of oxygen in a breathing gas can lead to drowsiness and loss of mental acuity in many persons, and to death with more severe reductions. On the other end of the scale, it is peculiar that O_2 is toxic to humans at too high a partial pressure. Although partial pressure danger limits in these two cases are approximately 0.1 and 2.0 atma, respectively, with a margin of safety a better range is 0.2 to 1.4 atma. Strictly speaking however, even exposure to the latter pressure, if prolonged, could be fatal.

Any diver, whether skin diving or using some means of air supply, experiences a steadily increasing pressure as he goes deeper in water. It turns out that the pressure increases by approximately one atm for each 33 ft (10 m) of depth increase by the diver. For a free dive, the high pressure existing some distance beneath the surface can be a real source of *squeeze*, wherein the diver is acutely aware of being compressed. Nevertheless, I have seen men free dive to 100 ft (30 m) with no apparent discomfort—and free divers have reached over twice this depth (Kenny, 1972). But the principle behind diving with an air supply is that the diver is at equilibrium with the ambient pressure at his depth. He breathes air (or another breathing mixture) that is at the same pressure as that of his surroundings. As long as he is pressure-equilibrated, there is no more discomfort from pressure than normally experienced on land when loaded down by miles of atmosphere.

But herein lies danger in various forms. If the depth reaches the equivalent of approximately 9 atm (10 atma with the atmospheric pressure), an average person is in danger of convulsions from O_2 poisoning. And yet I know a young diver who spent 10 minutes at 300 feet (90 m), breathing air, with no convulsions. This shows clearly that there are individual tolerances different from the average.

Nitrogen

There are two diving problems concerning "inert" nitrogen. *Nitrogen narcosis* is caused by the increased partial pressure of N_2, which causes the gas to act like an intoxicant or, depending upon a person's point of view, an anaesthetic. In some cases the diver can display ridiculous behavior such as offering his mouthpiece to passing fish or scrubbing himself with a piece of rough coral. On the other hand, he may simply pass out. In either case the affected diver is in danger of injury or death unless he can cope with the symptoms. Not only may different divers be affected in different ways, but the onset of problems may occur at different depths (pressures). I have vivid memories of my own first experience with this situation, sometimes known as "Martini's law," at 130 ft (40 m). Both construction and inspection divers, diving up to 200 ft (60 m) on the Barbers Point outfall project (#49 in Table 10-1), experienced repeated problems due to N_2 narcosis.

The most dreaded feature of N_2 poisoning for divers is the effect of nitrogen leaving the system too rapidly. Once gas enters the blood stream it is carried throughout the body tissue where it is absorbed. The amount of gas taken up by the tissue depends upon the type of tissue, the difference in concentrations of the gas between the blood and tissue, and time. As the diver descends, more N_2 (and O_2) enters his tissues; as he ascends, the N_2, which is not, of course, used in metabolic processes, must leave the tissues. The rate of ascent must be sufficiently slow that no N_2 bubbles form as that gas leaves the tissues. If bubbles do form, they can become trapped in body tissues, leading to excruciating pain and the very real possibility of disability, paralysis, and death. This situation is called the *bends* or *decompression sickness* (Type I). It killed one diver during the construction of the Point Loma outfall at San Diego, California

(#18). Another diver on this same job was left with permanent disability because of the bends, and a veteran diver on the Barbers Point project incurred partial paralysis through the same effect, after working at 210 ft (64 m) and breathing air. A second Barbers Point construction diver was also left with permanent damage from the bends as was one man from Mokapu (#47).

Decompression

The time taken by an average diver to decompress depends on the nature of his diving activities during the preceding twenty-four hours—how much time was spent at what pressures. Tables have been developed that present the average diver with information on his maximum allowable rate of ascent and decompression schedule. Since divers cannot be expected to move around with bulky tables under their arms, many use special decompression meters that provide the requisite (average) information. But many divers also simply plan their dive and associated decompression beforehand, knowing how long their tank of air will last. But this can be dangerous since many dives do not go as planned. A spear dropped over a ledge, an aggressive shark, deeper water than intended—these and many other factors mean alterations in plans.

The standard U.S. Navy (no-decompression) rate of diver ascent is 60 ft (18 m) per min. Decompression stops are made, if required, progressively at 50, 40, 30, 20, and 10-ft (15, 12, 9, 6, and 3-m) depths. According to the U.S. Navy decompression tables, which are for the diver breathing air, decompression stops are not necessary for the following maximum stays at the depths indicated: 50 min at 70 ft (21 m); 40 min at 80 ft (24 m); 30 min at 90 ft (27 m); 25 min at 100 ft (30 m). A diver at a depth of 100 ft (30 m), for example, for 40 min, should, besides the standard no-decompression ascent rate, spend 15 min at the 10-ft (3-m) decompression stop. The tables are good but not infallible. I know an experienced Honolulu diver who got "bent" with resulting permanent lower body paralysis on a recreational dive to 100 ft (30 m) when the tables were religiously followed and the diver was rushed into a recompression/decompression chamber when he displayed indications of central nervous system decompression sickness. It is wise not to push the tables to the limit.* Dive after dive to shallower depths can lead to problems, and the tables allow for such accumulation. I know of mild cases of the bends after repeated work dives in 70 ft (21 m) of water.

A diver complaining of itching or joint twinges or pains, displaying unusual rashes and swelling or extreme fatigue or loss of appetite, should be suspected to be suffering from decompression sickness and should be treated by recompression in a special chamber immediately, and there decompressed according to treatment tables. Medical personnel experienced in hyperbaric medicine should be present if possible.

---

*Beckman (1976) recommends use of the more conservative British Royal Naval Physiological Laboratory air diving tables rather than those of the U.S. Navy.

A recompression-decompression chamber (Chapter 10) serves functions other than treating "bent" divers, but it is the only reasonable way of dealing with an attack of the bends. The diver in pain should not and could not be reoutfitted and sent back down into the water. Piping, valves, dials, and meters on the outside of the chamber permit an operator outside it to control internal conditions. Voice communications between the operator and patient can be maintained.

One way of hastening the decompression process is to use pure O_2 in the last stages. N_2 can be eliminated about twice as fast when the diver breathes O_2 rather than air. This point is expanded upon in Chapter 10.

Breath-Holding

Besides everything else discussed thus far, the diver must be wary not to swim too rapidly to the surface because of the air trapped in his lungs. The air that occupies the lungs of the diver at a depth of 33 ft (10 m), for example, would occupy about twice the lung volume at the surface. A diver ascending rapidly without allowing the lung air to be expelled could suffer from one of the following disorders. *Pneumothorax* involves air being forced into the space between the lung and the chest wall. This air, with no escape route, must expand, and it does so by collapsing the nearby lung and pushing the lung and heart off to one side. The air can be released by means of a needle inserted into the cavity, but the results of pneumothorax can be fatal.

Air embolism, which can also result in death, involves air sacs and blood vessels which rupture under the action of the expanding air in the lungs. Bubbles of air are then forced into the capillaries within the lungs, thence to the heart and then on into the arterial circulation system. The bubbles can block arteries and cut off the blood supply to the tissues beyond. Often it is the brain that is cut off, and irreversible damage to brain tissue denied oxygen can occur in several minutes. Unless the victim is recompressed promptly to reduce the size of the bubbles and to permit blood to flow again, severe damage and ultimately death can follow.

Long-Term Effects of Diving

The acute effects of diving have been considered in the preceding paragraphs, but there are also long-term effects. One of these concerns hearing damage, thought to be due to the rushing sound of air entering the helmets of "hard-hat" divers (Section B-5). I know an "old pro," for example, who keeps one hand cupped behind his ear toward anyone talking to him.

Even mild cases of the bends can lead to permanent crippling at the joints and nerve impairment leading to loss of feeling. A very serious chronic hazard of diving without acute warning is *aseptic bone necrosis* (ABN) thought to be due to repeated compression and decompression, even to moderate depths. "Aseptic" means occurring in the absence of infection; "necrosis" refers to the pathological death of one or more cells, or a portion of tissue or organ, resulting from irreversible damage. When the structural damage associated with ABN extends to the joints, the results can be crippling. ABN has led to the early retirement of at least one veteran outfall construction diver.

B-3   SCUBA DIVING

Standard SCUBA Diving

Although this type of diving is by no means the oldest, it will be detailed first here since it is the simplest form of duration diving and is perhaps the most familiar to the public. In SCUBA diving the diver carries with him his source of air rather than being dependent upon a surface source and a hose to supply him and perhaps another hose to remove the exhaust gases. This source of air is a steel or aluminum tank filled with air under pressure (see Figure B-1). Although tank sizes vary, the most common is the 72-ft^3 (2.0 m^3) steel tank that is initially filled with air to a pressure of approximately 2500 psi (1724 N/cm^2). The volume of the tank is of course not 72 ft^3 (2.0 m^3). This number corresponds to the volume the amount of air in the tank would occupy at atmospheric pressure. The weight of air in the standard tank is about 5 lb (22 N).

How long a standard SCUBA tank filled with air at 2500 psi (1724 N/cm^2) will last depends on the person using it, the amount of work being done, the dive depth, and other factors. Let the volume rate of air usage per minute equal a factor times the pressure in atma. A reasonable range on the factor for (atmospheric) volume in ft^3 is about 0.5 to 1.5; for m^3 about 0.015 to 0.04. Many divers monitor their use of air through use of a pressure gage on the end of a hose connected to the top of the tank.

When empty, SCUBA tanks are usually neutrally buoyant in fresh or salt water. Almost all tanks have negative buoyancy when filled. They are carried in a frame that fits over the shoulders by means of a pair of curved bars (Figure 9-1) or around the shoulders by means of a pair of straps. In both cases a waist strap is a further link.

A valve at one end of the tank is designed to mate with a yoke assembly that is part of the air demand system supplying air to the diver. A proper seal between the yoke assembly and the tank valve is provided by a rubber O-ring gasket. The reduction of air pressure from that in the tank to that required by the diver is accomplished by the *regulator*. This usually has two stages. In the first stage, within the yoke assembly, the tank pressure is reduced through a spring-loaded piston to ambient pressure plus about 100 psi (69 N/cm^2). The second stage may be in the same casing or, in the common single-hose regulator, may be in the second assembly connected to the first stage by a suitable hose. It

Figure B-1 SCUBA divers working underwater

is the second stage of the regulator assembly that provides the "demand" function of the demand system and reduces the breathing gas pressure to ambient so that the diver remains properly pressure-equalized through changes in depth during the dive.

The standard form of SCUBA is open-circuit SCUBA. In this case, the gas exhaled by the diver is vented directly into the water. Since we tend to regard air as free, this is not a wasteful process. The great advantage to SCUBA is the maneuverability it affords the diver. The fundamental disadvantage is the limit imposed on a diver's time at depth imposed by the limited volume of air carried along.

There are various things about SCUBA diving, most mentioned in Section B-2, that make it mandatory that a prospective diver take a formal course to become certified. The National Association of Underwater Instructors (NAUI) is one certification body in the United States.

Equipment for SCUBA Diving

A diver needs more than just a tank for an underwater sortie. To permit vision underwater he wears a special mask consisting of a rubber frame with a plastic or glass face plate. This mask fits around the eyes and the nose. In conjunction with the mask many divers also carry a snorkel, a plastic or stiff rubber tube that enables a swimmer to breathe while swimming face down at the surface. If a diver runs out of air and must swim on the surface to a boat or to shore, a snorkel is indispensable, especially if the water is at all rough.

In particularly cold waters, a rubber hood is worn over the top and back of the head. Even in relatively warm waters, such as the 70 to 80°F encountered in Hawaii, most divers wear a "wet suit." This outfit, made of rubber in various thicknesses and worn directly over the skin, is not designed to prevent the penetration of water underneath the suit, but the gap is sufficiently small between the skin and rubber that the body can keep this water warm. A properly fitting wet suit should stop the circulation of water as a diver moves around. Both wet suit jackets and pants can be worn, but the latter is not as usual as the former. A "dry suit" prevents (ideally) any penetration of water into the area between the rubber suit and the skin. Such outfits make use of tight-fitting cuffs and are employed in very cold water.

Over the wet suit, it is advisable for divers to wear an inflatable life vest or *buoyancy compensator*. Such vests can be filled by blowing into a special tube or by pulling, in an emergency, a cord that releases the contents of a CO_2 cartridge. Air from the first stage of the regulator is also a possibility. When a diver is tired and out of air, and perhaps in the grip of a persistent current, it is a tremendous plus to be able to deploy a buoyancy arrangement.

A diver entering the water wearing a wet suit and SCUBA tank would probably find himself very "positive"—positively buoyant. To permit the diver to submerge more easily, it is normal that he wear a weight belt made of fiber passed through special flat lead weights that fits around the waist. A quick-release buckle at the front of the belt permits the diver to discard it quickly if a subsurface emergency should arise and he needs to maximize his rate of ascent to the surface—within the constraints of avoiding pneumothorax or air embolism.

It is advisable for divers to carry a knife. This is normally a leg knife whose scabbard is tied to the diver's leg with two belts. It should be tied inside the leg so that a weight belt, dropped in an emergency, does not become trapped in it. The knife is not a defensive weapon, but is very useful if one becomes entangled in rope or fine wire. It can also be used as a measuring device.

Flippers are worn by the diver to provide a more powerful means of propulsion through the water than his own feet. Some flippers have solid portions under the heel and some do not. In the latter case the diver will often wear "booties" inside the flippers. One design of rubber booties has a thick and ridged bottom. These booties can then be used for walking across rocks or a reef before or after diving, or for protecting the feet on the bottom when it is more convenient to walk around while working than to swim. See Figure B-1.

Depth gages, watches, and gloves are other items worn by some divers.

B-4 SYNTHETIC MIXED GASES FOR BREATHING

It is the oxygen in air that is used by the body in respiration. The nitrogen carrier accomplishes no useful purpose *per se*; in fact, as has been mentioned, its presence creates problems for the diver. Helium (He), another inert gas, has been successfully substituted for N_2 in a diving gas mixture known as *heliox*. Use of this mixture successfully eliminates nitrogen narcosis; in fact, there is no problem of this nature for helium down to a depth of over 1000 ft (305 m). Longer no-decompression times at all depths are also possible with heliox as long as the oxygen percentage is above 21%. For a heliox mixture with 60% He and 40% O_2, allowable no-decompression times are as follows: 85 min at 70 ft (21 m), 60 min at 80 ft (24 m), 45 min at 90 ft (27 m), and 35 min at 100 ft (30 m). But there is a price to be paid. For one thing, a diver breathing heliox speaks virtually unintelligibly (the Donald Duck effect). For another, helium removes body heat far faster than nitrogen; diver heating may thus become necessary.

Standard SCUBA rigs containing heliox *could* be used, but every diver exhalation would waste a quantity of expensive helium carrier. For this reason heliox is used in a system where all or part of the exhaust gas is recirculated to *scrubbers* that remove from it the carbon dioxide exhaled by the diver and then return it to the diver after a metered addition of O_2 or of fresh heliox. The *semi-closed system* is the one in which part of the exhaled gas is vented to the water. The *closed-circuit system* returns all of the exhaled gas to a breathing bag or canister. The standard SCUBA system, as mentioned earlier, is an open-circuit system.

The three systems listed in the previous paragraph are all "exhaust" systems. The gas *supply* is one of two types, a *demand supply system* or a *continuous flow system*. The former, used in SCUBA rigs, supplies air only when the diver inhales. The latter (Section B-5) provides a constant stream of breathing gas past the diver's face.

B-5 TETHERED DIVERS

Introduction

When a diver is dependent upon a supply of breathing gas from a source that he does not carry with him, normally on a vessel on the surface, he is said to be *tethered*, since he is at the end of a line through which his breathing gas passes. Another very likely line is an electrical cable, so that voice communication can be maintained between the diver and his assistant (*tender*) on the surface. The tender may be an apprentice diver, but the job of tender is a special classification. I will briefly discuss two separate tethered systems. With one system, the diver has great mobility and low stability; with the other, the diver does not have great mobility but is very stable. The latter case will be considered first.

Hard Hat Diving

The traditional type of tethered diving is *standard* or *hard hat diving*. Standard diving equipment ("heavy gear") is detailed in the U.S. Navy Diving Manual (1975). The diver wears a watertight suit that covers the whole body with the exception of head and hands. This outfit is made of vulcanized sheet rubber between layers of cotton twill. Overalls made of light canvas are worn over this dress to protect it against wear and chafing. Rubber cuffs at the wrists make waterproof seals. Gloves, too, are often used.

A heavy rubber gasket is fitted around the neck of the dress. A breastplate made of copper slips over the head of the diver and inside the flap provided by the gasket, and threaded studs around the edge of this breastplate protrude through reinforced holes in the gasket. The breastplate is shaped to fit comfortably over the shoulders, chest and back, and the neck portion has a threaded ring that mates with an interrupted ring inside the diver's helmet. Wing nuts are used to link the breastplate and diving dress. The helmet is of spun copper. The U.S. Navy version is shown in Figure B-2. Figure B-3 shows two tenders putting a hard hat on a construction diver. A watertight seal between helmet and breastplate is provided by a leather gasket set in a recess in the breastplate ring.

Figure B-2 U.S. Navy diving "hard hat" (Courtesy, Desco Corp.)

Figure B-3 Tenders putting "hard hat" on outfall construction diver

A safety lock on the back of the helmet prevents accidental separation of the helmet from the breastplate.

For vision the diver has three or four ports in the helmet. The port in front, the faceplate, is hinged and sealed over a rubber gasket by a swing bolt and wing nut. Brass guards protect all four ports in Figure B-2 from breakage.

There are two gooseneck (angled) fittings on the back of the helmet (see Figure B-3). The fitting on the diver's left is for attaching the intercommunication amplifier and the diver's lifeline cable. That on the right receives the safety air nonreturn valve, which in turn receives the diver's air hose. The valve prevents loss of air from the suit if the air hose bursts or air-supply system fails.

The diver's air hose is negatively buoyant; it is made of cotton-reinforced vulcanized rubber. This hose consists of two parts; a 3-ft (1-m) length runs from the nonreturn valve, under the diver's left arm (Figure B-4), then to an air-control valve hung on the diver's chest. A much longer air hose, secured to the breastplate by a rope lanyard extends from the air-control valve to the air

Figure B-4   Outfall construction diver leaving diving platform on trestle

supply. The lifeline and intercommunication cable are also secured to the breastplate by a lanyard before passing under the diver's right arm to its gooseneck fitting.

Air is pumped continuously to the diver. It is vented into the water through a regulating exhaust valve in the helmet. Because the suit and the large helmet are filled with air, the diver must wear a considerable amount of weight to assure negative buoyancy. There are two standard sources of weight, a belt weighing 84 lb (374 N) and shoes weighing 35 lb (156 N) a pair. Lighter boots and ankle weights are also possible.

If the breathing gas is heliox, then suitable tanks of gas, plus valves, regulators, tubing, and meters must be maintained on the support vessel. In helium-oxygen diving, the standard helmet is modified to recirculate part of the gas through a CO_2 scrubber in a canister on the outside of the helmet.

The communication cable links up the diver with a receiving-transmitting station topside. In some cases ("Northwest style") the diver communication link is in his breastplate rather than his helmet. Although general messages can be passed between a diver and his tender by using tugs on the hose, an audio communication link should be maintained in good working order. Signals by jerking on the diver hose are by no means failsafe. I have been told of one instance in shallow-water outfall construction work when the tender forgot to engage the air compressor feeding air under pressure to the tank from which the diver received his air supply. The diver's tugs on the hose were misinterpreted as a call for more hose. Because of this confusion, the diver was killed.

Band Mask and Hat Diving

The cumbersome nature of hard-hat (helmet) diving outfits has led manufacturers to develop lighter head systems that can be used in conjunction with a wet suit, weight belt, and fins, as with SCUBA diving.

The *band mask* is a full-face mask. Such a mask, with rubber body, is shown in Figure B-5. Another possible body material is fiberglass. The mask is worn against a face seal set in the front of a hood worn by the diver over his head, neck and near-neck areas of chest and shoulders. A "spider," consisting of straps extending over the back of the head draws the mask up tight against the seal. Inside the main mask is an oral-nasal mask providing another line of defense against water entry, and also providing for the possibility of vocal communication with the surface and, finally, reducing the possibility of CO_2 buildup in the mask.

Figure B-5   Band mask (Courtesy, Desco Corp.)

Divers have three general complaints against the band mask:

1. communication is marginal;
2. there is no head protection; and
3. water penetrates into the mask from under the hood.

These problems have largely been eliminated by the development of new types of lightweight diving hats.

Around his neck the diver places a yoke that sits on his shoulders and is secured around his body by straps which can be part of a special chafing suit. Shortly before the diver enters the water, his tender sets a special hat (Figure B-6) on the yoke. This hat, covering the head completely, is turned and locked in place on an interrupted thread ring or secured in some other manner. A favorite material for such hats is fiberglass, strong but light (see Figure B-7). Examples are the Advanced Diving (Gretna, Louisiana) air and helium hats, and the Advanced Diving faceplate, made of Lexan plastic, is literally bulletproof.

Gas supply can be either demand or continuous. If air is used it is vented to the water. If heliox is the breathing mixture, part of the waste gas can be scrubbed by using canisters mounted on the diver's back before being rerouted to the diver with additional heliox. A communications link between diver and tender is normal. This cable would be bundled with a lifeline and the breathing gas supply hose, which is customarily made of flexible reinforced rubber.

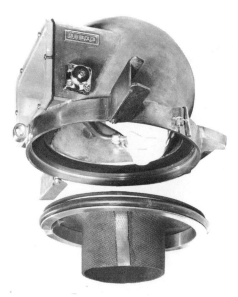

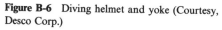

Figure B-6   Diving helmet and yoke (Courtesy, Desco Corp.)

Figure B-7   Com Hat I (Courtesy, Commercial Diving Division of U.S. Divers Co.)

A two-stage air compressor driven by a small diesel engine is normally used to supply air. It is used with a storage, or volume, tank, of minimum volume 60 ft^3 (1.7 m^3) to serve both as a steady air-flow supply and a reserve supply should the compressor fail. Pressure in the system is maintained at least 50 psi (34.5 N/cm^2) above the ambient pressure at the diver's depth. Filtering at the outlet from the tank purifies the air, and a check valve at the mask or hat prevents flooding or squeeze should the air hose become severed or the air flow stopped.

B-6 REFERENCES

BECKMAN, E. L. (1976): "Recommendations for Improved Air Decompression Schedules for Commercial Diving," University of Hawaii, Honolulu, Hawaii, Sea Grant Program, *Technical Report 76-02* (October).

KENNY, J. E. (1972): *Business of Diving*, Gulf Publishing Co., Houston.

U.S. Navy Department (1975): *U.S. Navy Diving Manual*, Washington, D.C.

B-7 BIBLIOGRAPHY

DOAK, W., and J. McKENNEY (1970): "Getting to Know Your Decom Meter," *Skin Diver*, Vol. 19, No. 4 (April), pp. 22–26, 28.

EDMONDS, C., C. LOWRY, and J. PENNEFATHER (1976): *Diving and Subaquatic Medicine*, Diving Medical Centre, Mosman, N.S.W., Australia.

HUTCHINSON, W. R. (1976): "Aseptic Bone Necrosis: A Potential Sport Diving Problem," *Skin Diver*, Vol. 25, No. 4 (April), pp. 40–41.

McKENNEY, JACK (1976): "The Ins and Outs of Buoyancy Compensators," *Skin Diver*, Vol. 25, No. 4 (April), pp. 42–47.

MOUNT, TOM (1970): "The Decom Meter vs U.S. Navy Tables: A Comprehensive Comparison of Two Methods of Computing Decompression," *Skin Diver*, Vol. 19, No. 11 (November), pp. 32–35.

SHILLING, CHARLES W., and MARGARET F. WERTS (1971): *An Annotated Bibliography on Diving and Submarine Medicine*, Gordon and Breach Science Publishers, New York.

SHILLING, CHARLES W., and MARGARET F. WERTS (1973): *Underwater Medicine and Related Sciences: A Guide to the Literature*, 2 vols., IFI/Plenum, New York.

Glossary

Table C-1: Symbols Used in Chapters 7 and 8

Symbol	Description	Equation or Figure	English System	SI System
a_e	area of jet at vena contracta	—	ft^2	mm^2
a_H	horizontal acceleration	(8-46)	ft/s^2	m/s^2
a_j	minimum area of jth port	—	ft^2	mm^2
a_o	minimum area of specific port	—	ft^2	mm^2
A	projected area of immersed object perpendicular to approaching velocity vector	—	ft^2	m^2
A'	cross-sectional pipe area	—	ft^2	m^2
A_j	cross-sectional diffuser area at jth port	—	ft^2	m^2
b	buoyancy flux per unit length of diffuser	(7-30)	ft^3/s^3	m^3/s^3
b_0	shortest distance between two wave orthogonals in deep water	—	ft	m
b_*	shortest distance between two wave orthogonals in any water depth	—	ft	m
B	width of source	—	ft	m

566

Table C-1 (cont.): Symbols Used in Chapters 7 and 8

Symbol	Description	Equation or Figure	Units English System	Units SI System
$\mathbf{B_e}$	wastewater disposal parameter	(7-11)	—	—
$\mathbf{B'_e}$	energy-ratio parameter	(7-8)	—	—
B_0	total width of slot(s) in diffuser	(7-40)	ft	m
$\mathbf{B_0}$	wastewater disposal parameter	(7-12)	—	—
c	pollutant concentration	7-23	—	—
c_{max}	pollutant concentration along centerline of surface plume	7-23	—	—
c_0	initial pollutant concentration	7-23	—	—
C_c	contraction coefficient	(7-2)	—	—
C_D	drag (force) coefficient	(8-8) and (8-23)	—	—
C_H	horizontal force coefficient	(8-13)	—	—
C_I	inertia (force) coefficient	(8-25) and (8-27)	—	—
C_j	discharge coefficient	(7-17)	—	—
C'_j	contraction coefficient	(7-16)	—	—
C_L	lift (force) coefficient	(8-10)	—	—
C_m	added mass coefficient	(8-27)	—	—
C_{max}	maximum-horizontal-force coefficient	(8-28)	—	—
C'_{max}	maximum-horizontal-force coefficient for angled pipe	(8-35)	—	—
C_V	vertical force coefficient	(8-14)	—	—
d	diameter of sediment particle or rock	—	ft	mm
d_e	effective diffuser port diameter	(7-3)	ft	mm
d_j	diameter of jth port from outer end of diffuser	—	ft	mm
d_0	diffuser port diameter	—	ft	mm
D	outside diameter of outfall pipe	—	ft	mm
D_j	inside pipe diameter at jth port	—	ft	m
e	natural base for logarithms (2.71828…)	—	—	—
E_j	total net head at diffuser port	(7-15)	ft	m
$f\{\ \}$	function of	—	—	—
f	Darcy-Weisbach friction factor	—	—	—

Table C-1 (cont.): Symbols Used in Chapters 7 and 8

Symbol	Description	Equation or Figure	Units English System	SI System
f_C	Coriolis parameter	(8-4)	s^{-1}	s^{-1}
f_j	Darcy-Weisbach friction factor	(7-22)	—	—
F	horizontal wave-induced force on object parallel to velocity vector	(8-22)	lb	N
F	current flow Froude number	(7-31)	—	—
F_D	drag force	(8-8) and (8-23)	lb	N
F$_e$	effective densimetric Froude number	(7-5)	—	—
F_H	horizontal force, perpendicular to pipe, for angled current	(8-13)	lb	N
F_I	inertia force	(8-25)	lb	N
F_{max}	maximum horizontal force on pipe	(8-28)	lb	N
F'_{max}	maximum horizontal force on angled pipe	(8-35)	lb	N
F$_0$	densimetric Froude number	(7-6)	—	—
F$'_0$	modified densimetric Froude number	(7-50)	—	—
F_N	normal force between pipe and boundary	—	lb	N
g	acceleration due to gravity	—	ft/s^2	m/s^2
g'	adjusted gravitational acceleration	(8-38)	ft/s^2	m/s^2
g'_0	buoyant body acceleration	(7-1)	ft/s^2	m/s^2
G	relative pipe clearance	(8-12)	—	—
G_A	Airy theory acceleration factor	8-16	—	—
G_V	Airy theory velocity factor	8-16	—	—
h	undisturbed water depth	—	ft	m
h_j	vertical distance between receiving water surface and center of jth port	—	ft	m
h_1	head loss	(7-22)	ft	m
h_0	water depth at continental shelf break	—	ft	m
H	wave height	—	ft	m
H_0	wave height in deep water	—	ft	m
H_*	depth of diluted wastewater cloud	—	ft	m

Table C-1 (cont.): Symbols Used in Chapters 7 and 8

Symbol	Description	Equation or Figure	English System	SI System
j	index ($j=1,2,\ldots,N$) for numbering diffuser ports from offshore end	—	—	—
J	ratio of upstream velocity head to total net head	(7-19)	—	—
J'	ratio of downstream velocity head to total net head	—	—	—
k	surface wind stress parameter	—	—	—
k_0	exponential decay coefficient	(7-70)	s^{-1}	s^{-1}
k_1	minor energy loss coefficient	(7-13)	—	—
k_*	wave number	$k_*=2\pi/\lambda$	ft^{-1}	m^{-1}
K	bottom friction parameter	(8-5)	$ft^{1/3}$	$m^{1/3}$
K_H	wave height change coefficient	(8-18)	—	—
K_{max}	coefficient for maximum vertical wave-induced force	(8-30)	—	—
K'_{max}	coefficient for maximum vertical wave-induced force on angled pipe	(8-36)	—	—
K_r	wave refraction coefficient	(8-19)	—	—
K_s	wave shoaling coefficient	—	—	—
l	average distance between diffuser ports	—	ft	m
l_*	specified length of pipe	—	ft	m
L	scale of patch of diffusing substance	—	ft	m
L_1	outfall length	—	ft	m
L_2	length of diffuser	—	ft	m
L_x	width of surface plume	7-23	ft	m
m	sea floor slope	—	—	—
m_0	momentum flux parameter	(7-42) and (7-52)	—	—
M	ratio of vertical momentum of source to horizontal momentum of ambient current	—	—	—
n	Manning's roughness factor	—	$ft^{1/6}$	$m^{1/6}$
N	number of diffuser ports	—	—	—
p	pressure within pipe	7-2	lb/ft^2	N/m^2
p_a	ambient pressure	—	lb/ft^2	N/m^2
P_L	lift force	(8-10)	lb	N

Table C-1 (cont.): Symbols Used in Chapters 7 and 8

Symbol	Description	Equation or Figure	Units English System	SI System
P_{max}	maximum vertical wave-induced force on pipe	(8-30)	lb	N
P'_{max}	maximum vertical wave-induced force on angled pipe	(8-36)	lb	N
P_V	vertical force for angled pipe	(8-14)	lb	N
q	discharge per unit length of diffuser	$q = Q/L_2$	ft^2/s	m^2/s
q_j	discharge through jth port	7-2	ft^3/s	m^3/s
q_0	discharge through specific port	(7-4)	ft^3/s	m^3/s
Q	total diffuser discharge	—	ft^3/s	m^3/s
Q_j	pipe discharge upstream of the jth port	7-2	ft^3/s	m^3/s
r	horizontal coordinate	7-23	ft	m
R	radius to maximum winds in a hurricane	—	miles	km
$\mathbf{R}$	steady flow Reynolds number	(8-11)	—	—
$\mathbf{R}_j$	steady pipe flow Reynolds number	(7-23)	—	—
$\mathbf{R}_w$	Reynolds number for wave situations	(8-24)	—	—
$\mathbf{R}_*$	boundary Reynolds number	(8-40)	—	—
s	specific gravity of water	—	—	—
s_s	specific gravity of solid material	—	—	—
S	vertical coordinate with origin at sea floor	8-17	ft	m
$\bar{S}$	ratio of unit current and effluent discharges	(7-32)	—	—
S_c	centerline dilution of jet or plume	—	—	—
S_{Ct}	Coriolis tide	(8-2) and (8-3)	ft	m
S_m	minimum effluent dilution at water surface	(7-35) and (7-37)	—	—
S_{pt}	pressure tide	—	ft	m
S_{st}	storm tide	(8-1)	ft	m
S_t	terminal jet dilution	(7-45), (7-47) and (7-55)	—	—
S_{wt}	wind tide	(8-2) and (8-3)	ft	m
S_*	storm surge less pressure tide	(8-2) and (8-3)	ft	m

Table C-1 (cont.): Symbols Used in Chapters 7 and 8

Symbol	Description	Equation or Figure	Units	
			English System	SI System
t	time	—	s	s
t_{90}	elapsed time for 90% coliform dieaway	—	min	min
T	wave period	—	s	s
T_0	stratification parameter	(7-41)	—	—
T_0'	modified stratification parameter	(7-51)	—	—
T_*	dimensionless shear stress	(8-39)	—	—
u	current speed	—	ft/s	m/s
u_e	effective jet velocity through port	—	ft/s	m/s
u_j	jet velocity through jth port	7-2	ft/s	m/s
u_{max}	theoretical peak horizontal wave-induced flow speed	—	ft/s	m/s
$\dot{u}_{max}$	theoretical peak horizontal wave-induced flow acceleration	—	ft/s^2	m/s^2
u_0	nominal jet exit velocity	(7-7)	ft/s	m/s
u_s	peak horizontal flow speed under solitary wave	(8-36)	ft/s	m/s
U	Ursell parameter	(8-33)	—	—
U_{max}	actual near-bottom peak horizontal flow speed under wave	—	ft/s	m/s
$\dot{U}_{max}$	actual near-bottom peak horizontal flow acceleration under wave	—	ft/s^2	m/s^2
V	mean flow speed for steady flow in pipe or channel	—	ft/s	m/s
$\forall$	volume of pipe	(8-26)	ft^3	m^3
$\bar{V}$	depth-averaged alongshore current speed at any time	(8-6)	ft/s	m/s
V_f	forward speed of hurricane	—	ft/s	m/s
V_i	minimum steady flow speed required to cause sediment movement	(8-17)	ft/s	m/s
V_j	flow speed in diffuser upstream of jth port	7-2	ft/s	m/s
V_{st}	steady-state, depth-averaged alongshore current	(8-7)	ft/s	m/s
w_t	terminal width of jet or plume	(7-49)	ft	m
W	wind speed at 10-m height above sea surface	—	ft/s	m/s

Table C-1 (cont.): Symbols Used in Chapters 7 and 8

Symbol	Description	Equation or Figure	Units English System	Units SI System
W_B	buoyant weight of pipe	—	lb	N
W_*	weight of pipe	—	lb	N
x	horizontal coordinate	7-23	ft	m
y	depth of flow in channel	—	ft	m
z	vertical coordinate with origin at still water surface	—	ft	m

Table C-2: Greek Symbols Used in Chapters 7 and 8

Symbol	Greek Letter	Description	Equation or Figure	Units English System	Units SI System
α	alpha	coefficient	(7-58)	$ft^{2/3}/s$	$cm^{2/3}/s$
α_H	alpha	breaking wave factor for peak horizontal pipe forces	—	—	—
α_V	alpha	breaking wave factor for peak vertical pipe forces	—	—	—
β	beta	dimensionless parameter	(7-63)	—	—
β_1	beta	empirical factor	(8-31) and (8-32)	—	—
β_1'	beta	regression estimate of empirical factor	(8-34)	—	—
β_3	beta	empirical factor	(8-21)	—	—
β_s	beta	factor	(8-44)	—	—
γ	gamma	specific weight of water	—	lb/ft^3	N/m^3
γ_a	gamma	specific weight of receiving water	—	lb/ft^3	N/m^3
γ_d	gamma	specific weight of effluent	—	lb/ft^3	N/m^3
γ_s	gamma	specific weight of solid material	—	lb/ft^3	N/m^3
γ_*	gamma	specific weight of receiving water at level of diffuser port	—	lb/ft^3	N/m^3
δ	delta	factor	(7-46)	—	—
Δ	delta	clearance of edge of pipe from bottom	—	ft	m
Δh	—	change in pressure head	—	ft	m

Table C-2 (cont.): Greek Symbols Used in Chapters 7 and 8

Symbol	Greek Letter	Description	Equation or Figure	English System	SI System
Δm	—	infinitesimal mass	—	slugs	kg
Δp	—	central pressure reduction in a hurricane	—	—	mb
Δt	—	time increment	—	s	s
$\Delta \forall$	—	volume storage	(7-28)	ft^3	m^3
ϵ	epsilon	diffusion coefficient	(7-59)	ft^2/s	cm^2/s
ϵ_0	epsilon	reference diffusion coefficient	(7-62)	ft^2/s	cm^2/s
ϵ_*	epsilon	average height of pipe wall roughness	—	ft	m
ζ	zeta	angle wind makes with perpendicular to coast	—	—	—
η	eta	wave ordinate	8-17	ft	m
η_c	eta	ordinate at wave crest	8-17	ft	m
η_t	eta	ordinate at wave trough	8-17	ft	m
θ	theta	angle of flow or wave orthogonal to line of a pipe	—	—	—
$\bar{\theta}$	theta	average angle of attack	—	—	—
λ	lambda	wave length	8-17	ft	m
$\bar{\lambda}$	lambda	average wave length over certain length of outfall	—	ft	m
λ_0	lambda	wave length in deep water	—	ft	m
μ	mu	dynamic viscosity of water	—	lb-s/ft^2	N-s/cm^2
μ_0	mu	volume flux parameter	(7-43) and (7-53)	—	—
μ_s	mu	coefficient of static friction	—	—	—
μ_r	mu	dilution parameter	7-19 and 7-22	—	—
ν	nu	kinematic viscosity of water	$\nu = \mu/\rho$	ft^2/s	cm^2/s
ξ	xi	vertical coordinate, positive upwards, with origin at center of port	7-16	ft	m
ξ_{max}	xi	maximum height of rise of jet in stratified fluid	(7-44), (7-48) and (7-54)	ft	m
π	pi	3.14159...	—	—	—
ρ	rho	(mass) density of water	—	slugs/ft^3	kg/m^3
ρ_a	rho	receiving water density	—	slugs/ft^3	kg/m^3
ρ_d	rho	density of wastewater	—	slugs/ft^3	kg/m^3

Table C-2 (cont.): Greek Symbols Used in Chapters 7 and 8

Symbol	Greek Letter	Description	Equation or Figure	English System	SI System
ρ_*	rho	density of receiving water at level of diffuser port	7-16	slugs/ft^3	kg/m^3
σ	sigma	standard deviation of concentration profile	—	ft	m
σ_r	sigma	local standard deviation in horizontal direction	—	ft	m
σ_z	sigma	local standard deviation in vertical direction	—	ft	m
Σ	sigma	summation	—	—	—
τ	tau	shear stress exerted by steady flow on boundary	(8-41)	lb/ft^2	N/m^2
τ_d	tau	minimum shear stress required to initiate movement of particle of size d	—	lb/ft^2	N/m^2
ϕ	phi	latitude	—	—	—
ϕ_t	phi	height of rise parameter	(7-44)	—	—
ψ	psi	adjusted period parameter	(8-20)	—	—
ω	omega	radian frequency	$\omega = 2\pi/T$	s^{-1}	s^{-1}
Ω	omega	angular velocity of the earth	—	s^{-1}	s^{-1}

Table C-3: Abbreviations for Units

Abbreviation	Units	Abbreviation	Units
atm	atmospheres (of pressure)	gpcd	U.S. gallons per capita per day
atma	atmospheres absolute	gpd	U.S. gallons per day
BTU	British thermal units		
cal	calories	hp	horsepower
cc	cubic centimeters	Hz	Hertz (cycles per second)
cfs	cubic feet per second		
		in.	inches
cm	centimeter	kg	kilogram
fpm	feet per minute	kHz	kiloHertz
fps	feet per second	km	kilometer
ft	feet	ℓ	liter
gm	gram(s)	lb	pound

Table C-3 (cont.):

Abbreviation	Units	Abbreviation	Units
m	meter(s)	ppt ($^o/_{oo}$)	parts per thousand by weight
mb	millibar		
mg	milligrams	psf	pounds per square foot
Mgd	million U.S. gallons per day		
		psi	pounds per square inch
MHz	megaHertz	psia	pounds per square inch absolute
mℓ	milliliter		
mm	millimeter	rpm	revolutions per minute
mmho	(conductivity in) millimhos	s	seconds
		μ	microns
N	Newtons	μs	microsecond
n.m.	nautical miles		
ppm	parts per million by weight		

Index

A

ABN, 555–56
Abyssal depths, 17
Accelerometer, 173, 232
Acoustic velocity:
  of rock, 227
  of sea water, 206, 223–24, 227
  of sediments, 227
Activated carbon, 77
Activated sludge treatment, 65, 69–70
Acute toxicity bioassay, 110–16
Acute toxicity, sources of information, 117
Added mass coefficient, 387–88
Adsorption, 77
Advanced primary treatment, 62
Advanced waste treatment, 75–78
Advection, 42, 318
Aeration in wastewater treatment, 62, 69–70
Aerobacter aerogenes, 58
Aerobic processes, 55
Aerobic sludge digestion, 74
Aesthetics, 79, 96
"A" frame:
  on anchor scow, 472
  on hydraulic dredges, 457
  on oceanographic ships, 169
Agreement, construction (see Contract, construction)
Agulhas Current, 31

Aids to navigation, floating, 547–48
Air:
  compressor, 565
  as diver breathing gas, 554
  embolism, 555
  lift, 494, 502
  major constituents, 552
  use by body, 552
Airy wave theory:
  major assumptions in, 382
  near-bottom water motion, 382–84, 391, 393
  pressure predictions, accuracy of, 177, 302
  water motion predictions, accuracy of, 382–83, 393
Alaska, Gulf of, 175
Alenuihaha Channel, 45
Algae:
  blue-green, 123, 243
  brown, 27, 243
  golden-brown, 243
  green, 27, 243
  red, 27, 243
  sampling, 256–57, 263
  yellow-green, 243
Aliquot, 256, 260
Alongshore currents, 37
American Society for Testing and Materials, 348 (*footnote*), 427 (*footnote*)

American Water Works Association, 348
Ammonia stripping, 77
Anaerobic digestion of sludge, 72–74
Anaerobic processes, 56
Analog instrument, 159
Anchor-first mooring, 168
Anchor-last mooring, 168
Anchoring, sub-bottom information for, 229
Anchors:
  clump, 163
  for crane barge, 472
  Danforth, 162–63
  deploying, 168–69
  dragging, 502–03
  for dredges, 457–58
  for moored buoy systems, 162–63
  for pipe bottom-pull barge, 465, 467
  pipe damage due to, 356–57, 365, 520
  scow, 458, 472
Angle of attack, 386, 391
Animals, marine, 423–44
Anode systems:
  galvanic, 354
  impressed current, 354
Anodes, sacrificial:
  on moored buoy systems, 165
  on outfalls, 354

Anticyclonic disturbance, 30
Application factor, 117
Aquaculture, wastewater, 84
Armor rock:
    classification of, 489
    danger of dropping, 489
    danger of pipe damage from, 353
    grading of, 490, 492
    impracticality of some plans and
        specifications, 491–92
    loading of rock barge, 494
    need for, 362, 392–93
    placement methods, 490–94
    provisioning of crane barge, 494
    removal of, 494
    sizes of, 489–90
    stability of, 393–99
    weights of, 489–90
Artificial sea water (*table*), 18
Aseptic bone necrosis, 555–56
Association of Diving Contractors,
    495–96
ASTM (*see* American Society for
    Testing and Materials)
Astronomical distances:
    earth-moon, 540
    earth-sun, 540
Atoll, 42
ATP, 22
Autotape:
    accuracy, 545
    approach, 544–45
    in bathymetric surveys, 224, 544
    cost, 545
    in side-scan sonar surveys, 226
    soil sampling stations, 544
Autotroph, 22
AWT (*see* Advanced waste treatment)

B

Backfill, pipe trench:
    erosion of, 364–65, 377
    materials, optional, 343, 362
    by natural means, 498
    pipe damage from rock, danger
        of, 353
Backshore, 38
Bacteria:
    coliform (*see* Coliform bacteria)
    definition, 57
    pathogenic (*see* Pathogens)

Bacteriophages, as indicator
    organisms, 131
Baja California, 23, 529
Ballast rock:
    configuration of matrix, 342
    danger of dropping, 489
    for diffusers, 493
    grading of, 490
    as a matrix, 399
    need for, 392–93
    placement methods, 490–94
    sizes of, 489–90
    weights of, 490–91
Band mask, 563–64
Bangalore, 461
Bar, 38, 42
Barbers Point, Hawaii:
    Hawaiian Independent Refinery
        Inc., oil pipelines, 240, 362, 485,
        502
    Standard Oil Co., oil pipelines,
        358–59, 392, 395, 519
Barbers Point outfall:
    bedding stone, placing of, 489
    the bends, diver problems during
        construction, 554
    bid bond, 433
    bid opening, 436
    bids, 436
    blasting beyond trestle, 460–61
    construction delay requirement,
        441
    construction progress payments,
        441
    corers, use for, 232
    cost of obtaining plans and
        specifications, 433
    diving survey and probing (*foot-
        note*), 234
    documents for bidders, 432
    EIS, review of, 138–39
    electronic positioning, use for
        barge, 544
    format for estimated costs, 431
    hydraulic energy dissipation, 300
    Local Notice to Mariners, 462
    lump-sum contract statement, 435
    major characteristics of, 452
    money left on the table, 434
    nitrogen narcosis problems during
        construction, 553
    pipe damage during construction,
        500

Barbers Point outfall (*cont.*):
    pipe flow velocities, predicted,
        296–98
    port discharges, predicted, 296–98
    pre-bid meeting, 432
    problems with beveled pipe sec-
        tions, 500
    public hearing, 140
    removal of pipe sections, 500–01
    seismic profiling surveys, 228
    side-scan sonar surveys, 226
    sizing of protective rock, 397
    submerged drilling rig, 235–36
    time of project completion, 433
    trestle, construction, 478
Barges, construction:
    crane, 471–77
    mooring of, 465, 472
    pipe-lay, 470–71
    pipe-pull, 465, 467
    rock, 492, 494
Barrier reef, 42
Bathymetric chart (*see* Charts, nauti-
    cal)
Bathymetric surveys for outfalls, 224
Bathymetry:
    definition, 222
    echo-sounding, determination by,
        223–24
    electronic positioning during
        survey, 544–45
    by fathometer, 223–24
    by lead-line, 222–23
    by side-scan sonar, 225
    storm surge, effect on, 368, 370–71
    wave breaking, effect on (*footnote*),
        392
    wave propagation, effect on, 380
Bathythermograph, 201–02, 526
Bays, 38
Beach:
    berms, 38
    characteristics of, 38–40, 239–40
    profiles, 241
Beach erosion:
    cycles of, 40, 239–40
    physics of, 40
    problems for outfalls, 240–41
    surveys for outfalls, 240
Bearing of object, 548–49
Beaufort numbers (*table*), 35
Bedding stone:
    dropping of, 488–89

Bedding stone (*cont.*):
  dumping of, 475
  placing, 488–89
  sluicing of, 475–76, 488
  for spreading pipe load, 366
  through tremie pipes, 489
  washing under pipe, 475
Bell, diving:
  Horse, on the, 476–77
  spherical, 493–94
Bell, pipe, 345, 401, 471, 477
Bends, the:
  manifestations of, 533–54
  outfall construction divers, effects on, 553–54
  source of, 553
  treating in chamber, 555
Beneficial uses of marine waters, 96
Benthos, 23
Berm, beach, 38
Bidding, competitive:
  bid opening, 435–36
  bid preparation, 432–35
  disqualification of bidders, 436
  nature of, 423
  notice inviting bids, 426, 432
  possible problems with low bidder, 436
  pre-bid meeting, 432
  successful bidder, 435–36
Binary fission (*footnote*), 22
Bioassay, toxicity, 110–18
Bioassays:
  acclimation of test organisms, 113
  acute toxicity, 110–16
  batch, 112
  choice of test organisms, 110, 117–18
  continuous-flow, 112
  static, 112
Biochemical oxygen demand:
  removal in wastewater treatment processes, 61–62, 68, 70, 78
  test, 55
  of wastewaters, 55
Biogenous sediment, 38
Biological:
  aeration, 69–70
  filtration, 65
  tower, 68–69
  treatment, 65–70, 124
Biomagnification, 24

Biomass:
  algal, 253, 256, 260, 263
  definition, 22
  per species, 256
Biostimulation, 123–24
Blasting, underwater:
  debris, removal of, 459–60
  deep cuts for outfalls, 228
  direct effects on marine biota, 516
  dynamite, 460–61
  indirect effects on marine biota, 516
  regulations (*footnote*), 442
  shaped charges, 458–61
  for submarine pipelines, 458–64
Blocks (*see* Pulleys)
Bloom:
  algal, 123–24
  plankton, 25
Blow count, 230
BOD (*see* Biochemical oxygen demand)
Bolts, draw, 349, 472
Bond:
  bid, 433
  contract, 434
  payment, 434
  performance, 434
  surety, 434
Boston Whaler, 170, 249, 257
Bottles, sampling:
  collection of water sample, 202–04, 264
  contamination of samples, 203, 263–64
  Nansen, 203–04
  rod and reel, handling by, 264
Bottom samplers, 231
Bottom-pull, pipe-laying method:
  Annacis outfalls, 299
  layout of pipe strings, 465
  for longitudinally-post-tensioned concrete pipe, 351
  for non-straight pipes, 469
  outfalls laid, 469
  pipe route, 467
  pipe stability, 467–68
  potential problems, 465, 467–69
  protection of pipe, 467
  pulling head, 468
  releasing sources of buoyancy, 467
  required land area, 355, 464
  submerged pipe weight, 467–68

Bottom-pull, pipe-laying method (*cont.*):
  types of pipes, 355
  winch power for, 465, 467
Boulders (*table*), 39
Boundary layers, 396
Box samplers, 253
Breaker:
  plunging, 36
  spilling, 36
  surging, 36
Breaking waves, 36
Breaking wave forces on pipes, 393
Breakwaters, 394–95
Breathing gas, diver:
  air, 554, 556, 561, 564–65
  filtering, 565
  heliox, 559, 562, 564
  purifying, 565
  scrubbing, 559, 562, 564
Bristle worm (*see* Polychaete worms)
British Columbia ocean water quality control plan, 106–09
Brooks' dispersion models, 321–25
Brown algae, 27, 243
BT (*see* Bathythermograph)
Budding, coral, 25
Budget estimating, construction (*see* Cost estimating, construction)
Buoyancy of steel pipe (*table*), 346, 361
Buoyancy flux, unit, 303
Buoyant spreading, 287, 317
Buoyant weight, 376
Buoys, marker, 163–64
Buoys, subsurface, 170–71
Buoys, surface:
  life of, 163–67, 174–75
  notification regarding, 167
  troubles with shipping, 166
  waverider, 173–75
Buoy systems, moored, 161–72
Burial, outfall:
  backfill, trench, 343, 362
  configurations, 342
  need for sub-bottom information, 229

C

Cable, torque-balanced, 166
CAFE-1, circulation computer model, 327
Calcarenite sediments, 38

California Lambert zones, 541
California ocean water quality control plan, 104–07
California State government:
  Marine-Estuarine Technical Committee, 195
  Navigation and Ocean Development, Dept. of, 177
  Regional Water Quality Control Boards, 119–20, 525
  State Water Resources Control Board, 119–20, 195
California waters:
  minimum initial wastewater dilutions, 105, 317
  wave data for, 177
Cameras, underwater, 234, 256–57
Capitella capitata, 108, 528
Capstan, 191
Carbohydrates, 21
Carbon dioxide, dissolved in sea water, 18
Carnivores, 26
Carrier gas, 552
Cathodic protection, 354–55
"C" clamp, 482
Celerity, wave, 35
Centrifuges, use in dewatering sludge, 71
CERC (see Coastal Engineering Research Center)
Chamber:
  decompression, 496, 555
  recompression, 555
Change orders, construction, 428, 442–43
Chart recorders, 158–59, 161, 179, 227
Charts, nautical:
  coastal, 542–43
  coordinates, 542
  definition of, 542
  distribution of, 542
  length scale, 542
  Mercator, 378
  navigational hazards, 543
  position from horizontal sextant angles, 547
  preparation for marine survey, 545
  projection distortion, 542
  scale of, 542
  sea floor, nature of, 543

Charts, nautical (cont.):
  use of, 5, 209
  water depth contours, 542–43
  water depths, 542–43
Charts, synoptic, 377–79
Check valves, 519
Chemical analyses, possible contamination, 264
Chemical treatment of wastewater, 62–63
Chemostat, 112
Chloramines, 64
Chlorination of wastewaters:
  procedures, 63–64
  sources of chlorine, 64
  validity of, 131–32
Chlorine residual, total, 64
Chlorinity of sea water, 205
Chlorophyll:
  a-biomass relationship, 260
  definition, 21
  link with turbidity, 267
  types, 243
Chronic toxicity of wastewaters, 117–18
Circles:
  great, 378, 540
  small, 540
Clarification, 60
Clarifier:
  primary, 61–62
  secondary, 66, 70
Clarity, water, 266–68
Clarke-Bumpus sampler, 261
Clay:
  engineering characteristics of, 366
  particle size characteristics (table), 39
Clean Water Act of 1972 (footnote), 103
Clean Water Restoration Act of 1966, 99
Clump anchor, 163
Coastal depths, 17
Coastal Engineering Research Center:
  information on flow-induced sediment instability, 396
  wave data gathering, 173
  wave data presentation, 180
  wave data processing, 179
Coastal Zone Management Act of 1972, 102 (footnote), 139
Cobbles, size of (table), 39

Coefficient of static friction, 376
Coliform bacteria:
  die-away in sea water, 126–29
  disappearance in sea water, 126–29
  standards for receiving waters, 131–32
Coliphage, 131
Comminutor, 60
Compass:
  deviation, 542
  gyro, 542
  hand-bearing, 209
  magnetic, 541
  rose, 542
Compensation depth, 22
Composting, sludge, 75
Computer, for diffuser design calculations, 295–96, 298
Concentration, contaminant, 318–27
Concrete, attack by acids, 348–49
Concrete-covered steel pipe, 347–48, 358–59, 502
Concrete, tremie, 342, 362, 521
Concrete pipe, longitudinally-post-tensioned, 351
Concrete pipe jacket, 347–48, 358–59, 502
Concrete-steel cylinder pipe, 344, 348, 350
Concrete weight coating, 347–48, 376, 465
Conductivity of sea water, 205–06
Conglomerate rock, 38
Conservative substance, 317, 323–25
Construction cost escalation, 430
Construction cost index, ENR, 430
Construction effects on the marine biota, 516–18
Construction equipment:
  rental costs, 428–29
  rental of, 428–29
Construction manager, 424–25
Construction projects:
  engineering, 426
  heavy, 426
  insurance, 442
  payment for progress, 427
  safety, 442
Construction Safety Act of 1969, 442
Consultant, outfall (see Designer, outfall)
Contact basin, 64

Continental shelf:
  definition, 17
  subdivision of, 38–39
Continental slope, 17
Continental terrace, 17
Contract:
  construction (*see* Contract, construction)
  design, 422–23
  fixed-price, 423
  negotiated, 422–23
Contract, construction:
  award, 423
  lump-sum, 433, 435
  signing of, 435
  unit price (*footnote*), 435
Contract documents, 432
Contract forms, 433–35
Contraction coefficient, port, 288, 293
Contractor, construction:
  problems confronting, 423
  qualification statement, 433–34
  responsibilities of, 434
  selection of, 423
Contractor, general, 423
Control, 113
Convection, 42
Convergence, 31, 44
Copenhagen sea water, standard, 205
Copper, as pollutant, 109, 118–19
Coralline algae, 41
Corals, reef-building, 24–25, 41–42
Corers, bottom:
  gravity, 231–32
  piston, 231–32
  use in monitoring receiving waters, 528
Coriolis acceleration (*see* Coriolis force)
Coriolis force:
  effects on currents, 33, 44, 46
  factor in storm surge, 371–72
  influence on global winds, 29–30, 378
  introduction, 28–29
Corps of Engineers:
  beach sand replenishment, 5
  EIS review, 138
  permits, 99, 101, 134–36
Corrosion:
  fatigue, 166

Corrosion (*cont.*):
  in moored buoy systems, 165
  protection, 354–55
  protective coatings, 347, 502
  stability of plastics to, 351
  of steel pipe, 347
  of submarine electrical cables, 177
Cost estimating, construction:
  difficulties for marine work, 429
  need for, 427–28
  plans, need for, 422, 427
  procedures, 427–31
  specifications, need for, 427
Costs:
  outfalls (*table*), 126
  of receiving-water monitoring, 525
  treatment, wastewater (*table*), 126
Costs, outfall construction
  alignment tolerances, effect of (*footnote*), 425
  backfill, 431
  contingency, 431
  contractor profit, 431
  excavation, 431
  grade tolerances, effect of (*footnote*), 425
  mobilization/demobilization, 431
  overhead (*footnote*), 431
  pipe and pipe-laying, 431
  structures, 431
CPM, 437–42
Crab traps, 113, 250
Crane barge pipe-installation method, 471–77
Crest, wave, 34
Critical path method:
  activities, 437
  advantages (*footnotes*), 440
  arrow diagram, 437
  control-monitor phase, 439
  crash time duration, 438
  critical activity, 438
  critical path, 438
  early finish, 438
  early start, 438
  events, 437
  float, 438
  late finish, 438
  late start, 438
  need for, 437
  network diagram, 437
  normal time duration, 437–38

Critical path method (*cont.*):
  overall advantage of, 437
  planning phase, 437
  in practice, 440–42
  resource levelling, 439
  scheduling phase, 437–39
Crown, pipe, 339
Current data:
  frequency distribution, 192
  histogram, 192
  use in outfall projects, 191
  persistence in, 192
  presentation of, 181, 191–93
  processing of, 190
Current forces on exposed pipe:
  current perpendicular to pipe, 372–74
  drag coefficient (*see* Drag coefficient, current)
  drag force, 372–74
  lift coefficient (*see* Lift coefficient, current)
  lift force, 373
  pipe at angle to current, 375
  sliding instability, 376
  vortex-shedding, effect of, 374
Current forces on sediment, 377
Current measurements:
  Eulerian, 181, 186–93
  flow, 181, 186–93
  Lagrangian, 181–86, 191, 195–96, 200
  path, 181–86, 191, 195–96, 200
Current meters:
  Aanderaa, 187
  biological fouling of, 187–88
  ducted-impeller, 188, 386
  effect of mooring system motion, 187
  Ekman (–Merz), 186
  electromagnetic, 188–89
  Savonius-rotor, 186–87
  tilting, 172, 189–91
  ultrasonic, 188–89
Current rose, 181, 191
Currents:
  alongshore, 37
  coastal, 320–21
  data presentation, 191–93
  design, 371–72
  ebbing, 44
  flooding, 44
  hurricane-generated, 371–72

Currents (*cont.*):
  littoral, 37
  longshore, 37
  oscillatory, 303, 320–21
  reversing, 44
  rip, 37, 40
  set, 44
  strong, 45, 364, 371–72
  tidal, 44, 193
  unidirectional, 303, 320–21
  vertical variation, 181
  wind-driven, 44, 193
Current systems, variation in, 193,
    320–21
Cyclone:
  Althea, 371, 392
  Tracy, 368
Cyclonic disturbance, 30, 46, 367–68
Cysts, amoebic, 57, 130

D

Damage, submarine pipeline:
  abrasion, due to, 353, 358
  anchors, from, 356–57
  armor rock displacement, from,
    361
  boulders, from, 361
  causes of, 356
  chipped concrete, 356, 358
  currents, from, 359
  construction practices, from, 361
  differential erosion of sea floor,
    from, 359
  earthquakes, due to, 361
  fishing trawls, from, 356–57
  ice, due to, 361
  joint pullout, 359–60
  laying, during, 356
  manhole structures, 359–60
  pipes on wave-cut terraces, 359
  pollution due to, 356, 360–61
  prevention by burial, 357, 362
  during rocking, 500
  sea floor slides, from, 359
  threat from dredges, 365
  tsunamis, from, 360
  vortex-shedding, from, 359
  waves, from, 356, 358–61
Darcy-Weisbach pipe friction factor,
    294, 398, 522–23
Data presentation:
  chemical parameters, 206, 266

Data presentation (*cont.*):
  currents, 191–93
  for geophysical surveys, 225, 227
  marine biota, 244, 250, 252, 255,
    256, 261
  soil gradation, 238
  waves, 179–80
Datum, map versus chart (*footnote*),
    367
Davy Crockett, 472–77, 516
Dechlorination (*footnote*), 63
Declination:
  magnetic, 541–42
  of moon, 32–33
  of sun, 32–33
Decompression, diver:
  breathing air, 496–97
  breathing heliox, 559
  breathing oxygen, 496–97
  British Royal Naval Physiological
    Laboratory tables (*footnote*),
    554
  in deck decompression chamber,
    496–97
  stops, 554
  time taken, 554
  U.S. Navy tables, 554
  in water, 496
Decompression chamber, 496, 555
Decompression sickness (*see* Bends,
    the)
Deep water for waves (*footnotes*),
    35
Demobilization, construction, 430–31
Denitrification, 77
Density of sea water, 18–19, 310–12,
    315, 317
Depth, water, 35
Depth gage, diver, 492
Depth-limited design criterion, 385,
    392
Depth pay, diver, 497–98
Depth sounder (*see* Fathometer)
Depuration (*footnote*), 130
Design report:
  availability to bidders, 432
  final, 343
  preliminary, 343
  utility for construction, 422
Design wave:
  computations for, 384–85
  derivation of, 381–82
  recurrence interval, 381

Design wave (*cont.*):
  use of, 381
Design wave-induced force:
  horizontal, 390–91
  vertical, 390–91
Designer, outfall:
  choice of, 422–23
  functions of, 423–25
Detention time in wastewater treat-
    ment, 61
Detergents, 54, 77
Detonating cord, 460
Detritus, 26–27
Deviation, compass, 542
Diatoms (*footnote*), 21, 243
Die-away:
  of coliform bacteria, 57, 126–29,
    325
  of fecal streptococci, 57
Diffuser, outfall:
  alignment, 291
  at angle to outfall, 469
  buried, 298
  definition, 9, 284
  design of, 290–300
  design flows, 296–98
  factors of influence, 363
  internal hydraulics, 290–300
  internal settlement of particulates,
    291
  location, choice of, 363
  painting for diver visibility, 520
Diffuser discharges:
  into fresh water, 308–09
  wave effects on, 298, 301–02
Diffusion:
  eddy, 318
  molecular, 318
  oceanic, 193–201, 318–20
Diffusion coefficient (*see* Dispersion
    coefficient)
Digital instrument, 159
Diluters, 112
Dilution, effluent:
  definition, 285
  minimum surface, 305–06, 317
  secondary (*see* Dispersion, sec-
    ondary)
  solutions for, 286
Dilution, initial:
  centerline, 308–09, 312–16
  definition, 286
  equations for homogeneous receiv-

Dilution, initial (*cont.*):
  ing water, 303–09
  in homogeneous receiving water,
    303–09
  limitations of equations, 287, 303
  of round jets, 307–09, 312–14
  of round plumes, 307–08
  in stagnant, density-stratified
    waters, 309–16
Dilution, wastewater disposal by,
  78–80
Dinoflagellates, 21 (*footnote*), 243
Direction, determination of, 541–42
Disappearance, coliform, 57, 126–29,
  325
Discharge coefficient:
  port, 293
  riser, 299–300
Discharge, port, 288, 292–95, 304,
  307
Diseases, water-borne, 57–58
Disinfection, 63–64
DISPER-1, dispersion computer
  model, 327
Dispersion, wave, 34, 379
Dispersion coefficient:
  basic concept, 318
  effects on, 194, 318–20
  horizontal versus vertical, 321
  tests to determine, 200–01
  use of drogues in determining,
    194–95
Dispersion models, accuracy of, 287
Dispersion, oceanic, 193–201, 318–20
Dispersion, secondary:
  Brooks' model, 321–25
  defined, 287
  volume source model, 326
Dissolved gases in sea water, 18
Dissolved oxygen:
  in sea water, 18
  in wastewater, 56
  tests, 264–66
Distillation, 78
Diurnal tide, 32
Diver breathing systems:
  closed-circuit, 559
  continuous flow, 559
  demand supply, 559
  semi-closed, 559
Divergence, 31, 44
Divers' activities related to outfalls,
  551

Divers, construction:
  in bell on Horse, 476–77
  dangers for, 495
  drilling the sea floor, 461
  in heavy wave surge (*footnote*),
    461
  hose burial, possibility of, 489
  in-water work with the Horse, 474,
    476
  nitrogen narcosis problems, 553
  pay format, 497–98
  pay rates in Hawaii, 498
  pay rates in Southern California,
    497–98
  problems with tugger cables, 480
  sluicing bedding stone, 488
  standby, 495
Divers, inspector:
  in bell on Horse, 476–77
  checking joint gaps, 475
  nitrogen narcosis problems, 553
  problems with tugger cables, 480
  problems of determining pipe
    station, 492
  surveys of armor rock, 492–93
  surveys from diving bell, 493
  tour of damaged Mokapu outfall
    trestle, 484
Diversity:
  equitability, 244
  evenness, 245
  index, 244–46
  richness, 244
  variety, 244
Diving:
  activities related to outfalls, 551
  air compressor, 565
  band mask, 563–64
  buddy, 495
  chronic effects, 555–56
  dangers, 495, 552–56
  divers' joints, problems with, 556
  free, 553
  hard hat, 560–63
  hats, 564–65
  hearing damage, 555
  heavy gear, 560
  lifeline, 562, 564
  no-decompression times at depth,
    554, 559
  optimum exposure time, 497
  problems with breath-holding,
    555

Diving (*cont.*):
  safety regulations, U.S., 495–96
  standard, 560–63
  storage tank, air, 565
  volume tank, 565
DO (*see* Dissolved oxygen)
Doldrums, 30
Downwelling, 31
Drag coefficient, current:
  definition, 372
  design value of, 374
  factors of influence, 373–74
  for a sphere, 395
Drag coefficient, wave, 387
Dragging (*see* Trawling, bottom)
Dredges:
  bucketline, 453
  cable-tool chipper, 459
  clamshell, 454–55, 459
  of contaminated bottom material,
    517
  cutterhead, 456–58, 463
  cutter suction, 456–58, 463
  dipper, 453
  dragline bucket, 453–54, 459, 462
  dustpan, 456
  grab, 454
  grapple, 454
  hopper, 455–56
  hydraulic, 455–58
  mechanical, continuous, 453
  mechanical, repetitive, 453–55
  spoil area, 455–56
  suction, 453, 455–58
  trailer suction, 455–56
Dredging:
  effects of marine biota, 516–18
  unavoidable adverse effects, 518
Drift bottles, 182
Drift cards, 182
Drifter:
  bottom, 182–83
  current, 182
  seabed, 182–83
  Woodhead, 182–83
Drilling the sea floor:
  notification regarding, 230–31
  from a platform, 236–37
  from the sea floor, 235–36, 464
  from the water surface, 230,
    237
Drogues:
  circular, 183

Drogues (*cont.*):
cruciform, 183–84
in dispersion studies, 194–95
function of, 181–82
marking, 185
parachute, 183–84, 195
plane, 183
in t_{90} studies, 128
tracking, 184–85
wind effects on, 185, 195
Dunes, 38
Dyes, fluorescent:
continuous releases, 199
in dispersion measurements,
195–201
fluorescein, 196–97
pontacyl pink, 196–97
rhodamine B, 196–97
rhodamine WT, 196–97
for sediment motion, 241
slug releases, 199
Dye studies:
continuous dye release, 199
dyes, 195–97
instantaneous dye release, 198–99
instruments, 195–99
organizing results, 199–201
positioning by sextant, 198, 209
track, vessel, 198
Dynamics, sea floor, 241–42
Dynamite, 460–61

E

EA (*see* Environmental assessment)
Earth:
angular rotation, rate of, 371
coordinates on, 540–41
diameter through poles, 540
equatorial diameter, 540
geographic meridians, 541
magnetic meridians, 541
map projections, 542
spherical representation, 540
Earthquakes in outfall design, 400–01
Ebbing current, 44
Echo-sounding, 223–24
E. coli (*see* Escherichia coli, as
indicator organisms)
Eddies:
erosion due to, 392
oceanic, 318–20

Eddies (*cont.*):
from water motion past pipes, 392
Eddy diffusion coefficient, 200, 318
Effective time, median, 116
Effluent:
definition, 9
standards, 79, 106–07
EIS (*see* Environmental impact
statement)
Electrical cables, submarine, 161,
176–77
Electrodialysis, 78
Electronic positioning methods,
544–45
Electroplating industries, pollutants
from, 56, 119
Elutriation, 72
Emergency Temporary Standard,
495–96
Encroachment, salt water, 82–83
Endangered Species Act of 1973
(*footnote*), 138
Energy, hydraulic:
equation, 292–94
head, 292–94
kinetic, 289, 292
mechanical, 292
potential, 289, 292
pressure, 292
useful, 292
Engineer-designer, outfall (*see* Designer, outfall)
Engineering News-Record construction cost index, 430
Enrichment, nutrient, 123–24
Enteroviruses, 129
Environmental assessment, 133, 134,
136–37
Environmental impact statement:
draft, 138–39
final, 138
notice of intent, 134
origin, 99
preparation of, 134, 137–38
review, 138
slowness in preparing, 102
Environmental Protection Agency
(*see* U.S. Environmental Protection Agency)
EPA (*see* U.S. Environmental Protection Agency)
Epifauna:
motile, 23

Epifauna (*cont.*):
sessile, 23
Erosion, sea floor:
over buried pipes, 393
depth of, 394
differential, as outfall problem,
520
due to eddies, 392
behind pipes, 358–59, 392
problems during pipe-laying, 482
risk of baring outfall, 519
from waves, 40–41
Error function, 323
Escalation, construction cost, 430
Escherichia coli, as indicator
organisms, 57–58
Established flow, zone of, 287
Estuaries, 1, 38, 286
Euphotic zone, 22
Eutrophication:
artificial, 76
cultural, 76
natural, 123
Excavating:
by blasting, 458–64
by cutter-head vehicle, 498
using dredges, 447, 453–58
using fluidization, 499
during pipe burial, 498–99
by ship propeller slip stream, 463
for submarine pipelines, 447,
453–64, 498–99
by submersible vehicles, 463
by water jets, 499
Excavator vehicles, 463
Explosives, use of, 458–64
Exponential decay, 126–27
Extended-bell concrete pipe, 349
Extreme value distribution, 381

F

Fathometer, 223–24, 545
Fatigue:
corrosion, 166
metal, 166–67
Fauna, 242
Fecal contamination, 57
Fecal streptococci, 57, 130–31
Feces, 54
Federal Water Pollution Control
Act Amendments of 1961, 98

Federal Water Pollution Control Act Amendments of 1972 (*see* Public Law 92-500)
Federal Water Pollution Control Act of 1956, 98
Fertilizers, 74
Fetch, 34, 377–79
Filter:
cake, 72
press, 72
Filter fluorometer, 195–98
Filtration of fine suspended matter, 77
Fin rot in fish, 527 (*footnote*), 529
Fish:
bite, 166
characteristics, 246–47
counting, 251–52
demersal, 248, 250
groundfish, 248
length-weight relationship, 246–47, 252
Fishing:
longline, 250
rig, 526–57
using traps, 527
by trawling, 248–49
Fish and Wildlife Coordination Act of 1958 (*footnote*), 138
Five freedoms, 79
Fix, navigational, 549
Flavor impairment of organisms, 118
Flexural stresses, pipe, 391–92
Float gage, 161
Flocculation, 60, 63
Flooding current, 44
Flora, 242
Florida:
alongshore sediment transport, 37
currents, 30, 37
wastewater disposal in the ocean, 140
Florida Current, 30
Flotation, wastewater treatment by, 62
Flotation pipe-installation method, 469
Flow capacity of pipe, 290
Flow establishment, zone of, 286
Flow-induced pipe oscillations, 520
Fluid, inviscid, 382, 387–88

Fluorescence:
defined, 195
factors of influence, 197–98
Fluorimeters, 195–98
Fluorometers, 195–98
Fog:
laser alignment system, influence on, 545
radar, influence on, 549
Folsom splitter, 260
Food and Agriculture Organization of the United Nations, 6
Foraminiferans, 245, 262
Force, pipe:
due to currents (*see* Current forces on exposed pipe)
from earthquake motion, 400–01
normal, 376
due to waves (*see* Wave forces on exposed pipes)
Forecasting:
storm surge heights, 368
wave conditions, 377
Foreshore, 38
Formalin, 255, 260
Four-thirds power law, 318–20
Fréchet probability distribution, 381
Free available chlorine, 64
Friction coefficient, sliding, 376
Friction factor, Darcy-Weisbach, 294, 398, 522–23
Fringing reef, 41
Front:
cold, 377
definition, 30
occluded, 377
wave (*footnote*), 386
Froude number:
adapted densimetric, 315–16
current, 303–07
densimetric, 288–89, 312–13
effective densimetric, 288–89, 307–08
Fully-developed sea:
definition, 34
wave characteristics (*table*), 35

G

Gale force winds, 35 (*table*), 368
Gantry crane, 457, 480
Gas diffusion, 552

Gases:
heliox, 559, 562, 564
synthetic mixed, 559
Gaskets, pipe joint, 349–53, 473, 500–02
Gaussian probability distribution (*see* Normal probability distribution)
General conditions of specifications, 425–27
General contractor, 423
General provisions of specifications, 425–27
Geomagnetic electrokinetograph, 181
Geophysical investigations, 222–28
Geophysical surveys, 225–28
Geostrophic wind, 378
Grab samplers:
characteristics of, 231–32, 254–55
desirable features of, 254–55
Ekman, 254–55
orange peel, 254–55
Petersen, 254–55
Smith-McIntyre, 254–55
van Veen, 254–55, 257
Granules (*table*), 39
Gravel:
bedding (*see* Bedding stone)
particle size characteristics (*table*), 39
pea, 488
Gravity coring sampler-tubes, 231–33
Gravity flow, of outfall, 300, 302
Grazers, 22
Great Barrier Reef, 42
Great Circle, 378, 540
Green algae, 27
Grit, 60
Grit chambers, 60
Groundwater recharge, 81–83
Groundwater replenishment, 81–83
Gulf Islands, British Columbia, 45
Gulf of Mexico:
design hurricane, 369
hurricanes, 368
pipeline failure by sliding, 392
submarine pipelines in, 362
tidal range, 367
Gulfs, 38
Gulf Stream, 30–31
Gusts, wind, 43
Gyre, 31

H

Hadal depths, 17
Halocline, 20
Hard hat diving:
    advantage of, 560
    air hose, 561
    clothing, 560
    disadvantage of, 560
    helmet design, 560–61
    safety fittings, 561
    signals by tugs on hose, 563
    watertight seals, 560
    weights, 562
    wire communications with surface,
    562–63
Hat diving, 564–65
Hawaiian Independent Refinery Inc.
    oil pipelines, 240, 362, 485, 502
Hawaiian Islands:
    sea life, 26
    winds, 30, 42
Hawaiian Plane Coordinate Grid
    System, 541
Hawaii State government:
    Agriculture, Department of, 139
    Harbors Division, 136, 139
    Health, Department of, 136, 139
    Land and Natural Resources,
    Department of, 136, 139
    Planning and Economic Develop-
    ment, Department of, 139
    Social Services and Housing,
    Department of, 139
    Transportation, Department of,
    139
Hawaii, University of,
    Environmental Center, 139
    oceanographic research, 168–69,
    187–88
    Water Resources Research Center,
    139
Head:
    elevation, 292–94
    loss, 292–94, 301
    loss coefficient, minor, 293
    pressure, 177, 292–94
    total, 292–94, 364
    velocity, 292–95, 301
Headland, 38
Heavy metals, in wastewater, 56
Helicopters:
    for bathymetric surveys, 222–23

Helicopters (*cont.*):
    for dye studies, 199
Heliox, 559, 562, 564
Helmets, diving, 560–61, 564
Herbivores, 22, 26
Hermatypic corals (*see* Corals,
    reef-building)
Heterotrophic organisms, 22
High tide, 31
High water, 31
Higher high water, 32
Higher low water, 32
Hindcasting, wave, 377–79, 381
Holdfast, 27, 243
Holding tanks for bioassays, 113–14
Holoplankton, 23
Honolulu, City and County of:
    Board of Water Supply, 139
    Finance, Department of, 425
    General Planning, Department
    of, 139
    Land Utilization, Department of,
    139
    Public Works, Department of, 138,
    140, 346, 426
    Transportation Services, Depart-
    ment of, 139
Honolulu, Hawaii:
    astronomical tidal levels, 367
    raw sewage effluent, 95
    tides (*tables*), 33
    wastewater treatment, 125
Horse, the:
    changes to, 474
    cost of, 472
    design of, 472–73
    lifting of, 474
    need for, 472
    operation of, 473
    Orange County No. 2 outfall, 472
    pipe acquisition, 473
    pipe installation procedures,
    474–75
    positioning of, 474
    rate of pipe-laying, 477
    Sand Island No. 2 outfall, 472–77
    use in outfall repair, 521
"H" pile, 478, 483, 484
HRS (*see* Hydraulics Research
    Station)
Hurricane:
    Camille, 369
    Carla, 46, 368

Hurricane (*cont.*):
    central pressure reduction, 368,
    484
    design, 368, 369, 382
    exiting (*footnote*), 369
    Kate, 483
    landfalling, 369
    minimum sustained wind speed
    (*table*), 35, (*footnote*) 46, 367
    radius to maximum winds, 368,
    484
    track, 368
    velocity of forward movement,
    368
    wave generation models, 382
    wind speed variation, 368
Hydraulic design:
    change of pipe direction, 401
    diffuser flows, 290–98
Hydraulic jump, 300
Hydraulics Research Station, 393,
    395, 396
Hydrometer, 237–38
Hydrophone, 223, 226–28
Hyperbaric medicine, 554
Hyperion outfalls, 14, 109, 120

I

Ideal fluid (*see* Inviscid fluid)
Imhoff cone, 54
Incineration, 74
Incipient lethal level, 115
Incompatible pollutants:
    definition, 101
    limits on, 121
    list, 119
Indicator organisms:
    concept, validity of, 127, 130–31
    definition, 57
Indirect costs (*footnote*), 438
Inertia coefficient:
    added mass coefficient, link with,
    387
    for pipe in ideal fluid, 388
    for rock matrix, 399
    for trapezoidal shape, 399
Inertia force:
    earthquakes, during, 401
    for flow-rock interaction, 396
    under solitary wave, 398
    in wave force model, 387

Infauna, 23
Initial dilution zone (*footnote*), 79
Injection wells, 82
Inshore zone, 464
Inspection divers (*see* Divers, inspector)
Inspection of operational outfall:
  diversion of flow, 402
  frequency of, 519–20
  methods, 520
  need for, 519–20
Instruments for chemical parameters, 266
Intertidal zone, 39
Intrusion, salt water, 82–83
Invert, pipe, 339
Invertebrates, marine, 244–45
Inviscid fluid, 382, 387–88
Ion exchange, 78
Irrigation, wastewater, 81
Isaacs-Kidd midwater trawl, 261
Isobar, 377, 379
Isobath, 249
Iterations, computational, 295

J

Jet:
  buoyant, 284
  effluent, 284
  line, 285–86, 309
  round, 307–09, 312–14
  slot, 285–86, 309
  terminal height of rise, 312–16
  total width, 313
  velocity of, 289
  zone of established flow, 287
  zone of flow establishment, 286
Jetting, by divers, 235–36
Joint pressure test, 501

K

Kaneohe Bay, Hawaii:
  dye study, 199
  fishponds bordering (*footnote*), 84
  wastewater disposal in, 84
Kelp, 27, 243, 361
Kinks, cable, 165
Kuroshio Current, 31

L

Lagoon, 42
Lagooning, sludge, 75
Lagrangian current measurements, 128, 181–86, 195–96
Lake Erie, 123
Lake Michigan (*footnote*), 132
Lambert Zones, 541
Landfills, 74–75
Laser alignment system, 460, 474, 501, 545–46
Latitude:
  definition, 541
  north, 541
  parallel of, 541–42
  south, 541
Lay barge pipe-laying method, 470–71
Lead line:
  for pipe trench elevation, 480
  in water depth determinations, 222–23
Lethal concentration:
  definition, 111
  median, 111
  threshold, 115
Lethal level, incipient, 115
Lethal time, median, 116
Lift coefficient, current:
  definition, 373
  design values for, 374
  factors of influence, 374
Light-dark bottle technique, 262
Limiting nutrient, 22
Linear wave theory (*see* Airy wave theory)
Lineup (*see* Range, bearing)
Lipids, 21
Liquefaction, soil:
  causes of, 358
  consideration in pipeline design, 358
  definition, 358
  factor in pipeline instability, 357–58, 366
  failure of pipe during laying, 487–88
Lithogenous sediment, 38
Littoral:
  currents, 37
  transport, 40, 241
  zones, 39

Long Island beaches, closing, 105
Long Island Sound, New York:
  currents in, 45
  swimming in (*footnote*), 132
Longitude:
  definition, 541
  east, 541
  meridian of, 541–42
  west, 541
Longlines, fishing, 250
Longshore currents, 37
Los Angeles, California, outfalls (*see* Hyperion outfalls)
Los Angeles area:
  wastewater flow, 14, 83
  water supply, 83
Los Angeles City:
  Board of Public Works, 120–23
  source control program, 118–23
Lower high water, 32
Lower littoral zone, 39
Lower low water, 32
Low tide, 31
Low-tide terrace, 38
Low water, 31

M

Magnetic tape recorder, 158, 161, 174, 179, 266, 545
Maintenance, outfall:
  cleaning, 522–23
  costs of, 521
  diversion of flow, 402
  manual, 519
Mangrove, 27
Manholes, pipe, 359–60, 401–02
Manifold, 284
Manly Beach, Australia, 94–95
Manning's roughness factor, 290, 295–96, 346, 371
Manual, operation and maintenance, 519
Map, 367 (*footnote*), 542
Marine biota, construction effects on, 516–18
Marine biota, effects of sediments on, 516–18
Marine Protection, Research and Sanctuaries Act of 1972, 102 (*footnote*), 138
Maritime Services Board of New South Wales, 174–75

Martini's Law, 553
Mask, band, 563–64
Massachusetts Institute of Technology:
  circulation computer program, 327
  dispersion computer program, 327
Mass coefficient, 387–88
Mass emission rates of pollutants (*table*), 119
Mean sea level, 32
Median tolerance limit (*footnote*), 111
Membrane-filter technique for coliforms, 58, 128
Mercator chart, 378
Mercenaria mercenaria, 124
Meridian:
  definition, 540
  of longitude, 541–42
  prime, 540
Meroplankton, 23
Messengers:
  definition, 203
  for sampling bottles, 203–04
  for trawls, 261
Metabolism, 21
Metal-finishing industries:
  cost of pretreatment (*footnote*), 117
  wastewaters from, 56, 118–19
Methane, 56, 72
Methemoglobinemia (*footnote*), 81
Middle littoral zone, 39
Midwater trawl, Issacs-Kidd, 261
Mile:
  nautical, 541
  statute, 541
Minor constituent of sea water, 18
Mixed layer:
  definition, 19
  extent of, 303, 309, 311
Mixed liquor:
  definition, 69
  suspended solids, 69
Mixed tide, 32
Mixing, effluent, 303–27
Mixing zone, 79
Mobilization, construction, 430–31
Mokapu outfall:
  autotape positioning system, use of, 544
  blasting for, 459

Mokapu outfall (*cont.*):
  change orders during construction, 443
  construction diver injury, 495, 554
  dredging offshore, 454
  dredging from trestle, 455, 459
  joint pressure testing, 501
  longitudinal trestle beams (*footnote*), 484
  major characteristics of, 452
  movement of sea floor materials, 482–83
  pipe joint design, 350
  pipe material option, 355
  temporary end bulkhead, 483, 485
  trestle, construction, 478, 483–84
  trestle piles, 484
  underwater inspections, during construction, 442, 484
  wave damage to trestle, 483–84
Momentum flux parameters, 312–16
Monel:
  manhole covers, 402
  straps, 358
Money left on the table, 434
Monitoring, effluent, 523, 525
Monitoring, receiving-water:
  abundance of organisms, 529
  the benthos, 526–28
  bottom sampling, 526
  chemical water column parameters, 263
  costs, 525
  by County Sanitation Districts of Orange County, California, 525–27
  discoloration of receiving water, 526
  diseases in the benthos, 529
  diseases in fish, 527 (*footnote*), 529
  diversity index, calculation of, 526
  documentation of, 527, 529
  enrichment by effluent, 529
  fish-spearing, 527
  fish traps, 527
  floatables, 526
  fundamental purpose of, 523
  grab samples of sediment, 526
  improved conditions with effluent cleanup, 529
  nature of, 524
  number of species, 529

Monitoring, receiving-water (*cont.*):
  pollutants in organisms, 526, 528–29
  pollutants in sediments, 526, 528
  procedures for negative reports, 524
  procedures for positive reports, 524–25
  by SCCWRP, 528–29
  Southern California waters, 528–29
  stations, selection, of, 526 (*footnote*), 529
  trawling, 527, 529
  types of species, 529
  water column parameters, 526
Moody diagram, 294
Mooring systems:
  anchor-first, 168
  anchor-last, 168
  deploying, 168–70
  retrieving, 170–72
  slack-line, 162, 173–75
  taut-line, 162–64, 168–69
  for tilting current meters, 172, 191
  vessels for deploying, 169–70
Morison equation, 387–88
Most probable number technique for coliforms, 58, 128
Motile epifauna, 23
Mud, drilling, 230, 235
Mudstone, 38, 229 (*footnote*)
Multimedia beds, 77
Muskie hearings, 125
Mutualism (*footnote*), 25

N

Nansen bottle, 203–04
Narcosis, nitrogen:
  elimination through use of heliox, 559
  inspection divers, problems for, 553
  manifestations of, 553
  nature of, 553
  outfall construction divers, problems for, 553
National Commission on Water Quality, 103
National Environmental Policy Act of 1969 (*see* National Environmental Protection Act of 1969)

National Environmental Protection
Act of 1969:
application of, 133, 136–37
major elements, 99
National Historic Preservation Act
of 1966, 138
National Oceanic and Atmospheric
Administration:
administration of laws, 138–39
charts from, 542
Environmental Data Service, 160
storm surge predictions, 369
National Pollutant Discharge
Elimination System permit
program, 101, 132–34
National Technical Information
Service, 368
Nautical chart (*see* Charts, nautical)
Nautical mile, 541
Navigation aids, floating, 547–48
Neap tides, 32
Negative declaration, 134
Nekton, 23
NEPA (*see* National Environmental
Protection Act of 1969)
Neritic zone, 17
Nets, plankton:
bongo, 261
flowmeters for, 257–58, 261
frames, 258
materials in (*footnote*), 257
mesh sizes, 258
use of, 257–58
Neuston, 258
New York Bight, 75, 105
Niskin:
bottle, 204
sampler, 264
Nitrates:
in drinking water, 81
as limiting nutrients, 22, 76
as nutrients, 76
as source of nitrogen, 265
Nitrification, 56
Nitrification-denitrification, 77
Nitrogen:
dissolved in wastewater, 56
Kjeldahl, total, 265
total, 265
Nitrogen narcosis (*see* Narcosis,
nitrogen)
Nonconservative substance, 317,
325

Normal force, 376
Normal probability distribution:
for bioassay test results, 115
for distribution of contaminant
concentrations, 200, 321–27
North Equatorial Current, 31
North Sea:
diver injury and mortality, 495
pipelines in, 242
sea floor dynamics, 241–42
unsupported pipeline spans (*footnote*), 359
North Star, 541
Notice of intent, 134
Notice of intention to bid, 435
Notice to proceed, 440
NPDES (*see* National Pollutant
Discharge Elimination System
permit program)
NTIS (*see* National Technical
Information Service)
N_2 (*see* Nitrogen)
Nutrient, limiting, 22
Nutrients:
definitions, 22
determination of levels of, 265
important examples, 76, 265
limiting, 76, 85 (*footnote*)
removal from wastewaters, 76–77

O

Oahu, Hawaii:
fringing reefs, 41
large swell, 380–81
tidal currents, 44
tsunami, 45
Occupational Safety and Health Act
of 1970, 442, 495
Ocean Dumping Act (*footnote*), 102
Oceanographic data-gathering:
approaches in, 157–59
arrangements for, 157–58
Oceanographic vessels, requirements
for, 208
Ocean station vessels, 381
Offshore, the, 38
Offshore zone, 464
Oligotrophic waters, 123
Operation, outfall:
costs of, 521
manual, 519

Operation, outfall (*cont.*):
need for inspection, 519–20
pipe wall deposits, problem of,
522–23
Optimum diver exposure time, 497
Orange County No. 2 outfall:
autotape positioning system, use
of, 544
bends in, 401
borings for, 236
camera evaluation of diffuser
performance, 520
characteristics (*table*), 15, 451
damage to concrete pipe sections,
499–500
diffuser performance, evaluation
of, 520
dredging, offshore, 454
dredging from trestle, 463
dropping of pipe after release, 482
egress structure, 403
the Horse, use of, 472
manhole covers, 402
manholes, 402
manhole spacing, 402
monitoring of performance, 525–27
pipe sizes considered, 364
placing of armor rock, 490
placing of bedding stone, 489
sizing of protective rock, 397
studies by SCCWRP, 528
surge tank, 301
time of project completion, 433
trestle, construction, 478
wastewater flow in, 14
Orange County Sanitation District:
chlorination of wastewaters (*footnote*), 131
wastewater reclamation plant, 83
Organic materials, removal from
wastewater, 77
Organic sediments, 38, 517
"O"-rings:
in instrument housings, 189
in pipe joints, 349–51, 473, 500–01
in SCUBA regulators, 556
Orthogonal, wave, 36, 380, 391
OSHA (*see* U.S. federal government,
Occupational Safety and
Health Administration)
O_2 (*see* Oxygen)
Outfall:
coatings (*see* Pipe coatings)

Outfall (*cont.*):
  construction (*see* Outfall construction)
  definition of, 1
  inspection after completion, 519–20
  joints (*see* Pipe joints)
  linings (*see* Pipe linings)
  operation (*see* Operation, outfall)
  pipe details (*see* Pipe, submarine)
  pipe dimensions (*see* Pipe dimensions)
  pipe materials (*see* Pipe materials)
  pipe sizes (*see* Pipe dimensions)
  protection (*see* Protection, submarine pipeline)
  repair, 521
Outfall construction:
  coordination of pipe laying and rocking, 493–94
  costs, 429–31
  detailed specifications for, 427
  excavating, 447, 453–64
  general provisions for, 426–27
  manholes, marking, 502
  notice to proceed, 440
  pre-bid meeting, 432
  progress reports, 441
  removal of pipe sections, 494
  sea forecasts, use of, 502–03
  specifications for, 425–27
  temporary end bulkhead, 483
  weather forecasts, use of, 502–03
Outfall design:
  against earthquakes, 400–01
  need for construction methods input (*footnote*), 425
  range of pipe sizes, 364
  water levels, 366–71
  against waves, 386–93
Outfall names:
  Agaña (Guam), 298, 360–61
  Agat (Guam), 298, 360
  Annacis Island (Vancouver, B.C., Canada), 298–99
  Ashbridge's Bay (Toronto, Ontario, Canada), 448
  Baglan (England), 451
  Barbers Point (Ewa, Hawaii)(*see* Barbers Point outfall)
  Barking Sands (Hawaii), 360
  Bowery Bay (New York), 298
  Brooklyn (New York), 298
  Broward County (Florida), 228

Outfall names (*cont.*):
  Burwood Beach (Newcastle, Australia), 397
  Cannes (France), 451
  Chula Vista (California), 359
  Clover Point (Victoria, B.C., Canada), 354, 355
  Coney Island (New York), 298
  Crown-Simpson No. 1 (Humboldt Bay, California), 356
  Deal (New Jersey), 448
  Depoe Bay (Oregon), 464
  Doctor's Gully No. 2 (Darwin, N.T., Australia), 236
  Five Finger Island (Nanaimo, B.C., Canada), 451, 467
  Georgia-Pacific Paper Co. (Newport, Oregon), 298, 355, 362, 364, 454, 472, 482, (*footnote*), 485
  Gentofte (Denmark), 373
  Gisbourne (New Zealand), 351, 450
  Glenelg (S.A., Australia), 448
  Goleta (California), 354, 361, 520
  Grimsby-Lincoln District (Grimsby, Ontario, Canada), 298
  Hammermill Paper Co. (Erie, Pennsylvania), 298
  Harmac Pulp Ltd. (Nanaimo, B.C., Canada), 452, 467, 469
  Hastings (England), 298, 450, 469
  Hawaii Kai (Hawaii), 298, 301, 320, 483
  Hilo No. 1 (Hawaii), 360
  Hilo No. 2 (Hawaii), 360, 521
  Hollywood (Florida), 450, 472, 482 (*footnote*)
  Humboldt Bay (California), 240, 345, 520
  Huntington Beach (California), 494
  Hyperion No. 1 (El Segundo, California), 15
  Hyperion No. 2 (El Segundo, California), 15
  Hyperion No. 3 (El Segundo, California), 15
  Hyperion No. 4 (El Segundo, California), 15
  Hyperion No. 5 (El Segundo, California), 15, 356

Outfall names (*cont.*):
  Hyperion No. 6 (El Segundo, California), 15, 356, 373, 448
  Hyperion No. 7 (El Segundo, California), 15, 355, 449, 472, 477, 519, 520, 528
  Hyperion Sludge (El Segundo, California), 75, 354–55, 357, 448, 469, 520, 522
  ITT Rayonier Inc. (Port Angeles, Washington), 298
  Ingoldmells (England), 450
  International Paper Co. (Gardiner, Oregon), 449, 469
  Ipanema Beach (Rio de Janeiro, Brazil), 452
  Kahe Thermal Power Plant (Waianae, Hawaii), 442, 516
  Kahului No. 1 (Hawaii), 360
  Kahului No. 2 (Hawaii), 360
  Kapaa (Hawaii), 360
  Kewalo Basin (Honolulu, Hawaii), 345–46
  Lion's Gate STP (First Narrows, Vancouver, B.C., Canada), 298, 365, 451, 520
  Macauley Point (Victoria, B.C., Canada), 450, 470
  Marsden Thermal Power Station (New Zealand), 458, 471
  Menasha Corporation (Coos Bay, Oregon), 298, 348, 356, 519–20
  Midway Island No. 1, 360
  Midway Island No. 2, 360
  Mokapu (Kailua, Hawaii)(*see* Mokapu outfall)
  Monmouth County (New Jersey), 451, 469
  Mornington (Victoria, Australia), 448
  Napier (New Zealand), 436
  Nassau County (Long Island, New York), 451
  Nelson (New Zealand), 351, 450, 462
  Newport (Oregon), 298, 364
  North Miami Beach (Florida), 449
  Orange County No. 0 (Newport Beach, California), 15
  Orange County No. 1 (Newport Beach, California), 15, 301, 448, 478, 523, 528

Outfall names (*cont.*):
  Orange County No. 2 (Newport Beach, California)(*see* Orange County No. 2 outfall)
  Otter Rock (Oregon), 344, 361, 362, 365, 461, 469
  Oxnard (California), 520, 528
  Palm Beach (Florida), 448, 478
  Pearl City (Hawaii), 360
  Pittsburgh (California), 449
  Point Loma (San Diego, California)(*see* San Diego outfall)
  Port Fairy (Victoria, Australia), 451
  Port Lincoln (S.A., Australia), 449
  Portsmouth (New Hampshire), 449
  Powell River (B.C., Canada), 451, 470, 520
  Prince Rupert (B.C., Canada), 357, 451, 470
  Rainier Nuclear Power Plant (Washington), 373
  Rochester (New York), 233, 358, 451
  Saipan, No. 1, 360
  Saipan No. 2, 360
  Salem (Massachusetts), 522
  San Francisco (California), 237
  San Francisco Bay (California), 450
  Sand Island No. 1 (Honolulu, Hawaii), 298, 345, 359–60, 448, 477
  Sand Island No. 2 (Honolulu, Hawaii)(*see* Sand Island No. 2 outfall)
  Sandy Bay (Hobart, Tasmania, Australia), 449
  Santa Barbara (California), 452, 478, 487
  Scheveningen (The Netherlands), 109
  Straight Point (England), 450
  Swanbourne (W. A., Australia), 449
  Tipalayao (Guam), 361, 373
  Ventura No. 1 (California), 359
  Waianae (Hawaii), 298, 301
  Wanganui (New Zealand), 235–36
  Waitara (New Zealand), 369, 452, 469
  Watsonville No. 1 (California), 448

Outfall names (*cont.*):
  Watsonville No. 2 (California), 449, 469
  West Point (Seattle, Washington), 450, 519
  Weyerhaeuser Co. (Longview, Washington), 373
  Whites Point No. 1 (Los Angeles, California), 15
  Whites Point No. 2 (Los Angeles, California), 15, 448
  Whites Point No. 3 (Los Angeles, California), 15
  Whites Point No. 4 (Los Angeles, California), 15, 449, 459, 478, 528
  Woodman Point (W.A., Australia), 449
  Wrightsville (North Carolina), 359
Outfall route:
  considerations involving, 364–65
  down cliffs, 365–66
Outfalls:
  alignment problems, 501
  blockage by debris, 522
  blockage by sand, 356
  on charts, 543
  cleaning of, 522–23
  differential settlement of, 366
  erosion of supporting material, 359
  external inspection, need for, 519
  flow capacity, decrease of, 522–23
  hydraulic capacity tests, 345–46, 522–23
  installation, 464–88, 498–99
  laying, 464–88
  length, 363–64
  natural channels for (*footnote*), 363
  new pipe, insertion of, 523
  placement on charts, 520
  pre-design seafloor reconnaissance, need for, 363
  predicted operation and maintenance costs, 521
  relining of, 523
  settlement of, 366
  shore exit point, choice of, 363
  termini, 403
  across tidal flats, 487
  wall deposits, 522

Outfall sewer, definition, 1
Overhead, 431 (*footnote*), 441 (*footnote*)
Overrocking, 492
Oxygen:
  as breathing gas, 555
  dissolved in sea water, 18, 22
  dissolved in wastewater, 56
  minimum human requirement, 552
  solubility of, in water, 265
  toxicity to humans, 552–53
Ozone, as disinfectant, 63

P

Paper, underwater-writing, 208, 252, 396
Paper tape recorder, 174, 179, 545
Passes, reef, 42
Pathogenic bacteria (*see* Pathogens)
Pathogenic viruses (*see* Pathogens)
Pathogens:
  control by treatment, 130
  definition of, 57, 129
  destruction of, 63–64
  examples of, 129–30
  indicator organisms for, 58, 130
Pea gravel, 488
Pebbles, sizes of (*table*), 39
Pelican release hook, 171
Penetrometer, 233
Peru, 26, 31
Pesticides, 55
pH:
  definition, 55
  determination of, 265–66
Phenols, 54
Phosphates:
  as limiting nutrients, 76
  as nutrients, 76, 265
Photic zone, 20
Photometer, 268, 526
Photosynthesis, 21
Physical sediments, 38
Physicochemical treatment, 78
Phytoplankton:
  counting of, 260
  definition, 21
  identification of, 260
  sampling, 257–59, 261–62
Pig, pipe, 522–23

Pile bent, 478, 484
Pile driver, 485
Piles, 228, 478–79, 483
Piling, sheet:
  cutting of, 483
  driving of, 478, 487
  excavation inside, 480
  focusing of wave energy, 480–81
  leaving in place, 483
  trestle on, 486–87
  use for stabilizing pipe-laying,
    480–81
Pingers:
  on outfalls, 520
  for pipe pigs, 523
  on releases, 172
Pipe coatings:
  coal tar enamel, 347
  concrete, 347–48, 358, 465
  for corrosion protection, 347, 354
Pipe, concrete:
  damage during joint makeup, 500
  damage to sections above water,
    499–500
  longitudinally-post-tensioned, 351
  reinforced, 348
  repairing damage, 500
Pipe dimensions:
  concrete pipe, 348–49 (table)
  fiberglass-reinforced pipe (table),
    353
  high-density polyethylene (table),
    352
  steel pipe (table), 346
Pipe joints:
  ball, 345, 348
  ball-and-socket, 345
  bell, 345, 401, 471, 477
  beveled, 401
  checking gaps, 475–77
  in concrete pipe, 349–51, 471, 477,
    500–01
  desirable characteristics, 344, 500
  double-gasket, 349–51, 475–76,
    500–01
  double-gasket pressure test, 475,
    482, 501
  flanged, 344, 348, 352–53
  flexible plastic gaskets, 501–02
  fused, 352–53
  gaskets, 349–53, 473,
    500–02
  "O" rings, 349–51, 473, 500–01

Pipe joints (cont.):
  pulled, 401
  pullout, 520
  rigid, 345
  source of pipeline damage, 359
  spigot, 345, 401, 471, 477, 500
  welded, 348, 353, 465–66
Pipe linings:
  cement mortar, 344, 347, 356
  coal tar enamel, 347
  epoxy, 347
  polyethylene, 344, 347
  polyvinyl chloride, 347
  stainless steel, 348–49
  tar epoxy, 347
  vinyl, 348–49
Pipe materials:
  cast iron, 344–46
  choice of, 355
  concrete, 344, 348–51, 355
  corrugated iron, 344
  ductile iron, 344–45
  fiberglass-reinforced (pipe), 344,
    351–53, 365
  high-density polyethylene, 344,
    351–53, 355
  iron, 344–46
  plastic, 344, 351–53
  polyvinyl chloride, 344, 351–53
  PVC (see polyvinyl chloride)
  reinforced concrete, 344, 348–51,
    355
  steel, 344, 346–48, 355, 358–59
  vitrified clay, 344
  wood stave, 344
  wrought iron, 344
Pipe roughness:
  absolute, 290
  Manning's n, 290, 295–96, 346
  relative, 294
Pipe sizes (see Pipe dimensions)
Pipe, submarine:
  alignment of, 500–01
  anchored, 342, 345–46, 351,
    463–64
  ballasted, 342, 393, 399
  bends, 401
  beveled sections, 500
  buried, 342, 376, 393
  clearance, relative, 373
  damage to, 355–62
  egress structure, 403
  elevation of, 482

Pipe, submarine (cont.):
  failures of, 355–62
  flapgate structure, 403
  jetted-in, 342, 498
  manhole covers, 401–02
  manholes, 359–60, 401–02
  mitered sections, 401
  overstressing, 520
  pile-supported, 359
  repair, 521
  skunk line, 475
  sliding instability, 376, 392
  stapled, 342, 376, 463–64
  stresses during laying, 470
  support by concrete-filled burlap
    bags, 362
  supported, 342
  support of suspensions, 520
  with weights, 351
Pipe, tremie, 489–90
Piston corers, 231–33
Plane coordinate systems, 541
Plankton, 21–23, 257–61
Plankton samples:
  sorting, 260–61
  splitting of, 260
Plankton tow:
  surface, horizontal, 258, 261
  vertical, 257–58, 262
Plans, construction:
  definition, 422
  standardized drawings, 426
Plans and specifications:
  alterations to, 442–43
  change orders, 442–43
  definition, 422
  need for changes, 443
Plants, marine, 242–43
Planula, coral, 23, 25
Platforms, offshore, 477
PL 92-500 (see Public Law 92-500)
Plume:
  definition, 284, 289
  line, 305–07
  plane, 307
  round, 307–08, 312–13
  slot, 305–08
  terminal height of rise, 312–16
  total width, 313
  two-dimensional flow regimes,
    304
  width, 313
Plunging breaker, 36

Point Loma outfall (*see* San Diego outfall)
Pneumothorax, 555
Poles:
  geographic, 541
  magnetic, 541
  north, 540
  south, 540
Pollution:
  definition, 6
  forms of, 96
Pollution indicators:
  approaches with marine biota, 528
  chemical parameters, 527
  coliform bacteria, 57–58, 129–32
  integrative nature of benthic animals, 527
  polychaete worms, 528
Polychaete worms, 23, 85, 108, 262, 528
Ponds, stabilization, 65
Pony, the, 481–82
Port diffuser:
  bell-mouth, 285, 293, 298
  blockage by ballast rock, 493
  blockage by sea life, 519
  check valves, fitted with, 519
  covers during construction, 493
  definition, 9, 284
  design of, 285, 291
  flow through, 288–89, 290–91, 293–95, 301–02
  numbering of, 292
  spacing, 289, 294, 307, 309
  sum of areas of, 291, 295
Positioning of marine vessels:
  celestial method, 543
  charts, 542–43
  by electronic means, 544–45
  floating navigation aids, using, 547–48
  link with outfalls, 539–40, 543
  need for, 543
  range-azimuth approach, 546
  sextant angles, horizontal, 198, 209, 546–47
  by theodolites, 546
Potentiometer, 178
Preaeration, 60
Pre-bid meeting, 432
Prechlorination, 63
Pressure:
  absolute, 552

Pressure (*cont.*):
  atmospheric, 552
  increase with water depth, 552–53
  partial, 552
  sensors, 176–78, 206–07, 380
  squeeze, 553
Pressure sensors, for waves, 176–78, 380
Pretreatment, 58, 118–23
Pretreatment standards, 102, 121
Primary treatment of wastewater, 59
Primer cord, 460–61
Probability distributions:
  Fréchet, 381
  Gaussian, 115, 200, 321–27
  normal, 115, 200, 321–27
Probing, by divers, 235–36
Probit axis, 115
Production by plankton, 22, 25–26, 262
Protection, submarine pipeline:
  by ballast rock, 340–42, 393, 399
  by burial, 362, 364–65, 393
  by concrete-filled burlap bags, 362
Proteins, 21
Protractor, three-arm, 208–09, 547
Public health:
  considerations for outfalls, 126–32
  divers in polluted waters, 95
  illnesses to swimmers (*footnote*), 132
Public hearing, 135 (*footnote*), 140
Public Law 80–845, 98
Public Law 84–660, 98
Public Law 91–190, 99
Public Law 91–596, 495
Public Law 92–500:
  elements, 100–02
  introduction, 97
  progress under, 102–03
  secondary treatment, minimum requirement of, 124–25
  treatment works construction, 102
Public Law 92–532 (*footnote*), 102
Public Law 92–583 (*footnote*), 102
Public Law 93–207 (*footnote*), 103
Public Law 93–243 (*footnote*), 103
Public Law 93–254 (*footnote*), 102
Public Law 93–592 (*footnote*), 103
Public meeting (*footnote*), 135

Public notice, 134–35
Pulleys, 465, 481
Pulling pipe:
  from the land, 351, 355, 464–69
  to the land, 355
Pulp and paper mills, effluents from, 109
Pumps:
  for dredges, 455–58
  in fluorescent dye studies, 198
  to obtain water samples, 264
  for wastewater flows to outfalls, 291, 300, 366, 523
  for water-jet nozzle, 488
Punched paper tape, 158–59
Pure death process, 126
Putrefactive materials, 59
Pynocline, 20

Q

Quadrat, 252–53, 263
Quality control, construction, 426, 499, 502

R

Rack, bar, 59
Radar:
  antenna, 548
  dependability of, 549
  for determining position, 549
  distance scale, 548
  plan position indicator, 548–49
  screen, 548–49
  sea return, 184, 549
  source of word, 548
  target, 548–49
  in tracking drogues, 184–85
Radioactive wastes, 97
Radiography, 465–66
Rams, hydraulic, 473, 481
Range, bearing, 170, 547
Range, distance, 224, 545, 547–49
Raw sewage:
  definition, 6
  effluents, 94–95
Raw wastewaters, 6, 52–56
Readout, data, 158

Receiving waters:
  classification for effluents,
    286
  definition, 1
  density-stratified, 309, 311
  homogeneous, 303
  monitoring, 523–25
  standards, 79
  wastewater disposal by dilution,
    78–80
Recharge, groundwater, 81–83
Reclamation, wastewater, 82–84
Recompression chamber, 555
Recorder:
  chart, strip, 158, 159, 161, 179,
    227
  magnetic tape (see Magnetic tape
    recorder)
  paper tape, 174, 179, 545
Recurrence interval, 369, 381
Red algae, 27
Reef:
  barrier, 42
  fringing, 41
  passes, 42
  platform, 42
Refraction, wave, 35–36
Refractory materials (footnote),
  76
Refuse Act, 99, 101
Regional wastewater plan, contents
  of, 2–4
Regression, linear, 246
Release hook, pelican, 171
Releases:
  acoustic, 172
  timed, 171–72
Remote sensing:
  of currents, 181, 185
  of outfall effluents, 534
  of shoreline changes, 240
Replication, sampling, 258
Residue:
  dissolved, 53
  filtrable, 53
  fixed, 53
  nonfiltrable, 53
  total, 53
  volatile, 53
Resolution of measurement system,
  226
Resource leveling, 439

Respiration, definition of, 21
Reuse, wastewater:
  controlled direct, 84
  indirect, 83
  uncontrolled indirect, 83
Reverse osmosis, 78
Reversing currents, 44
Reversing thermometers, 202, 204
Reynolds number:
  boundary, 397
  for current forces, on exposed pipe,
    373
  for diffuser discharges, 305
  influence on drag coefficient, 373
  for rock movement, 395
  wave, 387, 389
Rhumb line, 378
Richardson's Law, 318–20, 322–23
Rig fishing, 526–27
Rip current, 37, 40
Rip head, 37, 40
Risers:
  blockage by sand, 519
  check valves for, 519
  definition, 284, 298
  discharge coefficients, 299–300
  external conditions, effect on flow,
    299
  hydraulic design of, 299–300
  risk of damage to, 298, 356–57
  use of, 298
Rivers and Harbors Act of 1899,
  99, 101
Roaring forties, 30
Rock, armor (see Armor rock)
Rock, ballast (see Ballast rock)
Rocking, pipe, 500
Rock matrix:
  grading of, 400
  sliding of, 399
Rock skips, 491–92
Rope:
  dacron, 163
  nylon, 163
  polyethylene, 163
  polypropylene, 163, 166, 257
  synthetic fiber, 163
Rose:
  current, 181, 191
  wind, 43
Roughness, pipe, 290, 295–96, 346,
  371

S

Salinity of sea water, 17–20, 205–06
Salt water encroachment, 82–83
Salt water intrusion, 82–83
Samplers:
  bottom, 231
  box, 253
  grab, 231–32, 254–55, 257
Sampling biological:
  by camera, 256–57, 263
  data presentation, 250, 255
  density of organisms, 252–53, 263
  by divers, 251–53
  by grab sampler, 254–56, 257
  notification of, 253
  plankton, 257–61
  by raking, 250
  sample handling, 250, 255
  by trawl, 249–50, 257
Sampling bottle, Nansen, 203, 204
Sampling bottle, water, 202–04
Sampling the sea floor:
  with corers, 231, 232–33
  coring, 230
  by diver probing, 234–36
  drilling from a barge, 230, 237
  from drilling platform, 236–37
  with grab samplers, 231–32,
    254–55, 257
  by submerged drilling rig, 235–36
  by vibracore, 233–34
Sand:
  erosion, 40–41, 239–42, 359–60,
    364–65
  particle size characteristics, 39
    (table), 238–39
  waves, 241–42
San Diego, California:
  biological surveys, 262–63
  outfall (see San Diego outfall)
  sludge disposal, 75
San Diego outfall:
  blasting for, 459
  blockage of, 522
  concrete pipe wall erosion, 349
  hydraulic energy dissipation, 300
  inspection by submersible, 520
  liners in pipe, 348
  lives lost during construction,
    442
  loss of pipe liners, 348–49

San Diego outfall (*cont.*):
  major characteristics of, 449
  repairing chipped pipe, 500
  studies by SCCWRP, 528
  tower use during construction, 477
  trestle use during construction, 478
  wastewater flow in, 14
Sand Island No. 2 outfall:
  allowable pipe gaps (*footnote*), 475
  armor rock, nominal sizes of, 341–42
  armor rock plans and specifications, 491–92
  author's experience with, 11, 516–17
  blasting for, 460
  borings for, 236–37
  bottom inspection by submersible, 234
  burial of, 394
  completion date of (*footnote*), 95
  construction cost estimates, 430–31
  construction delays due to waves, 438
  construction diver injury (*footnote*), 495
  construction performance schedule, 440–42
  construction progress payments, 441
  corer use for, 232
  cross-sections of buried pipe, 341
  delay to activity on critical path, 494
  depths along, 235
  design water density profiles, 310
  design wave, 381–82
  diffuser characteristics, 340
  diver decompression during construction, 496
  dredging, offshore, 454, 460
  dredging from trestle, 455
  dropping of pipe after release, 482
  flapgate structure, 340, 403
  grading of armor rock, 490
  Horse, use of the, 472–77
  impaired water visibility during rocking, 516–17
  inertia coefficient for rock matrix, 401

Sand Island No. 2 outfall (*cont.*):
  inspection openings in diffuser, 402
  joint pressure testing, 501
  joints in diffuser, 501
  major characteristics of, 340–42, 452
  manhole covers, 402
  manholes, 402
  manhole spacing, 402
  money left on the table, 434
  mooring of crane barge, 472
  OSHA inspection during construction, 442
  pipe damage during construction, 494, 500
  pipe flow velocities, predicted, 295–96, 298
  pipe-laying sequence, 482
  placing of armor rock, 490–91
  placing of bedding stone, 489
  Pony, the, 481–82
  port discharges, predicted, 295–96, 298
  removal of high pipe sections, 494
  repairing chipped pipe, 500
  rock barge, 492
  rock skips, 491–92
  sheet piling, 479–81, 483
  size of bedding stone, 488
  sizes of armor rock, 489–90
  sizes of ballast rock, 489–90
  sizing of protective rock, 397
  successful bid (*table*), 431
  terminal structure, 340, 403
  trench for, 480
  trestle, construction, 478
  trestle pile penetration, 478
  trestle piles, 483
  underwater inspections, 442
  wall in pipe, 483
  wave damage to trestle, 483
  weights of armor rock, 490
  weights of ballast rock, 490
San Francisco Bay:
  biological sampling in, 249
  outfalls in, 123
Saprophyte, 58
SCCWRP (*see* Southern California Coastal Water Research Project)
Scope, anchor line (*footnote*), 163

Screening wastewater, 59
Screens, 59
Scripps Institution of Oceanography, 177
Scrubbers, gas, 559, 562, 564
SCUBA divers:
  in biological sampling, 251–53, 262–63
  gathering sea floor data, 234–36
SCUBA diving:
  advantage of, 557
  air tank, 556
  air usage, rate of, 556
  booties, 558
  buoyancy compensator, 558
  buoyancy of diver, 558
  depth gages, 558
  disadvantage of, 557
  dry suit, 558
  face mask, 557
  flippers, 558
  hood, 558
  knife, 558
  life vest, inflatable, 558
  regulator, 556–57
  snorkel, 558
  source of word (*footnote*), 551
  weights, 558
  wet suit, 558
Sea, fully-developed, 34–35
Sea floor:
  dynamics, 241–42
  erosion, 241–42
  sampling (*see* Sampling the sea floor)
Sea floor conditions:
  design need for, 229
  by side-scan sonar, 225–26
  by television, 234
Sea grasses, 27
Sea level, mean, 32
Sea return, radar, 184, 549
Sea state (*table*), 35
Secchi depth, 267
Secchi disc, 257, 266–68
Secondary dispersion, effluent, 317–27
Secondary treatment:
  as minimum for wastewaters, 124–26
  processes, 65–70

Secondary treatment (*cont.*):
   requirement for wastewater
      effluents, 103
Sedimentation tanks:
   primary, 61–62
   secondary, 66, 70
Sediment instability:
   from currents, 377
   embedded particles, 377, 396–97
   nonembedded particles, 396
   from Shields' curve, 397–98
   from wave motion, 393–94
Sediments:
   biogenous, 38
   calcarenite, 38
   lithogenous, 38
   organic, 38, 517
   physical, 38
   terrigenous, 38
Sediment transport, alongshore, 40
Seismic reflection profiling, 226–28
Seismic sea waves, 45
Self-monitoring, 122–23
Semidiurnal tide, 32
Sensors:
   for chemical parameters, 266
   conductivity, 206
   defined, 158
   dissolved oxygen, 206
   pH, 206
   pressure, 176–78, 206–07
   salinity, 207
   temperature, 206–07
Sessile epifauna, 23
Set, current, 44
Settling tanks, 61–62, 66, 70
Setup wind (*see* Wind, setup)
Sewage:
   biochemical oxygen demand, 55
   characteristics, 53–56
   definition, 1
   dissolved gases in, 56
   field (*see* Wastewater field)
   flow, 52
Sewer systems:
   combined, 52
   infiltration, 52–53
   pumping, 52
   separate, 52
Sextant:
   in celestial positioning, 543–44
   definition, 543

Sextant (*cont.*):
   dye studies, positioning in, 198,
      209
   for measuring horizontal angles,
      546–47
   positioning by, 198, 209, 546–47
Shackle pins, 166
Shackles, 163–66
Shallow-water wave, 35 (*footnote*),
      45
Shannon (-Weaver) diversity index,
      245–46
Shaped charges, 458–61
Shear strength, of soil, 239
Shear stress, boundary, 397–98
Shellfish:
   bioassays on, 110, 118
   depuration (*footnote*), 130
   diseases from consuming, 130
   standards for meats of, 108, 131
   water quality for, 104–05, 108,
      131
Shields' curve, 397
Ship observations:
   of waves (*footnote*), 377
   of winds (*footnote*), 377
Shredding of wastewater screenings,
      59–60
Sidecasting of dredged material, 454,
      516–17
Side-scan sonar survey systems:
   for pipeline surveys, 520
   in preoutfall-design surveys,
      226
   technique, 225–26
Sieves, 237, 255
Significant wave height:
   annual maximum, 381
   maximum, 381
   definition, 34
Silt, size range of (*table*), 39
Simple harmonic motion, 382,
      394
Skimming tanks, 60
Skips, rock, 491–92
Skunk line, pipe, 475
Slack-line moorings, 162, 174–75
Slack tide, 44
Sleepers, 482
Slings, pipe, 473, 480–82
Sludge:
   aerobic digestion, 74

Sludge (*cont.*):
   anaerobic digestion, 72–74
   barge disposal, 75
   cake, 71
   chemical conditioning of, 72
   composting, 75
   concentration, 71
   conditioning, 72
   definition, 60
   dewatering, 71–73
   disposal, 74–75
   heat treatment of, 72
   lagooning, 75
   moisture content, 70–71, 73,
      75
   outfall, 75
   removal from tanks, 62, 68, 70
   solids content, 71, 73
   thickening, 71–73
   treatment, 70–74
   yield from wastewater treatment,
      71, 74
Small circle, 540
Soffit, pipe, 339
Soil:
   analysis, 237–39
   bearing capacity of, 366
   cohesion, 239, 366
   conditioners, 74–75
   friction angle, 239, 366
   gradation, 238
   settlement of pipe on, 366
   shear strength, 239
   sieve size, 400
Solids:
   settleable, 53
   suspended, 53
   total, 53–54
Sound, 38
Source control, 80, 118–23
South Equatorial Current, 31
Southern California Coastal Water
      Research Project:
   cost figures for wastewater treat-
      ment and outfalls, 125–26
   current-measuring station setup,
      191
   grab sampler tests, 254–55
   monitoring outfall performance,
      528–29
   view on secondary treatment of
      marine effluents, 124

Specifications, construction:
  definition, 422
  detailed specifications, 425, 427
  general conditions, 425–27
  general provisions, 425–27
  standardized components, 425
Spectrophotometer, 268
Spider I:
  dimensions of, 485
  in Hawaiian waters, work in, 485, 487
  longitudinal speed of, 487
  operation of, 485–87
  origin of, 485
Spider II:
  borings for San Francisco outfalls, 237
  in Hawaiian waters, work in, 485
  Santa Barbara outfall, work on, 487
Spigot, pipe, 345, 401, 471, 477
Spilling breaker, 36
SPLASH, 368–69
Splash zone, 39
Split-spoon soil sampler, 230
Splitter, Folsom, 260
Spoil, dredge, 455–56
Spread, construction, 428
Spreading:
  buoyant, 287, 317
  surface, 287, 317
Springline, pipe, 298, 339, 349
Spring tides, 32
Spuds, dredge, 457
Squeeze, pressure, 553
Stabilization ponds, 65
Standard deviation:
  of biological data, 256
  of contaminant concentration profiles, 322
Standards:
  effluent, 79
  effluent versus receiving-water, 523–24
  receiving-water, 79
  stream, 79
Standing crop, 22
Stand, tidal, 31
State Plane Coordinate System, 541
Static friction, coefficient of, 376
Statute mile, 541
Step aeration, 70

Sterilization, 63
Stiffleg, 471
Still water level (footnote), 34
Stinger, 470
Stone, bedding (see Bedding stone)
Storm drains, 52
Storm surge:
  bathystrophic model, 369–71
  computation of, 368–71
  definition, 367–68
  frequency information, 369
Strain gage, 176
Stratification parameter:
  for slot jets, 315–16
  for round jets, 312–13
Stream standards, 79
Strobe light, 184
Strongback, 471
Sub-bottom conditions, obtaining:
  by divers, 234–36
  by drilling and sampling from the water surface, 230–31
  by probing and jetting, 235
  using seafloor drilling rigs, 235–36
  by seismic profiling, 226–28
Subcontractor, 423
Sublittoral zone, 39
Submarine pipelines, installation:
  by bottom-pulling, 355, 464–69
  in inshore waters, 478–88
  by lay barge, 470–71
  by towers, 477
Submersibles:
  in gathering preoutfall-design data, 234
  for outfall inspections, 520
Subpolar low, 29
Subtropical high, 29
Surety, 434
Surface-loading rate, 61, 66
Surface-pulling pipe installation method, 470
Surface spreading:
  wastewater on land, 82
  wastewater on receiving water, 287, 317
Surfactants, 54
Surf zone:
  definition, 36, 343, 464
  need for outfall inspections within, 519–20
  pipe damage in, 358–61, 364–65

Surf zone (cont.):
  wave patterns approaching, 380
Surge, storm (see Storm surge)
Surge tanks, 300–01
Surge, wave:
  effect on damaged corals, 516
  influence on biological sampling, 256
  influence on seafloor drilling rigs, 236
  influence on underwater experiment, 396
  pipe-laying difficulties due to, 438
Surging breaker, 36
Survey instruments:
  autotape, 224, 226
  distance-measuring, 546
  interrogator, 545
  rental of, 545
  responder, 545
  sextants, 543–44, 546–47
  theodolites, 546
Suspended solids, removal in waste-water treatment processes, 61–62, 68, 70, 78
Swell:
  as design wave, 381
  large, 35, 380–81
  nearby storms, from, 35, 381
  origin of, 34, 379
  typical surface profile, 383
Synoptic chart, 377–79
Sydney, Australia:
  Botany Bay, 174–75
  North Head, 41
  Port Jackson, 94
  raw sewage effluents, 94–95
  wave data, 174–75
Symbiont, 25
Symbiosis (footnote), 25
Synergism, 110
Synodic month, 32

T

Tahiti, 30
Tanks:
  activated sludge, 69–70
  anaerobic sludge digestion, 73
  holding, for bioassays, 113–14
  primary sedimentation, 61–62

Tanks (*cont.*):
  test, for bioassays, 114
  trickling filter, 66
Tape recorder, magnetic (*see* Magnetic tape recorder)
Taut-line moorings, 162–64, 168–69
Taxonomy:
  defined, 242
  selected marine invertebrates (*table*), 245
  selected marine vertebrates (*table*), 244
  system of classification, 243–44
Television, underwater, 234, 465
Temperature of sea water, 18–20
Tender, diving, 495–96, 559, 563–64
Terrace, low-tide, 38
Terrigenous sediments, 38
Tertiary treatment of wastewaters, 75–78
Tethered divers, 559–65
Thermal pollution, 97
Thermistor, 206
Thermocline, 19
Thermometers:
  protected, 204
  unprotected, 204
Threshold lethal concentration, 115
Tidal currents, 44, 193
Tidal range, 32
Tidal wave, 45
Tide:
  astronomical (*see* Tides, astronomical)
  Coriolis, 370–71
  pressure, 370
  slack, 44
  storm (*see* Storm surge)
  wind, 369
Tides, astronomical:
  diurnal, 32
  high, 31
  information available on, 33, 161
  low, 31
  measurement of, 161
  mixed, 32
  neap, 32
  semidiurnal, 32
  spring, 32
Time of outfall construction completion, 433
t_{90}:
  choice for design, 129

t_{90} (*cont.*):
  definition, 127
  determination of, 127–28
  factors of influence, 128–29
  values obtained, 128–29
  variability in, 128–29
Tolerance limit, median (*footnote*), 111
Tolerances, outfall specification:
  alignment (*footnote*), 425
  armor rock surfaces, 491–92
  grade (*footnote*), 425
Towers, offshore, 477
Toxicity:
  acute, 110–11, 113
  bioassay, 110–18
  chronic, 110–12
  concentration, 116
  curve, 115
  definition, 63
  emission rate, 116
  of heavy metals, 109–10
  long-term, 110
  short-term, 110
  tests, 110–18
  units, 116
Toxic residues, 118
Trace constituent of sea water, definition of, 18
Tracer, definition of, 195
Tracers, radioactive:
  in dispersion measurements, 195, 198
  for sediment motion, 241
Tradewinds, 30
Transducer, 223, 226
Transect, 251–53, 262
Transients, hydraulic, 300–02
Transmissometer, 268
Transport, sand, 40, 241
Traps, fish, 527
Trawling, bottom:
  for biological sampling, 250–51
  efficiency of, 250–51
  in receiving-water monitoring, 527
  results of, 250
  standardized procedures in, 249–50
  treatment of organisms, 250
Trawls:
  bottom (*see* Trawls, bottom)
  cod end, 248–49, 257–58
  cod line, 248
  flopper, 248

Trawls (*cont.*):
  Isaacs-Kidd midwater, 261
  otter (*see* Trawls, otter)
  uses of, 247–48
Trawls, bottom:
  beam, 248
  for biological sampling, 249–50
  details of, 248–49
  groundrope, 248
  headrope, 248, 250
  otter (*see* Trawls, otter)
  uses of, 247–48
Trawls, damage from:
  moored buoy systems, 174
  submarine pipes, 356–57
Trawls, otter:
  for biological sampling, 249–50
  bobbins, 249
  doors, 249
  kites, 249
  mud lines, 249
  otterboards, 248–50
  towing speeds, 249
  uses of, 247–48
  warps, 248, 250
Tremie concrete, 362, 521
Tremie pipe, 489–90
Trencher, travelling pipeline, 498–99
Trenches, outfall:
  excavating, 447, 453–64
  filling in by sand, 240–41
Trenching, outfall:
  after laying, 447, 453, 498–99
  pretrenching, 447, 453–64
Trestles:
  construction of, 478–79
  damage from blasting, 463
  excavation from, 454–55, 463, 478, 480
  failure of, 483–85
  height above sea, 483
  need for, 478
  piles of, 478
  pipe-laying from, 480–82
  removal of, 478
  sub-bottom information for, 228
  use of tugger winches, 480–81
  utilities for, 478
  wave damage to, 483–84
  wave surge under, 438, 480–81
Trickling filter process, 65–69
Trophic pyramid, 21, 26
Tropic of Cancer, 32

Tropic of Capricorn, 32
Trough, sublittoral, 38
Trough, wave, 34
Try net (*see* Trawls, otter)
Tsunami:
  definition, 45
  erosion from, 239
  at Hilo, Hawaii, 45, 360
  on Oahu, Hawaii, 45
  pipe damage from, 360
Tubercles, pipe, 344
Tugs, 458, 470–72
Tumors, fish, 527 (*footnote*), 529
Turbidity, water:
  definition, 20
  determination of, 266–68
Turbulence:
  anisotropic, 326
  influence on current flow force
    coefficients, 373
  influence on wave force
    coefficients, 389
  isotropic, 326
208 plan (*footnote*), 2
Typhoon:
  Jean, 360
  June, 360
  Pamela, 360

U

Underwater operations, 494–98
United Brotherhood of Carpenters
  and Joiners of America, 495
United Nations Food and Agriculture
  Organization, 6
Upwelling, 26, 31, 309 (*footnote*), 529
Ursell parameter (*footnote*), 390
U.S. Coast Guard:
  boat certification, 208
  inspection of offshore construction
    projects, 442
  Local Notice to Mariners, 167,
    230–31, 253, 462, 521, 547–48
  notification of, 134, 231, 253, 462
  permits, 134
  regulations, 426
U.S. Environmental Protection
  Agency:
  origin of, 99
  permits for outfalls, 132–36
  regional offices, 133

U.S. Environmental Protection
    Agency (*cont.*):
  value engineering requirements, 443
  work under PL 92–500, 100–03,
    132–34
U.S. federal government:
  Advisory Council on Historic
    Preservation, 138
  Agriculture, Department of, 138
  Army, 138
  Army Corps of Engineers (*see*
    Corps of Engineers)
  Atomic Energy Commission, 139
  Coastal Engineering Research
    Center (*see* Coastal Engineering
    Research Center)
  Coast Guard (*see* U.S. Coast
    Guard)
  Commerce, Department of, 138, 542
  Corps of Engineers (*see* Corps
    of Engineers)
  Council on Environmental Quality,
    100
  Court of Appeals, 496
  Federal Insurance Administration,
    369
  Federal Power Commission, 139
  Federal Water Pollution Control
    Administration, 84
  Fish and Wildlife Service, 138
  Health, Education and Welfare,
    Department of, 99, 139
  Housing and Urban Development,
    Department of, 139, 369
  Interior, Department of the, 138
  Labor, Department of, 495
  National Climatic Center, 160
  National Oceanic and Atmospheric
    Administration (*see* National
    Oceanic and Atmospheric
    Administration)
  National Ocean Survey, 4
  National Oceanographic Instru-
    mentation Center, 159
  National Technical Information
    Service (*see* National Technical
    Information Service)
  National Weather Service, 43, 160,
    369, 502
  Navy, 138, 173, 554
  Occupational Safety and Health
    Administration, 442, 495–96

U.S. federal government (*cont.*):
  Office of Economic Opportunity,
    139
  Public Health Service, 99, 139
  Soil Conservation Service, 138
  State, Department of, 139
  Transportation, Department of, 139
Unit processes, 59
Upper littoral zone, 39
Urine, 54
User charges, sewer system, 102

V

Vacuum filtration, 71
Value analysis, 443–44
Value engineering, 424, 443–44
Valve, check, 519
van Dorn bottle, 204, 264, 526
Vane-shear tester, 236
Variation, magnetic, 541–42
Velocity, diffuser pipe flow, 292–97
Vena contracta, 288
Vertebrates, marine, 244
Vibracore, 233–34
Victoria, B.C., raw sewage effluent,
    95
Viruses:
  definition, 57
  enteric, 129
  pathogenic (*see* Pathogens)
Visibility, water:
  impairment by fine bottom sedi-
    ments, 474
  impairment during rocking opera-
    tions, 516–17
Volume flux parameters, 312–16
Volume source dispersion model,
    326–27
Vortex energy-dissipation system,
    300

W

Walking construction platforms,
    485–87
Wastewater:
  biochemical oxygen demand, 55
  characteristics, 53–56
  chronic toxicity of, 117–18

Wastewater (*cont*.):
  definition, 1
  disposal (*see* Wastewater disposal)
  dissolved gases in, 56
  flow, 52
  screening, 59
Wastewater disposal:
  by barge, 80
  in injection wells, 82
  on land, 80
  by outfall, 80
Wastewater field, submerged, 309
Wastewater field, surface:
  depth of (*footnote*), 308
  formation of, 317
Wastewater plan, 2–4
Wastewater treatment plant, bypass-
    ing of, 522
Water-borne diseases, 57–58
Waterhammer, in outfalls, 301
Water jet nozzle for bedding stone,
    475–76, 488
Water-pollution control:
  definition, 79
  legislation in the United States,
    98–103
Water Pollution Control Act of 1948,
    98
Water Quality Act of 1965, 99
Water Quality Improvement Act
    of 1970, 99
Water quality plans, 103–09
Water tunnel, pulsating, 393
Wave:
  celerity, 35
  decay, 379
  deep-water (*footnote*), 35
  design (*see* Design wave)
  diffraction (*footnote*), 389
  energy dissipation, 379–80
  generation, 34, 377–79
  height (*see* Wave height)
  height changes, 379–80, 391
  hindcasting, 377–79, 381
  kinematics, 37–38
  length, 34
  orthogonal, 36, 380, 391
  period, 34, 380
  pressure height, 302
  refraction, 35–36, 379–80, 391
  refraction coefficient, 380
  refraction by computer, 380

Wave (*cont*.):
  refraction diagram, 380, 391
  shallow-water, 35, 45
  shoaling, 379–80, 391
  shoaling coefficient, 380
  short-term variability, 384
  specific energy, 34–35
  spectrum, 34, 177–78
  surge (*see* Surge, wave)
Wave-cut terrace, 41
Wave data:
  collection of, 173–79
  importance in outfall projects,
    172
  joint frequency table, 180
  presentation, 180
  processing, 179–80
  recording, 179–80
  sources of, 172–73
  zero-upcrossing method, 180
Wave force distribution along pipe,
    392
Wave force phenomena:
  experiments in the ocean, 386–87
  modelling of, 386, 393
Wave forces:
  on buried pipes, 400
  pipe in trench, 400, 484
  on pipes during construction, 400,
    484
Wave forces on angled pipe, 391
Wave forces on exposed pipes:
  drag coefficient, 387
  drag force, 387
  inertia coefficient, 387
  inertia force, 387
  maximum-horizontal-force
    coefficient, 388–89
  maximum horizontal forces,
    388–89
  maximum-vertical-force coefficient,
    390
  maximum vertical forces, 390
  Morison equation, 387–88
Wave height:
  annual significant, 381
  definition of, 34
  maximum significant, 381
  significant, 34
Wave-induced near-bottom water
    motion:
  under breaking waves, 385, 392–93

Wave-induced near-bottom water
    motion (*cont*.):
  under non-breaking waves, 37–38,
    382–85
  prediction by Airy theory, 382–84
  prediction by solitary wave theory,
    398
  prediction by stream function
    theory, 384–85
  strength of, 386–87
Wave mast, tiltable, 178–79
Wave-pipe research by the author,
    343, 359, 386–87, 392
Waverider:
  calibration of, 174
  data recording, 179
  mooring system, 173–75
  principles of operation, 173–74
  redesign of, 175
  reliability, 174
Wave-rock experiments by the author,
    396
Waves:
  ability to move large objects, 395,
    484–85
  bathymetric surveys, effects on, 224
  breaking, 36, 379, 394, 483–84
  characteristics of, 34
  data presentation, 179–80
  depth of influence, 241, 386
  erosion by, 40–41
  influence on outfall construction,
    381–82
  water motion beneath, 37–38
Wave staff:
  continuous resistance, 178–79
  resistance, 178–79
  step resistance, 178–79
Wave theories:
  Airy, 177, 302, 382–85
  deterministic (*footnote*), 384
  higher order, 384
  linear, 177, 302, 382–85
  probabilistic (*footnote*), 384
  solitary, 398
  stochastic (*footnote*), 384
  stream function, 384–85
Weight coating, concrete, 347–48,
    376, 465
Weir, overflow, 62
Well, injection, 82
Westerlies, 30

Whaler, Boston, 170, 249, 257
Whites Point outfalls, 14
Winches:
  for anchors, 169
  bathythermograph, 201
  for bottom-pulling pipe, 465, 467
  plankton sampling, use in, 257–58
  for sensor cables, 205
  for surface-pulling pipe, 470
  tugger, air-powered, 480–81
  underwater (*footnote*), 465

Wind:
  anemometers, 160
  data, 160
  rose, 43
  setup, 46
  stress, 370–71
  waves (*see* Waves)
Wind-driven currents, 44, 193
Windhoek, South West Africa, 84
Winkler test, 265–66
Wire-angle meter, 198, 202, 261
Wire rope, 163, 165

Z

Zone of mixing, 79
Zones, coastal:
  intertidal, 39
  littoral, 39
  lower littoral, 39
  middle littoral, 39
  sublittoral, 39
  upper littoral, 39
Zooplankton, definition of, 22
Zooxanthellae, 25, 518